Robotic Exploration of the Solar System
Part 4: The Modern Era 2004–2013

Paolo Ulivi with David M. Harland

Robotic Exploration of the Solar System

Part 4: The Modern Era 2004–2013

 Springer

Published in association with
Praxis Publishing
Chichester, UK

Dr Paolo Ulivi
Toulouse
France

Dr David M. Harland
Space Historian
Kelvinbridge
Glasgow
UK

SPRINGER–PRAXIS BOOKS IN SPACE EXPLORATION

ISBN 978-1-4614-4811-2   ISBN 978-1-4614-4812-9 (eBook)
DOI 10.1007/978-1-4614-4812-9
Springer New York Heidelberg Dordrecht London

Library of Congress Control Number: 200792775

Cover design: Jim Wilkie
Project copy editor: David M. Harland
Typesetting: BookEns, Royston, Herts., UK

Printed on acid-free paper

Springer is a part of Springer Science+Business Media (www.springer.com)

Contents

Foreword

When Paolo Ulivi asked me to write the Foreword to this, the fourth volume in his series about the robotic exploration of the solar system, I was much honored. He has done an amazing job of capturing the pioneering spirit, and the books are both broad in scope and rich in detail.

Our solar system was born over four and a half billion years ago. Blobs of gas, dust and rock became a number of planets orbiting a newly born star. Although life soon developed on one of those planets, the Earth, it remained at the microbial level until macroscopic life appeared several hundred million years ago. And although the human species evolved a million years or so ago, it is only in the last several thousand years that we became intelligent enough to develop the means to explore our planet. Among the first exploration machines were boats to discover new lands. Later, submarines enabled us to explore the underwater realms. For the last fifty years we have been exploring the other objects of the solar system. Although brief, less than a human life span, this pioneering exploration has required us to develop both the technology to send robotic probes to obtain data from far out in space and to broaden our knowledge of physical principles to understand the results, and it was important that someone devote time to documenting this initial phase of our exploration of the solar system. So, many thanks to you, Paolo for this huge effort.

Because the solar system is vast, our probes must travel tremendous distances. In general, each probe has a specific target. One of the first to receive our attention was the planet Mars. Our exploration of this fascinating world has raised many questions. Which processes shaped its surface? Why is its atmosphere so tenuous? Did life ever develop there? Might terrestrial life be the offspring of Martian life carried to Earth inside a meteorite? The solar system is full of interesting places to explore beside Mars. There is a mysterious ocean beneath the icy crust of Europa, one of Jupiter's satellites. And there is the atmosphere of Titan, one of Saturn's satellites, whose composition resembles that of the early Earth. The more we explore the solar system, the better we understand our own world. And we have barely begun. There are many adventures ahead.

My particular interest is Mars. The third volume in the series ended with the twin rovers *Spirit* and *Opportunity* starting their explorations of that planet a decade ago.

Each vehicle weighed 185 kg, with 8 kg of scientific instruments. This fourth volume describes the arrival and initial activities of the *Curiosity* rover, which weighs around 900 kg with 80 kg of scientific instruments. In only eight years, the mass and volume of scientific instruments carried on Martian rovers increased by a factor of ten. In a sense, I, myself, live on Mars. I joined the Centre National d'Études Spatiales (the French space agency) in Toulouse in 1994 and it has been a tremendously satisfying experience. First, I spent eight years building mechanisms for robotic applications in several exploration missions. Then I taught astronauts to use scientific equipment for research in the neurosciences field to learn how the brain reacts to weightlessness, as a step towards having humans venture out into the solar system. Then from 2009 to 2012, I organized the French Instruments Mars Operations Centre (FIMOC) for the two instruments on *Curiosity* that were developed with French participation: SAM (Sample Analysis at Mars) and ChemCam (Chemistry & Camera).

Each night, work starts in Toulouse around 5 p.m. It is then 8 a.m. at the Jet Propulsion Laboratory (JPL) in Pasadena, California, the main operations center that coordinates activities and drives the *Curiosity* rover. But each scientific instrument is operated by several centers around the world, which must follow the JPL time zone. I am more involved with ChemCam than SAM. Each night, my team writes the sequences to have the rover fire laser shots at rocks to identify their chemical compositions. As a result, I have the very strong sense that I am living on Mars.

As I write these lines, we have just achieved 100,000 laser shots. Even after more than a year on Mars, when I see the laser impacts on a Martian rock carried out the previous day, it is an emotional moment. I feel that I am contributing to the first steps in exploring the planet by digging holes like a pioneer. I hope, in my old age, to see my younger colleagues operating robots in more complex exploration, and eventually to see human beings walking on the surface of Mars.

Eric Lorigny
CNES, Toulouse, France,
December 2013

Author's preface

The third part of Robotic Exploration of the Solar System closed with the many Mars launches of 2003 that delivered to the Red Planet the European Mars Express orbiter and the long-lived American Spirit and Opportunity rovers. This fourth (and for the moment, final) part of the series will cover all the planetary and deep-space missions launched in the 9-year period running from 2004 to 2013. The US program continued on the inertia and large budgets that it had accumulated over the previous years, prior to a slow-down and then a shocking contraction. It now looks like the second half of the 2010s and at least the start of the 2020s will be only marginally richer in NASA missions than the 1980s, a period that was described in the second part of this series. However, other agencies and nations have flown a large number of missions, with mixed success. As I write these words, the European Space Agency has just awakened its Rosetta comet orbiter and lander, possibly one of the most interesting missions of the decade. It has also apparently settled on the future of its ExoMars mission, and is planning a Jupiter orbiter for the 2020s and 2030s. India is flying its first mission to venture beyond the Moon. China is still in contact with its first asteroid mission, and is planning others, in addition to an amazingly successful lunar exploration program. And hopefully future Russian missions will fare better than the poor Phobos-Grunt.

As in the third part of the series, we made some arbitrary choices regarding which missions to include. We have omitted all the solar observers and other missions that are stationed at the Lagrangian points of the Sun-Earth system. On the other hand, we have included the Kepler space telescope because it was inserted into solar orbit like a deep-space mission and it was part of the Discovery series of planetary missions.

So this brings the story of the robotic exploration of the solar system up to January 2014. Perhaps further updates will be provided in future volumes...

Paolo Ulivi
Toulouse, France
January 2014

Acknowledgments

I must thank all of those who generously provided documentation, information and images for this volume, including Michael A'Hearn, Alessandro Atzei, Tibor Balint, Rui Barbosa, Jens Biele, William H. Blume, Philippe Coué, Nick Cowan, Jacques Crovisier, Paolo D'Angelo, Dwayne Day, Haroldo Fraga de Campos Velho, Qiao Dong, Dan Durda, Thérèse Encrenaz, Peter Falkner, Giancarlo Genta, Brian Harvey, Raymond Hoofs, Ji Wu, Horst U. Keller, Erik Laan, Geoffrey A. Landis, Mingtao Li, Paulett C. Liewer, Stephen C. Lowry, Aleksander Lyngvi, Steven E. Matousek, James V. McAdams, Ralph McNutt, Elsa Montagnon, Masato Nakamura, Catherine Olkin, Dmitry Payson, Pierpaolo Pergola, Ettore Perozzi, Andy Phipps, Colin Pillinger, David S. F. Portree, the late Patrick Roger-Ravily, Jean-Jacques Serra, Robert Shotwell, Alan Stern, Makoto Taguchi, Jan van Casteren, Pierre Vernazza, Victor Vorontsov, Vu Trong Thu, Hiroshi Yamakawa, and Zou Xiaoduan.

I also offer my thanks to user 'pandaneko' of the unmannedspaceflight.com forum for his Japanese translations. It and the NASASpaceflight forum proved invaluable as sources of information.

My special thanks go to Eric Lorigny, who enthusiastically wrote the Foreword, to Marc D. Rayman for proof reading and first-hand comments on the section about the Dawn mission, and to Corby Waste for providing the background illustration for the front cover, which is a customized depiction of a proposed spacecraft about to fly through plumes of water vapor emitted by the south polar region of Saturn's moon Enceladus. Philippe Kletzkine and the Solar Orbiter team at ESA kindly provided the very latest mission schedule just in time for publication. Thanks also go to my father Carlo for translations from Russian, and to my brother Federico for keeping me up to date with the landing of the Curiosity rover via SMS when I was cut off from the rest of the world somewhere in Mongolia.

And my very special thanks go to David M. Harland, who reviewed and edited all of the volumes of this series, and to Clive Horwood of Praxis and Maury Solomon of Springer in New York for believing in what proved to be a decade-long project.

Although I have managed to identify the copyright holders of most of the drawings and photographs, in those cases where this was not possible and I deemed an image to be essential to the story, I have used it with as full a credit as possible; I apologize for any inconvenience this may create.

10

New frontiers in the solar system

THE DECADAL SURVEY

By the early 2000s, most solar system bodies had received at least a preliminary inspection by spacecraft, making it difficult for the US National Aeronautics and Space Administration (NASA) to decide upon the next steps in its program and to prioritize objectives. Hitherto, the agency had decided the targets for its missions based on strategic planning by its Office of Space Science. It relied on input from the scientific community, in particular from the Space Science Board of the National Research Council, part of the National Academy of Sciences. In 2001 therefore, NASA asked the National Research Council to assess the current state of US planetary exploration and recommend which missions scientists believed to be the most significant and should be attempted in the 2003–2013 period. The Council interviewed some of the top planetary scientists, not only in the US but also abroad, and produced a report which, although the first of its kind in this field, was similar to the 'decadal surveys' that NASA had previously used for its astronomy, astrophysics and solar physics programs. While the survey was underway, NASA announced the creation of a new series of medium-class planetary missions, which it decided to call New Frontiers, to complement the Discovery and flagship-class missions. Discovery missions would fly every 1.5 to 2 years, New Frontiers every 5 years on average and flagships every 10 years for comprehensive investigations of especially significant targets. Like Discovery, New Frontiers was to be a program of flights that would be selected on a competitive basis and led by principal investigators, but with budgets of up to $650 million to enable them to attempt more ambitious objectives. The first mission retroactively adopted was the Pluto-bound New Horizons (discussed later). The choice of future missions, however, would be based extensively on input from the National Research Council report. Published in 2003, *New Frontiers in the Solar System – An Integrated Exploration Strategy* addressed four topics: (1) The First Billion Years of Solar System History, (2) Volatiles and Organics as the Stuff of Life, (3) The Origin and Evolution of Habitable Worlds, and (4) How Planetary Systems Work. It argued for the development of a number of 'backbone' medium and large missions, or at least for their objectives to be included in future missions.

The Europa Geophysical Observer was proposed as a flagship mission to confirm the presence of an ocean of water beneath the icy crust of the eponymous Jovian satellite. This was felt to be the first logical step in assessing Europa's suitability to support life, and would also provide insights into how tides shaped major bodies in the solar system. It would exploit the 'Fire and Ice' studies, but was recognized to be a project on the scale of Galileo or Cassini.

Of course, scientists voiced interest in a Pluto flyby, canceled by NASA in 2000 and revived as the New Horizon mission with the expanded objective, in addition to studying Pluto and Charon, of providing the first look at one or more bodies of the Kuiper Belt. In fact, a 'Kuiper Belt Pluto Explorer' was deemed the highest priority medium-sized mission for the decade. Although not one of the priority missions, interest was also expressed in flying a similar Trojan and Centaur Reconnaissance mission to start by encountering at least one of the Trojan asteroids that occupy the Lagrangian points of Jupiter's orbit, on the way to a hybrid comet-asteroid Centaur in an orbit between Jupiter and Saturn. Another flyby probe could be sent to obtain a second look at Neptune and its moon Triton, prior to a flyby of a Kuiper Belt object. Other objectives could be satisfied by Earth-based telescopes, in particular defining the true extent of the Kuiper Belt.

Cratering processes and the evolution of planetary surfaces would be addressed by a sample-return from the south pole of the Moon, where the deep Aitken basin is believed to have excavated through the crust into the mantle without prompting an upwelling of magma to mask the basin floor.

Another proposal addressed the gas giants, in particular Jupiter. Although Galileo answered many questions, there were still mysteries concerning the giant of the solar system. For example, does it have a central rocky core? Whilst Saturn, Uranus and Neptune were all believed to possess one, the evidence for Jupiter was in dispute. A Jupiter Polar Orbiter with Probes would address various issues. It would drop three probes to continue studies of the atmosphere initiated by Galileo, and in particular precisely determine its composition in order to place constraints on the region of the solar system in which the planet condensed out of the solar nebula. And by having a periapsis of less than 0.1 Rj (Jovian radii) above the cloud tops, this spacecraft would also advance the study of the planet's magnetosphere. In fact, the discovery of many planets of at least Jupiter's mass orbiting around other stars at distances even closer than the orbit of Mercury from the Sun raised the prospect that gas giants condensed much further out and subsequently migrated inward owing to gravitational interactions with the protoplanetary disk. If this happened to Jupiter, we are fortunate that it stopped at its current position, since it would otherwise have disrupted the inner solar system.

A Venus In-Situ Explorer (VISE) would address the evolution of the atmosphere of that planet in order to clarify why Venus, Earth and Mars, seemingly so similar to one another, evolved so differently. Moreover, the mission would yield data on the runaway greenhouse effect as a point of comparison in the context of Earth's global warming. After collecting a sample off the surface, the lander would use a balloon to study it in the cooler upper atmosphere, where it could be analyzed over a longer period of time than was feasible on the surface – on which the Veneras had survived for at most several hours. And whereas the Veneras had been able only to determine the chemical composition of the surface material, VISE would study the mineralogy, texture and possible stratigraphy of the sample to yield much greater insight into the planet's geological history. A near-infrared camera would provide panoramas of the surface below 10 km of altitude, and these would be correlated with Magellan radar data to determine the geological context. Of course, the mission would provide data on winds, etc, during its descent, surface time, ascent and balloon flight.

A Comet Surface Sample Return would provide materials to study the history and evolution of volatiles, and in particular of primitive organic molecules understood to be part of a comet's composition. As utilizing a high-speed sampling method such as that of Stardust would not permit volatiles or organics to be collected and preserved while flying through the coma, a sample would be taken directly from the nucleus. A cheap version of this mission would obtain a surface sample, but a more expensive option would involve drilling to obtain a sample of the pristine material beneath the 'evolved' surface. An asteroid sample-return would obtain material from the surface of a near-Earth asteroid, either a well-characterized one like Eros or a taxonomic type rich in organics. Such a mission could also use a small mobile robot (a rover or a hopper) to explore the surface. And asteroids, in particular near-Earth ones, would also be studied by large ground-based survey telescopes with the goal of locating 90 per cent of those larger than 300 meters in order to determine more precisely the risk that they pose to Earth. Continued ground support for space missions was called for at a variety of electromagnetic wavelengths. In the past, NASA's Infrared Telescope Facility in Hawaii had shown that the atmospheric probe released by the Galileo spacecraft sampled an anomalously dry 'hot spot' on Jupiter. In the future, the New Horizons Kuiper Belt mission would require terrestrial telescopes to discover and localize its targets.

Of course, the National Research Council recommended that the Cassini mission continue beyond the end of its primary mission in July 2008. Apart from urging that they continue to be launched every 18 months, the report made no recommendations for Discovery missions. In fact, given the quick-response character of many of these missions, they were not actually suitable for 'strategic' planning.

Recognizing the stand-alone status of the Mars exploration program the National Research Council made specific recommendations, including support for NASA's creation of a Mars Scout series of inexpensive missions; a cheap upper-atmosphere mission; a network of landers to obtain seismic evidence to determine whether the planet has a liquid core; a 'smart lander' that would obtain the ground-truth to assess the identification from orbit of sites which appeared to have been modified by water, and also to validate sample-return technologies; and, of course, a series of precursors that would lead to a sample-return mission after 2015, since this would be the only way to determine conclusively whether life once existed or still exists on the planet.

A number of other outer solar system missions were deemed interesting, and the National Research Council urged that these be studied for possible implementation in the 2010s. These included a Saturn Ring Observer, a Uranus Orbiter, a Europa Pathfinder that would deliver a small payload to the surface using airbags to cushion

The priorities defined by the 2003 National Research Council decadal survey *New Frontiers in the Solar System.*

"Crosscutting Themes"	Key questions	Recommended missions or facility
The first billion years of the solar system	What processes marked the first stages of planet and satellite formation?	Comet Surface Sample Return Kuiper Belt-Pluto Explorer South Pole-Aitken Basin Sample Return
	How long did it take Jupiter to form, and how was the formation of Uranus and Neptune different from that of Jupiter and Saturn?	Jupiter Polar Orbiter with Probes
	How did the flux of impactors decay during the early history of the solar system, and in what ways did it influence the emergence of life on Earth?	Kuiper Belt-Pluto Explorer South Pole-Aitken Basin Sample Return
Volatiles and organics	What is the history of volatile compounds and water?	Comet Surface Sample Return Jupiter Polar Orbiter with Probes Kuiper Belt-Pluto Explorer
	What is the nature of organic material in the solar system?	Comet Surface Sample Return Cassini Extended
	What mechanisms affect the evolution of volatiles?	Venus In-Situ Explorer Mars Upper-Atmosphere Orbiter
The origin and evolution of habitable worlds	What processes generate and sustain habitable worlds, and where are habitable zones in the solar system?	Europa Geophysical Explorer Mars Smart Lander Mars Sample Return
	Does life exist beyond Earth? Or did it exist? Why have the terrestrial planets evolved so differently?	Mars Sample Return Venus In-Situ Explorer Mars Smart Lander Mars Long-Lived Lander Network Mars Sample Return
	What are the hazards to life on Earth?	Large-Aperture Synoptic Survey Telescope

"Crosscutting Themes"	Key questions	Recommended missions or facility
How Planetary Systems Work	How do processes that shape planetary bodies operate and interact?	Kuiper Belt-Pluto Explorer South Pole-Aitken Basin Sample Return Cassini Extended Jupiter Polar Orbiter with Probes Venus In-Situ Explorer Comet Surface Sample Return Europa Geophysical Explorer Mars Smart Lander Mars Upper-Atmosphere Orbiter Mars Long-Lived Lander Network Mars Sample Return
	What does the solar system tell us about the development and evolution of other planetary systems?	Jupiter Polar Orbiter with Probes Cassini Extended Kuiper Belt-Pluto Explorer Large-Aperture Synoptic Survey Telescope

the impact, and a complex, ambitious (and expensive) Europa Astrobiology Lander that would, amongst other things, collect and analyze deep samples. Titan Explorer would involve an orbiter and some kind of aerial robot – a balloon, an airship or an aircraft. Detailed planning and identification of scientific objectives for this mission would, however, have to await the results of Cassini and Huygens. An Io Observer would involve a Jupiter orbiter making repeated flybys of that volcanic moon, while a Ganymede Orbiter would be similar in concept to the Europa radar orbiter. A most important mission to the outer planets in the 2010s would be a Neptune Orbiter with Probes, and it was urged that preliminary studies be started. It would investigate the planet, its magnetosphere, its ring system (which Voyager 2 had shown to be rich in fine structures) and its satellites – in particular geologically active Triton to measure its effect on the planet's magnetic field in order to determine whether, like some satellites of Jupiter and Saturn, it possesses a subsurface ocean of liquid water. The report also suggested that NASA invest in new technologies for future missions, in particular for sample returns, aerocapture, technologies to collect, store and examine samples at cryogenic temperatures, and optical communications to boost data rates far above those of conventional radio systems.[1,2]

DECODING THE SOLAR SYSTEM

As related in Part 2 of this series, in 1984 the European Space Agency (ESA) started to plan a revolutionary mission to land on the nucleus of a comet, collect samples and return them to Earth. In 1986 this became a joint project with NASA, which would provide some of the key technologies. This Comet Nucleus Sample Return (CNSR) mission was renamed Rosetta, after the basalt stone discovered by French troops in 1799 in Rashid (Rosetta) in the Nile Delta and now on exhibit at the British Museum in London. The hope was that, just as the stone enabled archaeologists to unlock the Egyptian language of hieroglyphs, the cometary material would unlock the secrets of the origin of the solar system. However, owing to the financial and programmatic difficulties that NASA experienced in the early 1990s, that agency had to withdraw from the project; at the same time stopping work on the Mariner Mark II planetary mission bus which Rosetta was to use.[3] ESA started to look at possible alternatives in 1992, with the primary objective of defining a scientifically meaningful cometary mission that could be performed using only European technology and within its own financial resources. As redefined, the mission lost its capability to return samples to Earth. This was traded for a longer period of exploration, unconstrained by the need to start the return leg of the flight. As a further bonus, it would now be able to study the evolution of the nucleus from 3 to 1 AU, as it approached perihelion. Given the increase in interest in the minor bodies of the solar system after Galileo's flybys of Gaspra and Ida, the mission was to encounter one or two asteroids on its way to the comet. The spacecraft for the sample-return mission was to have been powered by plutonium-fueled Radioisotope Thermal Generators (RTG) supplied by NASA, but owing to advancements in low-temperature solar cells which could now yield sufficient power even at Jupiter's distance from the Sun, the redefined spacecraft would use solar power. It would be launched using an enhanced Ariane 5 and use planetary flybys to gain the additional energy required to reach its primary target. Although it would be tracked by a European network of antennas and ground stations, centered initially on the 30-meter-diameter Weilheim dish, NASA's Deep Space Network would be available in a backup capacity during critical phases and maneuvers.

A number of comets were considered as the target of the revised Rosetta mission. They were all members of the Jupiter family which had a perihelion near 1 AU and an aphelion around 5 AU, a low inclination with respect to the ecliptic and could be reached within 9 years of a 2003–2005 launch using the Ariane 5, assisted by flybys of Venus, Earth and/or Mars. Also briefly considered were hybrid asteroids that had shown evidence of being extinct comets, notably (2201) Oljato and (4015) Wilson–Harrington. Usually, each launch opportunity also permitted flybys of one or more main belt asteroids.

The scientific objectives at the comet were to characterize the nucleus in terms of its shape, orientation and spin; investigate its morphology; determine the chemistry, mineralogy and isotopic composition of the volatile and refractory materials on its surface; investigate the activity and processes occurring on the nucleus and in the coma; and monitor how the comet interacted with the solar wind as it approached perihelion. One or more landers were to address some of the scientific objectives of the sample-return mission, in particular performing in-situ analyses. Surface samples would provide material chemically unaltered and unchanged, unlike the gas and dust collected by flyby missions and by Stardust, which evolved as it escaped the coma and was modified by high-speed impacts. This would enable more meaningful analyses to be performed. In the initial design, the lander was to be essentially passive, with no autonomous capability for landing, and would rely only on dampers and shock absorbers to cushion its uncontrolled and unstabilized fall. Various options were considered, including press-down thrusters primed to fire as soon as contact was detected, or Russian-style spherical landers that would deploy spring-loaded petals in order to right and stabilize themselves once on the surface. The original design was to be battery-powered to facilitate several hours of operations, after which the lander would expire. Because of this, no command link was intended from the orbiter to the lander, which was to execute a pre-programmed sequence of activities.[4] In a redesign, the orbiter was to carry one or two longer-duration landers. One, called ROLAND (Rosetta Lander), was to be made by a European consortium led by the German Max Planck Institute and the DLR (Deutsche Zentrum für Luft-und Raumfahrt; German center for flight and spaceflight). The second, aptly called Champollion in honor of Jean François Champollion, the French anthropologist who decoded the Rosetta stone, was to be supplied by NASA's Jet Propulsion Laboratory (JPL) and the French Centre National d'Etudes Spatiales (CNES; national center for space studies). In fact, unlike the main spacecraft, the landers were not to be funded directly by ESA but by national space agencies.

The redesigned Rosetta mission was approved by ESA in November 1993, with launch expected in the early years of the new century. Its total cost, including flight operations, was expected to be 770 million euros.

But financial and programmatic difficulties soon forced NASA to withdraw from Champollion, whose payload was in part merged with ROLAND, and CNES joined that project. Nevertheless, JPL decided that the Deep Space 4 (DS4) project, part of its New Millennium program, would land Champollion's scientific instruments on a comet, perform sample collection and return, and demonstrate technologies for use on sample-return missions. After launch in April 2003 Deep Space 4 was to use its xenon propulsion to match the orbit of comet 9P/Tempel 1 and in December 2005 become its satellite. After 4 months of mapping the surface of the nucleus, the scientists would choose a landing site and the JPL lander would be released to conduct a 3.5-day mission. Its software would have sufficient autonomy to land at a specified site at a vertical rate no greater than 0.25 m/s. It had a slim cylindrical body with a harpoon to secure itself to the surface and a 75-cm-diameter conical 'snow shoe' for stability in-situ. The scientific instruments on the lander would include a microscopic camera and an infrared spectrometer, a panoramic camera, gamma-ray and neutron detectors, a gas chromatograph and mass spectrometer, and a probe to determine the mechanical and thermal properties of the surface. Once a 1-meter-long miniaturized drill had taken a subsurface sample, the ascent stage would lift off and rendezvous with the orbiting cruise stage and return capsule, leaving the 60-kg platform on the surface with most of the instruments. The return to Earth would be accomplished at the next perihelion in May or June 2010. Atmospheric entry speed would be greater than 15 km/s; faster than any previous returning craft. (For example, the Apollo, Zond and Luna returns from the Moon all entered at less than 11 km/s, and Genesis and Stardust did so at less than 13 km/s.) The design of the sample-return capsule would therefore require extensive development. In particular, during entry it had to maintain the cometary material at a very low temperature to preserve its pristine state. The technologies to be tested included lightweight solar arrays which would be deployed by an inflatable boom, multi-engine ion propulsion, autonomous and precise landing, anchoring and sampling mechanisms, high-performance electronics, UHF transceivers for mother-to-daughter vehicle communication, small transponders, and automatic rendezvous and docking techniques. The orbiter would have four Deep Space 1-derived ion engines, powered by solar panels capable of providing up to 12 kW. Two of the engines would operate during most of the cruise.[5,6,7,8] After 3 years of study, NASA threatened to cancel the mission unless its cost could be capped at $158 million. A revision involving canceling the sample-return portion of the mission and spending a longer time in the vicinity of the comet was rejected, and in July 1999 the project was terminated on budgetary grounds in order to pay for overruns of other programs. As ESA director of space science, Roger Bonnet put it, Europe, with Rosetta, would seize "the opportunity the USA always misses".[9]

Meanwhile, Daimler–Chrysler (later Airbus Space Germany), leading a consortium of European partners, mainly in Germany, France and Italy, finished the development of Rosetta. The orbiter resembled a geostationary communication satellite. The 2.8 × 2.1 × 2.0-meter aluminum honeycomb body supported a pair of gimbaled 14-meter wings with a total area of 64 square meters that gave the spacecraft a span of 34 meters. Each wing had five solar panels. The low-intensity low-temperature cells would provide 8,700 W at 1 AU, 395 W at aphelion beyond the orbit of Jupiter, and about 850 W at comet encounter. They were supplemented by four batteries. It would be possible to provide power to all of the instruments within 3.2 AU of the Sun. Of the remaining sides of the box: one carried the launcher interface; another (the upper surface) held the payload support panel used to accommodate all of the scientific instruments and navigation cameras; another had the attachment for the lander; and the last held the 2.2-meter-diameter carbon-composite high-gain antenna which had two degrees of freedom in pointing. There were also two 0.8-meter-diameter medium-gain antennas on fixed mounts, and two omnidirectional low-gain antennas. Thermal radiators and louvers were mounted on the panels which would almost always face away from the Sun. A careful thermal design kept the hardware and instruments from freezing at large distances from the Sun. To support Rosetta and the other deep-space missions that ESA was planning, the agency decided to establish a small European network of antennas. This initially comprised two 35-meter-diameter dishes situated 120 degrees apart in longitude at New Norcia in Western Australia and at Cebreros in Spain, and provided coverage of missions for two-thirds of each day. In 2012 a third antenna was added near Malargüe in Argentina to provide continuous coverage.[10,11]

The main corrugated aluminum thrust tube inside the Rosetta bus gave it

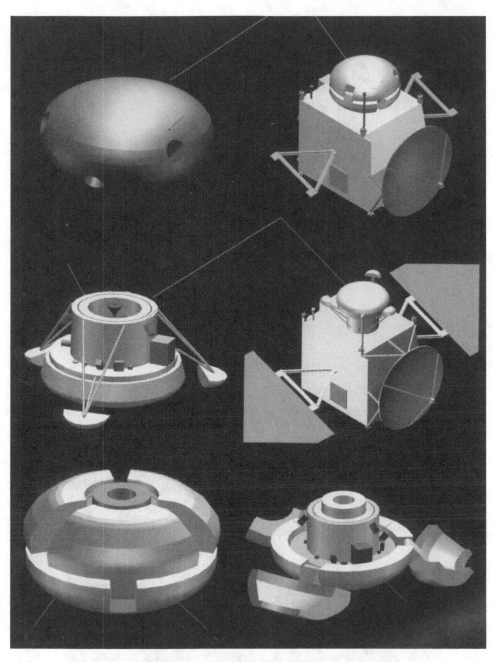

Some early designs for the Rosetta cometary lander. (ESA)

NASA's Deep Space 4 comet lander and sample-return.

strength. Also carried internally were two 1,106-liter nitrogen tetroxide and hydrazine tanks, and four smaller tanks with 140 liters of helium pressurant. The 3-axis-stabilization system used reaction wheels and a total of sixteen 10-N bipropellant thrusters set in pairs at the corners of the bus. There were also eight identical thrusters for trajectory control and maneuvers near the comet. Attitude determination was by star cameras, laser gyroscopes and four Sun sensors, two of which were mounted on the bus and two on the solar panels. In addition to providing navigational information, two CCD cameras could monitor up to 10 stars to provide a backup source of data for attitude determination. Including 165 kg of scientific payload, the lander and propellant, the spacecraft's launch mass was 3,065 kg. To provide a margin for contingencies, the propellant load of 1,900 kg was more than sufficient for the total velocity change of about 2.2 km/s required to conduct the mission.

No fewer than 11 scientific instruments were mounted on the bus, three of which, together with components of a fourth, were provided by American scientists through NASA funds.

The imaging system consisting of a wide-angle and a narrow-angle camera was to image the asteroids and the nucleus of the comet. The wide-angle camera had a focal length of 14 cm and the narrow-angle camera had a focal length of 717 mm and an aperture of 90 mm, yielding fields of view of 12 and 2.2 degrees respectively and resolutions of 10.1 and 1.9 meters at a distance of 100 km. Both were equipped with 2,048 × 2,048-pixel CCD sensors, which were twice as large as detectors flown on other planetary missions. The narrow-angle camera had two filter wheels with eight positions each for combinations of 11 color filters, four re-focusing plates for different wavelengths, and one neutral filter. A similar mechanical arrangement gave the wide-angle camera 12 narrow-band and two wide-band filters. The cameras had their structures wholly made of silicon carbide (like the optics) to ensure consistent thermal expansion of the entire instrument.[12] Two spectrometers operating in the far-ultraviolet, visible and infrared were to analyze the gases in the coma and tail of the comet, measure the production rates of water, carbon monoxide and carbon dioxide, determine the composition of the nucleus and map the temperature of its surface. This data would also assist in identifying a suitable site for the lander.[13,14] An ion and neutral particle spectrometer was to identify the composition of the coma and measure the velocity of charged particles.[15] A microwave radiometer was to measure the temperature of the surface and subsurface of the nucleus, identifying the main gases and measuring their production rates. Between them, these instruments would analyze the gases in the coma, detect water and oxides of carbon and measure their rates of production.[16] An ion mass analyzer and a grain-impact analyzer were to determine the number, composition, size, mass, velocities and other characteristics of dust from the nucleus which impacted at speeds of up to several tens of meters per second, while a microscope would give information on particle size, volume and shape.[17,18,19] A 'plasma consortium' consisting of Langmuir probes, a dual-fluxgate magnetometer, a mutual impedance probe, ion and electron spectrometers and an ion composition analyzer would monitor cometary activity, examine the structure of the inner coma, and investigate the comet's interaction with the solar wind.[20,21,22,23,24,25] A sounding experiment would use an external 1.5 × 1.5-meter H-shaped antenna to transmit long-wavelength radio. The plan was to measure the reflection, scattering, time delay and attenuation of the signal between the orbiter and the lander to map the structure of the nucleus to a great depth, in particular to determine whether it was a 'rubble pile' or a solid body.[26] And, of course, radio tracking of the spacecraft would yield the mass and other gravitational data on the comet, and of the asteroids passed on the way to the primary target.[27,28,29] A rarely mentioned engineering instrument would monitor the radiation environment of the spacecraft and provide a continuous measurement of the ionizing particles encountered during the lengthy flight. Prior to replay to Earth at rates of up to 20 kbits per second, data from the scientific instruments would be stored in 25-Gbit solid-state memories.[30,31]

The Rosetta lander was carried on the rear of the orbiter – i.e. on the side that was maintained facing away from the Sun in normal circumstances. The German Space Agency played the leading role in the project, with significant contributions from the space agencies and major scientific institutions of France, Italy, Hungary, the United Kingdom, Finland, Austria and Ireland. ESA itself joined the consortium at a later date, and contributed over half of the lander's total cost of 220 million euros. The 97.9-kg lander comprised an octagonal carbon-fiber and aluminum honeycomb box,

The Rosetta orbiter during solar panel deployment tests. (ESA)

The Rosetta orbiter, minus solar panels, during thermal vacuum tests. The Philae lander is mounted on top in this picture. (ESA)

1 meter square and 80 cm tall. Apart from the 'balcony' side which housed most of the 26.7-kg scientific payload, it was covered with solar cells capable of generating 10 W in almost any orientation at a heliocentric distance of 3 AU. The power system was supplemented by sets of non-rechargeable primary batteries and rechargeable secondary batteries. The lander had no means of directly communicating with Earth. All commands and data transfers had to be relayed via the orbiter. A 6.5-W radio system provided data rates of up to 16 kbits per second at distances of 150 km from the orbiter. Three long legs extended from the body, connected by a universal swivel

joint to overcome a surface slope and to allow full rotation of the lander. There was a twin-sole damping system and an 'ice screw' on the tip of each leg to engage with the nucleus. Like the original Rosetta sample-return and Champollion landers, to avoid rebounding after touchdown in the feeble gravity, a harpoon would be fired at about 100 m/s on contact. A cable tensioning mechanism would reel it in. Instruments on the harpoon included accelerometers and temperature sensors. There was a second harpoon as a backup. Attitude control during the descent would be by a single reaction wheel and by a cold-gas thruster that could also double as a 'hold down' system at the moment of landing. It was decided to do without isotopic heat sources and electrical heaters, and instead to use a thermal insulation 'hood' to keep the lander relatively warm in the shadow of the orbiter and at large heliocentric distances.[32]

The lander carried an extensive suite of miniaturized instruments. A sampling and distribution system included an integrated drill and sampler. It was derived from the original sample-return proposal, and had a hollow steel auger with a retractable drill bit to remove the cylindrical sample. It was equipped with synthetic polycrystalline diamonds to enable it to penetrate to a depth of 23 cm in materials having textures ranging from fluffy to hard rock whilst drawing less than 15 W of power. Moreover, the system was itself a scientific instrument because the drilling speed, kinematics, power drain, etc, provided data on the strength and other mechanical and structural characteristics of the cometary material. The sample distribution system consisted of a rotating carousel with 26 mini-ovens for 20 mm^3 (i.e. 3 mg) of material. Ten of the ovens were for medium-temperature analyses, and the others, which were capable of being heated to 800°C, were for high-temperature analyses. The drill, its motors and the sample distribution system were contained inside a column-shaped carbon-fiber fairing on the 'balcony'.[33,34] Both a gas analyzer and a gas chromatograph and mass spectrometer were to make elemental, molecular and isotopic composition analyses. The instruments also had the capability to measure the chirality (i.e. the left/right-handedness) of complex organic molecules to assess their relevance to pre-biologic chemistry. The Ptolemy gas chromatograph and mass spectrometer (named after one of the three hieroglyphs contained in cartouches used to decipher the Rosetta stone) was to measure the isotopic ratios of hydrogen, carbon, nitrogen and oxygen in the surface material, as well as in the 'atmosphere'. It had originally been planned that the orbiter and a second lander would be equipped with similar instruments named Cleopatra and Berenice.[35,36] A miniaturized alpha-particle and X-ray spectrometer based on instruments carried on Mars 96, Mars Pathfinder and the two Mars Exploration Rovers would be lowered onto the surface and use radioactive curium to perform an elemental analysis of the surficial layer.[37] A spike-penetrometer and thermal probe on a boom would swing down onto the surface 1 meter from the lander, and then use a hammer to drive the spike up to 32 cm into the surface to determine its physical characteristics, mechanics and thermal conductivity. The penetrator was initially to have been equipped with a gamma-ray densitometer also, but this was deleted as a cost-saving measure. The boom would later be retracted in order to allow the lander to rotate on its swivel joint, while cables maintained the data and power connection. The thermal characteristics of the surface were also to be measured by an infrared radiometer on the lander.[38] The plasma environment near the surface of the nucleus, and the manner in which it interacted with the solar wind, would be determined by a fluxgate magnetometer at the end of a 60-cm-long boom, an electron analyzer and Faraday cup, and two sensors for low pressures. The magnetometer could make the first detection of the intrinsic field of a cometary nucleus, if one exists.[39] A suite of three instruments were to map the near-surface structure of the nucleus by means of acoustic sounding, its electrical characteristics (strongly correlated with the presence of water) and the dust environment.[40]

The Rosetta lander was fitted with various cameras, in two different groups. One, a 1,024 × 1,024-pixel camera pointing downward was to document the descent, with the final frame being taken less than 5 seconds before touchdown. Thereafter, it was to activate light-emitting diodes to provide illumination while it imaged the lander's belly and the operation of the other instruments, in particular the X-ray spectrometer targets, the drilling work and the resulting borehole.[41] The second set included five micro-cameras and a stereoscopic imager capable of taking a 360-degree panorama with a resolution of 1 mm, and two optical microscopes operating in the visible and infrared ranges capable of resolving detail as fine as 7 micrometers. The 1,024 × 1,024-pixel panoramic cameras were mounted on top of the lander. The microscopes viewed samples inside the medium-temperature ovens through a sapphire window.[42] Finally, a pair of perpendicular antennas would measure radio waves transmitted by the orbiter to investigate the internal structure of the nucleus.[43,44,45,46]

It was decided that Rosetta should sample the periodic comet 46P/Wirtanen. This was discovered by Carl A. Wirtanen of the Lick Observatory in January 1948, and except for 1980 (when it remained close to the Sun in the sky) it had been observed at every return since.[47] It was closely scrutinized in preparation for the mission. In particular, observations by the Hubble Space Telescope in August 1996 showed its nucleus to be only 600 meters in size and to spin with a 6-hour period. Moreover, it was also known that at times a large fraction of its surface was active, accounting for large rocket-like 'non-gravitational' changes to its orbit.[48] The spacecraft was to be launched during a 19-day window in January 2003 using an Ariane 5 G +, a version of the most powerful European rocket with the capability of starting its second (and final) stage in weightlessness. Rosetta would fly past Mars in August 2005, and then Earth in November 2005 and 2007 to build up the energy required to reach Wirtanen shortly after the spacecraft's aphelion in 2011. On the way, it would fly by asteroids (4979) Otawara in July 2006 and (120) Siwa in July 2008. These were selected in preference to (3840) Mimistrobell and (2703) Rodari in order to gain the opportunity to inspect Siwa, which, at 110 km, was one of the largest C-class asteroids – in fact, it was about thrice the size of Mathilde, which was inspected by the US Near-Earth Asteroid Rendezvous (NEAR) mission in 1997. Although Otawara was an unremarkable 4-km-sized S-class object, its rapid 3-hour rotation would enable its surface to be fully mapped during a flyby. After aphelion, Rosetta would rendezvous with Wirtanen in late November 2011. It would release its lander in the summer of 2012, and accompany the comet to its perihelion in July 2013. The Rosetta orbiter and lander were shipped to Kourou in September 2002 for integration. Remarkably, this would be only the second deep-space mission to depart from French Guiana and, like the first, Giotto in 1985, the target was to be a comet.

A model of the Rosetta Philae lander showing the 'balcony' side, where most instruments are mounted. The white cylindrical object is the housing of the sampler drill.

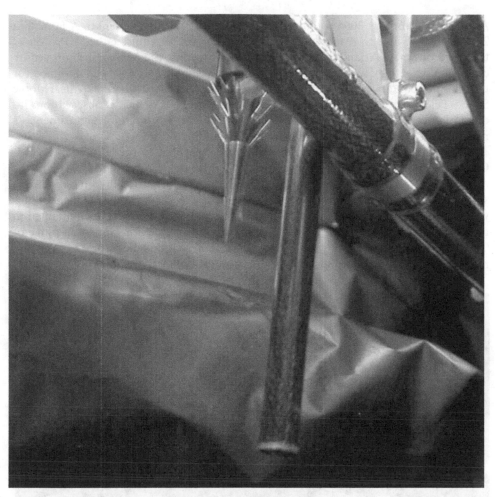

One of the two Philae anchoring harpoons protruding from the spacecraft after integration with the orbiter. (ESA)

On 11 December 2002, as Rosetta and its launcher were being prepared, disaster struck. The maiden flight of the more powerful ECA version of the Ariane 5 failed, destroying the two satellites that it was carrying. With the closure of the window for Rosetta looming on 31 January, ESA and Arianespace, the supplier of the launcher, reviewed the options and, even although the launcher for Rosetta differed from that which had failed, decided not to attempt to launch Rosetta in 2003. In fact, the cause of the disaster was soon found to be the uprated version of the first stage engine used by the Ariane 5 ECA.[49,50,51,52] Owing to the postponement, the Rosetta science team was asked to review the target. Over 150 candidates were examined, nine scenarios were presented in February to ESA's Science Program Committee, and three were selected for further study. The first option was to switch to a Proton and launch in January 2004 in an effort to reach Wirtanen. Alternatively, launching on the original

Ariane 5 in February 2004 could reach 67P/Churyumov–Gerasimenko. A launch in 2005 would be able to reach this same target, but would require either a Proton or the more powerful Ariane 5 ECA. The option of trading the Mars flyby for a more efficient Venus flyby was rejected, because the spacecraft had not been designed for the thermal conditions in the inner solar system. As ESA deliberated, the Hubble Space Telescope and the European Southern Observatory made observations to determine at least the principal physical parameters of Churyumov–Gerasimenko's nucleus and assess the chances of successfully landing on it.[53,54,55]

The 67th periodic comet to be cataloged was discovered by the Ukrainian Klim I. Churyumov on images taken by Svetlana I. Gerasimenko in September 1969 at the Almaty Observatory in Kazakhstan, and had been observed on each of its six returns since. Studies of its orbit established it to be a relatively fresh comet. An encounter with Jupiter in 1840 had reduced its perihelion from 4 AU to 3 AU. It remained in this orbit until it approached Jupiter again in February 1959, this time at a range of 0.052 AU, and the gravitational perturbation cut its perihelion to 1.28 AU and reduced its period to 6.6 years. When discovered in 1969 it was making its second

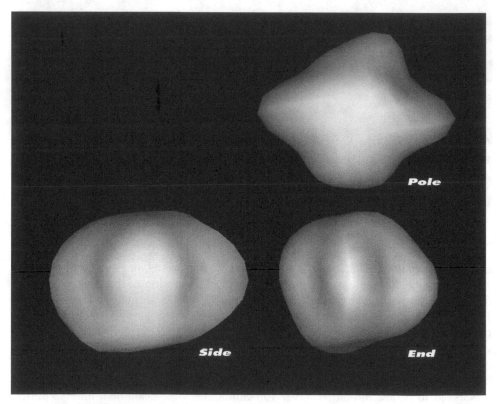

The 'starfish' nucleus of comet 67P/Churyumov–Gerasimenko reconstructed from Hubble Space Telescope observations. More recent analyses indicate that the nucleus may actually be a broad, slightly flattened object. (NASA, ESA and Philippe Lamy)

perihelion passage in this orbit.[56] Projecting forward, a more distant Jupiter encounter in 2007 would reduce the perihelion to 1.24 AU. Although relatively active for a Jupiter-family comet, and sporting a bright inner coma and a faint tail, Churyumov–Gerasimenko appeared to produce only one-fortieth of the dust issued by Halley. Indeed, at its 2002 perihelion it appeared to be issuing a mere 60 kg of dust per second and about twice as much gas. However, in rare outbursts its rate of dust production would increase by a factor or four. The Hubble Space Telescope took no fewer than 61 images of Churyumov–Gerasimenko between 11 and 12 March 2003, whilst the comet was receding from perihelion, and these revealed it to have a 3 × 5-km 'starfish' nucleus that rotated in about 12 hours.[57,58,59]

While ESA considered the mission options, the Rosetta spacecraft was stored in a clean room in Kourou. For safety, the harpoons were removed from the lander. The high-gain antenna and solar panels were removed from the orbiter. The 660 kg of hydrazine was offloaded in order to reduce the risk of an explosion, but the nitrogen tetroxide was left as a precaution against corrosion of the tanks and plumbing. Five instruments were unloaded for refurbishment. Various modifications were made to tailor the software and validate it for a mission to Churyumov–Gerasimenko. As the revised mission would probably require the spacecraft to pass slightly closer to the Sun than previously planned, reflective surfaces were added to some of the thermal blankets in order to prevent overheating.

On 14 May 2003, realizing there was little prospect of modifying the spacecraft in time for a Proton launch in January 2004 to retain Wirtanen as the target, ESA chose to launch on an Ariane 5 in February 2004 and shoot for Churyumov–Gerasimenko. The storage of the spacecraft, late launch, and longer mission would cost the agency an additional 80 million euros. A Proton launch in February 2005 remained a backup option, but given the need to modify the spacecraft, which could not start until it was evident that it would not be launched by Ariane 5, it was unclear how realistic this was. In order to reach Churyumov–Gerasimenko, the spacecraft would need to make flybys of Earth in March 2005, November 2007 and November 2009, and of Mars in March 2007. If the propellant reserve was sufficient to ensure the planned cometary mission, then the vehicle would detour to inspect targets of opportunity. There were several asteroid candidates, but the decision to attempt any encounters would not be made until after launch.[60]

Because Churyumov–Gerasimenko was three to four times the size of Wirtanen, its surface gravity would be at least 30 per cent greater and the speed of the lander at touchdown would be at least thrice that envisaged by its designers. This could cause problems of stability upon landing – it had been designed not to bounce on making contact, and a fast impact on a hard surface could cause it to topple over. After low-gravity pendulum tests showed that the addition of a simple 'tilt limiter' bracket to the landing gear would ensure a successful landing at rates of up to 1.5 m/s, this tiny bracket was installed on 30 September. Tests and simulations established criteria in terms of the mean density, size and surface slopes of the comet and approach speeds for various degrees of risk in attempting a landing. Another possible downside of the nucleus of Churyumov–Gerasimenko being significantly larger than Wirtanen was that radio waves for the sounding experiment might be completely absorbed, making

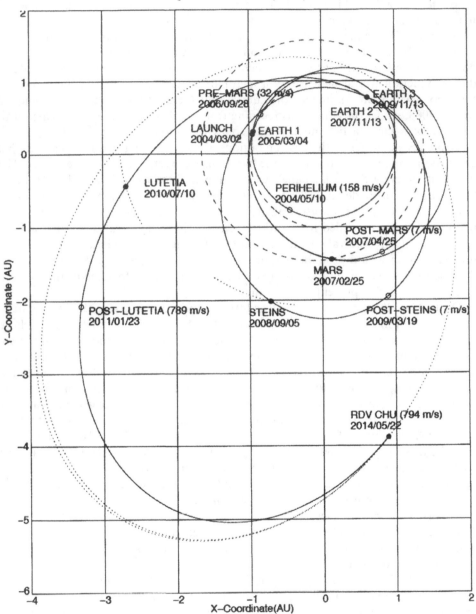

Projection to ecliptic plane

The circuitous road of Rosetta from launch in 2004 to rendezvous with Churyumov–Gerasimenko in 2014. (ESA)

it impossible to study the deep interior. Other issues concerned the higher radiation dose that the spacecraft would endure as a consequence of spending a longer period in space around the time of solar maximum.

The 21-day launch window for Churyumov–Gerasimenko opened on 26 February 2004, but the first two attempts were scrubbed: the first time owing to high altitude winds, and again because insulation on the cryogenic tank of the first stage became detached and the launcher had to be returned to the assembly building for necessary repairs. Rosetta finally lifted off on 2 March. The second stage of an Ariane 5 would normally ignite immediately after the first shut down, but for this mission it was to make a free-flight lasting several hours prior to firing. The hardware and operational modifications required for this technique would also be useful for future commercial flights and for resupply flights to the International Space Station using this vehicle.[61] The first stage inserted the stack into a 57 × 3,849-km orbit that would theoretically cause Rosetta to re-enter the atmosphere at its first perigee, but 2 hours after launch (as planned) the storable-propellant second stage fired on the descending leg to place the spacecraft into a 0.885 × 1.094-AU heliocentric orbit. Shortly after the launch, ESA announced that the lander had been named Philae, after an island on the Nile near Aswan where an inscription on an obelisk had confirmed the decryption of the hieroglyphics facilitated by the Rosetta stone. Some 8 hours after liftoff, the locks which had protected Philae from the rocket vibrations were released. After tracking had measured the accuracy of the launcher and the extent to which the spacecraft would have to maneuver to achieve the nominal trajectory had been calculated, the propellant margin was assessed and the asteroid encounters were announced. A team of European scientists had studied all possible launch scenarios and compiled a list of possible targets. The selection criteria included the possibility of determining the asteroid's mass by Doppler tracking of the spacecraft, and of investigating objects of either a hitherto unstudied taxonomic type or of as 'primitive' a type as possible. It had been decided to detour to (2867) Steins in September 2008 and (21) Lutetia in July 2010. Less than 10 km in size, Steins was discovered on 4 November 1969 by Soviet astronomer N.S. Chernykh at the Nauchnyj Observatory in Crimea and was named after Karlis A. Šteins, former director of the Latvian University Astronomical Observatory. Almost 100 km in size, Lutetia was discovered on 15 November 1852 by Hermann M.S. Goldschmidt in Paris and dedicated to the town itself, which was known to the Romans as Lutetia Parisiorum (Lutetia of the Parisii tribe).[62]

About a week into the mission, the instruments were switched on one at a time for calibration. Other than payload commissioning and several course corrections, few activities were scheduled for the first circuit of the Sun. But in April and May 2004 the opportunity arose for scientific observations of the long-period comet C/2002T7 LINEAR and also to study the tail and meteor stream of Giacobini–Zinner using the magnetometer. In a test in May, the lander took some delightful pictures from its attached position showing the rear of one of the solar panels illuminated by sunlight reflecting off the orbiter's body, proving the ability of its cameras to resolve details as fine as the honeycomb structure of the solar panels. Pictures of the empty ovens taken by the microscope verified the correct functioning of the oven carousel and its positioning system. The downward-looking camera successfully imaged the thermal

The Ariane 5 G+ named 'Ville de Colleferro' (after the town near Rome where the boosters are built) ready to launch Rosetta, one year late. (ESA/CNES/Arianespace, Photo Service OptiqueVidéo CSG)

blankets of the orbiter in the ghostly illumination of the light-emitting diodes. The orbiter's camera took calibration pictures, including stunning views of Earth and the Moon from a distance of 73 million km and also of the Orion Nebula. The imaging spectrometer and infrared radiometer had earlier taken calibration measurements of Earth from a range of 18 million km.[63],[64] After a deep-space maneuver of 152.8 m/s on 10 May, the spacecraft tweaked its trajectory on 16 May with a burn of by 5 m/s, then reached perihelion on 24 May.[65]

On 4 March 2005, one year into the mission, Rosetta returned to Earth as planned. It approached our planet almost exactly from the anti-Sun direction, a geometry that favored particles and fields studies but precluded observations by most of the other instruments. The point of closest approach was 1,954 km over the Pacific, just west of Mexico. Several instruments were activated as it receded over the

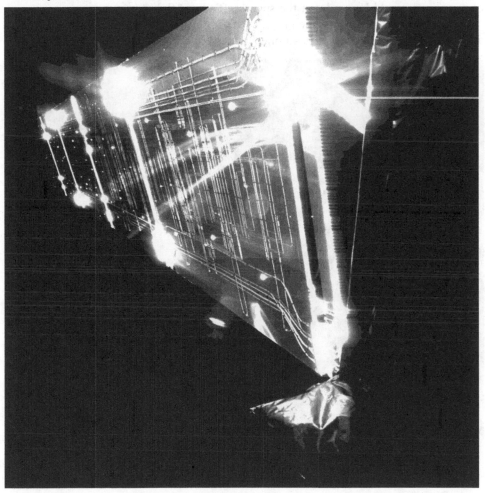

The rear of Rosetta's solar panels imaged by Philae during a calibration exercise in May 2004. (ESA)

dawn side, and multispectral images distinguished between vegetation over the Andes, the rainforest of Brazil and the pampas of Argentina. Infrared images also detected cloud patterns over the night-side.[66] About 16 hours after the flyby, Rosetta flew by the Moon at a range of 173,530 km and exercised software for the tracking of asteroids by slewing the spacecraft to maintain the navigation camera pointing at the Moon. Images of Earth were taken by the panoramic camera on Philae. The lander also calibrated its magnetometer against the measurements made by the orbiter's instruments in a well-known field. The science camera on the orbiter was unable to be used because of an unresolved problem with the cover of its optics, but this would soon be corrected by software patches. Doppler tracking detected the mysterious 'flyby anomaly' that had perturbed several spacecraft in making Earth flybys. For Rosetta, an unexplainable velocity change of 1.82 mm/s was observed. Additional observations would be made on the spacecraft's later flybys.[67]

At this point ESA had planned to put Rosetta into a low-power cruise mode, but at NASA's request this was postponed by several months so that the spacecraft

The Moon rising over the Pacific Ocean pictured by one of the Rosetta navigation cameras during the February 2005 flyby. (ESA)

could observe the collision of the Deep Impact projectile with the nucleus of Tempel 1. In contrast to ground and Earth-orbiting telescopes, Rosetta would be able to observe the comet continuously for days. It was first pointed towards Tempel 1 on 29 June and continued until 14 July, covering all of the impact phases on 4 July by taking pictures at a peak rate of one image per minute. Although the range was 80 million km, the view was at optimum solar elongation. The results revealed that the rate of water production increased 100,000-fold at impact, provided data on dust production rates and outflow speeds, and monitored how the comet's brightness changed during the collision. Rosetta found that the total amount of water released during the impact was about 4,600 tonnes, mostly as icy grains, while at least a similar amount (and in all likelihood much more) of dust was ejected. This confirmed Tempel 1 to be more of an 'icy dirtball' than a 'dirty snowball'. The data placed a lower limit on the volume excavated by the projectile, corresponding to a crater 30 meters in diameter; but it was probably much larger. The outburst in brightness, however, was extremely brief.[68],[69]

Afterwards, Rosetta entered a hibernation mode that was intended to last until the Mars flyby in February 2007. Onboard activity was minimized, and the spacecraft was tracked only once per week to ascertain its health and to download engineering data. In order to avoid excessive mechanical wear of the reaction wheels during such a long flight, the spacecraft used its thrusters for attitude control. Nevertheless, some scientific observations were made at solar conjunction in early 2006 when the line of sight of its radio signal passed through the corona, and then again in July when the spacecraft passed within 0.06 AU of the tail of comet Honda–Mrkos–Pajdušáková, although the data collected on that occasion was not retrieved until the conclusion of the period of hibernation. The fact that the instruments had successfully detected the tail prompted scientists to request that the plasma package be kept on during future comet tail crossings.

A large maneuver on 29 September 2006 and a series of minor burns refined the approach to Mars.[70],[71],[72] The mission plan for Wirtanen had included a gravity-assist at Mars, but unlike that opportunity this one offered only limited scope for scientific observations because the spacecraft would spend some 25 minutes in the shadow of the planet. As the first time the spacecraft lost sight of the Sun since launch, this was a situation for which it had not been designed. In addition, the line of sight to Earth would be lost for just over 14 minutes. Particles and fields observations were to be made in approaching the day-side of the planet and then receding over its night-side, in the process crossing the most important Martian plasma boundaries and spending a few minutes inside the planet's ionosphere. The results were to be correlated with those of the Mars Express orbiter.[73] The camera took global images at distances of the order of 240,000 km. The color filters highlighted dust clouds, and the ultraviolet filter chosen to detect cometary water revealed structures in the Martian atmosphere such as polar and limb clouds. A sequence of Phobos crossing the planet's disk was also obtained. The radiation monitor collected data on the Martian environment that may assist in the planning of human missions. The instruments on the orbiter were switched off for 3 hours for closest approach and the solar eclipse, but the lander's camera and magnetometer remained on battery power, wholly autonomous for the first time in the mission. The magnetometer recorded the Martian bow shock as the vehicle entered the ionosphere, followed by the crossing of the tail. Only 4 minutes before the moment of closest approach, the lander took one of the most stunning images in the history of solar system exploration, showing the black structure of the orbiter and one of the solar panels set against the disk of Mars, 1,000 km away, with Mawrth Vallis clearly visible. The time of closest approach was at 01:54 UTC on 25 February 2007, with the spacecraft passing some 250 km above a point at 43.5°N, 298.2°E at a relative speed of almost exactly 10 km/s.

As Rosetta receded over the night-side, it took a series of pictures of the crescent planet and Phobos at angles rarely seen by spacecraft and inaccessible from Earth. Long-exposure views were taken to look for atmospheric night-glow, but ended up being mostly overexposed.

Mars and the structure of the Rosetta orbiter in the foreground seen by Philae. (ESA)

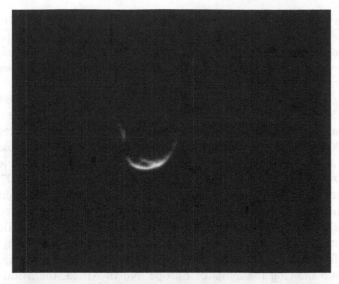

A crescent of Phobos seen by the Rosetta orbiter. (ESA)

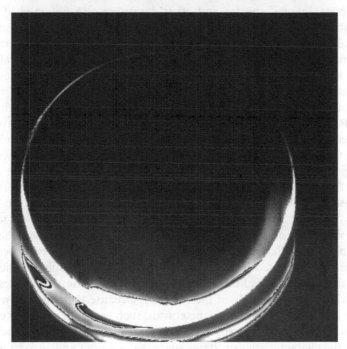

The night-side of Mars seen by Rosetta, with the overexposed crescent at bottom. This
is a geometry rarely used to image the planet. (ESA)

A few days after the Mars flyby, observations were made of Jupiter in support of NASA's New Horizons passage through the Jovian system, in particular using the ultraviolet spectrometer to monitor the Io torus. After these observations ended in May, a deep-space maneuver set up the Earth flyby on 13 November 2007 and Rosetta resumed its hibernation.

On making its second approach to Earth, Rosetta again approached the night-side and receded on the day-side. On 7 November, being faintly visible in the night sky, it was discovered by telescopes seeking dangerous near-Earth objects, cataloged as asteroid 2007VN84, and found to be about to make a shockingly close 0.000081-AU passage; a planetocentric distance of just 1.89 Earth radii! But its identity was soon recognized by a Russian astronomer and the designation 2007VN84 was canceled. Nevertheless, its detection proved that the search telescopes were alert, and amateur astronomers had an excellent opportunity to image the spacecraft.[74,75] It flew by at a relative speed of 12.5 km/s, passing 5,301 km over the Pacific southwest of Chile. A number of observations were made by Rosetta, this time using the science camera as well. In particular, the wide-angle camera took some amazing pictures of the Earth's limb showing the lights of Europe, North Africa, the Middle East and India at night. The camera also took views of Antarctica and searched for meteor trails from space. Images, spectra and other data were also obtained of the Moon and the Earth's magnetosphere. The flyby of Earth in 2007 yielded no measurable 'flyby anomaly' and stretched Rosetta's aphelion to 2.26 AU, within the asteroid belt. After making a course correction on 23 November, the spacecraft resumed hibernation, awaiting its appointment with Steins, which was to be Europe's first flyby of an asteroid.

Almost nothing was known about Steins when it was selected for the first asteroid flyby of the Rosetta mission, but it was starting to be characterised by investigations using the largest astronomical facilities on Earth, including the European Very Large Telescope in Chile and JPL's Table Mountain Observatory, and in November 2005 by the infrared Spitzer Space Telescope. Rosetta had itself observed the asteroid in March 2006 from a distance of 1.06 AU, using its science camera to obtain a 'light curve' at phase angles larger than geometry constraints ever allowed from Earth, and continuously for 24 hours, uninterrupted by day/night cycles. It took a total of 238 images spanning four rotations of the small asteroid. Although researchers initially listed Steins as S-class (like most previously visited bodies), the new observations of its spectral properties placed it in the E taxonomic class of reddish bodies having a high albedo that were believed to be thermally evolved owing to episodes of partial melting and differentiation early in their history. Because their spectra were similar to rare enstatite chondrite or aubrite meteorites, such asteroids could be expected to have surfaces of iron-poor (perhaps even iron-free) silicates. Some researchers noted that some of the characteristics of Steins suggested a young and very rough surface, at most a few million years old. Its rotation period was calculated to be 6.05 hours. Although asymmetries in the light curve confirmed it to have an ellipsoidal shape of 5.73 × 4.95 × 4.58 km, it did not appear to have any major irregularities. Fewer than 30 members of the E-class were known, and little was known of their evolutionary history. The class included (44) Nysa, which is the largest, and also the near-Earth objects (3103) Eger and (4660) Nereus, the latter of which was a recurrent candidate for space missions. The Earth-crossing orbits provided a geometry for enstatite and aubrite meteorites to fall on our planet.[76,77,78,79,80,81,82,83,84,85,86]

On 4 August 2008 Rosetta aimed its science camera at Steins, at that time some 26 million km distant. It imaged the asteroid for a month to refine the ephemeris, initially twice weekly, then daily through to a few hours before the encounter. A total of 340 images were taken in what was Europe's first optical navigation in deep space. Meanwhile, a more detailed light curve allowed the shape of the asteroid to be refined, revealing it to possess pointed features. Initially, Rosetta was to have flown by Steins at a range of about 1,745 km in order to rehearse the dynamical characteristics of the Lutetia flyby. No imaging was to be conducted around closest approach, as this would have meant exposing the 'cold face' of the spacecraft (and hence the radiators) to the Sun. But the scientists argued that this encounter should be devoted more to science than to an engineering rehearsal, and they managed to have the flyby range cut to the minimum which was deemed safe and to schedule observations for when the spacecraft was passing near the imaginary line between the Sun and the asteroid (in scientific parlance at 'phase angle zero'), at the time of closest approach, and also on the outbound leg. Not only would this pose a significant navigational challenge and expose the radiators to the Sun, the rapid slewing at closest approach would also drive the reaction wheels close to their limit. But the revised encounter sequence had been rehearsed in March, and confidence was high.

A maneuver on 14 August tweaked Rosetta's speed by a mere 12.8 cm/s to set up the flyby, and an 11.8 cm/s correction on 4 September refined the closest approach to 800 km, with an uncertainty of just 2 km. A total of 15 instruments were activated to perform imaging and spectrometry, measure particles and fields, analyze gas and dust, and measure magnetic fields using the lander's magnetometer. And, of course, the radio signal would facilitate Doppler tracking to measure the asteroid's mass. At best, Steins would span just over 300 pixels in the field of view of the narrow-angle camera with the smallest detail resolvable being about 30 meters across. With some 12 hours to go, the navigation cameras were allowed to take control of the spacecraft and of its orientation. This was a critical maneuver. When it was discovered that the cameras had difficulty in detecting and tracking Steins owing to its small angular size and thermal noise in the images, a series of camera settings were frantically tested until some settings were found which worked. The decision on whether to use autonomous tracking was deferred until 2 hours before the encounter. If Rosetta was unable to track Steins, it would have to execute a sequence of observations based on the team's best estimate of how the target would move in relation to the vehicle, and if this were inaccurate then some of the observations would be lost. It was decided to attempt autonomous tracking. The flyby required Rosetta to perform a flipping maneuver some 20 minutes prior to the encounter in order to hold Steins in the field of view of its cameras, while minimizing the exposure of the 'cold face' to the Sun. It started to slew to track the target 1 minute after executing this flip. These tracking activities resulted in the high-gain antenna losing its lock on Earth some 10 minutes before the flyby.[87]

Approximately 2 minutes before closest approach, Rosetta passed almost exactly between Steins and the Sun. At 18:38 UTC on 5 September 2008 Rosetta passed 802.6 km from the asteroid at a relative speed of 8.62 km/s.[88] Less than 2 hours after the spacecraft had severed its high-gain link, the Goldstone antenna resumed communications and confirmed that the encounter had been successfully completed and began to download engineering and scientific data. A total of 551 images had been obtained. The only major glitch was the narrow-angle camera switching to safe mode 9 minutes prior to closest approach, by which time Steins was still 5,200 km away, only to resume working several hours later. As a result, the highest-resolution pictures were lost. Only moderate-resolution color coverage was obtained. Still, all of the asteroid's surface was observed at low resolution and some 62 per cent of it at a resolution better than 200 meters. A resolution of 80 meters at best was achieved by the wide-angle camera, and a similar one by the narrow-angle camera before it went into safe mode.

Asteroid Steins proved to be grayish in color and to reflect about 35 per cent of the light it received from the Sun. It was 6.67 × 5.81 × 4.47 km in size and had a pronounced non-spherical shape. In fact, with a conical and a tronco-conical hemisphere it bore a striking similarity to a cut diamond. It was heavily cratered. One crater, 2 km across, was on the flat face of the 'diamond', corresponding to the south pole. A remarkable feature was a string of at least seven craterlets distributed along a meridian resembling the catenae present on much larger bodies. However, the mechanism believed to be responsible for such features (namely the tidal disruption of a small asteroid or comet that ventured too close and broke up) could never have been caused by an object as small as Steins, so perhaps it simply collected the debris of an already shattered object. Alternatively, it might be a series of collapse pits along a fault. Another elongated groove was located almost antipodal to the chain of craters or pits. The degraded appearance of the craters all across Steins hinted at the presence of a blanket of fine regolith on the surface, while the scarcity of small craters suggested that the surface of the asteroid was relatively

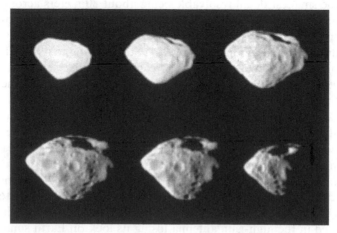

A mosaic of Rosetta wide-angle images of asteroid Steins. (ESA)

young.[89] The diamond shape was attributed to material sliding toward its equator as its rotation was spun up by the Yarkovsky–O'Keefe–Radzievskii–Paddack (YORP) effect, involving different radiation pressures over the night and day sides. If this is what really happened, this means Steins is a porous 'rubble pile' body like Itokawa. The spin-up and slippage of material would have obliterated the previously existing craters, thereby 'rejuvenating' the surface of the asteroid. It is also possible that as the shock wave from the impact that made the 2-km crater propagated through the ancient body, this caused its crust to peel off like the first layer of an onion, and thus reveal a fresh surface underneath. In fact, the age of the surface could be as little as a few hundred million years. Other instruments also delivered data on Steins. The imaging infrared spectrometer measured the surface temperature, which varied from –90°C down to a very chilly –240°C. The microwave radiometer measured a rock-like thermal inertia, implying a surface not covered by powdered regolith. The ultraviolet instrument collected spectra showing a low abundance of iron on the surface, confirming the association of Steins with the E-class asteroids, and searched for an exosphere of oxygen and hydrogen atoms. Finally, the magnetometers on both the orbiter and lander failed to detect any remanent magnetization.[90,91,92,93]

Just after the Steins flyby, Rosetta was involved in another experiment by taking images of the central region of the Milky Way at the same time as this was inspected by ground-based telescopes in an effort to detect gravitational 'microlensing' events. This effect occurs if a foreground star happens to cross the line-of-sight to a more distant star, and the gravitational bending of starlight by the intervening star creates a characteristic brightening of the more distant one. A certain number of foreground stars should be double or even be accompanied by planets, and these should leave a detectable 'fingerprint' in the way the background star brightens. Such events have been seen from Earth, but if two widely separated observers saw the same event then the mass of the foreground star and planet could be determined to a relatively high accuracy. It was estimated that a narrow-angle camera mosaic of the central region of the Milky Way should typically include about 50 microlensing events. Although several events are known to have been detected, the results of the experiment remain to be published.[94]

On 17 December 2008 Rosetta reached its aphelion at 2.26 AU, and was in solar conjunction at about the same time. It returned to Earth for its third flyby two years after its second. In preparation for this last Earth flyby, the spacecraft was awakened from hibernation on 8 September 2009, its systems and instruments were extensively checked out and its orbit was corrected. It once again approached the Earth from its night hemisphere, at a relative speed of 9.4 km/s. An amazing approach movie of the crescent Earth was taken over a period of 24 hours, as the spacecraft went from 1.1 million km to 322,000 km, just inside the Moon's orbit. Narrow-angle images of North America at night were obtained closer in, clearly showing city lights over most of the southern United States. Images of Tenerife Island at night were also obtained in an attempt to detect a laser beam fired by a telescope. Given the critical need to precisely target the flyby to ensure the rendezvous with comet Churyumov–Gerasimenko, the Earth and Moon observations received a relatively low priority. A major effort went into obtaining a very accurate orbit determination as the

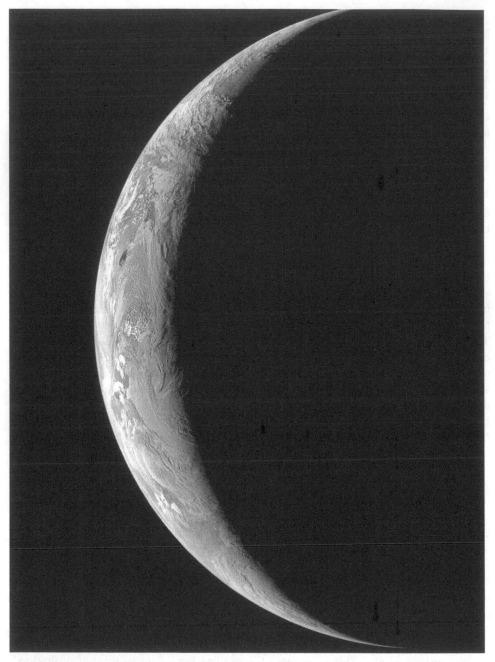

A wide-angle image of Earth taken by Rosetta from a distance of 350,000 km on 12 November 2009. The crescent includes part of South America and Antarctica. (ESA)

spacecraft approached Earth for a new measurement of the still unexplained 'flyby anomaly'. Therefore, instruments were operated only when this would not interfere with flyby operations and orbit determination. Closest approach was on 13 November 2009, at an altitude of 2,480 km above a point just south of the island of Java. No 'flyby anomaly' was detected this time either. Amazingly, therefore, an anomalous speed change was detected on the first Earth flyby, but not on the second or the third. More pictures of an almost fully illuminated Earth were obtained on the outbound leg by the navigation cameras as a test to rehearse operations in imaging and tracking an object filling the whole field of view, as asteroid Lutetia would do in 2010. Some of the pictures taken on the outbound leg fittingly showed Egypt and the Arabian Peninsula. A little over 12 hours after closest approach to Earth, Rosetta passed 233,000 km from the Moon and the microwave radiometer was trained on it in an attempt to detect the OH radical discovered by the Indian Chandrayaan orbiter and confirmed by Cassini and Deep Impact observations.

Then, just after the flyby, Rosetta hit its last perihelion inside the orbit of Earth, at 0.98 AU. This gravity-assist stretched its aphelion out to the orbit of Jupiter at 5.09 AU. The spin-stabilized deep-space hibernation mode that Rosetta would use during most of the flight to aphelion was rehearsed in January 2010. On penetrating the asteroid belt for the second time, the vehicle would pass Lutetia on 10 July 2010. The data from the IRAS infrared astronomy satellite enabled the average diameter of this object to be roughly estimated during the 1980s at 95.5 km, but light curves obtained since, most recently by the Hubble Space Telescope, established it to be asymmetric and to rotate in just over 8 hours. It also appeared to be covered with a fine-grained regolith. In 2008 and 2009, high resolution near-infrared images were obtained using adaptive-optics cameras on two of the world's largest telescopes: the Keck Observatory in Hawaii and the European Southern Observatory's Very Large Telescope in Chile. These images clearly resolved the shape of the asteroid and even gave some idea of its surface features. The shape was "well described by a wedge of Camembert cheese", and large depressions, probably craters were located near the north pole. The asteroid could be approximated as an ellipsoid roughly 124 × 101 × 80 km in size. Remarkably, the spin axis proved to be almost in the orbital plane, as is so for Uranus. As a consequence, each hemisphere would experience long seasons of continuous sunlight followed by continuous darkness. In particular, it was found that the Rosetta flyby would occur when the northern hemisphere was in constant sunlight. In these conditions, it would be difficult to gain a precise reconstruction of the shape of Lutetia to correlate with mass determinations. The shape of the southern hemisphere would have to be inferred from infrared radiometer scans. Nevertheless, Lutetia is one of the best-characterized asteroids to receive a spacecraft inspection. Yet, despite many spectroscopic studies, its taxonomic class was disputed. For a long time it was listed as an M-class object that spectrally resembled some iron-rich meteorites. Thus far, no such body has been visited. But this classification was not confirmed by infrared observations by the telescopes in Hawaii and Chile, nor by the Spitzer Space Telescope, nor indeed by ultraviolet observations made by the Hubble Space Telescope. Hence Lutetia is now regarded as an anomalous C-class object which reflects more light than other members of the class, possibly implying that its surface is 'unweathered'. In fact, it appears similar to some metal-rich carbonaceous chondrite meteorites.[95],[96],[97],[98],[99] Rosetta made observations of the asteroid in January 2007, just before the Mars flyby, in order to assist in characterizing its light curve, rotation period and spin axis.[100],[101],[102]

On 16 March 2010, just after performing a rehearsal of the Lutetia flyby, Rosetta was turned to image 'comet' P/2010A2. Discovered in January, this object orbited entirely within the main asteroid belt and sported a dust tail but no coma. Images of its 'head' by the Hubble Space Telescope showed a small nucleus, several hundred meters in size plus a mysterious X-shaped dusty structure. Scientists suspected that the comet-like appearance was due to dust released in the collision between two small asteroids, the impactor probably being only a few meters across. By virtue of being well above the object's orbital plane, Rosetta had a unique perspective on this. Its observations facilitated a reconstruction of the true 3-dimensional shape of the dust tail, or more correctly 'trail', establishing that it was the result of a single, brief event and was not created by long-duration cometary activity.[103]

Two months later, on 10 May, the probe provided support to the American Dawn mission by observing its target asteroid, (4) Vesta, over a period of two rotations, from a distance of over 40 million km.[104]

Rosetta started to track Lutetia using its navigation and narrow-angle cameras on 31 May, at first twice weekly and then from 28 June on a daily basis. A total of 272 images were acquired for optical navigation purposes. Five slots were available for course corrections, but only the first, on 18 June, was used to increase the minimum approach distance by some 500 km. The errors were negligible, and the spacecraft achieved a flyby distance within 27 km of that planned. In particular, no corrections were needed 40 and 12 hours before encounter. Images taken starting 9 hours 30 minutes before the flyby resolved the asteroid. Four hours before closest approach, Rosetta performed a flip maneuver to target the looming asteroid. The flyby mode

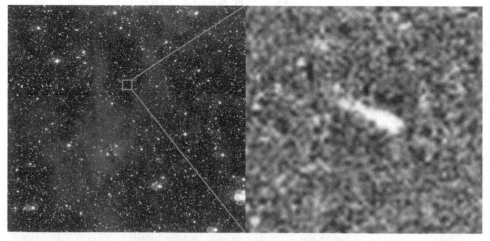

The asteroid belt 'comet' P/2010A2 imaged by Rosetta. This remarkable object was found to be a cloud of dust created by the collision between two small objects in the belt. (ESA)

was activated with 1 hour to go, allowing one of the navigation cameras to guide the probe so that Lutetia would remain within its field of view. Seventeen instruments were collecting data during the encounter: not only the remote-sensing imagers and spectrometers, but also the Ptolemy mass spectrometer that was attempting to detect a faint atmosphere surrounding the asteroid.[105]

About 18 minutes and 16,400 km from closest approach, Rosetta passed almost precisely between the Sun and Lutetia, then upon reaching the maximum latitude of 84°N it looked directly down at the large polar depressions. As expected, the high-gain antenna broke its lock on Earth with 5 minutes to go. Contact was maintained through the low-gain antenna, with Doppler tracking attempting to measure the mass of the asteroid. The 70-meter antenna of NASA's Deep Space Network near Madrid in Spain was tasked with these high-signal-to-noise-ratio measurements. Finally, at 15:45 UTC Rosetta flew by Lutetia at a range of 3,172 km and at a relative speed of 15 km/s. This was just a few kilometers from the closest distance the mission could pass lest the asteroid fill completely the field of view of the navigation cameras, complicating the autonomous tracking maneuvers. At that time, Rosetta and Lutetia were just over 3 AU from Earth and 2.71 AU from the Sun. Imaging continued for about 18 minutes after closest approach and then, 5 minutes later, the high-gain antenna re-established contact. Some amazing images were taken when Rosetta was 36,000 km away and 40 minutes out, as Saturn, some 6.5 AU distant, transited the field of view of the narrow-angle camera, placing it just alongside Lutetia.

Unlike at Steins, this time the performance of the payload was faultless and a total of 462 images of Lutetia were obtained, in addition to other data. They covered slightly in excess of 50 per cent of the surface, mostly on the northern hemisphere. The Earth-based determination of the shape of the asteroid were confirmed. Lutetia was 121 × 101 × 75 km across and angular, with flat plains and a giant dent in a side. Although it was pitted with large, smooth bowl-shaped craters, there were vast swathes lacking craters. Relatively few small craters were present. One 55 km bowl-shaped depression stretched across much of the visible surface. Large craters had depth-to-diameter ratios compatible with hard rock, while smaller ones had different ratios, probably indicating that they formed on a thick layer of regolith and ground-up rock. This was confirmed by thermal inertia measurements by the microwave radiometer, resembling those of lunar regolith. Judging from the subdued, softened outline of some of the craters, this layer must be at least 600 meters thick. Such a layer would be substantially thicker than that seen on other asteroids. It could be explained by the smaller percentage of impact debris reaching escape speed on such a massive asteroid. The debris that could not escape would fall back and blanket the surface. Details as small as 60 meters were visible, including many large boulders tens or hundreds of meters across that projected dark shadows. In fact, almost 240 boulders larger than 100 meters were identified. Some of them appeared to have rolled downslope, which seemed unlikely; probably, they recorded some traumatic event in the asteroid's past. Seen close up, the surface appeared to be crossed by grooves, scarps and pit chains. In fact, some of the highest resolution pictures could readily be mistaken for views of Phobos or Eros (which, in any case are two bodies several times smaller and probably made of low-strength material). Features initially received informal names after Parisian landmarks such as Chatelet, Trocadero and Marais, but later received a completely different, official nomenclature. Craters were named for ancient Roman cities, a large region was named after Hermann Goldschmidt, the discoverer of Lutetia, and others for provinces of the Roman Empire. Finally, other features were named after rivers and cities of that Empire. The 55-km impact basin was called Massilia after the ancient name of the city of Marseille.

The imaging spectrometer made a quick, push-broom swath across the surface of Lutetia at closest approach. No silicates nor hydrated minerals were found, and the surface was finally determined to have a composition similar to certain metal-rich classes of meteorites. Its temperature was no warmer than about $-30°C$. A search was made for any satellites of Lutetia, but nothing larger than 60 meters was found. As expected, although a rather precise determination of the mass was obtained, the first density estimates were rather poor because, approaching pole-on, Rosetta had not been able to determine the shape and volume of the night hemisphere. More precise determinations would have to await Earth observations made half an orbit

This picture of Lutetia with Saturn in the background was taken by Rosetta's narrow-angle camera at a distance of 36,000 km from the asteroid. (ESA)

A view of Lutetia taken near closest approach. The hollow in foreground is crater Massilia. (ESA)

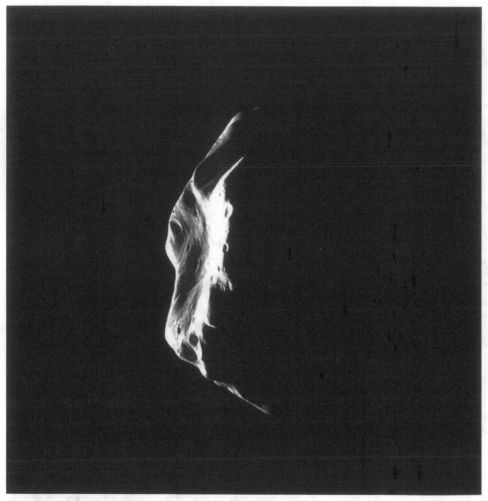

A view of Lutetia as Rosetta passed the terminator into the night-side. (ESA)

later, when the other side of the asteroid was illuminated. When these were finally available, a density more than three times that of water was able to be estimated; higher than for most of the asteroids previously visited by spacecraft but not enough to indicate that Lutetia was a metallic asteroid. Instead, it indicated that Lutetia was composed of rock with some metal, and could be even partially differentiated with a higher density metallic core that might have been molten at some time in the distant past. Unlike other asteroids visited by spacecraft, Lutetia appeared to be a primitive body that had never suffered destruction and re-aggregation during its existence and was thus definitely not a 'rubble pile'.[106],[107],[108],[109],[110]

Weeks after the flyby, a small leak was identified in the attitude control system thrusters piping. Because the system was not designed to be re-pressurized, future operations would have to be revised, but the team were confident that this would not impact the planned scientific objectives of the mission. Between 17 and 23 January 2011, Rosetta performed a series of four burns to produce a total velocity change of 778 m/s and target a rendezvous with comet Churyumov–Gerasimenko. Using only small, low-thrust engines, the critical deep-space maneuver would take a total of 17 hours to complete. On 18 January, one hour into one of the burns, small attitude disturbances were detected which caused the spacecraft to abort it and enter safe mode, pointing at Earth, and await instructions. Successive burns were made using the backup thrusters, putting them through a sort of operational test. In spite of the problem, the maneuver was completed one day late and achieved 98 per cent of the intended correction. On 15 and 26 March, Rosetta took the first sequence of 52 long-exposure pictures of its target. At that time the vehicle was 4.14 AU from the Sun, the comet was 5.1 AU from the Sun, the range was 1.09 AU, and the nucleus of the comet was no more than a dot in the star field. Rosetta was put into hibernation in June by turning off all its systems, including telecommunications, attitude control, propulsion, and scientific instruments. The only hardware that would be left on was the computer, clocks and heaters to keep electronics from freezing. Early on 8 June, Rosetta was oriented with its solar panels facing the Sun and made to slowly spin for stabilization. Finally, at 12:58 UTC the hibernation command was issued. Just over an hour later, there was loss of radio contact. It reached aphelion beyond the orbit of Jupiter on 2 October 2012 in this hibernation state and became the first solar-powered spacecraft to operate in that environment. But the energy in sunlight was 25 times less than at Earth, and the arrays provided only minimal power.

As the Rosetta mission progressed, our knowledge of its primary target significantly increased. In February 2004 (just before Rosetta lifted off) the Spitzer Space Telescope took 16 images of the nucleus of Churyumov–Gerasimenko when it was at a heliocentric distance of 4.48 AU, and then again in 2006 and 2007 around aphelion. Each such sequence spanned a full rotational period. The comet appeared to be inactive, but hundreds of days earlier it had ejected large dusty grains which, due to their mass, were relatively insensitive to solar radiation pressure and had traveled with the nucleus out as far as the orbit of Jupiter! Although this dust will pose a risk to Rosetta, the chance of an impact with one of these millimeter-sized grains during the approach phase is estimated at less than 1 per cent. Unlike other Jupiter-family comets which are active throughout the entire orbit, including at aphelion, Churyumov–Gerasimenko appeared to be quiescent when farthest from the Sun, giving Rosetta an opportunity to monitor the increase in activity as it approached perihelion. These infrared observations also indicated the nucleus to be ellipsoidal, with principal axes of 4.40 to 5.20 × 4.16 to 4.30 × 3.40 to 3.50 km. In fact, these observations yielded a different shape for the nucleus, which may be a broad and slightly flattened object resembling the nucleus of Tempel 1 instead of a 'starfish'. The fact that it reflects a mere 4 per cent of the light that it receives from the Sun is consistent with other cometary nuclei. The good news is that with a nucleus of that size and a low bulk density (inferred by extrapolating the Deep Impact findings at Tempel 1) the Philae lander is likely to make a successful touchdown. Of course, its chances will depend on the physical characteristics of the nucleus. Observations of the bare nucleus by the European Very Large Telescope in 2006 refined its rotation period to 12.7 hours.[111],[112],[113],[114]

In January 2014, after aphelion and with the illumination having increased enough to power its payload, Rosetta emerged from hibernation. A triply redundant clock woke up the probe at 10.00 UTC on 20 January 2014, after two and a half years in hibernation, in order to prepare for the rendezvous with the comet, still about 9 million km away. The wake-up process was carried out entirely autonomously without intervention from Earth. The spacecraft warmed up some avionics, stopped its spin and oriented the high-gain antenna to communicate with its mother planet. More than seven hours later, a 'sign of life' was sent to Earth, where it was received at 18:18 UTC. A thorough health check was then carried out. By early April, the solar panels were generating enough power to enable the instruments to be turned on and calibrated. The cameras were used to try to locate the comet. A second rendezvous maneuver will be made on 21 May 2014 to match the orbit of the comet and cut the relative speed to several tens of meters per second. By the time it is 4 AU from the Sun, the spacecraft will be flying in formation with Churyumov–Gerasimenko, some 10,000 km distant from it. The navigation camera will image the entire area of sky representing the ephemeris of the nucleus based on ground observations, which is expected to have an uncertainty of several tens of thousands of kilometers. Once the ephemeris is refined by optical navigation, the spacecraft will be allowed to reduce the separation to several thousand kilometers whilst also retaining a Sun-comet-spacecraft angle suitable for imaging. At 300 cometary radii or 1,000 km, Rosetta will reduce its relative speed to 1.5 m/s, which will eliminate any risk of damage through impact with dust released by the comet. As soon as the nucleus subtends several pixels in the camera's field of view the spin axis will be able to be determined, with the accuracy improving as the distance decreases. By the time the range is 50 to 100 cometary radii, precise tracking of the spacecraft will enable the mass of the nucleus to be measured by its gravitational influence. At such distances, the nucleus will span about 500 pixels, facilitating more detailed studies of its rotational state, precession, nutation, etc.

A series of maneuvers will allow Rosetta to close within 25 radii of the nucleus at a relative speed of several centimeters per second. On 6 August 2014 a 'capture burn' will insert it into an elliptic polar orbit ranging between about 5 and 25 radii with the periapsis near the equator on the sunward side of the nucleus. The actual parameters of the orbit will depend upon a variety of constraints, including the precise rotation period of the nucleus, the proximity to the surface at any moment, orbital stability, etc. Despite the comet having such a small mass, it should be possible for Rosetta to remain in orbit at perihelion, when solar perturbations, including radiation pressure, will be greatest and the gravitational 'sphere of influence' of the nucleus will shrink to a few kilometers; but it will be an extremely complex task – far more so than for the NEAR spacecraft orbiting Eros. The trajectory of Rosetta around the nucleus of Churyumov–Gerasimenko will bear little resemblance to the simple repeating ellipse of a Keplerian orbit. The first orbital phase will be dedicated to characterization, at a range of about 60 km, with Rosetta flying around the nucleus to determine its gross shape. This will be followed by a global mapping phase from about 20 km that comprises four half-orbits in two different orbital planes. These phases will be followed by orbits at lower altitude, starting off circular at 10 km and then dropping to a periapsis of 5 km in order to use low flyovers to obtain very-high-resolution imagery and characterize the candidate landing sites. These close passes will be designed to conform to the requirement that Rosetta must always be in sight of Earth and the Sun, must not crash on the nucleus if maneuvers fail, and must not cross dust or gas jets. After several weeks of looking for a suitable landing site, the lander will be released in November 2014, nominally on the 11th, at a heliocentric distance of about 3 AU and heading in-system. The ideal landing site would be flat and at a latitude ensuring day/night cycles for thermal balance.

   Rosetta will release Philae at an altitude below 1 km, and at a speed relative to the orbiter of between 5 and 52 cm/s. In the weak gravity, the descent will take up to an hour. Just after release, the lander's cameras will take a dramatic stereoscopic image of the orbiter above it. Only the cameras will be operated during the descent, but the other instruments will be in standby and ready to start their measurements following landing. The first analyses will then follow a pre-programmed sequence designed to ensure the return of at least a certain minimum amount of science, after which more refined schemes will be implemented. The lander is designed for a baseline mission of 65 hours but could survive for months, and possibly even to perihelion. Once the orbiter has delivered the lander and provided relay support, it is expected to conduct a series of observations to map most of the surface of the nucleus, then monitor the active regions and the build-up of the coma as the comet approaches and reaches its 13 August 2015 perihelion. The formal end of the mission in December 2015 will

Rosetta Rendezvous Maneuvers

Date	Velocity Change (m/s)	Distance from comet (km)
21 May 2014	321.8	938000
4 June 2014	263.9	389000
18 June 2014	88.9	163000
2 July 2014	66.1	44000
9 July 2014	24.6	19000
16 July 2014	10.5	8043
23 July 2014	5	3459
4 August 2014	2	400
7 August 2014	1	100

Rosetta Mission Phases

Date	Distance (km)	Phase
7–15 August 2014	60–70	Initial characterization
15–23 August 2014	20–40	Mapping
23 August–11 September 2014	20	Gravity mapping
11 September–18 November 2014	5–10	Close observations
11 November 2014	5–0	Landing

mark 30 years from its conception as the Comet Nucleus Sample Return. Of course, if the spacecraft is still in good shape, and if the propulsion leak does not have a serious impact, it could perform extended investigations, including (for example) deep excursions down the comet's tail.[115,116,117,118,119]

RETURN TO THE FORGOTTEN PLANET

While space agencies were deeply involved in exploring Mars and Venus, and in planning and undertaking missions to the outer solar system, only one spacecraft had ever reached the innermost planet. Mariner 10 performed three flybys of Mercury in 1974 and 1975 and revealed it to be heavily cratered.

Although some follow-up mission studies had been drawn up, very few scientists were studying the planet.[120] Still, several discoveries had been made. Mariner 10 had looked for a tenuous planetary atmosphere and found only an evanescent one made mostly of helium and atomic hydrogen, with possible traces of atomic oxygen. This is more properly called an exosphere, with a density so low that the molecules and atoms of gas stand little chance of colliding with each other and will either combine with the surface or escape to space. But in the mid-1980s continuing spectroscopic research by terrestrial observatories found strong emission lines of sodium, and this appeared to be the principal constituent of the exosphere. Whereas the hydrogen and helium would be captured from the solar wind, the sodium was most likely sputtered out of surface material by the impact of energetic solar wind particles. Furthermore, images of Mercury in the sodium emission band revealed that this component of the atmosphere was confined to high latitudes in both hemispheres, and varied in time. The mechanism for this phenomenon was probably related to the transportation of ions along the lines of force of the planet's magnetic field to the poles. Later studies also found potassium and calcium.[121,122] These results revealed something about the surface composition but, remarkably, despite specific searches, no one succeeded in detecting common rock-forming elements such as silicon, iron, titanium, aluminum, etc.

The most amazing post-Mariner 10 discovery concerning Mercury was made in August 1991 by two teams attempting to make radar maps of the hemisphere which faced Earth at inferior conjunction – the hemisphere that Mariner 10 was unable to view. One team used the 70-meter antenna of the Deep Space Network at Goldstone to transmit a beam of microwaves towards the planet, and then used the 27 antennas of the Very Large Array in New Mexico to collect the echo. In addition to producing a map with a resolution of 100 km, this study discovered a spot at the north pole that was highly reflective to radar. Another team used the 300-meter-diameter antenna of the Arecibo Observatory in Puerto Rico, the world's largest radio-telescope, as both the transmitter and receiver, and in March 1992 identified a similar spot at the south pole. The characteristics of these spots resembled those 'seen' by radar on Mars and the Galilean satellites of Jupiter, and strongly suggested the presence of water ice. A study published alongside the discovery report argued that since the spin axis of the planet lies within 1 degree of the perpendicular to its orbital plane, the polar regions will receive only grazing insolation, the floors of the craters there may have been in shadow for eons, and temperatures of –200°C could have allowed a deposit of water ice to accumulate. The radar-reflective spots showed a good correlation with craters at the north pole and an even better correlation at the south pole, where they seemed to be inside the crater Chao Meng-Fu.[123,124,125] Alternative explanations for the polar spots involved sodium deposits, and the behavior of silicates in the surface exposed to extremely low temperatures.

Other important advances in our knowledge of this small planet were provided by upgrades at Arecibo and the introduction of fast, low-noise digital video cameras on optical telescopes. Upgrades to the Arecibo radio-telescope had finally enabled radar astronomers to achieve kilometer-scale resolution on the surface of Mercury. They discovered a number of fresh craters on the hemisphere not viewed by Mariner 10; made better maps of the polar regions that Mariner 10 had imaged only obliquely, in particular establishing that a study of the spacecraft's data had placed the north pole some 65 km off its actual position; and imposed constraints on the extent of any ice. Meanwhile, video cameras on optical telescopes facilitated the production of large-scale maps of the hemisphere not viewed by Mariner 10 that placed radar-detected features into context and discovered a number of new features. The most interesting of these features was similar in size to the Caloris basin, which is one of the largest impact structures in the solar system. This was informally designated the 'Skinakas' basin, after the observatory in Crete which made the discovery. A bright spot at its center was initially presumed to be a mountain, but later revealed by radar to be a fresh crater. Since there was no evidence in the Mariner 10 imagery of hilly terrain antipodal to this structure corresponding to that opposite Caloris, its identification as an impact basin was disputed.[126,127,128,129]

All of these discoveries, and in particular that of ice at the poles, made a return to Mercury very appealing. Nevertheless, space agencies ranked the planet low on their agendas. Starting in the late 1960s ESA had considered a number of plans for flybys and orbiters, but only began seriously to consider an orbital mission in 1995. The result was the BepiColombo mission, scheduled for the 2010s. NASA investigated a number of missions for the Discovery program and in 1993 selected two for further study, but neither was implemented.[130] An orbiter was proposed by the Carnegie Institution and the Applied Physics Laboratory (APL) of Johns Hopkins University. It was called MESSENGER (MErcury Surface, Space ENvironment, GEochemistry and Ranging) because the planet had been named after the wing-footed messenger of the gods in Greek and Roman mythology. It was not selected when first proposed in 1996 as a Discovery mission, but was one of the finalist in the 1998 selection round. Also selected on that occasion were the Aladdin Mars sample-return, which was also reaching the shortlist for the second time, the Deep Impact mission to investigate the subsurface of a cometary nucleus, INSIDE Jupiter (INterior Structure and Internal Dynamical Evolution of Jupiter), and the Venus Sounder for Planetary Exploration that was known as Vesper. The INSIDE Jupiter mission proposed by JPL was to determine the internal structure of that planet by making high-resolution maps of the magnetic and gravity fields, determine the nature and extent of the motions in the atmosphere and interior, study the planetary dynamo, and investigate the interaction between the ionosphere and magnetosphere. The spacecraft would be solar-powered and be equipped with a magnetometer, energetic-particle detector and two scientific radio systems: one for gravity studies and the other for atmospheric and ionospheric soundings. It could be launched on a Delta II, but gravity-assists by Venus and Earth would be required to reach its target.[131] The Goddard Space Flight Center's Vesper was to orbit Venus for two local days and investigate the planet's atmosphere using infrared and ultraviolet cameras, a sunrise and sunset limb sounder, and an X-Band radio-occultation system capable of achieving a high spatial resolution.

As regards Mercury, MESSENGER was to map the entire surface of the planet at a resolution of several hundred meters and the north polar region at a resolution of about 5 meters, survey the elemental and mineralogical composition of the surface, in particular the radar-reflecting material of the 'polar spots', chart the gravitational and magnetic fields, and monitor how the planetary magnetic field interacts with the solar wind and the interplanetary environment. In fact, owing to Mercury's location so near the source of the solar wind, interactions with its weak magnetic field were expected to be particularly violent. Finally, the spacecraft would characterize neutral atoms in the exosphere and ions in the magnetosphere. One of the main objectives of the mapping activity would be to date the formation of the many scarps which cross the surface to gain insight into the thermal history of the planet, its molten core, and magnetic field. The mission would also address (as an aside) the relatively neglected but fascinating question of whether Mercury has any small satellites. Apart from an embarrassing false report, this was left unanswered by Mariner 10. But scientists are skeptical because the gravitational tidal effects of the Sun, the large eccentricity of the planet's orbit, and the Yarkovsky effect (a recoil caused by solar radiation which causes an orbit to evolve) would rapidly cause a satellite to strike the surface. If no satellites were found by the time MESSENGER finished mapping the planet's entire gravitational sphere of influence, this would impose an upper limit on the size of any such body at about 2 km.

In July 1999 NASA approved MESSENGER (and also Deep Impact) with launch scheduled for 2004. However, in 2003 it became evident that more time would be required to test the spacecraft and improve its redundancy. A launch postponement from March to May 2004 slipped the arrival at Mercury from April to July 2009. But then the launch was further postponed to August 2004, with the unfortunate effect of extending the cruise by 2 years and delaying orbital insertion to March 2011. After the loss of the CONTOUR (Comet Nucleus Tour) mission, NASA called for a more conservative design. This inflated the cost of MESSENGER by about 15 per cent and almost caused its cancellation. As we shall see, the Discovery program was increasingly subjected to such dilemmas following the abandonment of the 'faster, cheaper, better' method of management.

The design of the MESSENGER mission profile was complex, in that the 6.6-year cruise required no fewer than six gravity-assists while making 15 circuits of the Sun in order to achieve suitable conditions for insertion into orbit of Mercury. One of the flybys had to be with Earth, two with Venus and three with Mercury (if it had been possible to launch earlier in 2004, there would have been only two Mercury flybys).

A scheme for matching Mercury's orbit using a combination of powered maneuvers and flybys had been devised by JPL as early as 1985, and was to be implemented for the first time by MESSENGER. This scheme consisted of a series of circuits of the Sun between each flyby. These would resonate with Mercury and the ratio between them would gradually approach unity – i.e. between the first and second flybys the spacecraft would make two circuits in the same time as Mercury required for three orbits, giving a ratio of 2:3; between the second and third flybys it would be 3:4; and between the third flyby and orbital insertion it would be 5:6. Two months after each flyby of Mercury the spacecraft would make an orbit-shaping burn near aphelion to refine the next encounter. As the size and orientation of the vehicle's orbit became increasingly similar to that of Mercury, this would reduce the braking burn required for orbit insertion. Without these gravity-assists, the braking burn would have been close to 10 km/s; much greater than any conventional planetary mission to date. By accepting a protracted cruise, the burn could be reduced to 816 m/s. This was the exact reverse of the technique used by Galileo and Cassini, combining planetary flybys and deep-space maneuvers to increase the speed of a mission heading to the outer solar system. In this case, flybys and maneuvers would be used to reduce the speed relative to Mercury.[132] On arriving at Mercury, the spacecraft would be put in a 200 × 15,200-km orbit with a period of 12 hours that was inclined at 80 degrees with its periapsis near latitude 60°N. The orientation of the orbit would remain fixed in space, and over one planetary revolution a wide variety of surface illuminations would be encountered: from dusk to dawn and including night and noon. The latter would be a particularly challenging situation, as MESSENGER would fly between the Sun and Mercury. One side of the vehicle would receive heat directly from the scorching Sun, the other would receive heat radiated back to space by the super-hot surface of Mercury. As Mercury has an axial rotation period of 59 days and an orbital period of 88 days, a solar day (i.e. from noon to noon) lasts 176 Earth days. The baseline orbital mission was to last 12 months, or a little over two local days. The first day's observations would create a morphological base map covering more than 90 per cent of the surface at an average resolution of 250 meters. These images would have oblique illumination to provide shadows to reveal the topography and geological features. The second day was to be dedicated to targeted high-resolution observations and to the acquisition of a second full map with a slightly different perspective in order to allow a 3-dimensional reconstruction of the topography.[133]

The design of the MESSENGER spacecraft was largely dictated by its mission, in particular the thermal requirements. During the cruise its heliocentric distance would range between a minimum of 0.31 AU and a maximum of 1.08 AU, subjecting it to a 12-fold variation in solar irradiation. It was decided to equip the spacecraft with a titanium frame to support a 2.5 × 2.1-meter semicircular sunshade made up of layers of plastic insulation and a ceramic cloth outer cover. The temperature of the sunward side of this shield could exceed 350°C but the spacecraft in its shade would remain at about 20°C. As long as the sunshade was pointed within 12 degrees of the direction of the Sun it would protect the spacecraft. One of the operational constraints caused by this pointing requirement was that the timing of trim maneuvers while in orbit of Mercury would be limited to just two 'windows' for each

88-day local year. A pair of gimbaled gallium arsenide solar arrays 1.5 × 1.65 meters in size were to provide 390 W at Earth and up to 640 W at Mercury. The arrays had to capable of being rotated relative to the Sun because (for example) if they were held perpendicular to the Sun at perihelion their temperature would exceed 150°C and this would reduce their efficiency. For thermal protection, the panels were covered with two rows of mirrors for each string of solar cells. As in the case of Mariner 10, some degree of attitude control would be able to be achieved by varying the orientation of the panels to control the torque of solar radiation pressure. With the arrays fully deployed, the spacecraft had a span of 6.14 meters. The main body was a 1.42 × 1.85 × 1.27-meter graphite-composite structure whose low thermal expansion coefficient and thermal conductivity would be better suited to the operating environment than a frame made of metal. Some degree of structural support was provided by the three internally mounted lightweight titanium tanks, one being for nitrogen tetroxide and the other two for hydrazine – one each for the main engine and for the attitude control system. Attitude determination was by star trackers, Sun sensors and an inertial platform of accelerometers and gyroscopes. In addition to a single 672-N bipropellant engine for major maneuvers, the propulsion system had four monopropellant 26-N thrusters for attitude control during orbit insertion and other large maneuvers, and a dozen 4-N thrusters for making smaller maneuvers and for general attitude control. In order to minimize propellant usage there were also reaction wheels for attitude control. As the pointing constraints imposed by the sunshade ruled out the use of traditional steerable antennas, MESSENGER had two planar phased-array high-gain antennas capable of data rates up to 104 kbits per second (marking the first time this technology was used on a planetary mission), one mounted at the front and the other at the rear. It also had four low-gain antennas and two medium-gain 'fan beam' antennas. A pair of 1-Gbyte solid-state recorders (one backup) were to store data prior to transmission to Earth.

Although the quantity of propellant required to perform the mission restricted the mass of the scientific payload to 50 kg, MESSENGER carried no fewer than seven instruments, making it one of the most capable missions of the Discovery program. The imaging system consisted of a pair of 1.024 × 1,024-pixel CCD cameras, one a wide-angle camera with a 10.5-degree field of view fitted with 11 color filters and a clear position, and the other a narrow-angle camera with a 1.5-degree field of view for grayscale imaging. It was mounted on a pivoting platform which enabled it to be pointed without requiring the spacecraft to violate the constraints of the sunshield. The cameras would undertake the optical navigation to refine the flybys of Mercury, and then map the planet. An atmospheric and surface composition spectrometer was to measure the known and predicted atmospheric species. Its compact Cassegrain telescope illuminated two channels, one ultraviolet-and-visible and the other visible-and-infrared. An X-ray spectrometer had three separate sensors to detect aluminum, magnesium, silicon, calcium, titanium and iron in the uppermost few millimeters of the planet's surface as this was made to fluoresce by solar radiation. The data from these two instruments was intended to provide data on the composition of the surface to enable scientists to decide between at least three competing theories for why Mercury is so dense and apparently enriched in metals. If the neutron

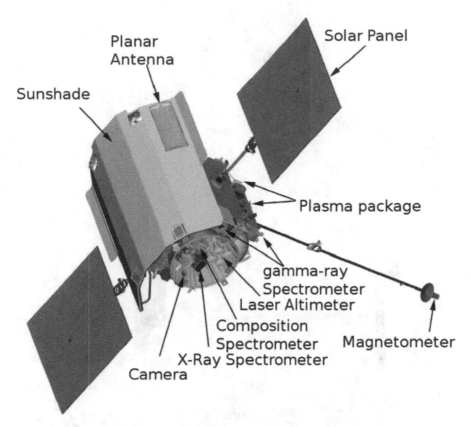

The MESSENGER Mercury orbiter showing the location of its scientific instruments.

spectrometer detected hydrogen from the radar-bright sites at the poles, this would confirm that they are water ice. An ultraviolet spectrometer and energetic-particle spectrometer were also to search for the signature of hydroxides. The spacecraft was equipped with an altimeter that consisted of a laser transmitter and an array of four refracting telescopes for detection. It was capable of a ranging resolution of 30 cm at a distance of 1,000 km. The planet's gravity field would be charted by combining altimetry and precise Doppler tracking, with the results indicating whether the planet's crust is decoupled from a liquid core. A 3-axis fluxgate magnetometer at the end of a 3.6-meter carbon-fiber boom (and protected by its own small sunshield) would be able to distinguish between 'fossil' magnetism and one still being generated by motions in a liquid core – both have been proposed as the source of the field discovered by Mariner 10. The payload was rounded off by an energetic-particle and plasma spectrometer, and the X-Band transponder for radio science.[134]

At launch, MESSENGER had a mass of 1,107 kg. The 599 kg of propellant that it carried was capable of a velocity change ability of about 2,200 m/s. At 54 per cent, this propellant fraction was unusually high but it was the only means of tackling the

MESSENGER (behind its extensive sunshield) being mounted on the final stage of its launcher.

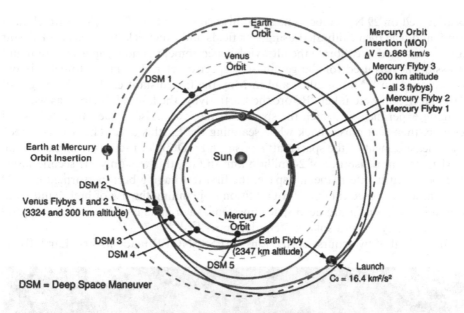

Earth
Orbit

Mercury Orbit
Insertion (MOI)
ΔV = 0.868 km/s

Venus
Orbit

Mercury Flyby 3
(200 km altitude
- all 3 flybys)
Mercury Flyby 2
Mercury Flyby 1

DSM 1

Earth at Mercury
Orbit Insertion

Sun

DSM 2

Venus Flybys 1 and 2
(3324 and 300 km altitude)

Mercury
Orbit

DSM 3
DSM 4

Earth Flyby
(2347 km altitude)

DSM 5

Launch
C₃ = 16.4 km²/s²

DSM = Deep Space Maneuver

The circuitous route of MESSENGER from Earth to Mercury orbit.

mission without using an ion engine. As a result, it required a Delta II 7925-Heavy, the most powerful launch vehicle allowed by the Discovery program, and even then an additional kick stage was needed. Including the spacecraft and its instruments, the launch vehicle, flight operations and data analysis, the cost of the mission was put at $466 million. Remarkably, when adjusted for inflation, this was almost the same as Mariner 10 had cost![135] The assembly of the spacecraft was completed in December 2003, and 3 months later it was shipped to Florida. The window opened on 11 May, but it was decided to delay the launch to early August in order to perform additional tests and (as noted above) accept the penalty of extending the cruise by 2 years. This window opened on 30 July but the earliest possible liftoff date was 2 August owing to competition for the facilities at Cape Canaveral. The window lasted only 12 days. A tropical storm that was developing over the Atlantic and along the rocket's flight path caused the first launch attempt to be scrubbed, but MESSENGER lifted off at 06:16 UTC on 3 August 2004. After a brief coast, the second stage of the Delta and the kick stage fired in succession over the Indian Ocean to place the spacecraft into a 0.923 × 1.077-AU solar orbit at the relatively steep inclination of 6.5 degrees to the ecliptic.[136]

Slowly spinning for stability, the spacecraft deployed its solar arrays and oriented the sunshade opposite the Sun to warm itself. In fact, because there was insufficient power to operate the heaters together with the science instruments at 1 AU from the Sun, while MESSENGER was operating its instruments early in the cruise it would face its back to the Sun to use solar heat to warm its electronics. On 24 August, as some of the instruments were being activated and calibrated, the spacecraft's first course correction adjusted its heliocentric velocity by 18 m/s. The imaging system was activated on 29 November, imaged a target mounted on the payload attachment fixture, and was then calibrated using star fields and other celestial objects. The star Regulus was used to calibrate the ultraviolet spectrometer, and a supernova remnant in Cassiopeia was used for the gamma-ray spectrometer. The magnetometer boom was deployed on 8 March 2005, and in May the first distant calibration images of Earth and Moon were taken. Before the Earth flyby, the laser altimeter was used for a ranging experiment in which the spacecraft received pulses issued by a terrestrial telescope, then sent pulses back while scanning the Earth's disk. The results agreed with a measurement of the spacecraft's position by Doppler radio tracking to within just 41 meters at a distance of 24 million km! In fact, this was only the second laser-link experiment in interplanetary space, the first test having been performed in 1992 using Galileo's science camera.[137] In addition to planning the use of high-bandwidth communications on future deep space missions, such links could yield fundamental insights into physics and solar system dynamics.[138,139]

After a total of five course corrections, MESSENGER made its first Earth flyby,

South America seen by MESSENGER after its Earth flyby.

passing 2,347 km above central Mongolia at 19:13 UTC on 2 August 2005, traveling at a relative velocity of 10.8 km/s. As on earlier missions, opportunity was taken to refine instrument calibrations. The camera imaged the Pacific region, and the north and south continents of America. Ultraviolet scans were obtained of the atmosphere. In addition, the X-ray spectrometer scanned the Moon and the particles and fields instruments took measurements while the spacecraft was inside our magnetosphere. On the outbound leg, the spacecraft compiled a stunning color movie showing Earth rotating. It experienced the smallest recorded anomalous flyby velocity, amounting to just 0.02 mm/s.[140]

This flyby left the spacecraft in an orbit ranging between 0.603 and 1.015 AU. On 12 December the first large deep-space maneuver set the date and geometry for the first Venus flyby, in the process raising the aphelion to 1.054 AU. This 524-second burn consumed about 18 per cent of the total propellant onboard. At 316 m/s, it was expected to be the largest burn of the interplanetary cruise. The next milestone was to be the flyby of Venus in October 2006. Weeks beforehand, and still 16.5 million km away, sequences of low-resolution navigation images discerned details of that planet's atmosphere. Although not strictly necessary for a Venus flyby (because the ephemeris was well known) this exercise was a useful rehearsal for the Mercury flybys. The spacecraft passed Venus at 08:34 UTC on 24 October at an altitude of 2,987 km and a relative velocity of 9.07 km/s. This high-latitude pass provided an eclipse of the Sun and an Earth occultation, but no scientific observations were made because Venus was within 1.4 degrees of solar conjunction and the bandwidth was insufficient. The 56 minutes that the vehicle spent in shadow would be the longest period of darkness of the entire mission, and put the batteries to a severe test. The flyby changed the vehicle's orbit to 0.526 × 0.901 AU, which was resonant with that

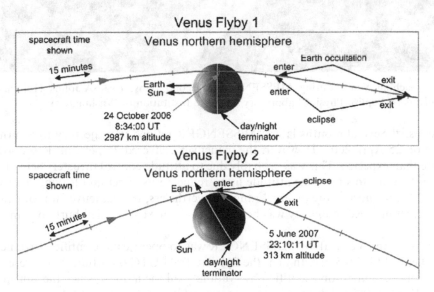

The geometry of MESSENGER's two Venus flybys.

Approaching Venus during MESSENGER's second flyby. (NASA/Johns Hopkins University Applied Physics Laboratory/Carnegie Institution of Washington)

of Venus.[141] Several months later, MESSENGER refined its trajectory to return to Venus on 25 April 2007. However, the maneuver delivered 26 per cent less velocity change than expected. This was probably due to attitude control thrusters being fired during the burn to stop some attitude jitter, reducing its effect and leaving the flyby about 200 km short of the planned range. Nevertheless, the maneuver was adequate and correcting the discrepancy was held over to the next scheduled burn, one month later.

On its second Venus flyby MESSENGER was to operate its scientific instruments in a rehearsal for Mercury. In fact, the flyby at 23:08 UTC on 5 June 2007 presented unique opportunities for scientific observations – all the more so since the European Venus Express (of which more later) was in orbit. Ground-based telescopes made complementary observations.[142] MESSENGER approached the planet from the

MESSENGER departing Venus. (NASA/Johns Hopkins University Applied Physics Laboratory/Carnegie Institution of Washington)

dark side, spending about 20 minutes in its shadow. The closest approach altitude of 316 km was within a mere 1.7 km of that intended. During the entire encounter, a large amount of data was collected and a total of 614 pictures were taken. Unfortunately, since the point of closest approach was over the Earth-facing hemisphere it was not possible to obtain additional atmospheric radio-occultation data. Color images of the night-side were obtained on the way in, and then black-and-white and color mosaics. High-resolution black-and-white mosaics and infrared pictures were obtained of the day-side, together with photometric data. In the near-infrared it had been discovered that after computer processing 'subtracted' the contribution from the surface, it was possible to make a rudimentary study of the composition of the lower atmosphere; and this technique was tested for the first time using images of Ovda Regio. Scans of the dusk and night hemispheres, as well as the terminator line, were obtained by the atmospheric spectrometer. It also scanned the outer atmosphere and the 'corona' of hydrogen that surrounds the planet. Ultraviolet and infrared spectra were collected to produce vertical profiles of the atmospheric chemistry. This complemented the data obtained at the same time by the visible and infrared imaging spectrometer on Venus Express. In fact, MESSENGER made its flyby near the boundary of Rusalka Planitia and Aphrodite Terra, which was observed by Venus Express only a few hours later, allowing almost direct comparison of their observations.[143,144] The other instruments also collected data. The plasma and particles instrument studied charged particles in the planet's bow shock and other plasma boundaries, plus interactions between the ionosphere and the solar wind. The laser altimeter obtained radiometric and cloud scattering data, and the X-ray spectrometer measured various species in the highest reaches of the atmosphere, in particular oxygen and carbon atoms. In fact, telescopes in orbit around Earth had revealed Venus to be a weak X-ray source, fluorescing in sunlight. On its outbound leg, MESSENGER took additional mosaics of Venus and compiled a movie. Although the spacecraft's velocity relative to Venus differed by just 3 m/s between the first and second flybys, the much closer approach deflected its trajectory more dramatically and put it into a 0.332×0.475-AU orbit that would intersect the orbit of Mercury. Shortly after the flyby, another experiment of laser ranging over a distance of 104 million km was attempted, but a series of mechanical problems as well as bad weather at the transmitter on Earth prevented successful acquisition.

The perihelion distance of MESSENGER's new orbit on 1 September was much closer to the Sun than Mariner 10 had ever ventured, and the solar arrays were tilted 70 degrees from the Sun to prevent their overheating. In fact, during its perihelion passage the spacecraft utilized radiation pressure to slowly refine its trajectory, using this solar sailing technique to save propellant. Mariner 10 had used solar pressure for attitude stabilization, but this was the first time it was used for orbit shaping. More solar sailing would be done between the next flybys. On 17 October the spacecraft performed its second large deep-space maneuver, and went through the longest solar conjunction of the mission. A 1.1-m/s correction on 19 December refined its aim for a 200-km flyby of Mercury the following month. Scientists had recently published an important paper about the planet reporting that radar observations on more than 20 different occasions had provided very accurate measurements of its rotation, and small variations which could not be explained by a completely solid interior implied the presence of a core that was at least partially liquid and decoupled from the outer mantle and crust. It was hoped that the gravity survey conducted by MESSENGER once it was in orbit of the planet would enable constraints to be placed on the size of this core and thereby provide insight into the planetary magnetic field. On the last day of 2007 it recorded a solar flare that spewed a large quantity of high-energy neutrons which were duly recorded by the gamma-ray and neutron spectrometer. This was the first time neutrons were detected from an average-size solar flare. More events were expected to be observed as the Sun's activity increased from its 11-year minimum.[145]

On 9 January 2008 MESSENGER began to take optical navigation images of the planet. These confirmed that it was on course for the desired flyby. This would be MESSENGER's first opportunity to provide first-class science. All the instruments would take data, including distant ultraviolet scans of the polar regions in search of hydrogen emissions from the putative ice. The camera was to provide color mosaics and high-resolution black-and-white images, including stereo pairs. One objective of particular interest was to obtain the first views spanning the diameter of the Caloris basin, because this had always been on the terminator when Mariner 10 flew by. The hope was that imaging the basin and its ejecta using multispectral filters would yield insights into the composition of the subsurface material. For particles and fields, the low-altitude flyby over the equatorial zone offered a way to sample the environment with a geometry that would be impossible to replicate once in orbit; all the more so since the flyby would sample the magnetic tail of the planet.[146] Because the antennas would not be pointing at Earth during the actual encounter, only a beacon would be transmitted as an indication of the spacecraft's state of health. The engineering and scientific data would be stored onboard for subsequent replay.

MESSENGER approached Mercury from the night-side, with regions imaged by Mariner 10 visible along the terminator. An abrupt increase in the magnetic field at 18:08 UTC on 14 January 2008 marked the vehicle's crossing of the bow shock to reach the magnetopause. The flyby trajectory was inclined at just 5 degrees to the equatorial plane. The moment of closest approach at 19:05 occurred 201.4 km above the night-side at 4°S, 38°E, at a relative velocity of 7.1 km/s. Around this time, the spacecraft spent 14 minutes in eclipse and the line of sight to Earth was occulted for about 47 minutes. Three minutes after closest approach, it emerged from behind the disk to re-establish the line of sight to Earth, and shortly thereafter crossed the dawn terminator. As it departed, the spacecraft was able to see the sunlit side of the planet, including 21 per cent of the area that had been hidden to Mariner 10 for all three of its encounters. At 19:19 MESSENGER again crossed the bow shock to re-enter the interplanetary magnetic field. Two days later, transmission began of the 500 Mbytes of data that had been stored during the encounter, including 1,213 images.

The previously unseen portion of Mercury's surface showed all the characteristics of that seen earlier, with lobate scarps hundreds of kilometers in length and wrinkle ridges evidently caused by a shrinkage of the planet. The scarp of Beagle Rupes was seen in its entirety, and found to extend for more than 600 km. As previously noted, many craters were substantially deformed by compression along scarps. An analysis revealed that the shrinkage of Mercury had been on average 30 per cent greater than previously believed. The new images indicated that some of the scarps were created prior to the emplacement of many of the smooth mare-like plains, with this process ongoing even after the creation of the youngest plains. One outstanding issue from Mariner 10 was whether the smooth plains were of volcanic origin or were blankets of basin ejecta. Mariner 10's flybys occurred a few years after the Apollo 16 mission to the Moon, when only impact-modified rocks were found in an area expected to be volcanic, and this biased scientists of that era against a volcanic interpretation of the plains of Mercury. In fact, Mariner 10's images had not been good enough to resolve the kind of features such as small domes and sinuous rilles that would argue in favor of volcanism. MESSENGER, however, was able to see volcanic vents and deposits

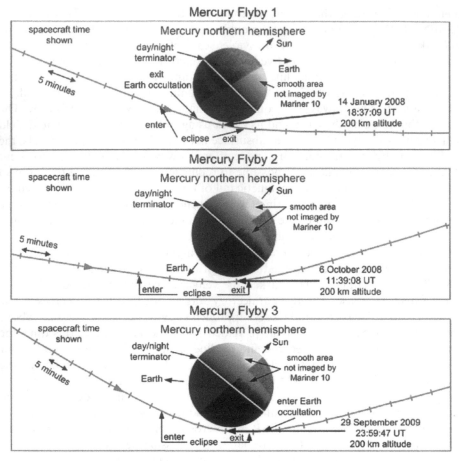

The geometry of MESSENGER's three Mercury flybys. The image also shows the extent of the hemisphere of the planet that was not seen by Mariner 10.

along the southern edge of Caloris. A kidney-shaped depression about 20 km across sitting within a broad smooth deposit was almost certainly a volcanic vent, as were a number of other small irregular depressions.

One focus of this first encounter was an inspection of the Caloris basin, where the crust was most deeply exposed and tectonism and volcanism were more evident. In addition to mosaics at a resolution of 200 to 300 meters, a full 11-color mosaic was obtained at 2.4-km resolution. These revealed the history of the basin from the initial impact to the upwelling of magma through deep fractures in the floor of the cavity to form the interior plains, the later formation of wrinkle ridges, radial and concentric grabens, and the imprinting of the impact craters. At 1,550 km, the diameter of the basin proved to be 20 per cent greater than previously believed. Near its center, a unique feature was seen in the shape of a 41-km-diameter crater situated at the 'hub' of a number of radial fissures resembling a spider's web. Pantheon Fossae, as it was

Departing Mercury after the first flyby in 2008. Only the rightmost quarter of the planet had been seen by Mariner 10. The brighter circular patch at upper right is the Caloris Basin, one of the largest impact craters in the solar system. (NASA/Johns Hopkins University Applied Physics Laboratory/Carnegie Institution of Washington)

named, seemed to be due to the stretching of the surface by an underlying volcanic plume. Dozens of similar structures had been documented on the surface of Venus. Grabens in Pantheon Fossae began radially and then transformed into a polygonal pattern which cut across wrinkle ridges. Such extensional features are extremely rare on Mercury. The crater at the center of Pantheon Fossae was named Apollodorus, in honor of the architect credited with conceiving the Pantheon in Rome. A newly discovered basin about 250 km in diameter which appeared to be relatively youthful, perhaps no more than 1 billion years old, was named Raditladi.[147] Multispectral data yielded details of the surface composition and also the degree to which it had been weathered by exposure to the space environment.

The remarkable 'spider' Pantheon Fossae and crater Apollodorus discovered by MESSENGER during its first Mercury flyby. (NASA/Johns Hopkins University Applied Physics Laboratory/Carnegie Institution of Washington)

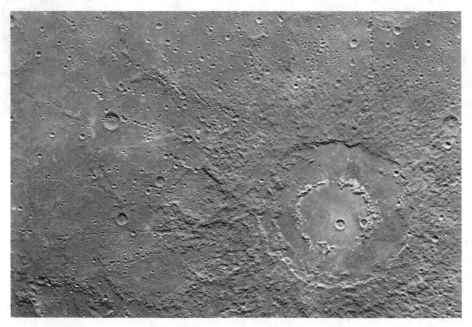

The young Mercurian basin Raditladi. (NASA/Johns Hopkins University Applied Physics Laboratory/Carnegie Institution of Washington)

As the spacecraft flew by Mercury, the laser obtained the first swath of altimetry ever recorded of the surface. This 3,200-km-long profile spanned 20 per cent of the equatorial region, but because it was over the night-side and terrain which had not been visible to Mariner 10 only Earth-based radar imagery would be available for correlation until MESSENGER was able to document this region in sunlight. The altimeter began to collect data 1 minute prior to closest approach, and took it for 10 minutes. Only the laser returns acquired at altitudes under 1,500 km were judged to be usable. The scan contained many depressions that were probably impact craters, and their profiles were typically shallower than their lunar counterparts, doubtless as a result of Mercury's gravity being stronger.

The spacecraft spent more than 1 hour inside Mercury's magnetosphere. The path was geometrically similar to the first one by Mariner 10, but with the closest point of approach at 201 rather than 327 km. This similarity facilitated a ready comparison of the two datasets. Serendipitously, near closest approach MESSENGER crossed the current sheet which separated the two polarities of the magnetic field. There was no indication of either the magnitude or orientation of the field having changed since 1974. Some minute deviations from a purely dipolar field were detected for the first time, but their reality was debatable. The most significant result was that there were no isolated crustal magnetic anomalies to indicate that the field was fossilized into the surface rocks. This strengthened the case for the field being created by an active dynamo in a liquid core. But this was only a flyby, and the interior of the planet and the nature of its field would be thoroughly investigated once the spacecraft was in orbit of the planet. As expected, the magnetosphere was permeated by sodium ions derived from the tenuous atmosphere. There were also oxygen, silicon, potassium, calcium, heavier ions related to sulfur, and even water-ice-group ions very probably sputtered from the ice located in the 'cold traps' at the poles. Neutron spectrometry showed that, at least in terms of its iron content, the Mercurian surface appeared to resemble the lunar regolith sampled by the Apollo and Luna landers. The ultraviolet spectrometer scanned the magnetic tail from 24,500 km out until the spacecraft was 4,500 km from the planet. Strong emission by sodium was detected all along the tail for a distance of at least 100,000 km. An asymmetry in the atmosphere between the northern and southern hemispheres could reflect conditions in the magnetic fields of either the planet or the interplanetary medium.[148,149,150,151,152,153,154,155,156,157,158]

As a result of the flyby, MESSENGER was in a resonant 0.313 × 0.700-AU orbit, the perihelion of which it reached on 23 January. A deep-space maneuver in mid-March and more solar sailing refined the second encounter with Mercury.

During a 9-day campaign in June the spacecraft snapped 240 images of the outer part of a putative belt of asteroids located between Mercury and the Sun. They were named Vulcanoids, after Vulcan, a planet nineteenth century astronomers inferred to exist inward of Mercury in order to explain some perturbations of its orbit which were subsequently realized to be relativistic effects.[159] Numerical simulations showed that stable orbits were possible only in a narrow band near the ecliptic at heliocentric distances between 0.08 and 0.21 AU. Objects closer to the Sun would be eroded by the sheer heat of sunlight, and those further out would be perturbed by the gravity of the inner planets. Coronagraph searches by solar observatories and telescopes on high-flying aircraft had failed to detect any object larger than 60 km. Better data could only be obtained by searches performed by spacecraft in the inner solar system. In the case of MESSENGER, objects down to 15 km in size would be detectable. The discovery of Vulcanoids would be extremely important in reconstructing Mercury's geological history. The process of dating a surface by 'counting craters' relies on assumptions about the population of impactors. As Vulcanoids could have hit only Mercury, their existence would have given rise to cratering that would require to be 'subtracted' in determining ages. In addition to Vulcanoids, such searches could also detect other putative objects like Mercurian Trojans (if any exist) or small asteroids having orbits confined entirely within that of Earth (several such objects are known, but they are very difficult to discover using ground-based telescopes).[160,161,162] MESSENGER's test evaluated the limiting brightness of detectable objects, checked whether known asteroids were visible, and rehearsed an observing technique that used the sunshield to ensure that no stray light entered the camera even when the field of view was very close to the Sun. However, it is unlikely that large Vulcanoids (if they ever existed) could have survived to the present day. Dynamical simulations showed that frequent high-speed collisions in this belt would have reduced such objects to kilometer-sized debris that would be more exposed to the Yarkovsky effect and would either fall into the Sun or undergo gravitational interactions with Mercury in a relatively short time. If the belt were now reduced to only 100 objects of this size, then these would be extremely difficult to detect.[163]

On 6 October 2008 MESSENGER had its second encounter with Mercury. As about 1.5 solar days had elapsed for the planet, the opposite hemisphere was now in sunlight. This allowed the spacecraft to inspect another 30 per cent of the surface that had never been visible to Mariner 10. In terms of geometry, the flyby was very similar to the first, approaching from the night-side, the closest point of approach occurring in eclipse, and emerging into daylight for the departure. However, as the pass was entirely over the Earth-facing hemisphere there was no Earth occultation this time. Previously unseen terrain would be visible across the terminator during the approach, so before the spacecraft adopted its encounter attitude and cut its high-gain link it returned some navigation images which showed this region. It crossed the bow shock at 07:19 UTC. The point of closest approach at 08:40 UTC was at an altitude of 199.4 km. (The desired value of 200 km made this the most accurate flyby of a solar system object by any mission.) It started to take pictures as soon as the illuminated terrain became visible. At 08:53 it crossed the bow shock back into the interplanetary medium. This time a total of 1,287 pictures were obtained.

The two MESSENGER flybys increased the detailed coverage of Mercury begun by Mariner 10 to just over 90 per cent. On this occasion the illuminated hemisphere included the bright crater Kuiper, one of the first small-scale features to be resolved as Mariner 10 made its first encounter. This hemisphere was seen to be smoother on average, somewhat analogous to the near-side of the Moon being generally smoother than its far-side. This could indicate either a difference in internal structure between the two hemispheres or a bias in the arrival direction of impactors. This hemisphere hosted a bright and evidently fresh 85-km crater, later dedicated to the Japanese painter Hokusai, that had already been inferred from radar observations. A complex system of thin rays of ejecta extended 4,500 km over much of the globe, almost to its antipodes, which was twice as far as the spectacular rays from Tycho on the Moon. In fact, the brightest part of this ejecta, in closest to the crater, had already been glimpsed telescopically. Another large crater visible at this same time was also known from radar observations, when it had been interpreted as a shield volcano.[164] The 'Skinakas' basin ought to have been plainly visible, but its absence proved that its earlier identification had been false. However, a basin about 715 km in diameter was found on the terminator of the hemisphere visible on the inbound leg. This was named Rembrandt. It was only moderately flooded by volcanic plains, and thus bore

A view of Mercury after the second MESSENGER flyby. Note the bright rays that appear to originate from the crater at the top. (NASA/Johns Hopkins University Applied Physics Laboratory/Carnegie Institution of Washington)

An inbound image of MESSENGER's second Mercury flyby with the Rembrandt basin straddling the terminator. (NASA/Johns Hopkins University Applied Physics Laboratory/Carnegie Institution of Washington)

The 290-km double-walled Rachmaninoff impact basin that was first seen during MESSENGER's third Mercury flyby. (NASA/Johns Hopkins University Applied Physics Laboratory/Carnegie Institution of Washington)

a resemblance to Mare Orientale on the Moon. Images from MESSENGER's first flyby enabled the full extent of Rembrandt to be mapped, revealing a remarkable network of 'wheel-and-spoke' ridges produced by overlapping phases of contraction and extension. Superimposed on this basin was part of a 1,000-km scarp, the longest yet found on the planet.[165] With so much of the surface finally documented, it was possible to make some preliminary generalizations. In particular, the smooth plains were globally distributed and cover 40 per cent of the planet, with the majority being of volcanic origin. The multispectral data also revealed that the magmas differed in composition. A major proportion of the surface appeared to be a low-reflectance bluish terrain which was probably iron-rich and titanium-rich material excavated by impacts.[166]

Some nice examples of volcanic pits and depressions on the floor of Mercurian craters. (NASA/Johns Hopkins University Applied Physics Laboratory/Carnegie Institution of Washington)

The new imagery enabled the laser altimetry from the first flyby to be placed into topographic context. Further, the 4,000-km-long swath from the current flyby was of terrain that had been previously imaged by the spacecraft. This happened to include a 1-km-tall scarp whose profile satisfied scientists that it had indeed been produced by compression as the interior of the planet shrank. The results placed constraints on the structure of the crust.[167]

The surface and atmosphere composition instrument obtained some 380 spectra, starting just before the spacecraft crossed the terminator through to when the planet departed the instrument's field of view. In addition to mapping the distribution of known atoms and ions in the exosphere and tail, the results identified for the first time the presence of magnesium in the tail.[168],[169]

The bright halo surrounding this probably volcanic irregular depression on Mercury had already been noted in high-resolution images taken from Earth. (NASA/Johns Hopkins University Applied Physics Laboratory/Carnegie Institution of Washington)

The different longitudes of MESSENGER's two equatorial flybys, taken together with one polar and one equatorial by Mariner 10, permitted a crude reconstruction of the structure of Mercury's magnetic field. This found the field strength to be almost equal on opposite sides of the planet, implying a dipole with its axis aligned within a few degrees of the rotation axis. This supported the hypothesis that the field was the result of motions in a liquid metallic core. The second flyby allowed a particularly detailed study of the structure of the planet's magnetosphere. The results suggested a 'perforated' magnetopause in which the interplanetary medium was able to reach the planet through multiple open paths, with 'reconnection events' occurring when the interplanetary field connected with the planetary one. All these phenomena have also been observed on Earth, but on Mercury they are evidently more frequent, probably influencing also the evolution of the surface and of the exosphere. A magnetically confined bubble of plasma was also encountered for a period of 4 seconds, making Mercury the fourth planet where such 'plasmoids' have been observed – the others being Earth, Jupiter and Saturn.[170,171] No significant magnetic anomalies were found closer to the planet, although by analogy with the Moon (where magnetic anomalies occur opposite large impact basins and maria) they had been expected to be present antipodal to the Caloris basin. Unfortunately, this region would be difficult to survey once MESSENGER was in orbit since it was in the southern hemisphere and far from the northerly latitude of periapsis and the spacecraft would be at high altitude while in line of sight. Since the magnetic anomalies on the Moon protect the surface in those areas from the darkening effect of the solar wind, these are usually marked by bright 'swirls'. A search of MESSENGER imagery for swirls not only proved negative, but also established several possibilities present in the Mariner 10 pictures to be of a different origin.[172] Small trajectory perturbations on both of MESSENGER's flybys may have been caused by localized mass concentrations, probably (by analogy with the Moon) due to the magmatic infill of large basins. These potential anomalies would be further investigated during the orbital phase of the mission.[173,174]

As a result of this second flyby MESSENGER's orbit had shrunk again, and now ranged between 0.302 and 0.630 AU. A two-part maneuver in December altered the heliocentric velocity by 247 m/s to target the third and final flyby. While at periapsis in February 2009, a 5-day campaign obtained 256 more pictures of areas to the east and west of the Sun for a second search for Vulcanoids.

The third encounter would be very similar to the second, occurring at the same orbital position and almost exactly 2 solar days later, so that the same hemisphere was on view. It would however, allow a small, hitherto unseen portion of the planet to be viewed on the inbound leg and thereby almost complete the reconnaissance of the planet, apart from the polar regions. The almost identical circumstances of the second and third encounters allowed scientists to plan several focused observations. Long exposure images were to be taken from 8 days prior to until 21 days after the encounter in order to construct a more precise curve of how Mercury reflected light at a number of wavelengths and to search for satellites as small as 100 meters in size. The narrow-angle camera was then to obtain high-resolution imagery of the southern hemisphere on the outbound leg to complement the northern coverage obtained by the second flyby. Moreover, as the gravity of Mercury bent the trajectory through almost 50 degrees, it would allow cameras to view some portions of the surface on both the inbound and outbound legs of the encounter. At the same time, the surface composition instrument was to study eleven preselected targets on the planet. This would also be the last opportunity to explore the planet's magnetic field at equatorial latitudes.[175]

The approach phase went without a hitch. A bright spot seen in telescopic images and on the limb for the second encounter was resolved and revealed to be a rimless depression with an irregular shape some 30 km across, surrounded by a bright halo. The origin of this feature remained unexplained, although the halo was suspected of being volcanic. An impact basin 290 km across was seen, similar to Raditladi. Its floor had remarkably few craters superimposed, suggesting that it had been flooded by magma after its formation, which would make it one of the most recent examples of volcanism on the planet. An irregular depression to the northeast could mark the site of the volcanic vent. The basin was later named Rachmaninoff and its age was estimated at 1 billion years. Rachmaninoff had a complete second wall and portions of a third ring were visible to the southwest. The Rembrandt basin was also visible, straddling the terminator. Further evidence of volcanism was found within several craters that had irregularly shaped pits on their floors, some of which were quite spectacular. MESSENGER crossed the bow shock at 20:56 UTC and entered the magnetopause at 21:28. In transiting the magnetic tail, the magnetic field was seen to increase several fold over intervals of mere minutes. Plasmoids traveling tailward were encountered on several occasions.[176,177] The vehicle entered the shadow cone of the planet 14 minutes before closest approach, and was to run on battery power for 18 minutes. However, after 10 minutes in the shadow, and just 4 minutes before closest approach, the signal from the spacecraft was unexpectedly lost. Apparently, its fault management system had noted a problem with the battery power supply in eclipse, interrupted all activities, including science observations, and put the spacecraft into a 'safe mode', pending intervention from Earth. Meanwhile, it passed over the Mercurian surface at a height of 228 km at 21:55 UTC on 29 September 2009. Since the spacecraft's heliocentric orbit was by now a close approximation of that of Mercury, the relative speed of the encounter was a mere 1.5 km/s. Just four minutes after MESSENGER emerged from eclipse, it was occulted by the disk of Mercury from the perspective of Earth, and so attempts to regain control could not start until after it had emerged from occultation 52 minutes later. About 6.5 hours after the problem appeared, MESSENGER was returned to normal operations and data stored on board was transmitted to Earth to determine what had happened. All of the observation on the inbound leg had been successfully made and their data stored, but the outbound observations were lost, including distant satellite searches. Good neutron spectrometer data was returned up to the safing event. Together with spectra obtained during the first flyby, this data revealed the surface of the planet to be unexpectedly rich in iron and titanium. It was already known that iron silicates were remarkably rare on the planet, and hence this element must be present in some other form. In fact, until then, Mercury had been a mystery, because it was known to be the terrestrial planet with the largest metallic core in terms of relative volume, yet without iron being detected on its surface. The ultraviolet spectrometer mapped sodium, magnesium and calcium in the planetary exosphere including at both poles, unseen by the previous encounters. A much weaker neutral sodium tail than on the second flyby was crossed, caused by the changed conditions in the interplanetary environment. However, the densities of calcium and magnesium were seen to increase. Remarkably, the concentration of magnesium and calcium over the two poles was asymmetrical, while that of sodium was not. Calcium ions were observed for the first time several planetary radii downwind, anomalously concentrated in small areas close to the equatorial plane.[178] Despite the safing event, the primary objective of the flyby was successfully achieved, namely easing the spacecraft into its final solar orbit of 0.303×0.567 AU.

In November 2009 a 3.3-minute, 177-m/s burn scheduled a return to Mercury in March 2011, when the spacecraft would enter orbit of the planet. Mercury orbital operations were rehearsed in early 2010. Meanwhile, the mission continued to seek Vulcanoids around perihelia. In some of these observations, its cameras happened to image known solar system objects, including the Earth-Moon system.

On 3 November, MESSENGER's cameras scanned the ecliptic plane and took a portrait of the solar system "from the inside out", complementing that taken from beyond the orbit of Neptune by Voyager 1 in 1990. A few additional pictures taken on 16 November completed a whole scan of the ecliptic. The portrait consisted of 34 wide-angle frames plus insets taken by the narrow-angle camera. All of the planets of the solar system were visible except Uranus and Neptune, which were too faint and distant. Also visible were the Earth's Moon and the Galilean satellites of Jupiter. A small portion of the Milky Way crossed the mosaic.

MESSENGER returned to Mercury for orbit insertion in March 2011. About 48 hours prior to the burn, the spacecraft turned to point its high-gain antenna to Earth, while also turning off all its instruments except the gamma-ray spectrometer, which was placed in standby. Unlike most planetary orbit insertions, this one would occur in sight of Earth, without occultations and loss of signal. But communications during the burn would be difficult due to the attitude of the vehicle, with the signal received on Earth being weaker than was usual. For this reason, antennas of the Deep Space Network station at Goldstone, California, were operated as an array comprising the largest 70-meter dish and three smaller ones. On 18 March, MESSENGER aligned its thrusters to the direction of travel and then started its engine at 00:45 UTC. The 15-minute burn consumed about 31 per cent (185 kg) of the fuel supply initially loaded onboard, leaving less than 10 per cent for use during the orbital mission. It reduced the spacecraft's speed by 862 m/s and braked it into a 207 × 15,261-km orbit with a 12.07-hour period that was close to that desired. As planned, the 82.5-degree orbit crossed the equator at an altitude of about 4,900 km northbound and 1,200 km southbound, and its plane was initially almost perpendicular to the line to the Sun, ensuring that the probe would fly over the thermally benign terminator and be constantly in sunlight.

After an interplanetary cruise of six and a half years, MESSENGER thus became the first artificial satellite of Mercury. By this point, all of the planets known since antiquity had received at least one orbital explorer. Ten minutes after the end of the burn, the spacecraft turned toward Earth and sent back the engineering telemetry that it had recorded during the maneuver, confirming that it was indeed in orbit around Mercury and that there was no need for a clean-up maneuver to refine the nearly perfect orbital parameters. Starting on 23 March, instruments were turned on, checked and calibrated. On 29 March, over a 6-hour period starting at 09:20 UTC, the cameras took 364 images. Fittingly, this was the same date as Mariner 10's first flyby. These first orbital images of Mercury covered the south pole and the craters Matabei and Debussy, with its extensive rays, as well as an as-yet-unseen area nearby. Scientific sequences were designed as soon as MESSENGER's orbit had been determined and its future position could be computed to define instrument pointing. Routine science observations began on 4 April. Early in the mission, the

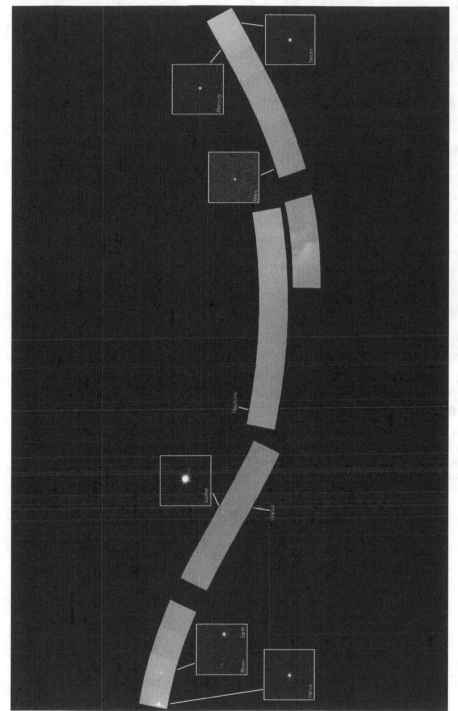

A family portrait of the solar system 'from the inside' taken by MESSENGER in November 2010. (NASA/Johns Hopkins University Applied Physics Laboratory/Carnegie Institution of Washington)

This is the first image of Mercury taken by MESSENGER after its orbit insertion. It was taken by the narrow-angle camera on 29 March 2011 with a resolution of about 380 m/pixel. (NASA/Johns Hopkins University Applied Physics Laboratory/Carnegie Institution of Washington)

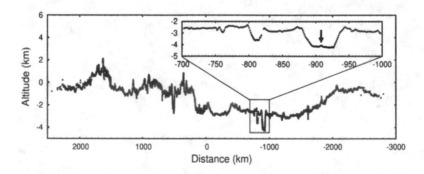

This is MESSENGER's first orbital laser altimeter swath on Mercury. The inset shows the detail of a crater with a prominent central peak. (NASA/Johns Hopkins University Applied Physics Laboratory/Carnegie Institution of Washington)

This image was taken by the wide-angle camera on MESSENGER on 17 May 2011 near apoapsis. Compared with close-ups of the surface, it clearly shows the range of altitudes of the elliptical orbit. (NASA/Johns Hopkins University Applied Physics Laboratory/ Carnegie Institution of Washington)

the cameras had their first look at the geologically complex terrain antipodal to Caloris, unseen since the Mariner 10 flybys. This area represented the spot where seismic waves from the formation of the giant basin had been focused and created a jumble of hills and fractured terrain.

MESSENGER began its 100th orbit on 6 May, at about the same time as it passed for the first time the challenging phase of the mission in which its periapsis occurred at local noon, facing the warmest terrain on the planet. During this time, the probe had made more than 70 million magnetic field measurements, and taken 300,000 infrared, 12,000 X-ray and 9,000 gamma-ray spectra of the surface as well as 16,000 images. The laser altimeter had mapped most of the northern hemisphere. In fact, its very first swath had captured one of the radar-bright permanently shadowed polar craters and measured the depth of its floor. On 13 June, as it was going through a 4-day solar conjunction, the probe completed its first 88 days in orbit, corresponding to a full planetary year. Two days later, the first orbit correction maneuver lowered the periapsis from 506 back to 200 km. To prevent the periapsis from drifting above 500 km, corrections would be needed once every local year. Due to pointing constraints, these maneuvers could be made only during two windows per Mercurian year, each lasting only a few terrestrial days. A second correction one month later, on 16 July, re-established the 12 hour period. In fact, every maneuver to lower the periapsis would cut the orbital period by about 15 minutes. As a result, it would need to be followed by a second maneuver to restore the 12-hour period. A total of three such maneuver pairs were to be carried out during the primary mission.

X-ray and gamma-ray spectra showed the composition of Mercury to be geologically unique and different from either Earth or the Moon, confirming Mariner 10's results that its surface had relatively low iron and titanium abundances in spite of the probable presence of a large metallic core. A few per cent was made of sulfur, possibly sourced from explosive volcanic vents. Minerals containing sulfur and iron were formed in the absence of oxygen, which is logical since oxygen-bearing molecules would be rare that close to the Sun. Overall, the composition of the surface resembled some metal-rich chondrite meteorites. Gamma-ray spectrometry measured the relative proportions of potassium and thorium to obtain information on the temperature of the solar system when Mercury formed. The potassium-thorium ratio was similar to that of the other terrestrial planets, and an order of magnitude greater than that of the Moon. This proved Mercury was not depleted in volatile elements, like potassium, and indicated that extreme conditions were not needed in order to explain the formation of such a small planet with a proportionately large iron core, and in particular that no high-temperature process such as an early gigantic collision or an exceptionally violent young Sun were needed.

Images of the north polar regions mapped relatively young volcanic plains which covered almost 6 per cent of the whole surface of the planet. The smooth flow of basaltic lava had filled in a topographic low, completely or partially burying older craters and displaying large flow fronts as well as teardrop-shaped 'islands' similar to Martian flood terrains. The presence of large buried craters indicated that the layer of basalt could be up to 2 km thick. Clusters of pits several kilometers across, adjacent to the plains, were evidently the vents from which the lava had erupted. The volcanic flow was less cratered than surrounding terrains and its age was estimated at about 3.8 billion years, which was similar to that of the Caloris basin. During the first 9 months of operations, over 4 million laser altimeter points were taken and revealed Mercury's northern hemisphere to be much flatter than either the Moon or Mars, with an altitude range of less than 10 km. Moreover, the northern plains were located in a depression 2 km deeper than the surrounding terrain. The location of the lowlands at the pole may be due to a reorientation of the planet's spin axis to place it coincident with the maximum inertia axis. The same phenomenon is responsible for asteroids and comets spinning along their shortest axis, and possibly for Enceladus having its tiger stripes and liquid subsurface water at the south pole. Located within the plain was a 950-km-wide rise and gravity anomaly which had no corresponding topography, and may be evidence of an intrusion of magma that deformed the crust. Likewise, altimetry showed that the northern portion of the floor of the Caloris

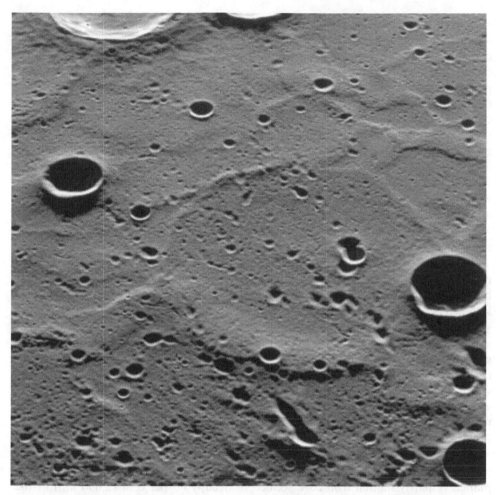

A 29 March 2011 image of the northern polar smooth plains, near crater Hokusai. (NASA/Johns Hopkins University Applied Physics Laboratory/Carnegie Institution of Washington)

basin bulged upward, in places exceeding the height of the rim. The laser altimeter peered into the permanently shadowed polar craters, taking more than 4 million topographic and 2 million reflectivity measurements over the polar regions during the primary mission alone. Craters at intermediate northern latitudes in which ice deposits were suspected from radar observations yielded only weak returns that were incompatible with the presumed presence of slabs of water ice. This led to the conclusion that the ice deposits must be buried under radar-transparent ice-poor dark material rich in organic molecules similar to the lag of a cometary surface. In fact, at such latitudes ice could be stable, and if buried beneath a layer of insulating material it would not sublimate. At higher latitudes, exposed ice would be able to survive as the floors of craters became sufficiently cold. As a result, laser-bright regions in large craters like Prokofiev (108 km across) and Kandinsky (62 km), both located around 85°N, were interpreted as exposed ice. The cameras provided further proof of the presence of ice at the poles by verifying that all known radar bright deposits discovered from Earth matched steep-walled polar craters that had at least part of their floors in permanent shadow. Definitive evidence of ice was provided by the neutron spectrometer, which detected several per cent fewer fast neutrons present over the north pole than over the equator indicating the presence of hydrogen from the dissociation of water molecules. Because of the relatively high altitude of the northern polar passes, at up to 600 km, the individual deposits could not be identified but were taken to be exposed patches of ice from the impact of comets and water-rich asteroids onto the planet, followed by migration and cold-trapping of water molecules in the polar craters. Ice deposits would be several million years old, and their long-term preservation would be aided by the relative stability of Mercury's spin axis orientation, which is a side-effect of the planet's spin-orbit resonance.

Bright areas on the floors of some craters, observed in Mariner 10 imagery, were resolved as clusters of irregular, shallow depressions which lacked well-defined rims but had bright interiors and halos, ranging from several tens of meters to several kilometers in size. These hollows appeared relatively crater-poor, implying they were youthful, and as such they would represent areas of recent activity. These included some of the brightest spots on the whole planet. Shallows were interpreted as the sites of surface collapse following the sublimation or release of volatiles brought to the surface from great depth by impacts and exposed to solar heat. Mesas and hills within the areas could represent the uneroded remnants of the original surface. In a sense, shallows could be the Mercurian analog of Martian annual 'Swiss-cheese' terrain, although their formation would be a much slower process. Not a single such hollow was located on the northern volcanic plains.

Radio tracking of the spacecraft combined with altimetry data probed the interior of Mercury, revealing a complex structure unlike that of any other rocky, terrestrial planet. A partly molten iron core seemed to occupy 83 per cent (2,030 km) of the radius of the planet, with the mantle and crust together making up the remaining 400 km. Moreover, this outer shell appeared to have a high average density, but because heavy elements like iron and titanium were rare on the very surface a 'reservoir' of high-density material was implied and assumed to consist of a dense layer, possibly involving iron sulfide, sandwiched between the core and mantle and up to 200 km thick. Interestingly, such a layer would be an electrical conductor that would shield and weaken the otherwise strong magnetic field of the molten core.

MESSENGER mapped the magnetic field of the northern hemisphere in detail, finding its axis to be tilted only a few degrees relative to the spin axis. Moreover, the field appeared to be displaced some 480 km to the north, resulting in a stronger field at northern latitudes than in the south. This characteristic could reveal something of the mechanism generating the field in the first instance. A core dynamo-generated field like that of Earth, in fact, would be symmetrical. Not unlike the displaced non-axial fields of Uranus and Neptune, that of Mercury could arise not in the core but somewhere near the boundary between the core and mantle. On the other hand,

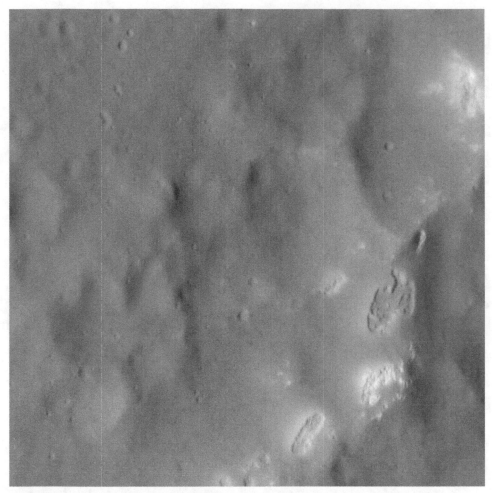

Hollows on the central ring of a 170-km crater. The picture has a resolution of about 15 meters. (NASA/Johns Hopkins University Applied Physics Laboratory/Carnegie Institution of Washington)

energetic electrons in the vicinity of the planet showed that the magnetic field was too weak to create true radiation belts, which made Mercury the only 'magnetic' planet not to possess Van Allen-like radiation belts. But the presence of a quasi-trapped particle belt at half the planet's radius from the surface is suspected from this data and from simulations of the planet's dynamics. Given its shape, Mercury's magnetosphere, like that of Earth, only gives access to high latitudes to solar wind ions, and in particular, owing to the displaced magnetic field over the southern hemisphere which MESSENGER could observe only at long range. On reaching the surface, ions would erode rocks and release atoms into the exosphere. As a consequence, the north pole was identified as an important source of sodium and oxygen, as well as water-related ions in the exosphere. Helium was also present,

A large field of hollows on the floor of 97-km crater Tyagaraja. (NASA/Johns Hopkins University Applied Physics Laboratory/Carnegie Institution of Washington)

possibly after being implanted in the surface by the solar wind, to which the planetary magnetosphere provided only a weak barrier, and subsequently released.[179,180,181,182,183,184,185,186,187,188,189,190,191,192]

At the conclusion of the primary mission, MESSENGER was given a 12-month extension to March 2013 in order to undertake more targeted observations and study the Mercurian environment's response to the changing solar activity as it headed to a new maximum. During the primary mission, the gamma-ray spectrometer produced more than 38,000 spectra, over 15,000 of which were acquired at low altitude, the cameras took 34,834 images covering 99.9 per cent of the planet at a resolution of 160 meters as well as stereoscopic coverage of 92.5 per cent of the surface. A global multispectral map at a resolution of 880 meters was obtained. And almost 12 million laser altimeter points were obtained. The spacecraft appeared to be in good shape overall, except for the gamma-ray spectrometer, which had to be turned off when its cooler failed.

In March and April 2012 MESSENGER performed a sequence of three maneuvers to modify its orbit, lowering the periapsis back to 200 km and reducing the period to 8 hours in order to provide the instruments with more time to observe at low altitude and target in particular the permanently shadowed polar areas. Also to be addressed were the processes that create the mysterious hollows, in particular by searching for changes over time and by imaging them at higher resolutions. Moreover, the cameras were to map the planet at a similar resolution to the primary mission but at different geometries and incidence angles. The idle time around apoapsis could be devoted to further searches for Vulcanoids and Mercurian satellites. For the second maneuver of this series, on 16 April, all of the oxidizer was intentionally burned to depletion, the 53.5-m/s maneuver reducing the period by almost 3 hours. This meant that any further maneuvers would have to be undertaken using the monopropellant thrusters. A final 4-minute maneuver on 20 April established the 278 × 10,134-km, 8-hour orbit which would resist periapsis decay caused by solar gravitational perturbations better than the 12-hour one. About 12.5 kg of propellant remained in the tanks. By the end of the first extended mission, perturbations had shifted the periapsis latitude from the original 60°N to 84°N, essentially over the pole. This enabled close-range studies of the permanently shadowed craters.

When the extended mission ended in March 2013, NASA, facing budgetary issues, was undecided about whether to approve a second extension or not. Until a formal decision was made, the team managing the mission was asked to continue routine operations of the spacecraft and scientific instruments. Scientific observations were thus allowed to continue. MESSENGER went into superior solar conjunction in the first half of May. A solar coronal mass ejection happened to cross our line of sight to the planet on 10 May and radio signals were used to investigate the magnetic field of the solar corona. In February and again in July 2013, MESSENGER made searches designed to detect possible satellites of the planet as small as 100 meters at distances in the range 2.5 to 25 planetary radii. During the course of the second survey, the spacecraft also happened to capture six images of the Earth and Moon, almost 100 million km distant.

In November, MESSENGER took advantage of close passes by two comets to perform further opportunistic observations. On 18 November 2013 comet 2P/Encke passed Mercury at a range of 3.7 million km (0.025 AU), 3 days prior to the comet's perihelion, and was well placed for observations. Only one day later, the bright long-period comet C/2012 S1 ISON passed by at a range ten times greater, heading for a very close perihelion on the 28th. ISON was discovered in September 2012 by Russian observers using the International Scientific Optical Network telescope. Initially hailed as the "comet of the century", the object was torn apart by the intense solar gravity as it reached perihelion less than 1.2 million km from the photosphere. MESSENGER's spectral observations of ISON revealed the presence of abundant carbon from organic grains. The relative fragility of these grains could be one of the reasons why the comet did not survive its encounter with the Sun.

If left unattended, MESSENGER would impact on Mercury in late August 2014, but the remaining fuel was sufficient for three maneuvers in 2014 and a final one in 2015 to raise its periapsis from as low as 25 km to at least 100 km. After that final maneuver, the periapsis is to be left to decay, briefly stabilizing at about 15 km. The spacecraft will fly as low as 2 or 3 km on its last orbits and pass just 150 km over the north polar craters before finally crashing on or about 28 March 2015, only days after the start of its fifth year in orbit. The impact point, at a latitude of about 58°N, will unfortunately be on the side of Mercury not visible from Earth at that time.[193,194,195]

AVENGING THE DINOSAURS

In the 1990s engineers at JPL and Ball Aerospace, and scientists at the US National Optical Astronomy Observatory and at the University of Maryland, made a study of a cometary mission in which a 500-kg projectile would be smashed into the nucleus of a comet while its bus observed the impact from a safe distance. The pictures and spectra of the ejecta taken during the cratering process would provide insight into the structural properties of the nucleus and identify differences between the surface, which is subjected to outgassing, and the pristine interior. In fact, until then most of our understanding of the nature, structure and composition of cometary nuclei was derived from observations of the coma, and there was no certainty that this could be extrapolated to the surface and subsurface. This mission, named Deep Impact, was proposed to NASA in 1996 for the Discovery program, but was rejected. However, after some minor modifications it was resubmitted for the next round of proposals and was approved in July 1999, together with MESSENGER. Several targets were under consideration, including nine bona fide comets and two objects that seemed to be extinct nuclei; namely (3200) Phaethon, which was the target of the original study, and (4015) Wilson–Harrington. The main evidence for Phaethon is that it is the parent of a meteor stream, so the link is speculative. The case for Wilson–Harrington is stronger, because after it was discovered and catalogued as an asteroid, a study of its orbit revealed that it had previously been listed as a faint comet. Although in neither case was there likely to be volatiles on the surface, the impactor could expose subsurface material to reinitiate cometary activity. Other candidates, including Tuttle–Giacobini–Kresak, had to be rejected because the nucleus was too small, or the encounter would occur with an unfavorable geometry with respect to the Sun and the Earth, or that the object was already being considered as a target for the CONTOUR mission.

A collision with 9P/Tempel 1 in July 2005 would occur when the comet was both close to perihelion and intersecting the plane of the ecliptic, which had the attraction of minimizing the launch energy, and with a good geometry relative to Earth and the Sun for mission requirements and for ground-based and orbital telescopes to observe the comet in the evening sky. Furthermore, optical targeting would be eased by the fact that the nucleus would be more than half-illuminated from the viewpoint of the spacecraft. It was also hoped the lander-equipped Deep Space 4 would manage to reach Tempel 1 soon after the impact in order to investigate the new crater in detail, but unfortunately this mission was canceled several weeks before Deep Impact was selected.[196] Discovered on 3 April 1867, Tempel 1 was the second periodic comet to be spotted by Ernst Wilhelm Liebrecht Tempel, who at that time was working at the Marseilles Observatory. It was observed on three returns, and then lost after a close encounter with Jupiter modified its orbit. Upon recovery in 1972 it was identified on plates taken 6 years earlier. It had been seen at every return since, usually as a faint object with little or no trace of a tail.[197] As it was chosen as a target primarily for its orbit, very little was known of the nucleus and worldwide observations were needed to place constraints on its size, shape and rotation period, and on the density of the dust in its coma. These studies would characterize the state of the comet and provide a basis for determining the effects of the mission. Several years of observations by some of the largest and most sensitive telescopes on Earth and in space revealed the nucleus to be an ellipsoid of 14.4 × 4.4 × 4.4 km that took between 39 and 42 hours to complete a rotation. It was typically dark, reflecting 4 per cent of the light that it received from the Sun. At any time only a few per cent of the surface appeared to be active, and the dust environment seemed to be relatively benign. It appeared that the spacecraft would approach near the equatorial plane, where the longest axis should be, but it was not possible to predict precisely whether the impactor would strike it broadside or head-on.[198]

The initial orbit design envisaged launching in January 2004, performing a year-long calibration and test phase that would end with an Earth flyby during which the software designed to autonomously track the cometary nucleus would be tested with the Moon as the target. Unfortunately, the project suffered delays in the delivery of several key components, and also difficulties in developing the navigation software and in correctly implementing its interactions with the attitude control system. As a result, the launch was slipped to January 2005. When the costs rose from the initially projected $240 million to $328 million, NASA considered axing it. The calibration phase was deleted but the mission was otherwise unchanged, with the gravity-assist of an Earth flyby being compensated by using a more powerful form of the Delta II. An impact on 4 July 2005 would not only be one day before the comet's perihelion, but would also mark an American national holiday. This solution roughly optimized several parameters, including lighting conditions and launcher performance, and the impact was scheduled for a 1-hour window when it would be able to be tracked by Deep Space Network antennas in both California and Australia and the effect of the strike could be observed by the large telescopes in Hawaii and in space.[199,200]

The Deep Impact spacecraft comprised a rectangular box to house the propulsion systems and electronics, an external platform with scientific instruments, a 1-meter high-gain antenna on a tripod, and a 2.8-meter square solar array that folded against the bus for launch and could supply up to 620 W of power once deployed in space. It had reaction wheels and thrusters for attitude control. Course corrections were to be made by four 22-N thrusters. Hydrazine was used for both maneuvering and attitude control. Including 86 kg of hydrazine, the mass was 601 kg. Parts of the bus which would face forward during the encounter were protected by dozens of small Whipple shields of various designs. The main shield was fairly conventional, having sheets of aluminum separated by a 10-cm gap, but innovative oblique graphite-epoxy shields protected the solar array. As a precaution in case the spacecraft was disabled by dust in the coma of the comet, most of the data was to be returned in real-time via a fast 200-kbits per second radio link. A recess in the bus housed the impactor, which also doubled as the interface to the launch vehicle. Battery-powered for its 24-hour free flight, and incorporating many hardware components identical to those of the bus, including a two-way communication system, the impactor had 7.8 kg of hydrazine for the lateral thrusters which would steer its final approach. It was a six-sided vehicle, 1 meter in size. There was a 113-kg conical 'dead cratering mass' on the 'ram' side in order to increase the momentum at impact. This was a machined block of pure copper so that its spectrum could be easily subtracted from that of the comet, on which copper had never been detected, and it incorporated cutouts for the targeting system. In fact, half of the mass of the impactor, including the front dust shield, was copper to minimize the amount of aluminum. It was calculated that by smashing into the nucleus at a speed of 10.2 km/s the 372-kg impactor, depending upon the nature and strength of the comet, would excavate a crater up to 200 meters across and 50 meters deep. The bus and impactor would both use an autonomous navigation system which exploited the lessons learned from Deep Space 1 and Stardust, employing imagery to update their positions relative to the nucleus to ensure that the impactor struck its day-side while the bus maintained a clear line of sight to observe the crater.[201,202,203,204]

The bus had two instruments on its external platform. One was a 30-cm-aperture telescope with a focal length of 10.5 meters, making it one of the largest ever flown on a planetary mission. This fed a 1,024 × 1,024-pixel CCD capable of a resolution of 1.4 meters and an infrared spectrometer that was much more sensitive than that of the Soviet Vega missions. They were to determine the composition and temperature range of the nucleus, and identify chemical elements and molecules in the plume of freshly exhumed ices. The second instrument was a 12-cm-aperture 2.1-meter-focal-length medium-resolution camera for wide-field views. Both cameras were equipped with filter wheels for color imaging. The impactor carried only a camera similar to the medium-resolution camera on the bus, but without the filter wheel. In addition to supplying pictures to the targeting system, it was hoped that in the final seconds before the impact this camera would show details of the nucleus as fine as 20 cm in size. If the flyby imagery showed large clumps of ejecta falling back onto the surface, their trajectories would serve to measure the mass of the nucleus.

But a large fraction of the scientific objectives of the Deep Impact mission would be accomplished by telescopic and spectrometric observations by some of the largest ground-based telescopes in Hawaii, the Hubble Space Telescope, the Spitzer Space Telescope, and NASA's Chandra and ESA's Newton X-ray telescopes. A number of other astronomy satellites would also contribute. The gap in professional monitoring of Tempel 1 before, during and after the impact was to be addressed by a program in which moderately equipped amateur astronomers would monitor the evolution of the coma, the production of dust and gas, and the interaction of the coma with the

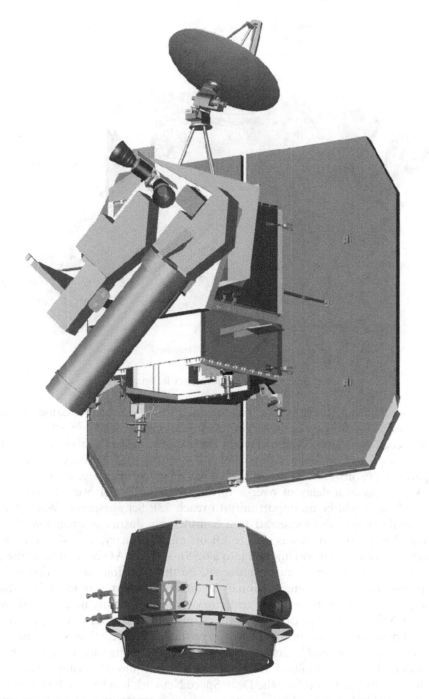

A CAD rendition of the Deep Impact spacecraft showing (top) the mothership and its telescopes and (bottom) the impactor.

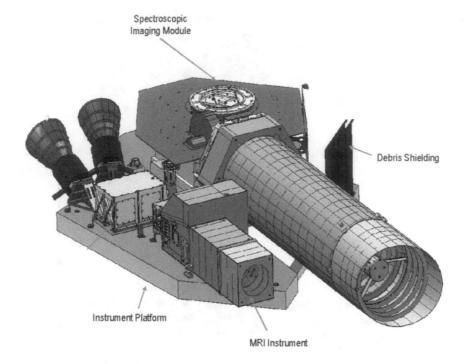

A CAD image of the Deep Impact imaging payload.

solar wind. Smashing the impactor into the nucleus was expected to cause the comet to undergo a 10-fold increase in brightness for days or weeks.[205,206,207,208]

The Deep Impact spacecraft arrived at Cape Canaveral in October 2004, with the launch window to Tempel 1 running from 30 December to 28 January. An issue with the spacecraft's software, and then the need to replace an interstage ring of the Delta II rocket, imposed a delay of over a fortnight. If the launch were to be delayed further, there would be an opportunity to reach 73P/Schwassmann Wachmann 3, but it would subject the spacecraft to a considerably dustier environment than its designers had presumed. It was able to lift off on 12 January, and after only a few minutes in parking orbit was injected into a 0.981 × 1.628-AU orbit around the Sun. In contrast to recent very accurate launches, on this occasion the injection was not very precise owing to a computational 'inconsistency' that was identified shortly beforehand. Although the resulting trajectory error could easily be corrected by the spacecraft's first maneuver, calculating the parameters of the burn first required that its actual position in space be identified. To further complicate things, the spacecraft adopted safe mode several hours after leaving Earth, faced its solar arrays to the Sun and spun itself up for stability. Fast intervention was required to bring it back online and pinpoint its position before the Deep Space Network had to switch stations. The safe mode had been triggered by incorrect readings from temperature sensors, and 20 minutes of the recovery procedure were assigned to orbit determination.[209] Over the

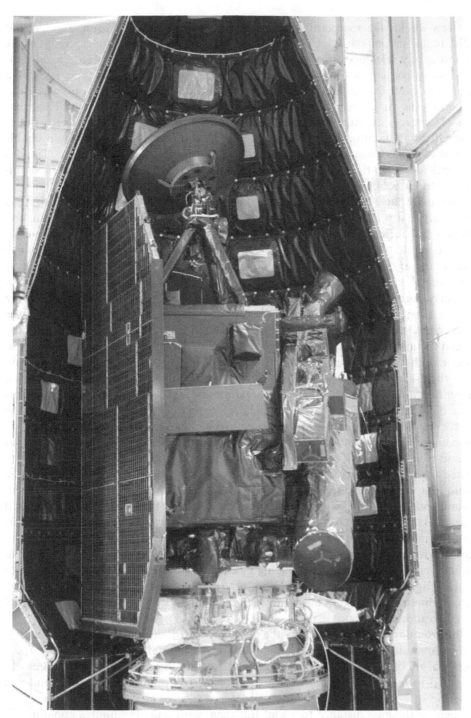

Deep Impact about to be closed inside the shroud of its Delta launcher.

next few weeks, the spacecraft took calibration images of the Moon, Jupiter, and individual stars and clusters. Tests revealed that after all the moisture present in the carbon-epoxy telescope tube of the high-resolution camera had been 'baked' out, the instrument could not be precisely focused. It was providing pictures of a resolution three to four times worse than predicted. The probable cause was a flat mirror in the optics which had become slightly warped during thermal vacuum tests. However, as with the NEAR and Stardust missions (whose camera optics suffered contamination) engineers were sure that image processing would enable the out-of-focus camera on Deep Impact to achieve all its scientific objectives. But the optical navigation would now have to be carried out using the medium-resolution camera.

The first course correction of 28.6 m/s was performed on 11 February 2005. The spacecraft first spotted Tempel 1 on 25 April at a range of some 40 million km, and detected the nucleus a month later. From 60 days to 7 days prior to the encounter, it would be limited to observing its target for 15 minutes every 4 hours, in order not to exceed temperature limits. Meanwhile, the comet was being intensely scrutinized by Earth-based telescopes, and it was the Hubble Space Telescope which first saw a cometary outburst on 14 June that caused a short-lived jet of dust to appear. A more massive event 8 days later temporarily expanded the size of the coma, and this was observed by Deep Impact itself. Course corrections on 5 May and 23 June not only refined the aim to exploit the latest ephemeris, but also rescheduled the encounter in order to enhance the opportunity for Hubble viewing, since this could observe for no more than 40 minutes per orbit. Starting one week before the encounter, imaging for navigation and science was to be almost continuous, except for when the spacecraft made a course correction or was occupied releasing the impactor. Despite having to use the medium-resolution camera, the optical navigation allowed the ephemeris of the comet to be computed much more accurately than was possible on the basis of ground observations alone. Indeed, the first orbit solution using in-flight imagery shifted the most likely position of the nucleus by more than 900 km! In the end, 922 ground-based and 3,956 spacecraft observations were used to refine the ephemeris to optimize the encounter. With 2 days to go, the spacecraft placed itself on a collision course. The following day, 12 minutes after pushing the impactor away at 35 cm/s, the bus made a 102.5-m/s burn both to establish a 500-km miss distance and also to slow down to ensure it would have 850 seconds of continuous line-of-sight viewing of the impact and of the rapidly evolving ejecta plume. If the impactor had failed to release, engineers would have had 12 hours to resolve the fault before the bus would have to make the deflection burn, in which case only observations of the unmolested comet would have been possible.

Just 16 minutes after its release, the impactor's attitude control system awakened and commanded a 17-cm/s burn to stabilize its attitude. But this moved the impact point 1.8 km off the nucleus. Two hours out, the autonomous navigation system was activated to refine the trajectory. However, owing to a small attitude determination error the 1.27-m/s maneuver half an hour later at a range of 53,600 km had the effect of opening the miss distance to 7 km! But then, with only 35 minutes remaining, a 2.26-m/s change in lateral velocity performed at a range of 21,600 km centered the predicted impact point on the nucleus. Finally, at 7,700 km, and with just 12 minutes to go, a correction based on a 'scene analysis' image-processing algorithm achieved a velocity change of 2.28 m/s which deflected the aim point 1.7 km to place it on the sunlit part of the nucleus.[210] Both vehicles were returning pictures, and these showed a varied topography which included smooth patches similar to the plateaus found on Borrelly, a long scarp (possibly layered) near the comet's equator, circular pits and sinkholes, and (for the first time on a cometary nucleus) bona fide craters rather than the circular depressions seen on Wild 2. In fact, the impactor appeared to be heading for a pair of craters. Overall, the nucleus gave the impression of being held together by self-gravity rather than the cohesivity of its material. The imagery also provided a direct measurement of the rotation period of the nucleus, and at 40.7 hours this was somewhat slower than implied by telescopic studies. Also contrary to expectation, it was moderately pear-shaped rather than elongated, and with a longest dimension of 7.6 km and a shortest dimension of 4.9 km it was considerably smaller. Temperature scans of the undisturbed nucleus indicated that the surface material was porous, with a low thermal inertia. The subsolar point was warmest, with a temperature of about 50°C. This data also indicated that the sublimation of water and carbon oxides that gave the nucleus its cometary appearance probably occurred below the surface. The most significant result from the bus was the identification of three small areas which covered a total of only 1/4,000th of the surface showing thermal characteristics and spectra compatible with the presence of 3 to 6 per cent of water ice. In fact, despite Tempel 1 being the fourth cometary nucleus 'remotely sensed' at close range, this was the first time that water ice had been detected on the surface of one. In any case, the data confirmed the belief that most of a comet's water and volatiles were held in subsurface reservoirs.[211]

A large dust particle hit the impactor 20 seconds before it struck, and turned its camera away from the nucleus. The attitude control system restored the line of sight with 10 seconds to go, but then there was another dust hit. The last blurry image was returned when the impactor was 4 seconds prior to impact, 40 km from the nucleus, but it was not of the targeted site. The last clear picture showed details only 3 meters in size, but by then the optics had evidently been 'sandblasted' by cometary dust and the quality was poor. At 05:44:36 UTC the projectile made an oblique impact on the day-side of the comet, striking near the edge of one of the two craters. The release of kinetic energy was equivalent to detonating 4,500 kg of TNT. As a result, the comet was slowed by an estimated 0.0001 mm/s and its perihelion distance was reduced by some 10 meters.[212,213]

The bus saw a small flash and then, after a short delay, a larger flash followed by a plume of dust and vapor. In fact, its images showed the *shadow* of the ejecta cone and expanding debris cloud as this sprayed out. Spectra of the incandescent plume showed the presence of water vapor, carbon dioxide, hydrogen cyanide and complex organics. Spectra taken minutes later showed a marked increase in the production of carbon dioxide, a less dramatic increase in the water, and traces of methyl cyanide. According to some estimates, about 20,000 tonnes of dust was ejected by the impact. This was 100 times more than expected. There were too many particles to have been produced by pulverization, suggesting instead that the impactor punched through an

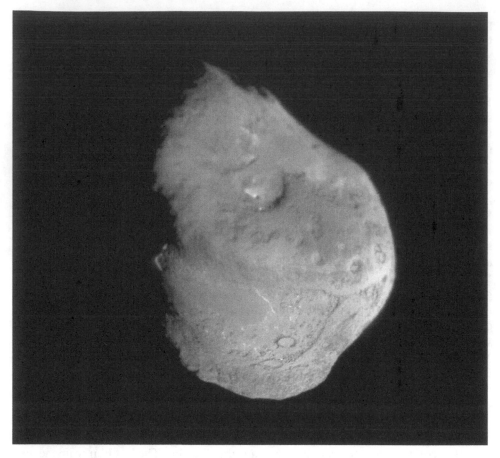

A mosaic of images of different resolution (best toward the bottom) of the nucleus of Tempel 1.

accumulation of dust several meters in thickness. The model that emerged indicated that the nucleus was very porous, and made of loose material held together by weak self-gravity. It seemed that the repeated sublimation and condensation at successive perihelia had produced a shielding cover of 'de-volatilized' dust with little or no water intermixed, below which was a layer of almost pure water-ice grains at least 10 meters thick.[214] The impact may well have dug through a layer of amorphous ice. The release of heat as ice changes phase from amorphous to crystalline would cause a chain reaction, and the expansion of gas would transport trapped dust. This process had been suggested (for example) for the anomalous brightening of comet Halley in 1991 some 5 years after its perihelion passage.[215] Modeling had suggested that the Tempel 1 dust would disperse in 200 seconds and reveal the new crater, but the very fine dust (likened to talcum powder) was still lingering over the site 13 minutes after the impact and denied the bus a view of the crater. After tracking the impact site for the scheduled 800 seconds, the bus had to rotate to face its dust

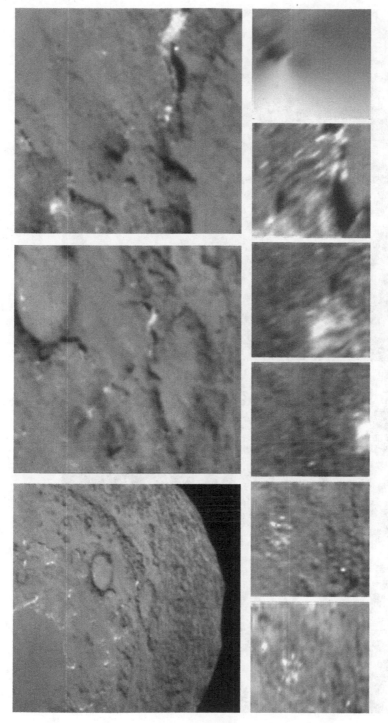

At top a sequence of images of increasing resolution taken by the targeting camera on the Deep Impact projectile. The impactor appeared targeted for the rim of one of the two craters. Note the bright material that could be freshly exposed. The bottom line shows the final sequence of images of degraded quality leading to the impact.

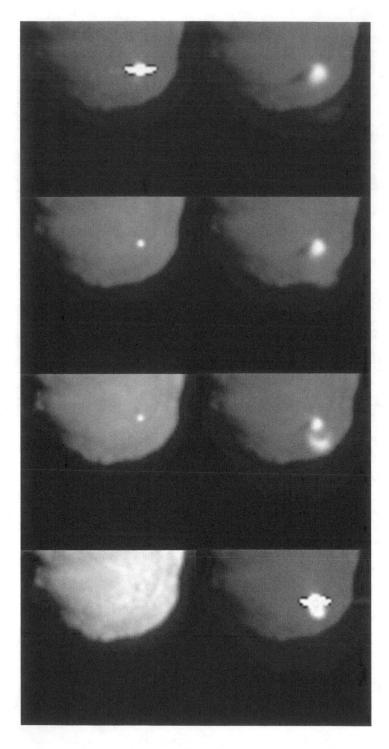

The impact as seen in images taken by the mothership about every 0.1 second. The shadow of the plume is clearly visible.

The nucleus of Tempel 1 after the impact seen by the Deep Impact mothership.

shields 'forward' for crossing the orbital plane of the comet – where most of the dust in the coma was expected to be. Only four large particles with masses in the microgram to milligram range hit Deep Impact around the time of closest approach. One hour later, it turned to begin to take pictures of the night-side of the comet. Surprisingly, the ejecta cloud was still attached to the nucleus. Its evolution yielded an estimate of the mass of the nucleus, establishing its density to be only 60 per cent that of water, consistent with a porous interior.

As seen from Earth, Tempel 1 developed a bright star-like nucleus within several tens of minutes of the impact, then returned to its usual faint and fuzzy aspect over a few days. Except for rare (and often disputed) observations of spacecraft striking the Moon, this was the first case of a human action having a visible effect on a celestial body. In many ways, the comet appeared to behave in the same manner as after a natural outburst. Seventeen orbits of the Hubble Space Telescope were dedicated to the project, with one monitoring campaign in June and another at the time of the impact. It first detected traces of the ejecta cloud 20 minutes after the impact occurred, and monitored its evolution. Initially semicircular, it was later forced by solar radiation pressure into the antisolar direction to join the tail. But the most important task for Hubble was to monitor the coma for the generation and evolution of volatiles like carbon monoxide. The Spitzer Space Telescope, in solar orbit, took infrared spectra which showed minerals, polycyclic aromatic hydrocarbons (not previously seen in a comet) and intriguing hints of compounds such as carbonates and clays which are produced in the presence of water. On the other hand, the presence of crystalline silicates was the first indication that material which formed in the high temperature environment of the inner solar system had been transported to more frigid regions and incorporated into a comet. (This would be proved conclusively within a year by analysis of the samples of Wild 2 returned to Earth by Stardust.) Other satellites in Earth orbit monitoring water production rates did not see any particular increase. In general, remote observations did not show any evidence of new chemical species being injected into the coma as a result of the impact; from this perspective only the ratio of dust to gas changed dramatically.[216,217,218,219,220,221,222,223,224,225,226]

The Tempel 1 encounter left Deep Impact with more than 40 kg of fuel to spare, and in an orbit with a period of 1.5 years that would return it to Earth 3 years after launch, which offered the opportunity of retargeting it for an extended mission. Five options were considered: three comets, and both of the hybrid objects Phaethon and Wilson–Harrington. The most favorable option in terms of propulsive requirements and flight time was 85P/Boethin. This comet was discovered by Leo Boethin in the Philippines in January 1975. It was observed on its return in 1986, but was too close to the Sun in the sky for viewing in 1997. This 11-year orbit made it a member of a small group of comets which complete one elliptical orbit in about the same time as Jupiter completes its circular orbit, with this resonance making the cometary orbits relatively stable. Calculations showed that Boethin had been pursuing essentially the same orbit for at least 700 years, and would continue to do so for another 1,000 years. As Boethin had not been seen in 20 years, a prerequisite to directing Deep Impact to this comet was that astronomers recover it by October 2007 and refine its ephemeris in time for the Earth flyby in December that would set up an encounter in December 2008. Astronomers hoped to have ESA's new Herschel infrared telescope in operation in time for the encounter.[227,228,229,230,231,232,233,234,235] As a second cometary flyby had never been a formal objective, the mission extension depended on NASA providing funding. On 20 July the spacecraft made a course correction to refine its return to Earth, and in early August, after recalibrating its instruments, it was placed into hibernation.

In 2006 NASA invited proposals for how to reuse the Deep Impact and Stardust spacecraft using 'Mission of Opportunity' funds allotted to the Discovery program. The selections were announced in July 2007. For Deep Impact no fewer than two non-overlapping missions received funding. First, in the early part of 2008, was the EPOCh (Extrasolar Planet Observation and Characterization) mission, for which

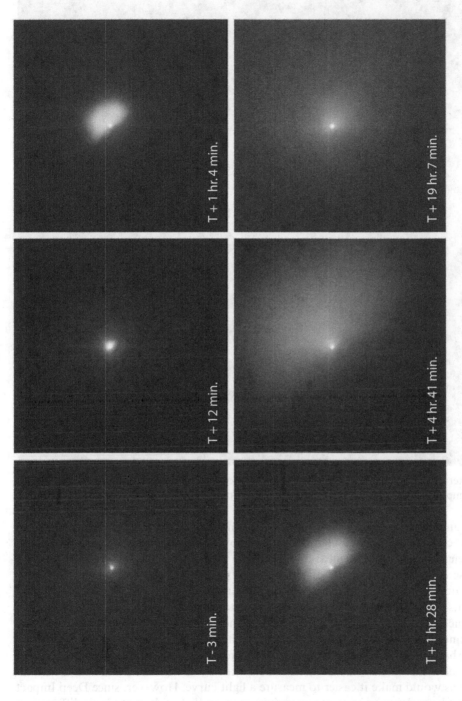

The evolution of Tempel 1 after the impact monitored by the Hubble Space Telescope. (NASA, ESA, P. Feldman of Johns Hopkins University, and H. Weaver of Johns Hopkins University Applied Physics Lab)

After closest approach, Deep Impact turned back to resume imaging the nucleus of Tempel 1. The cloud of debris raised by the impact was still present.

the main telescope (the largest ever flown on an interplanetary spacecraft, remember) would collect very precise 'light curves' of known extrasolar planets passing in front of their parent stars. The measurement of the dimming of the star would provide a precise determination of the diameter of the planet and this, combined with mass estimates derived from spectroscopic measurements by terrestrial telescopes, would yield its density. By obtaining more precise data using the Deep Impact spacecraft, it was theoretically possible to detect other planets orbiting further out from the star by minute variations in the timing of the transiting body due to gravitational perturbations. By an ironic twist, it was the inability of the camera to precisely focus that made this an attractive project, since spreading the light of a star over a number of pixels would make it easier to measure a light curve. However, since Deep Impact had not been designed to fixate on a given point in the sky it might have difficulty in maintaining its attitude sufficiently accurately.[236],[237] A rival idea had been to use the high-resolution telescope to make 'microlensing' observations of the type subsequently undertaken by Rosetta in 2008.[238] After completing its EPOCh observations, Deep Impact was to perform the Boethin encounter as the DIXI (Deep Impact eXtended Investigation) mission. The two projects were combined as EPOXI (EPOCh + DIXI) and financed with an overall budget of $30 million.

After 25 months in hibernation, Deep Impact was awakened on 26 September 2007 to refine the determination of its orbit and to check out its instruments in preparation for the Earth flyby. Meanwhile, comet Boethin was being sought using some of the world's largest telescopes, including some of the 8-meter 'giants', but in vain. A candidate was found within the search area, but the team was not sufficiently confident that it was comet Boethin to recommend targeting Deep Impact to it. The Spitzer Space Telescope was ideally suited to the task, but despite more than 15 hours of observation it found no sign of Boethin. In fact, not only was the comet not found by the October deadline, it remained elusive after it had been expected to be bright enough to be observed by moderately well-equipped amateurs. The DIXI team switched the target to 103P/Hartley 2, although this would add another 2 years and increase the cost. To reach Hartley 2, Deep Impact had to be placed into a CONTOUR-like orbit that was resonant with Earth. In fact, the flyby in December would achieve a 1-year orbit which was only slightly more eccentric than that of Earth and would provide returns in December 2008 and again in June 2010, at which time it would be directed to its target. The fate of Boethin's comet remains a mystery to this day. One possibility was that its position was well off that predicted, and too close to the star-crowded Milky Way to be spotted. Another possibility was that it had somehow switched off its cometary activity after 1986 and, since the nucleus was estimated to be at most a few hundred meters in size, become a dark asteroid that was difficult to see. The third (and most likely) possibility was that it broke up either in 1986, when it was 100 times brighter than expected, or during the unobserved return of 1997.[239] As demonstrated by Schwassmann–Wachmann 3, the target assigned to the ill-fated CONTOUR mission, fragmentation is a common fate for periodic comets.

On 1 November 2007 Deep Impact made a course correction to set up for its new target, although NASA did not formally approve the diversion to Hartley 2 until the end of that month. Meanwhile, the spacecraft made test observations of a binary star in Ursa Major in preparation for the EPOCh schedule. On approaching Earth, the cameras and spectrometer were trained on the Moon to recalibrate them. The Earth flyby on 31 December was above eastern Asia at an altitude of 15,566 km. This put the spacecraft into a 0.91 × 1.09-AU orbit inclined at just over 4 degrees to the ecliptic. A total of six known planets were selected for EPOCh observations during 4 months from January to May 2008, but some of these opportunities were lost to hardware problems. For example, observations of the largest transiting planet known were lost because the spacecraft was at perihelion and suffered overheating that put it into safe mode. There were also issues with the pointing accuracy of the telescope and with the downlink speed. It was decided to assign EPOCh additional observing time through to August. Four planets were targeted in this second round,

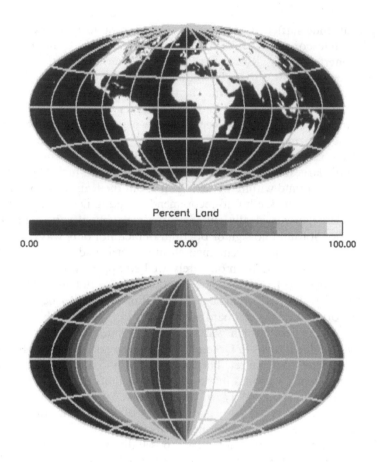

Earth as an alien planet. Distant imaging of Earth allowed scientists to test methods for mapping alien planets. This map shows the percentage of land as a function of longitude. (Courtesy of Nick Cowan)

of which three had not been observed earlier. An amazing total of 198,434 images cropped at 256 × 256 pixels around the expected position of the target star were taken, of which 97 per cent were successfully sent to Earth and 87 per cent were judged to be usable. Deep Impact also observed Earth several times from March to June 2008 in order to characterize how a distant habitable world would appear to a putative large orbiting telescope. On 29 May, at a range of 31 million km, it took a remarkable sequence of pictures at 15-minute intervals spanning a full axial rotation period and also caught the Moon passing in front of the Earth's disk. The EPOCh team shrank these images in order to produce a very-low-resolution map of Earth as an alien planet to simulate the view of a future 'planet finder' telescope. This resolved the oceans, land masses and, when infrared wavelengths were added, even the presence of vegetation. Better maps would be possible by disentangling the images produced by an observer who viewed the planet from a vantage point that was not in

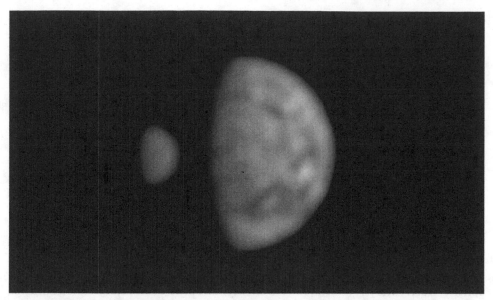

The Moon transiting in front of Earth from the point of view of Deep Impact on 29 May 2008.

the ecliptic plane.[240,241,242,243] In a technology demonstration experiment conducted in October, the spacecraft acted as a 'node' in the first experimental 'interplanetary Internet' and it simulated a network of Martian landers and orbiters by relaying dozens of images between different Earth stations.

After an Earth flyby over the southern Pacific Ocean at a range of 43,000 km on 29 December 2008, Deep Impact resumed hibernation. It was in an almost circular 1-year orbit inclined to the ecliptic that produced encounters at the nodal crossings in June and December. It repeated the 'Earth as an alien planet' experiment from the closer range of only 17 million km twice in 2009, viewing the north pole near the time of the spring equinox from well north of the ecliptic plane and the south pole near the autumnal equinox. Sequences caught sunlight glinting off seas and lakes. A similar technique may offer a way to detect large bodies of water on extraterrestrial planets. In between, the spacecraft made two relatively close approaches in late June and December, passing by at a range of about 1.3 million km on both occasions. In June, the spectrometer confirmed the presence of water molecules and ions on the Moon. Observations were made on 2 and 9 June and, although the range exceeded 5.9 million km, on both occasions the signal of water was clear. The data showed the strongest absorption bands over the north pole, and revealed dynamic processes driven by solar insolation by comparing spectra taken one week apart.[244] The final encounter on 27 June 2010 passed 30,480 km above the South Atlantic and gave the vehicle a 1.5-km/s boost and redirected it to Hartley 2.

We knew little about 103P/Hartley 2. It was discovered by Malcolm Hartley at the Siding Springs Observatory in Australia on 15 March 1986 and had been observed at every return since. Its present period is 6.41 years, but prior to 1875 it was in a 12-

year orbit resonant with Jupiter. It was placed into its present orbit by repeated close flybys with Jupiter in 1947, 1971 and 1982. It was studied in 1997 and 1998 by ESA's Infrared Space Observatory (ISO). In addition to measuring the production rates of water and other volatiles, this detected crystalline silicates in the coma. Furthermore, the data enabled the nucleus to be isolated from the coma, showing it to be at most 800 meters in radius and, remarkably, to be almost entirely active and issuing gas and dust from its whole surface. Moreover, telescopic pictures of the comet taken when it was far from the Sun showed the nucleus to be still active.[245,246,247,248,249] The Spitzer Space Telescope observed Hartley 2 in August 2008 at a heliocentric distance of 5.4 AU to measure the physical characteristics of the nucleus in preparation for the DIXI flyby. It was estimated to be 570 meters in radius (only one-fifth as wide as Tempel 1 and about one-hundredth as massive) and to reflect just 2 per cent of the light it received (making it dark, even for a comet). The observations also confirmed that when the nucleus was active, it was so across its entire surface.[250] A campaign to monitor Hartley 2 was mounted in preparation for the encounter. This involved no fewer than 51 telescopes in 10 countries, and was also allocated observation time on five astronomical satellites and on NASA's newly commissioned SOFIA (Stratospheric Observatory for Infrared Astronomy) airborne observatory. The campaign was in particular responsible for a preliminary determination of the comet's rotation period. The Spitzer observations were too brief to determine the spin period, but ground-based observatories and the Hubble Space Telescope found this to be about 16 hours. The Herschel space telescope made additional observations in support of the final encounter of the Deep Impact mission.

Deep Impact began imaging Hartley 2 on 5 September 2010, at a distance of 60 million km. An unusual phenomenon was noted between the 9th and 17 September, as the emission of cyanogen gradually increased 5-fold, then slowly decreased over the next several days. Usually these outbursts would be accompanied by an increase in the rate of dust production, but this time it remained steady. One possibility was that cyanogen was being released by dark grains made of complex organics and polymers like the CHON (carbon, hydrogen, oxygen and nitrogen-rich molecules) observed on Halley's comet by Giotto. After calibrating its instruments, cooling the spectrometer and making a course correction, on 1 October the spacecraft started to continuously image the comet. By the end of the month, it was observing the comet every 15 minutes using the camera, and every 30 minutes using the spectrometer for 16 hours per day. A course correction of 1.59 m/s was performed one week out, on 27 October, which was also the day that Hartley 2 was at perihelion. The comet was making its closest pass to Earth since its discovery, and had become a binocular object for amateur astronomers as it passed opposition to the Sun. At the same time, as it came within 17.7 million km of Earth, the radio-telescope at Arecibo obtained radar 'images' that resolved the nucleus. It appeared to be an elongated, probably double object akin to Borrelly's nucleus, with a longest axis of about 2.2 km. The rotation period was finally determined to be about 18 hours. Based on these observations, last minute simulations by JPL's engineers showed that the autonomous tracking software on the vehicle would probably decide on its own which of the two ends to lock on to. Radar observations also showed the presence near the nucleus of a cloud of particles centimeters across. In late October, jets were clearly seen 'sweeping around' the inner coma. Tracking of the jets also showed that the nucleus was not simply rotating end over end, it was probably also nutating and precessing. Hartley 2 was found to be producing carbon dioxide, itself a rare gas as comets go. Production of this gas was seen to vary periodically, and to match the dustiness variations of the coma, suggesting that the same area that was producing the former was also spewing out the latter. This was confirmed by the asymmetric distribution of carbon dioxide in the coma, matching that of dust grains and unlike the symmetrical one for water.

A final 1.4-m/s correction was made 2 days out. This was required in part because of the strong 'rocket effect' of the jets slightly perturbing the orbit of the comet. Fifty minutes prior to closest approach, the autonomous navigation system took control of the spacecraft, slewing it so that the nucleus remained centered in the field of view of the camera. Unlike at Tempel 1, no dust-protection attitudes were adopted for the encounter, but contact with Earth was still severed. Data collected by the mothership and impactor at Tempel 1 had shown the risk of a dust particle strike at the time of closest approach to be minor.

During 80 minutes around closest approach, the cameras took one picture every 4 seconds. About 37 minutes out, at a range of 27,350 km, the nucleus of Hartley 2 began to be resolved as an elongated, 'dog-bone' object that was casting a long shadow in the dusty coma. The spacecraft was approaching the comet almost perpendicular to the direction to the Sun, and over the terminator. It passed between the nucleus and the Sun, and at 14:00 UTC on 4 November flew by at a range of 694 km at a relative speed of 12.3 km/s. A range of 900 km had initially been targeted, but the low risk from dust had facilitated a closer flyby. After closest approach, the spacecraft turned to observe the opposite terminator and then the night hemisphere of the nucleus. Half an hour after closest approach, it turned back to Earth and began to download the pictures and spectra collected over the 18-hour period. Owing to the small range from Earth (just 0.156 AU or 23 million km) a fast data rate was able to be used.[251] About half of the nucleus was imaged in detail, at a best resolution of 7 meters, and the night-side was also mapped, in shape at least, thanks to the silhouette it projected against the bright coma background. It resembled Borrelly and Itokawa with two rough, boulder-strewn ends and a smooth middle section where material probably pooled in a gravitational low, as on Itokawa. But at 2,330 × 690 meters it was four times smaller than Borrelly. In fact, it was the smallest cometary nucleus yet imaged by a spacecraft. It was an extremely dark object, reflecting on average only about 5 per cent of sunlight. Dozens of jets were evidently emerging from it, all of which could be traced back to the rough areas on one end or another. Jets could even have caused the fields of boulders, ejecting them at a speed insufficient for them to escape the weak gravity. A large number of active jets were seen along the terminator of the larger of the two ends, the one closer to the camera, and jets were even faintly visible emerging from the night-side. Most emerged from the smaller end and from the bright mounds of the larger one, but they were seen to originate from every type of geological unit. Bright spots on the night-side marked jets oriented toward the spacecraft, and observed 'head on'. At this level of activity, it was estimated that Hartley 2 must be losing about 1.5 meters of material from its surface per orbit, so much so that it may disappear in just over a century. No obvious craters were visible, but there were several suspicious-looking circular depressions on the terminator. Surface features did not resemble those of Tempel 1 or Wild 2, with no depressions, craters, or flow markings. Rugged, sinuous narrow depressions were visible on one of the ends, as well as bright mounds and enigmatic dark smooth areas. Blocks 80 meters wide were visible on the larger, rougher end, as well as shiny blocks several times more reflective than the rest of the surface.

The high-resolution camera missed the nucleus on most shots but its images, once processed, revealed a number of star-like pinpoints around the nucleus that were determined to be small chunks of ice, centimeters in size, making the scene look like "one of those crystal snow globes". These were traveling at slow speed relative to the nucleus and were estimated to be fluffy aggregates of ice "like a dandelion puff" rather than lumps of "hail". Some fifty "snowballs" were tracked over sequences of images and their positions measured. All were within 30 km of the nucleus and most were within 10 km of it, moving at relative speeds of the order of 1 meter per second or less, yet in most cases faster than escape velocity. Scientists looked for evidence of impacts on the spacecraft. The attitude control telemetry recorded nine events within 10 minutes of the moment of closest approach that could have been due to strikes by particles of at most tenths of milligrams in mass; i.e. less than the mass of a

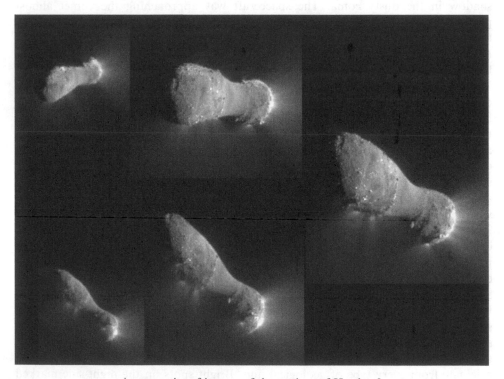

A composite of images of the nucleus of Hartley 2.

A view of space around the nucleus of Hartley 2. The two trails were left by 'snowballs' escaping the nucleus and flying close to the spacecraft.

snowflake. Drawing the observations to a conclusion, the spectrometer confirmed that Hartley 2 behaved differently from all the other comets that had been visited by spacecraft and was deemed to be a prototype for hyperactive comets. In particular, activity on Hartley 2 seemed to be dominated by carbon dioxide. Highly volatile carbon dioxide ice was seen to evaporate from the smaller end of the nucleus, drawing water and "snowballs" with it. In fact, Hartley 2 could be the first of a new, rarer class of comets visited by spacecraft. This class would not produce water by the direct sublimation of subsurface deposits, but from ice grains lofted by the sublimation of carbon dioxide. Signs of carbon dioxide-driven activity had been already detected during the flyby of Tempel 1, although it appeared localized in the hemisphere that was in darkness at the time of closest approach. In fact, it is possible that carbon dioxide is actually a common driver of cometary activity, and that it was not detected before because it was unobservable from the ground. The smooth waist region produced almost pure water vapor that was not visible as jets in visible-light images. Earth-orbiting satellites established that Hartley 2 was spewing up to 300 kg of water every second. An organic peak was also present in spectra. Remarkably, observations made by the Herschel infrared space telescope only days after the flyby showed a ratio of deuterium to hydrogen in the coma similar to that of the Earth's oceans. This provided the first evidence that water on Earth could indeed have been delivered by Jupiter-family comets. On the other hand, long-period comets such as Halley and Hale-Bopp which probably originated from the Oort cloud, are known to have different isotopic ratios.[252] Comets showing a similar hyperactivity to Hartley 2 included Wirtanen and Giacobini–Zinner. It will be remembered that the latter was the first comet to be visited by a spacecraft, the International Cometary Explorer, in 1985. Unfortunately, that had not been equipped with cameras and spectrometers for remote sensing and could not have observed this type of dynamics. Like all the other comets visited so far, Hartley 2 seems to be relatively young. It would be interesting if a future mission targeted a really old object such as Wilson–Harrington or Encke. Scientists predict that old comets could have more extreme surface morphologies.[253]

The fast flyby prevented any determination of the mass or density of the nucleus, but several considerations placed constraints on it. For example, the smooth waist and gravitational low provided an indirect way to measure the mass of the nucleus. It seemed to be an extremely porous object, with a density only 20 to 30 per cent of that of water. In fact, the nucleus had one of the smallest thermal inertias yet measured, with the warmest spot being right at the subsolar point (it would normally be shifted toward the afternoon hemisphere). There was evidence that the rotation status of the nucleus was actually more complex than initially believed, consisting, as in the case of Halley, of a rotation superimposed upon a fast precession or a 'roll' around the longest axis every 1.5 or 3 local days.[254,255,256,257,258,259,260]

Imaging of Hartley 2 continued with one picture every 2 minutes for 3 weeks after the encounter, until 26 November, increasing to about 125,000 the total number of images from the entire encounter.

Although EPOXI and Deep Impact were to cease operations in December 2010, NASA issued a call for ideas for a second extended mission. During 2011, the probe was used to take images of the deep sky, including galaxies and nebulas, as part of an operational exercise to keep its controllers proficient. Although the cameras were not designed for such targets, remarkably good pictures were obtained.

Deep Impact was left after the Hartley 2 encounter in a 0.98 × 1.22-AU orbit, with enough fuel for a velocity change of at least 18 m/s. A numerical search was carried out of the orbits of 10,000 near-Earth asteroids and comets to locate a suitable target for an additional flyby. Only six objects were found that were "encounterable", most of which were less than 1 km across. Only one asteroid matched all of the criteria of being sufficiently large to be detectable by the cameras in the approach phase, of having a sufficiently well-defined orbit, and of leaving a small fuel margin after the targeting maneuver. On the other hand, reaching it would require almost a decade in solar orbit. The target was to be the unnamed near-Earth asteroid 163249 (a.k.a 2002 GT), of which little was known apart from its orbit. Its diameter was estimated to be around 800 meters, and studies in support of the encounter revealed it to be a fast rotator, with a "day" lasting only 3.77 hours, and hinted at the possible presence of a small satellite. The object was also listed as an S-class (stony) asteroid. Better data were expected on the occasion of a late June 2013 close flyby of Earth during which 2002 GT would be imaged using radar and its orbit refined. It would be the last time the asteroid could be studied from the ground prior to the Deep Impact flyby, which would occur on 4 January 2020 with a closest point of approach of 1 or 2 km and a relative speed of 7.1 km/s. The spacecraft would approach from the night-side and leave over the day-side, so most of the imaging would be done on the outbound leg with a best resolution of about 4 to 10 cm from a distance between 20 and 50 km. Up to almost 80 km away, the asteroid would completely fill the field of view of the medium-resolution camera.

A single course correction burn would be needed in order to reach 2002 GT, but engineers wanted to accumulate more data on the quantity of fuel available and the maneuver was split in two. Therefore, on 29 November 2011 Deep Impact performed the first 140-second, 8.8 m/s correction. The second, 71-second, 1.9-m/s correction was made almost a year later, on 4 October 2012.[261] The mission extension however had yet to be formally approved by NASA. In the meantime, the medium-resolution camera was used to observe long-period comet Garradd in order to study outgassing and its periodicity over several weeks between February and April 2012. Images showed a uniform coma, but narrow-band uninterrupted imaging allowed scientists to precisely determine the rotation period of the nucleus to be 10.42 hours. The comet had an unusual behavior unlike other well-observed objects, in that carbon monoxide emissions did not correlate with water. A second window to observe the comet in July and August went unused, as the probe instead demonstrated the use of its telescopes for gravitational microlensing imaging. It was then to observe the bright comet C/2011 L4 PanSTARRS in late 2012 as this crossed the heliocentric distance at which water ice could be expected to sublimate. NASA Headquarters, however, ordered that the spacecraft be prepared for hibernation and the observations were canceled. After scientists pointed out that Deep Impact was a unique tool for observing bright comets, of which there were several expected in 2013, because it had an unobstructed view of them for days at a time, a Deep Impact Continued Investigations project, or "Deep Impact 3" was funded through the Discovery program. This would use minimal staff, including just one full-time engineer, reusable observing sequences which could be adapted for every cometary target and impose limited download on the Deep Space Network to return only a limited amount of data. First of all, the probe was to observe the "comet of the century" ISON in 2013, followed by the well-known short-period comet Encke, by C/2013 A1 Siding Spring inbound for a very close encounter with Mars in 2014, and by C/2012 K1 PanSTARRS, which would pass 0.12 AU from the spacecraft in August 2014. It was also to observe periodic comets 154P/Brewington and Borrelly and would be uniquely positioned to image Churyumov–Gerasimenko while Rosetta was orbiting it in 2015, at a time when the comet would be essentially unobservable from Earth.[262,263]

Starting this ambitious program as a cometary observer, Deep Impact took 146 medium-resolution images of ISON over a 36-hour period on 17-18 January 2013, when the comet was 793 million km from the vehicle. The comet already sported a tail some 65,000 km long, despite still being beyond the orbit of Jupiter. There was a second observation window in July and August at a time when the comet would not

A view of the spiral M51 galaxy with the supernova SN2011dh visible, taken by Deep Impact.

be observable from Earth due to being too close to the Sun in the sky, and also at the distance from the Sun at which water emissions were expected to "turn on". Infrared observations would thus monitor the production of water as well as emissions from carbon monoxide and carbon dioxide to determine the rotation period of the nucleus. Further campaigns would monitor the long-period comet as it receded from the Sun after its late November perihelion.

Unfortunately, while these observations of comet ISON were underway, contact was lost sometime after the final communication session on 8 August. Controllers believed by the end of the month that they had determined the cause of the problem. The onboard fault-protection software was unable to handle a date after 11 August, and this sent the computer into a state where it would be continuously and endlessly rebooting. Commands sent from Earth to reset the computer and place the vehicle in safe mode were evidently not acknowledged. Meanwhile, the situation onboard had probably disrupted attitude control, so much so that the orientation of the spacecraft was not known and the solar panels might no longer be pointing at the Sun, in which case the spacecraft would have only a few days before the batteries were exhausted. Without heating, the fuel would freeze and the electronics would fail. Having had no response for over a month, on 20 September 2013 NASA announced the official end of the mission.

EXPRESS TO VENUS

In March 2001, ESA, having identified the possibility of reflying the Mars Express bus for a low cost mission, issued a "call for ideas". This would be developed under a strict budget for launch in 2005. In response, the scientific community presented a variety of ideas, three of which were recommended for further study. Venus Express would use instruments from Rosetta and Mars Express to investigate aspects of that planet. Cosmic DUNE (Dust Near Earth) would spend 2 years orbiting around the Sun–Earth L2 Lagrangian point, 1.5 million km beyond the Earth's orbit, and utilize several instruments of Giotto, Vega, Cassini and Stardust-heritage to collect data on interplanetary and interstellar dust. SPORT Express, the Sky Polarization Observatory, would measure the polarization of the cosmic microwave background of the Big Bang to complement the results from the Microwave Anisotropy Probe and Planck missions operated by NASA and ESA respectively. In November, the agency's Space Science Advisory Committee recommended the adoption of Venus Express as a cheap mission of opportunity. It would be the first European mission to Venus, three decades after an orbiter was first proposed.[264,265] After a preparatory phase, Venus Express was started in July 2002, but it was subject to final approval from national authorities, in particular of Italy, whose participation would make money tight for that nation's other programs. This was given in November the same year. The spacecraft was to be delivered by the prime contractors Contraves, Astrium and Alenia Space in June 2005 for launch in November. In fact, to justify the 'Express' label, this was intended to be the agency's most rapid preparation of a science mission to-date.[266]

The differences between the Martian and the Venusian space environments obliged modifications to the bus. First and foremost, since Venus is on average half as close to the Sun than Mars, the solar thermal flux would be four times greater. And whereas Earth is an interior planet as viewed from Mars, requiring the thermal radiators to be carried on the opposite side to the Earth-pointing high-gain antenna in order to remain in shadow, this geometry would not apply in orbit of Venus. To retain the 'cold face' architecture, engineers decided to install a second, smaller, high-gain antenna pointing in the opposite direction to the main one. Hence, the main antenna (whose diameter was slightly reduced to 1.3 meters) would be used for about 75 per cent of the mission and the smaller 0.3-meter-diameter antenna would be used at times when Earth was nearest Venus. The greater strength of Venusian gravity in comparison to Mars meant the propellant load had to be increased by about 20 per cent for orbit insertion and shaping, and this resulted in a launch mass of 1,270 kg, which was some 50 kg more than Mars Express. Even so, this would facilitate only an orbit around the planet with a much longer period than employed at Mars. The thermal environment at the reduced heliocentric distance dictated the use of solar arrays of gallium arsenide instead of silicon cells. But the increased insolation meant the number of panels on each wing could be cut from four to two, and reduced in size to an area of 5.7 square meters. Furthermore, each panel would comprise alternating rows of cells and reflecting mirrors. The output of 1,450 W in Venus orbit would be much more than the minimum required to run the spacecraft's systems and payload. The arrays were designed to withstand the loads of aerobraking, although there were no official plans for this. As with Mars Express, thermal control would mostly be passive, but would use white paint instead of black paint, and highly reflective Kapton insulation instead of black blankets, etc.[267,268,269] These technologies were to benefit future ESA missions in the inner solar system, like the BepiColombo Mercury orbiter and the Solar Orbiter probe.

Despite more than three decades of Venusian exploration by American and Soviet missions, many questions remained in relation to (for example) the dynamics of the atmosphere, and in particular the super-rotation that caused the atmosphere to circulate tens of times faster than the planet rotated; the dynamics of the vortex discovered by the Pioneer Venus Orbiter at the north pole; the nature of the ultraviolet-absorbing substance that creates the dark markings in the atmosphere; the chemistry of the upper atmosphere; and the evolution of the paroxysmal greenhouse effect. Moreover, it was hoped that studies of the process of cloud formation and of the greenhouse effect on Venus might improve understanding of similar effects induced by human activity on Earth, and in particular put constraints on the temperatures likely to result from 'global warming'. The complex interaction of the Venusian atmosphere with the solar wind also merited further study. On becoming ionized, the upper atmosphere (more properly called the ionosphere) was known to develop its own induced magnetic field. Previous observations had not been able to establish whether this field was able to fend off the solar wind to prevent solar plasma from mixing with the atmosphere and influencing its evolution, especially at the time of solar minimum. The surface of Venus had been extensively mapped by radar and found to have been shaped by volcanism. There were hints that volcanism was ongoing, but this was far from certain. This was important, as volcanism could be a major source of transferring energy into the atmosphere. Infrared observations by Galileo and Cassini during their flybys had provided insight into the characteristics of the atmosphere, and Venus Express was to further this research.

One important factor in achieving the required low cost was the availability of a number of instruments left over from Mars Express and Rosetta that were suitable for studying Venus. The scientific payload comprised seven instruments with a total mass of 94 kg. A plasma and energetic atom analyzer with four sensors inherited from Mars Express was to study interactions between the solar wind and the upper atmosphere, and characterize the compounds lost by the atmosphere to space. The capability of this instrument far exceeded that of a simpler one carried by the Pioneer Venus Orbiter. An infrared Fourier spectrometer adapted from that of Mars Express was to characterize global temperature fields and high-altitude winds, identify gases inside and above the clouds, vertically profile the concentration of water, and seek gaseous emissions from volcanism. Another Mars Express-derived ultraviolet and infrared spectrometer was to chart the concentration of sulfur oxides and provide density profiles at high altitudes. This instrument would also exploit one infrared 'window' to study thermal emission from the surface on the night-side. Although the resolution would be limited to 50 km (at best) owing to scattering by liquid droplets in the clouds, such observations were particularly suited to detecting volcanic eruptions and fresh lava flows. It was also to 'sound' the atmosphere near the limb during stellar and solar occultations in order to use the strength of spectral absorption to profile the vertical distribution of hydrogen-bearing molecules at high altitudes (e.g. water, deuterated water, hydrofluoric acid and chloridric acid). The ionosphere and atmosphere would also be profiled by using radio occultations. The spacecraft would be able to undertake bistatic radar observations in collaboration with antennas on Earth. The Doppler effect would enable the planet's gravity field to be mapped. A Rosetta-heritage ultraviolet, visible and infrared imaging spectrometer would measure the composition of the lower atmosphere, the structure and dynamics of the clouds, the temperatures at medium altitudes, the temperature of the surface (in particular in search of volcanic 'hot spots') and also watch out for bolts of lightning. The high-definition color images of Venus provided by this instrument would raise public awareness of the mission. There was also a magnetometer derived from the Rosetta lander. Since reducing the magnetic signature of a spacecraft that had not been intended to carry a magnetometer was incompatible with the tight budget, the instrument was designed to distinguish artificial fields from natural ones by using two sensors: one mounted on the body and the other on a 1-meter-long boom. It would also serve as a receiver for low-frequency electromagnetic waves issued by lightning. And, of course, when Venus was at superior conjunction the spacecraft's radio signal would sound the solar corona.

In terms of scientific payload, the principal difference between Mars Express and Venus Express was the deletion of a lander (although a proposal was considered early on to deploy an inflatable atmospheric probe) and the subsurface radar with its whip antennas. When the project was initiated, a subsurface radar was included to sound the principal geological features of Venus down to a depth of 1 or 2 km, but this was soon canceled for financial reasons. The only wholly new instrument was the monitoring camera that replaced the camera used to document the release of the Beagle 2 lander. This was to take ultraviolet, visible and infrared images at resolutions varying between 200 meters at periapsis in the northern hemisphere and 50 km at apoapsis over the far southern latitudes in order to study atmospheric dynamics and cloud motions, monitor night-side airglow, and map the surface temperatures.[270,271,272]

The cost of the Venus Express mission, including 500 days of orbital operations, was 220 million euros, which was significantly less than most of the recently funded NASA Discovery missions.

To coincide with Venus Express, ESA inaugurated its second deep-space antenna, at Cebreros in Spain. In fact, the spacecraft's 24-hour operating orbit around Venus was designed so that this station would be in touch with it from 2 to 12 hours after periapsis, when most of the data collected on the previous orbit and periapsis passage would be downloaded. The spacecraft would record its data on a 1.5-Gbyte solid-state recorder and, depending on the distance from Earth and the high-gain antenna in use, replay it at between 19 and 228 kbits per second. The mission profile called for about 250 Mbytes of data to be returned every day.[273]

The Venus Express spacecraft. (ESA)

Due to the high level of reusability in hardware and instruments, the Venus Express development phase was the shortest ever for an ESA scientific mission; amounting to only a few months. Integration of the spacecraft began in April 2004, and testing was completed in July 2005. It was then delivered to Baikonur for fueling and launch on a Soyuz-Fregat. It was the first flight to set off for Venus from this cosmodrome for more than two decades. The launch window ran from 26 October to 25 November, with the best date being 5 November. However, when Venus Express reached the launch pad on 22 October, technicians found that the Fregat had been slightly damaged by hydrazine fumes and was shedding thermal insulation. For safety, it was decided to return the rocket to the assembly hall to be cleaned. The mission finally lifted off on 9 November 2005, right in the middle of the launch window, with its payload shroud adorned by Sandro Botticelli's *Birth of Venus*. After achieving parking orbit, the Fregat reignited to enter a 0.7 × 1.0-AU solar orbit. The 2-week launch delay facilitated almost the optimal minimum-energy

Venus Express and the Russian Fregat stage being prepared for integration with the launcher. Note at top the secondary high-gain antenna. (ESA)

trajectory, and the accuracy of the Fregat meant that Venus Express's propellant margin would permit a lengthy extended mission in orbit around its target. Signals acquired about 2 hours after launch by ESA's deep-space antenna at New Norcia in Western Australia confirmed that everything onboard was satisfactory. A course correction of 0.5 m/s was made the following day, in parallel with spacecraft commissioning activities. The instruments were turned on one by one and calibrated between mid-November and mid-December using the secondary high-gain antenna. The magnetometer boom was extended when the spacecraft was 3.5 million km from Earth, and test images of Earth and the Moon were taken by the imaging spectrometer and the monitoring camera in several spectral ranges. Apart from such tests, however, only magnetic field data was to be collected during the interplanetary cruise.

The main engine was employed for the first time by a course correction on 17

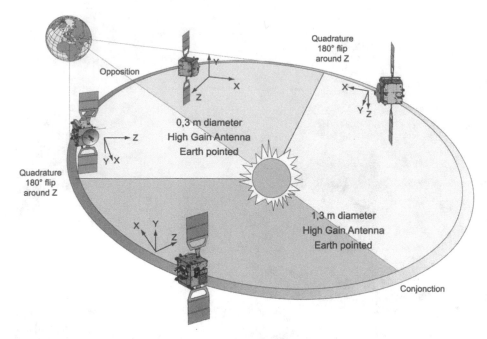

Adapting a spacecraft designed for Mars to Venus exploration required, among other things, the use of two different fixed high-gain antennas and 'flip maneuvers' twice on each orbit to prevent exposing the thermal radiators to the Sun. (ESA)

The Soyuz-Fregat 'Semyorka' launcher carrying Venus Express in the integration hall at Baikonur. (ESA)

February 2006. On 21 March Venus Express reached perihelion, and then on 11 April arrived at Venus. This marked the first operational use by the two ESA deep-space antennas of very precise navigation by determining the vehicle's position in the sky in relation to quasars. Supplemented by similar measurements using NASA antennas, this provided an uncertainty in the closest point of approach of just 3 km.[274] The spacecraft adopted the attitude required for orbit insertion and switched its communications to the low-gain antenna, which was to send a diagnostic carrier signal. The main engine ignited at 07:10 UTC for the 1,251.6-m/s braking maneuver. The spacecraft was occulted by the planet for 12 minutes midway through the maneuver. The burn lasted 3,163 seconds, which was almost twice as long as for Mars Express, and produced a preliminary 663 × 330,685-km orbit with a period of 9 days. By affording an unobstructed view of the south pole for several days centered on apoapsis, this orbit provided an early scientific opportunity. When the communications link was re-established, the telemetry showed Venus Express to be in excellent health. The primary mission called for observations over two local days, but with about 60 per cent of the propellant reserve remaining after insertion the vehicle would be able to operate for at least four, and possibly six. The successful arrival of Venus Express was a particularly significant moment for the long-neglected European exploration of the inner solar system, as it now had vehicles in orbit around the Moon, Mars and Venus, and was preparing a mission to Mercury. Most of the scientific payload was activated the day following orbit insertion to make some early observations of the previously poorly observed south pole. At a range of 206,000 km as Venus Express rose toward its apoapsis, it took the first-ever infrared images of the atmosphere over the pole. These revealed a dark vortex on the night-side which matched the analogous structure observed by the Pioneer Venus Orbiter at the north pole. An engine burn at apoapsis lowered the altitude of periapsis to 257 km. On making its first periapsis passage on 20 April, the spacecraft reduced the orbital period to 40 hours. A series of burns between 23 April and 6 May maneuvered into an orbit of 249 × 66,582 km with a period of 24 hours. Meanwhile, the commissioning and calibration of the instruments continued. The only major issue was that the scanner of the Fourier spectrometer locked in its closed position. Despite efforts to overcome this problem, the instrument remained unusable. On 23 May a bistatic radar experiment was performed in which the spacecraft aimed its high-gain antenna to bounce a radio signal off the surface of Venus at such an angle that the reflection could be detected by antennas on Earth. Such data enabled the nature of the surface to be characterised. Routine science observations began on 3 June, and during its primary mission Venus Express returned more data than any previous Venus orbiter except Magellan.[275,276]

The imaging instruments increased our understanding of the global atmospheric circulation, with observations in different spectral ranges providing measurements at different depths. As tracking features at the cloud tops several kilometers in size did not require a very high imaging resolution, this was usually done around apoapsis and during the descending portion of the orbit. The camera monitored the top layers of the atmosphere, keeping track of the temporal variations. A bright mid-latitude belt was seen to separate the polar regions (in which the atmosphere spirals like a

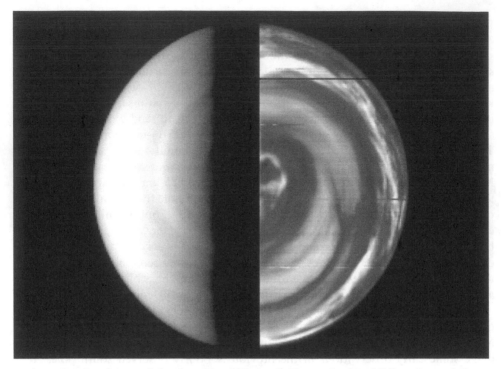

A composite picture of the day-side of Venus (left) seen in the visible range and the night-side (right) seen in the infrared. The infrared view penetrates the clouds to an altitude of about 55 km and clearly shows the south polar vortex. (ESA/CNR-IASF, Rome, Italy, and the Observatoire de Paris in France)

hurricane) from the equatorial region (with mottled and chaotic solar-driven convective clouds at the subsolar point). This belt probably marks the poleward limit of the 'Hadley cell' circulation pattern revealed by Mariner 10. The cloud patterns over the pole were seen to change from one orbit to the next. The highest resolution pictures revealed the convective cells at the subsolar point to be only 20–30 km in size, which was much smaller than inferred from lower resolution Mariner 10, Pioneer Venus Orbiter and Galileo imagery. They cannot be deeply rooted and hence probably do not contribute much to the transportation of solar energy across the atmosphere, or to 'driving' the super-rotating winds. Infrared data showed the cloud tops to be located on average at an altitude of 75 km, except toward the poles where they were at only 65 km. But at altitudes above 90 km the Sun is the principal factor driving atmospheric circulation, superimposing a day-side to night-side flow of warm air on the lower altitude super-rotation. Evidence for this flow was obtained by tracing hydrofluoric acid and chloridric acid, which are dissociated by solar ultraviolet on the day-side and their constituents are carried onto the night-side, where they reform. Maps of the circulation of oxygen over the night-side clarified the longstanding mystery of the airglow phenomenon. Carbon dioxide is dissociated over the day-side and the liberated oxygen atoms are carried onto the night-side, where they recombine to form oxygen molecules and emit photons at infrared wavelengths. Significant advances were made in understanding the behavior of the mysterious chemical that provides the ultraviolet absorption to create the distinctive ultraviolet-dark markings which are characteristic of Venus. This was revealed to vary with atmospheric conditions, in particular with the temperature at the cloud tops. The absorber seems to be present at latitudes dominated by strong convection, which draws it up from deep inside the clouds. Lower temperatures (for example in the cold collar at mid-latitudes) and stable air instead favor the creation of a bright haze of sulfuric acid. This link could be important, since it might one day assist in identifying the composition of the absorbing chemical.

As noted, while Venus Express was in its initial orbit it discovered a vortex at the south pole to match that discovered at the north pole by the Pioneer Venus Orbiter in 1979. This was a dipolar structure of two warm 'eyes' whose centers were separated by some 2,000 km and dropped down from the cloud tops to an altitude of 50 km, or possibly even lower. It appears to be caused by the heating of rapidly down-drafting masses of air. By tracking small features it was possible to measure wind speeds in the 'eyes' and in the 1,000-km-wide 'collar' of cold air which forms the outer part of the vortex. The eyes themselves made a full rotation in about 60 hours. It was possible to monitor how the vortex stretched, evolved and changed shape over a few days, often changing to bear little resemblance to the feature seen just one revolution before. The vortex had a center of rotation slightly displaced from the south pole and drifting around it in less than 10 days. This asymmetry, among its many effects, could even cause the winds at some latitude to change direction, against the remainder of the atmosphere.

Orbital geometry gave the first of three primary mission seasons of occultations of Venus Express by the planet in July and August 2006. This opportunity was used for 21 daily dual-frequency radio soundings at both ingress and egress. The different technical arrangement of the Pioneer Venus Orbiter had restricted occultation data to the ingress phase. Venus Express sounded the ionosphere and upper atmosphere in the 100–500-km altitude range, and the main bulk of the neutral-gas atmosphere down to about 50 km, at which level the signal was completely absorbed. The 3-dimensional profiles established the temperature at lower altitudes across both the day-side and the night-side to be fairly uniform. Complex structures were present above 60 km, in particular thermal inversions between the latitudes of 60 and 80 degrees marking the cold 'collars' of the polar vortices and hence probably caused by the down-drafting of the Hadley cell circulation flow. The most unexpected result was a 30–40°C difference between the temperature at high altitude on the day-side and on the night-side. This was surprising because the enormous mass of the atmosphere should distribute heat quite efficiently. Either other heating processes were at work or there was something present in the atmosphere that was partially absorbing the radio signal and yielding a misleading measurement. Solar and stellar occultations probed the chemistry of the high atmosphere, and gave some extremely interesting insights into the early history of Venus. In fact, this data established the deuterium in the upper atmosphere to be about 150 times more enriched relative to hydrogen than in the portions of the atmosphere sampled by descending landers –

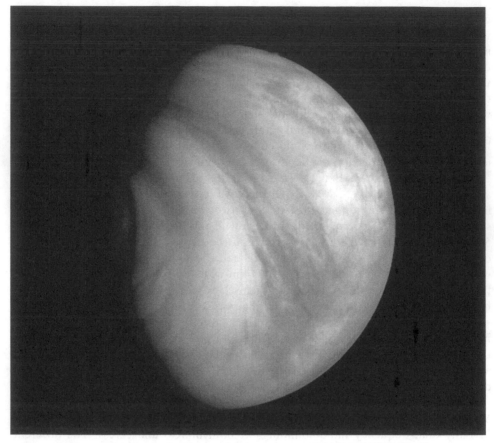

Venus clouds seen by the monitoring camera. (ESA/MPS/DLR/IDA)

and indeed than in the terrestrial oceans. This is so, apparently, because the heavier isotope of hydrogen is less readily able to escape to space than the lighter isotope, resulting in a gradual build-up of the concentration of deuterium. However, if solar heating were the only mechanism causing the escape of gas from the atmosphere, this would imply that the planet only had enough water in the past to form an ocean several meters deep. But if other processes have operated, including solar radiation ionization and erosion of ions by the solar wind, then there may once have been much more water present. Detailed analysis of ultraviolet occultation data revealed for the first time ozone in the Venusian atmosphere. The ozone probably forms from the dissociation of carbon dioxide molecules and recombination of oxygen atoms. Although its concentrations are too low to have an effect on the upper atmosphere temperatures, as it has on Earth, it would have a significant impact on the chemistry. Another unexpected discovery by the occultation instrument, which had collected data for 5 years, was the existence of a cold layer in the atmosphere at an altitude of about 125 km. In fact, this layer reached such a cold temperature that it would allow carbon dioxide to form ice crystals or even snow.

An ultraviolet view of the south pole of Venus. (ESA/MPS, Katlenburg-Lindau, Germany)

The imaging infrared spectrometer detected the hydroxyl (OH) radical by looking at sections of the limb, where such minor components are easier to detect through a thicker column of gas. Hydroxyl is extremely reactive, and therefore has a very short lifetime in any atmosphere. The fact that it was present meant that some mechanism must be producing it continuously.

The question of how the solar wind erodes the atmosphere could not be studied because magnetometer observations of the position of the bow shock upstream of the ionosphere, and its response to variations in the solar wind, revealed this to be at too great an altitude (at least at solar minimum) to allow the solar wind to penetrate the atmosphere, energize it and influence its evolution. Ions of hydrogen, helium and oxygen from the atmosphere were present beneath the bow shock, and the plasma analyzer showed that they escape into the planet's wake at a faster rate than was the case for neutral atoms. The measured ratio of escaping hydrogen to oxygen of 2.6 indicated that they were mostly derived from water (for which this figure would have been precisely 2.0).

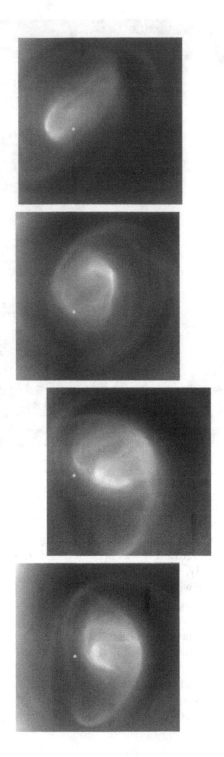

A sequence of images from the imaging infrared spectrometer shows the south polar vortex of Venus at an altitude of about 60 km. The second image was taken 4 hours after the first, the third 24 hours, and the last 48 hours later. The horizontally aligned white dots mark the position of the pole. (ESA/VIRTIS/INAF/Observatoire de Paris, LESIA/University of Oxford)

Gravity anomalies were studied during periapsis passes. With the periapsis at high northern latitude, this survey concentrated on Atalanta Planitia.

Venus Express also provided new data on the issue of whether lightning occurs. In principle there ought to be none, as small convective cells seem unable to produce electrical discharges and the clouds are made of smog-like aerosols which (on Earth) cannot prompt lightning. The first tentative detection was by Soviet probes in the late 1960s. The magnetometer on Venus Express 'listened' for electrical discharges for about 2 minutes around its periapsis passage, and detected brief but intense bursts of low-frequency electromagnetic waves having the characteristics expected of cloud-to-cloud discharges and at frequencies similar to terrestrial lightning. Cloud-to-ground discharges, however, which are common on Earth, seem not to be present on Venus, possibly because the cloud layers are at such a high altitude. The discrepancy between these results and the non-detection of lightning by the Cassini spacecraft during its flyby could be explained if Venus, like Saturn, undergoes long periods of atmospheric inactivity, or if Venusian radio emissions were restricted to certain frequencies.[277] Another mystery that remains to be solved, is whether there are active volcanoes. It had been hoped to trace concentrations of sulfur dioxide back to eruption sites. Although the concentration proved to be highly variable, it has not yet been possible to draw any conclusions in this regard. It is possible lightning occurs in the lower atmosphere only at the sites of active volcanoes. Good results were obtained by correlating surface emissivity of the southern hemisphere in the near-infrared with the topography revealed by the radar map produced by the Magellan orbiter. Averaging the data from the infrared imaging spectrometer, nine anomalous 'hot spots' were discovered in the three southern hemisphere topographic rises of Imdr, Themis and Dione. These zones of high thermal emissivity matched known volcanoes and coronae: Idunn, a large volcano in Imdr; Innini and Hathor in Dione; and Mielikki in Themis Regio. The higher temperature of these spots relative to the surrounding terrain was estimated at less than 20 degrees, meaning that they were unlikely to result from ongoing volcanic eruptions. It was felt, however, that they represented relatively recent flows of lava that had not yet been completely 'weathered' by the aggressive environment. Depending on the assumptions made, they could be between several hundred and several million years old. What matters, is that this discovery indicates that volcanism has not been limited to catastrophic global events like the one that resurfaced almost the entire planet half a billion years ago. Another suspect result was yielded by the occultation experiment by detecting an anomalous increase in sulfur dioxide concentration in the atmosphere above 70 km between 2006 and 2007 and then a 10-fold decrease until 2012. This behavior could be due to variations in the global circulation of the atmosphere, but could also be attributed to a volcanic eruption which had since stopped. On the other hand, the concentration of sulfur dioxide at lower altitudes apparently remained constant, and no other data corroborated the possibility of a recent or ongoing volcanic event.

A remarkable result of surface imaging was that features could only be matched with those seen by Magellan's radar almost two decades earlier if the Venusian day was on average 6.5 minutes longer. There was a similar discrepancy between long-

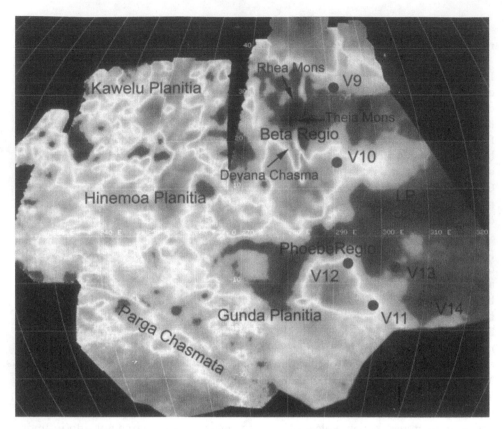

A thermal radiation map of the surface of Venus obtained by the monitoring camera. The map is able to distinguish between low altitude (bright) and high altitude (dark) areas. Dots show the landing sites of Soviet Venera landers and of the US Pioneer Venus Large Probe. (ESA/MPS/DLR)

term ground-based radar measurements and the Magellan data, which implied that the difference was real. It seemed that something – possibly friction from the dense atmosphere or a gravitational exchange of momentum between Venus and the Earth – was periodically slowing down the rotation of planet by a measurable amount. This issue will need to be solved before pinpoint landings of future probes can be planned.[278,279,280,281,282,283,284,285,286,287,288,289,290,291,292,293,294,295,296]

One little-known result of the Venus Express mission was its contribution to the study of the minor bodies of the solar system. In June and August 2006, soon after it entered orbit around Venus, the spacecraft made two encounters with periodic comet Honda–Mrkos–Pajdušáková, passing through the dust tail. It was well placed to seek associated interplanetary magnetic field disturbances.[297] Remarkably, instead, there were none of the magnetic field disturbances associated with asteroid (2201) Oljato and observed by the Pioneer Venus Orbiter. Possibly, the debris following the asteroid had cleared during the intervening years.[298,299]

In June 2007, just after Venus Express completed the first flip maneuver required (amongst other things) to aim the secondary high-gain antenna at Earth, it provided ionospheric data, multispectral images and other data in support of MESSENGER's flyby of the planet. At that time, it was descending from the apoapsis of its 410th orbit.[300]

As Venus Express completed its primary mission in September 2007, it received an extension until at least May 2009. In fact, the propellant reserve was considered to be sufficient to continue until 2013 and (with orbital maintenance) possibly even longer. Since orbit insertion, the altitude of periapsis had varied between 250 and 400 km, mostly due to solar gravity perturbations.

The MESSENGER flybys provided the first evidence of a rare magnetospheric phenomenon, a "hot flow anomaly", that was finally clearly detected by instruments on Venus Express on 22 March 2008. These are bubbles created by the solar wind's electrical field which trap ions from the planet's bow shock. Their detection made Venus the fourth planet to have hot flow anomalies after Earth (where they trigger auroras), Saturn and Mars. An even more interesting result was the detection, on 15 May 2006, of a "magnetospheric reconnection event". These occur when magnetic field lines break and reconnect, creating "plasmoids", or bubbles of magnetically contained plasma that travel down the magnetospheric tail. Plasmoids are frequently created in the vicinity of magnetic planets, but this was the first time one had been seen near a non-magnetic one.[301]

In late July 2008 the spacecraft initiated a 4-week campaign to lower its periapsis to 185 km in order to make deeper passes into the ionosphere and the induced magnetic field, for better particles and fields data. To maintain the period unchanged as the periapsis was lowered, the apoapsis was slightly increased in step. A small but measurable drag effect was measured at altitudes down to 185 km. In July 2009, Venus Express was able to document the evolution of a 'bright spot' that appeared in the southern hemisphere of the planet and was first noticed by a US amateur astronomer. In fact, the orbiter's monitoring camera had first imaged the 1,000-km-wide cloud for 4 days, which is one full atmospheric rotation, before it was discovered on 15 July. Four explanations have been offered for this unusual phenomenon: a large volcanic eruption, the effects of the interaction between the atmosphere and the solar wind, a comet or asteroid impact, or an unprecedented change in the cloud deck. Magellan radar images show Venus to possess relatively 'sedate' shield volcanoes, but only a powerful eruption could have issued a plume capable of rising so high into the atmosphere. The impact hypothesis would be a remarkable coincidence, as the spot appeared almost exactly 15 years after the impact on Jupiter of Shoemaker–Levy 9.

A second atmospheric drag experiment was carried out in October 2009. But the aerodynamic effects thereby measured were particularly dependent on the uncertain properties of the high atmosphere. Hence, a third campaign was carried out in April 2010 to measure the density of the Venusian atmosphere at altitudes in excess of 180 km by using a technique similar to that pioneered by the Magellan orbiter in 1994. In these tests, one of the solar panels was kept at a fixed orientation while the other was rotated in steps through a variety of angles during each periapsis pass. While the atmosphere introduced a torque on the orbiter, this could be measured by gyroscopes and counteracted by the reaction wheels. Finally, on 16 April, a proper 'windmilling experiment' was undertaken by orienting the two panels at 45 degrees in opposite directions. Whereas Magellan had suffered some serious failures at the time of the passes that required it to maintain its high-gain antenna pointed at Earth all the time, Venus Express was still healthy and the passes could be optimized for the greatest science return. Compared to atmospheric drag observations by the Pioneer Venus Orbiter and by Magellan, which sampled the equatorial regions, observations by Venus Express over the north pole showed a density about 60 per cent lower. A total of nine atmospheric drag campaigns had been carried out by September 2012.[302]

The mission was then extended three times, most recently to 2014, although the probe is expected to remain operational well into 2015, when its fuel should run out. In early August 2010, following a series of large coronal mass ejections, the solar wind dropped to a record low density. This enabled Venus Express to study the behavior of the Venusian ionosphere in more detail. As a result of the upstream solar wind conditions, the night-side ionosphere swelled and inflated to form a comet-like tail that may have extended for millions of kilometers. If pockets of the ionosphere became detached when the solar wind density returned to normal, this process could also affect the rate at which the atmosphere was eroded. The following October, the periapsis was reduced to a low 165 km for another atmospheric drag campaign.

Meanwhile, simultaneous and near-simultaneous data collected around Venus by Venus Express and around Earth by ESA's Cluster quartet of satellites allowed scientists to complete studies comparing the interactions between the solar wind, the non-magnetized Venusian environment and the terrestrial magnetosphere. In another joint campaign with Earth satellites in December 2010 and January 2011, Venus was monitored jointly by Venus Express and in a set of three observations by the Hubble Space Telescope, concentrating in particular on the distribution of sulfur dioxide in the planet's atmosphere. It was also planned to operate Venus Express in partnership with the Venus Climate Orbiter, which was to initiate observations of the planet in January 2011 but owing to a propulsion problem this Japanese spacecraft will not reach Venus before 2015, if at all.[303] On 23 January 2012 the periapsis was raised from 138 to 314 km, marking the completion of nine atmospheric drag campaigns. By then, a dozen Earth occultation seasons had yielded more than 500 atmospheric profiles. Venus Express was severely hit by solar flares as activity surged in early 2012. On one occasion in March, its star tracker was blinded for more than 2 days, scientific operations were interrupted, attitude control was switched to gyroscopes, and normal operations could not resume for a further 3 days. The spacecraft was then passing through a critical period that occurred every 19 months in which special care had to be taken in orienting the spacecraft in order to avoid exposing instruments to the Sun whilst communicating with Earth.

On 6 June 2012, Venus Express was in the unique position of being in orbit of Venus as the planet transited the disk of the Sun as seen from Earth. This rare event had happened only once during the space age, in 2004, when no spacecraft was in orbit, and will not repeat until 2117. During the transit there were no communication opportunities for 3 days since that would have required pointing radio-telescopes on Earth straight at the Sun. By observing the atmosphere during this period, Venus Express also served to calibrate studies of how monitoring the light of the parent star could discern the composition of the atmosphere of a transiting planet. This would have applications in the study of the hundreds of planets discovered to be transiting the disks of other stars. Moreover, combined observations from Earth and from orbit measured the temperature of the atmosphere over a wide range of latitudes, offering clues to solar heating and its link to the super-rotation.[304]

In addition to the routine collection of data, campaigns to investigate the sulfur chemistry of the atmosphere were planned for September and October 2013, and for the dynamics of the south polar vortex in December 2013 and January 2014. These studies targeted atmospheric changes and dynamics and the variation of sodium dioxide and nitrogen oxide concentration. After that, activities in 2015 will focus on aerobraking experiments and on low-altitude measurements.[305]

As an aside to Venus Express, in January 2010 a small asteroid was discovered, designated 2010AL30, that would fly by Earth on the 13th at only one-third of lunar distance. The object, assumed to be 10 meters across, had an unusual unstable 1-year orbit around the Sun crossing that of Earth and almost tangent to that of Venus. By backtracking the orbit based on an early determination, it was found that 2010AL30 made a close pass by Venus in spring 2006, preceded by one of Earth in late 2005. This, and the fact that Venus Express' trajectory prior to orbit insertion could have injected the spacecraft into a 2010AL30-like orbit, suggested that the 'asteroid' was the spent Fregat stage which dispatched the European mission. However, improved orbit determinations set the date of the encounter with Venus at 25 February 2006, some 6 weeks before Venus Express arrived at the planet. It would seem, therefore, that 2010AL30 is most probably just an unusual natural object in a very unstable and short-lived orbit.

TO PLUTO AT LAST

After being discussed throughout the 1990s and then attracting support as one of the 'Fire and Ice' missions, the cancellation of the Pluto–Kuiper Express in September 2000 prompted scientists to urge the US Congress to compel NASA to reinstate this mission.[306] In response, in January 2001 the agency yielded. But instead of resuming the previous development, it decided upon a Discovery-style competition and sought proposals from laboratories and institutions. The budget, which was capped at $505 million, was to cover construction of the spacecraft and payload, launch and flight operations through to the Pluto flyby. No funding was initially to be provided for investigating Kuiper Belt objects, as this would be covered by a mission extension if one was granted. As with the Discovery missions, the Pluto proposals had to follow a principal investigator-led management architecture. NASA would provide one RTG to the winner. Of the five proposals submitted, two were shortlisted for futher study. The Pluto and Outer Solar System Explorer (POSSE) was offered by the University of Colorado, JPL and Lockheed Martin, and would use an ion propulsion module to circumvent the constraints of a Jupiter launch window. New Horizons was proposed by the Applied Physics Laboratory of Johns Hopkins University and the Southwest Research Institute, and was broadly based on APL's CONTOUR cometary mission. After a 3-month assessment, New Horizons was selected on 19 November 2001. The launch was scheduled for December 2004, and the flyby of Pluto for as early as 2012. The final team, led by long-time 'Plutophile' Alan Stern of the Southwest Research Institute, included NASA's Goddard Space Flight Center, Stanford University and Ball Aerospace, with JPL providing tracking through its Deep Space Network. The launcher was to be a version of either the Delta IV or Atlas V, which were the two rockets under development for the joint Department of Defense and NASA Evolved Expendable Launch Vehicle program.[307],[308] In February 2002 NASA published its budget request for fiscal year 2003. This included strong support for continued low-cost Discovery missions, and proposed to start the New Frontiers series with New Horizons as the first mission of this type. Also in 2002, New Horizons received its greatest scientific endorsement when the decadal survey by the National Research Council assigned Pluto and Kuiper Belt objects the highest priority for solar system exploration.

The development of New Horizons faced several threats, including the lack of a US nuclear-qualified launch vehicle and a shortage of plutonium oxide to fuel the RTG; on several occasions President George W. Bush tried to delete funding, only for this to be promptly reinstated by Congress; a launch postponement from 2004 to 2006, and a resulting delay in arrival at Pluto; and even the tragic loss of some key engineers in an aircraft crash. The loss of CONTOUR had little impact, since the design flaws of that cometary mission were not part of the common spacecraft heritage. At one point, the problem with the RTG fuel seemed likely to slip the launch beyond 2006, but this was avoided. In fact, the January 2006 launch date was at the very end of a Jupiter gravity-assist window that had been optimum in 2004. If the launch were to slip beyond 2006, Jupiter would no longer be in position to slingshot New Horizons onto a fast 'trans-Plutonian' trajectory, and the mission would be obliged to pursue a direct flight that would reach Pluto no earlier than 2020, at which time a large portion of the surface would be in the deep darkness of winter.

New Horizons had an irregular six-sided 'grand piano' body, 0.68 meter tall, 2.11 meters wide and 2.74 meters long, inside which was a central aluminum thrust tube that enclosed the propellant tank and also supported the extra-thin aluminum honeycomb base and top panels – which in turn carried the side panels. The entire exterior was covered with thermal insulation blankets in order to retain as much as possible of the heat emitted by the electronics and so keep the systems warm in the frigid outer solar system. The propulsion system used four 4-N thrusters for course corrections and twelve 0.8-N thrusters for attitude control, all fed by a single tank with an initial load of 77 kg of hydrazine. The vehicle was to spin at 5 rpm for stability during most of the cruise, and because it had no reaction wheels it would employ thrusters to adopt 3-axis stability for its encounters. The communication system comprised a 2.1-meter high-gain antenna on the top for use during the Pluto flyby, a medium-gain antenna and a low-gain antenna on the feed tripod of the main

The 'naked' New Horizons spacecraft being prepared for launch. The dish-shaped object at left is the integrated camera, infrared detector an ultraviolet spectrometer.

dish, and a low-gain antenna on the base. From the launch vehicle interface to the tip of the antenna stack, the spacecraft was 2.2 meters tall. Like all recent missions, it had a solid-state data storage system, in this case with a capacity of 8 Gbytes to enable it to devote as much time as possible during its encounters to collecting data rather than returning it. In fact, the spacecraft would be so far from Earth that to transmit all of the data planned for the Pluto flyby at rates of 600 to 1,200 kbits per second would take nine months. The shortest side of the body held the flange to accommodate the RTG of the single-generator type used by Cassini. The plan was to provide an estimated 240 W at launch and 200 W for the Pluto flyby, but the shortage of plutonium fuel (despite purchasing an amount from Russia) will result in a shortfall of 30 W for the encounter. To raise reliability and maximize the likelihood of the spacecraft surviving 15 years in space, New Horizons incorporated standard APL architectural features such as having no mechanisms or moving parts other than the covers of the scientific instruments. On the downside, the entire vehicle must be rocked and steered to aim the instruments at their targets during an encounter.[309]

At 478-kg, New Horizons was much heavier than the miniaturized Pluto–Kuiper Express, and its 30 kg of payload ought to generate 10 times the scientific output. The payload of five instruments was intermediate between the 'Christmas tree' approach of the old Mariner Mark II Pluto flyby proposal and the 'minimalist' one of the Pluto–Kuiper Express. A multispectral remote-sensing package with a trio of

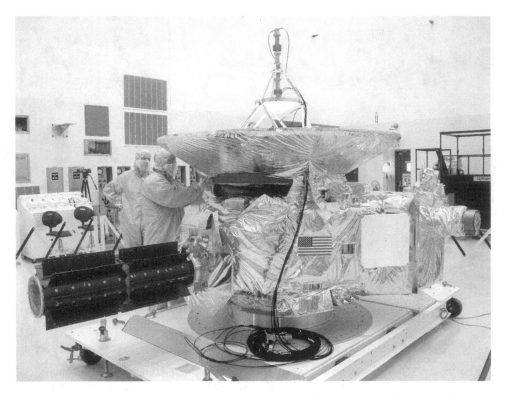

New Horizons after thermal blankets were applied and the RTG (at left) was installed.

black-and-white and color cameras, an infrared detector and an ultraviolet spectrometer based on a Rosetta system is to provide medium-angle mapping and spectroscopy of the atmosphere and surface of Pluto. A narrow-angle camera using a telescope with an aperture of 20.9 cm will supply high-resolution black-and-white images and global mapping in the weeks leading up to the encounter. A two-sensor particles package will investigate the solar wind in the outer solar system, and gases which escape from Pluto and are ionized by solar ultraviolet and 'picked up' by the solar wind. A student-built dust counter will characterize for the first time the dust at heliocentric distances exceeding 18 AU (the distance where the Pioneer Jupiter dust sensor cells failed) and extend the survey out into the Kuiper Belt. This instrument was renamed the Venetia Burney Student Dust Counter in honor of Venetia Burney Phair who suggested the name 'Pluto' for the newly discovered 'planet' in 1930, at which time she was aged eleven. In contrast to previous missions, where radio-occultation experiments were done in a 'downlink' mode with the spacecraft transmitting while Earth antennas listened, New Horizons will conduct atmospheric sounding by 'uplink' radio-occultations with the spacecraft listening to a signal sent in unison by two antennas of the Deep Space Network. By aiming the high-gain antenna at a solid body, this radio package will also be able to function as a microwave radiometer. The only typical 'first encounter' instrument missing is a magnetometer, but scientists hope to be able to infer the presence of a planetary magnetic field (if one exists) from the data supplied by the particles package.

In July 2003 it was decided to launch New Horizons on an Atlas V, and employ a STAR 48B solid-fuel kick stage to give the spacecraft sufficient speed to escape the solar system at Earth departure – in contrast to the Pioneers and Voyagers which used Jupiter gravity-assists to achieve this state.

Initially budgeted at $488 million (including reserves), it is now estimated that New Horizons (including launcher and Pluto operations) will cost about $700 million.[310,311,312,313]

Even as the spacecraft was being prepared, knowledge of the Kuiper Belt and its membership was rapidly increasing. After hundreds of objects had been discovered, it was realized that the belt seemed to have a rather abrupt outer edge. Also, some of the Kuiperoids were surprisingly large. The first of these was (50000) Quaoar, a 1,300-km object accompanied by a large moon. Next was (90377) Sedna, a large icy body that traveled between a perihelion of about 75 AU (which is beyond most of the Kuiper Belt) and an aphelion at 1,000 AU – an orbit which made this an inner member of the Oort Cloud, a much larger cometary reservoir that extends from several hundred AU to about 60,000 AU (i.e. 1 light-year), the frigid realm in which sunlight is less than one-billionth as intense as at Earth.[314] These discoveries put at risk Pluto's status as a planet. If its status as a planet had a cultural justification, this could no longer be held from the viewpoint of orbital dynamics and size. It was becoming evident that Pluto was merely 'King of the Kuiper Belt'. In the summer before New Horizons was to be launched, astronomers announced the discovery of Eris, a large Kuiper Belt object at 97 AU (the farthest distance that any member of the solar system had been observed). The first crude estimates of its size immediately showed Eris to be larger than Pluto. Had these objects been discovered shortly after Pluto (just as asteroids Pallas, Juno and Vesta were found soon after Ceres) they too would have been classified as 'minor planets' and the number of 'major planets' would have remained at eight – as it became with the discovery of Neptune in 1846. Even the presence of Charon as a moon of Pluto seemed less of an oddity because satellites of other trans-Neptunian objects were being discovered, some of which were large relative to their primaries. By the end of 2005, a total of 22 Kuiper Belt objects (including Pluto) were known to have at least one satellite.[315] A discovery announced just two months before New Horizons was due to launch restored to Pluto some of the status of 'King of the Kuiper Belt' that it had lost to Eris, and served to raise public interest in the mission. In analyzing very deep images by the Hubble Space Telescope, astronomers managed to find two specks of light orbiting Pluto at distances of 49,000 and 65,000 km, out beyond Charon. These new satellites were designated Nix and Hydra respectively, with the same initials as the New Horizons mission. A re-analysis of earlier imagery established that they had been detected in 2002. They are 45 to 130 km across, but where Nix, the innermost of the two, is slightly redder than Pluto, Hydra's color resembles that of Charon. It is considered likely that impacts of small Kuiper Belt debris on Nix and Hydra create short-lived faint icy rings. Since all the satellites orbit in the same plane and line up at 35-day intervals, when New Horizons approached the system it would have several opportunities to take well-illuminated family portraits.[316,317]

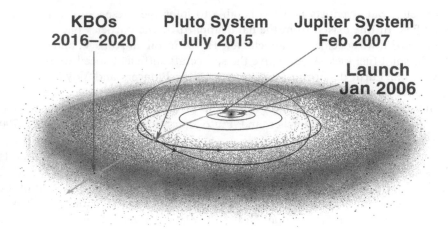

The New Horizons trajectory was unprecedented, requiring the spacecraft to be inserted into a solar system escape orbit at the time of launch.

The assembly of New Horizons was completed in June 2005 and it was delivered to Cape Canaveral in late September for mating with its RTG and final checks. The launch window opened on 11 January 2006 and ran to 14 February, but if the launch slipped beyond 27 January Jupiter would no longer be in position for the slingshot and the mission would have to attempt a flight direct to Pluto. There was a backup launch opportunity in 2007 for a direct flight, but this would require the propellant load to be reduced, and this would probably rule out encountering a Kuiper Belt object. In fact, the launch was delayed by several days when NASA decided to inspect the rocket's fuel tanks after a rupture occurred when a similar tank was undergoing structural testing. Weeks earlier, one of the five strap-on solid-fuel boosters had required replacement owing to damaged inflicted by hurricane Wilma. The first launch attempt on 17 January was scrubbed due to high winds at Cape Canaveral. A power shortage at APL's mission control center ruined the launch attempt the following day. But New Horizons lifted off smoothly on 19 January. The Centaur stage was abandoned in a solar orbit with its aphelion in the asteroid belt, but 44 minutes after launch the solid-fuel kick stage gave the spacecraft the greatest Earth-departure speed ever attained by a deep-space mission, surpassing even Ulysses. In fact, New Horizons was traveling so fast that it crossed the orbit of the Moon after just 9 hours – although its closest approach to that body was 184,700 km. As only minor course corrections would be needed to refine the aim for Jupiter, the excellent reserve of propellant would add flexibility in planning the Kuiper Belt phase of the mission. It made a two-part course correction on 28 and 30 January, and another on 9 March. The dust counter was switched on in March. A number of software patches were uploaded to fix some bugs and to implement additional capabilities.[318]

In early May, as instrument checkout and calibration activities continued, mission planners performed a search through the extensive database of known asteroids and

The Atlas V rocket used to launch New Horizons. Despite its name, it has little or no resemblance to the earlier versions of the rocket of that name.

realized that on 13 June the spacecraft would pass 2002JF56 at a range of 102,000 km and a relative speed of 26.6 km/s. Discovered by the LINEAR (Lincoln Near-Earth Asteroid Research) automated telescopes on 9 May 2002, this small main belt object was later renamed (132524) APL in honor of the institution which had developed New Horizons. It was decided to undertake observations to demonstrate the spacecraft's ability to track a nearby moving target as a rehearsal of encountering a Kuiper Belt object, even although the 'dust cover' of the telescopic camera was still closed and the resolution of the other camera would be comparable to the estimated size of the asteroid, at about 2 km. A call was issued to astronomers around the world to obtain as much data as possible on this object, which hitherto had been only a speck of light in a few telescopic pictures. In effect, all that was known about it was its orbit. One of the instruments that supplied new data was the 8-meter Very Large Telescope of the European Southern Observatory in Chile. Its observations a few weeks prior to the encounter showed APL to be of the S-class, 2.3 km across and fairly spheroidal.[319] New Horizons began to track the asteroid 35 hours out, and continued to do so until 8 minutes prior to the moment of closest approach, which was at an estimated range of 101,867 km. It obtained black-and-white and color imagery and, as expected, although the asteroid was not resolved, it was confirmed to be 2 or 3 km in size. Intriguingly, images taken within an hour of closest approach showed a second pinpoint source 2 pixels away from the asteroid, and traveling with it. This could be either a small satellite or an excrescence of the main body, but until the asteroid's rotational period is determined it will not be possible to decide which is the case.[320]

On 24 August 2006 the International Astronomical Union, recognizing that at least one object in the Kuiper Belt exceeded Pluto in size, voted to demote Pluto. Amongst other things, this resolution stated that for an object orbiting the Sun to be called a planet it must either have swept its orbit clear of debris or confined such material gravitationally, as Jupiter did with the Trojan asteroids. By this definition, Pluto was not a planet, since it occupied an orbit similar to the other members of the Kuiper Belt. The reclassification of Pluto as a 'dwarf planet' was irksome to the New Horizons team, but they were not going to abandon their target just because the mission could no longer be said to be investigating the outermost planet of the solar system. But one is left wondering what would have happened to the many Pluto mission proposals if this demotion in status had occurred earlier. The reduction in the number of planets to eight has only a cultural impact. Owing to this ruling (which has yet to be endorsed by the wider astronomical community) all more-or-less spherical bodies which have not swept their orbits clean became 'dwarf planets'. This class currently includes Ceres, Pluto, Eris and (136472) Makemake, the latter being another Kuiper Belt only slightly smaller than Pluto. On 11 June 2008 the Union ruled that dwarf planets that travel in orbits beyond that of Neptune constitute a family of 'plutoids'.

On 4 September 2006, shortly after its cover was opened and the first star field calibration pictures were obtained, the telescopic camera was aimed toward the next target, Jupiter, some 291 million km away. Its pictures rehearsed acquisition modes and exposure times for the forthcoming encounter, and managed to discern not only

Asteroid (132524) APL seen by the low-resolution camera on New Horizons. It is only 2 to 3 km across. The excrescence at lower right might be a small satellite.

details of the planet's atmospheric belts and clouds but also Io and Europa and their transiting shadows. On 21 and 24 September this camera proved its worth by taking its first pictures of Pluto as a faint speck of light moving against the background of stars, still 4.2 billion km away. By the end of September all of the spacecraft's instruments were ready for the Jovian encounter. Throughout this time, routine data was collected by the plasma instrument and the dust counter. New Horizons passed through solar conjunction in late November, but was able to continue to provide data during most of this 'blackout' period. The encounter program initiated on 8 January 2007 with distant imagery of the planet and infrared scans of Callisto. The atmosphere was remarkably quiescent, and a larger portion of the planet than usual appeared clear. It was hoped that these clearings would persist until late February. Although not an actual science objective, this flyby was an opportunity to rehearse techniques for data acquisition, in particular some of the outer irregular satellites. Although there were dozens of such moonlets, almost all that was known of each was its orbit, photometric properties and likely diameter. Given the approach trajectory, the only way the spacecraft would be able to make a close inspection of any of these objects would be to make a specific targeting maneuver, but it was decided to save the fuel. In January, distant optical navigation pictures were taken of Callirrhoe, an irregular moonlet discovered in 1999. Being less than 10 km in size, it appeared star-like. The observations simulated the detection and tracking of a Kuiper Belt object. Other images taken later during the encounter were used to reconstruct the 'light curves', shapes and sizes of Elara and Himalia, two of the larger outer satellites. Although New Horizons approached Jupiter closer than Cassini had, the minimum range was still 32 Rj (2.3 million km), outside the orbits of the Galilean satellites. The flyby occurred on 28 February at a relative speed of 21.2 km/s, slightly south of the planet's equatorial plane.[321,322]

Most of the activity of the encounter was in the second half of February. Starting in the early hours of 25 February, high fluxes of protons and electrons were detected that probably marked the crossing (or more likely multiple crossings) of the bow shock. However, this was impossible to confirm without an onboard magnetometer. The magnetopause was crossed later on the same day, at a range of 67.4 Rj. At the same time, the planet was being observed from Earth orbit by the Hubble Space Telescope and the Chandra X-Ray Observatory, and from the ground by a host of amateur and professional astronomers. This data enabled the New Horizons team to decide what their vehicle's camera should inspect. Thanks to the higher data rates made possible by the relative closeness of Jupiter to Earth, 34 Gbits of data was obtained during the encounter – more than was planned for Pluto. Some of this data was trickled back in almost real-time, but most of it was to be downloaded during a 3-month period starting in March.

In late 2005, several years after the decades-old 'white ovals' in a belt to the south of the Great Red Spot merged, the remnant began to take on a reddish hue, perhaps as a result of drawing up material from deep inside Jupiter. About half the size of the Great Red Spot to the north, it was nicknamed 'Red Spot Jr'. Interestingly, a similar but smaller spot had appeared at the time of the Pioneer 10 flyby.[323] High-resolution multispectral images of the spot by New Horizons showed the signature of ammonia swelling to the top of the atmosphere. An analysis of sequences of pictures taken 30 minutes apart measured the wind speeds within it at between 150 and 190 m/s; some of the highest speeds yet measured on Jupiter. As anticipated by distant imaging, the equatorial regions were experiencing a spell of unusually good weather, and thus were almost cloud-free. This assisted in the detection of deeper structures, in particular the bright and dark trains of atmospheric waves which spread at hundreds of meters per second. The turbulent wake of the Great Red Spot was also remarkably free of cloud. Thanks in part to the good visibility, infrared scans at diminishing distances detected short-lasting fresh clouds of ammonia ice which indicated active storms or regions of upwelling gas. It was possible to monitor their evolution over several rotations of the planet. When the imaging spectrometer began to suffer frequent resets on 25 February, the threshold of its radiation protection logic was increased and thereafter it operated satisfactorily. Finally, on 28 February, the day of closest approach, a sequence of color and methane-filter imagery resolved details of the Jovian atmosphere as fine as 45 km, which was comparable to the imagery obtained by Galileo while in orbit of the planet. The 11-km resolution of the New Horizons telescopic camera was comparable to the best imagery from the Voyagers during their much closer flybys.

Extensive medium-resolution imagery was taken of all the Galilean moons except Callisto. In fact, Ganymede and Callisto were on the opposite side of the planet from New Horizons in the approach phase, but it was possible to view the Jupiter-facing portions of these satellites where the surface composition had been poorly mapped (or in some areas not mapped at all) by Galileo owing to observational constraints. Maps of water ice on this portion of Ganymede showed spots of cleaner ice marking recent impact craters and their ejecta deposits. However, even the best images (obtained at a range of 3.5 million km) had a resolution of just 17 km. On 27 February the spacecraft started to take pictures of Europa. Especially interesting were the long-range views of its terminator which filled small gaps in the Galileo and Voyager coverage of such structures as broad arcuate troughs suspected of having been created by the wandering of the pole of the icy shell decoupled from the rocky core by an ocean. Such surface relief was best studied in images taken with the Sun at a low elevation above the local horizon.[324] Detailed maps of the 'non-ice' material on the surface of Europa were obtained by three scans of relatively good resolution. These confirmed a distribution consistent with Europa having swept up sulfur blown into space by the volcanoes on Io and then chemically altered on the surface by Jovian radiation. The light-scattering characteristics of the surface at various viewing angles were studied in greater detail than had been possible for previous missions, and found to be different from the solar system's other geologically active icy moons, namely Enceladus orbiting Saturn and Triton orbiting Neptune.[325]

Although New Horizons approached Io no closer than 2.24 million km, it obtained a total of 190 pictures employing its telescopic camera and 17 with its multispectral camera. Scientists expecting to capture the plume of Prometheus on the

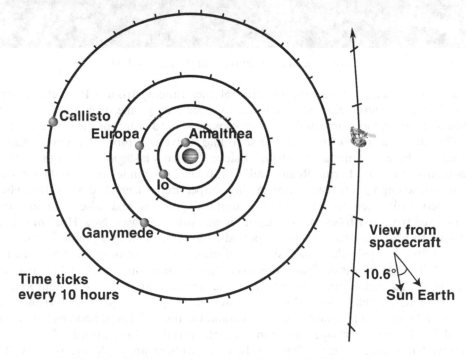

The New Horizons trajectory relative to the Jovian system.

A stunning image of Europa on the limb of Jupiter.

limb were amazed to observe 11 other plumes ranging from a Prometheus-style 'umbrella' of sulfur dioxide 80 km tall to a gigantic plume over Tvashtar in the north polar region. Some of the volcanoes were seen to erupt for the first time since 1979 when Voyager 1 discovered that Io had active volcanoes. No fewer than 19 sites on the entire moon had changed since Galileo's last global map of 2001. A new discovery in Lerna Regio had a 150-km-tall plume and a 400-km-wide concentric deposit on the surface. A ring-shaped deposit marked Tvashtar, where Galileo had taken its amazing 'fire curtain' pictures and where an eruption observed from Earth before the encounter was still underway. New Horizons made about 40 observations of it over a period of 8 days, viewing a plume 350 km tall and 1,300 km wide that enabled its evolution and morphology to be studied in detail. The brightness of Tvashtar proved that it was a source of basaltic lava, and the vent was issuing primarily gas that condensed during its ballistic flight and fell as discrete particles. New Horizons observed eclipses of Io in the visible and ultraviolet on four separate occasions around the time of closest approach, with the Hubble Space Telescope providing supplementary observations. The results showed dramatic auroral glows of Io's atmosphere and volcanic plumes. The glow was more intense at the sub-Jupiter point and at its antipode, possibly owing to local atmospheric density maxima. Brightness variations at eclipse entry and exit seemed to be related to surface sublimation. Maps of the glowing volcanic spots were also obtained.[326,327]

Having been tailored to the low light level of Pluto, the camera was able to give the "best-ever views" of Jupiter's dark rings. Images showed the presence of ripples in Jupiter's ring similar to, but less dramatic than those seen by Cassini at Saturn in 2009. In both cases the ripples are believed to have been formed by clouds of debris that swept by the rings and collided with the ring particles, moving them to slightly inclined orbits. Ripples were also seen in Galileo images from 1996 and 2000, and could be traced back to an event in the second half of 1994, specifically the impact of the comet Shoemaker–Levy 9 on the planet and its accompanying debris cloud. In

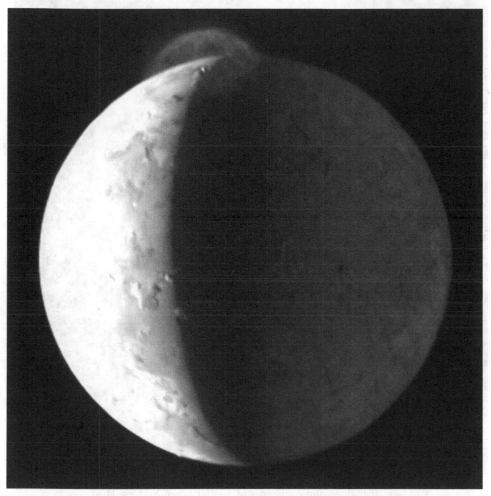

An image of Io taken by New Horizons. Note the huge plume of volcano Tvashtar at the top.

A mosaic of high-resolution images of Jupiter taken by New Horizons.

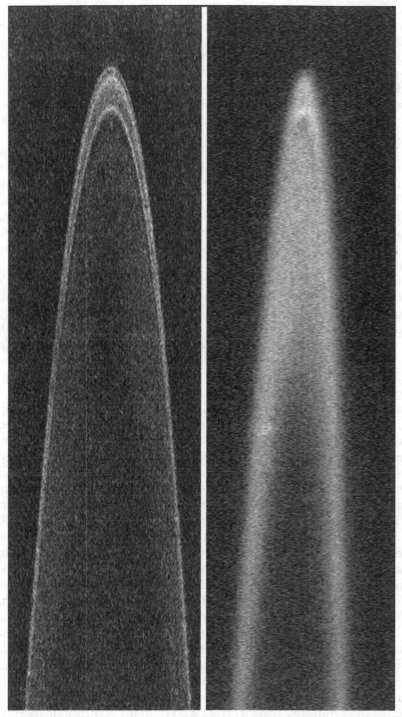

These images of the Jovian ring by New Horizons are the best ever taken, thanks to the camera being optimized for Pluto's low-light environment. The top image was taken during the inbound leg with the Sun behind the spacecraft and shows mostly large particles. The bottom image was taken after the flyby and shows the smaller ring particles forward scattering sunlight.

the case of New Horizons, the ripples appeared to have been caused by clouds that swept by in late 2001 and late 2003, implying that unseen objects disintegrate near the planet with a higher frequency than believed. This conclusion was supported by the unexpected appearance of fireballs from impacting asteroids or comets in the Jovian atmosphere in both 2009 and 2010.[328] Comprehensive searches for kilometer-sized embedded moonlets were made in an effort to determine whether such bodies were continuously replenishing the rings with dust. Until then, the smallest known moonlet was Adrastea, at 16 km across. Two 'movies' for a total of more than 100 frames were obtained on two consecutive days, each spanning a full rotation of the ring system. The main ring was clearly seen, as were density minima near the orbits of Adrastea and Metis, but no new moonlets were found. What was discovered, were unexplained families of clumps. One family was comprised of three clumps plus two others which were barely visible. Another was a pair of co-rotating clumps separated in longitude by several degrees. Such clumps may have been caused by interactions with the shepherd moons Metis and Adrastea (akin to the bright arcs of Neptune's rings), but if they formed when a small comet or asteroid crashed into the ring they must be short-lived features.

On 3 March, several days after the flyby, New Horizons turned to view Jupiter's night-time hemisphere. It took 16 long-exposure pictures of the polar regions, and, in addition to temperate latitudes where lightning was considered to be most likely and where flashes had been seen by the Voyagers, lightning was also seen at high latitudes for the first time – as six flashes in the northern hemisphere and seven in the southern hemisphere. These pictures also detected auroral phenomena and the 'foot' of the Io flux tube. Ultraviolet scans of the night-side detected a hydrogen glow that was much fainter than measured by Voyager 2 in 1979.[329,330,331,332] Withdrawing from Jupiter, on 19 March a memory error caused the main computer to reboot. This put the spacecraft into safe mode, which halted observations and 3-axis stabilization and set the vehicle spinning for stability. By then, however, the remote-sensing was essentially complete and the schedule called for New Horizons to enter spin-stability mode on 21 March. The plan called for particle observations to continue for 3 months as the vehicle made a particularly long excursion down Jupiter's magnetic tail – which is known to stretch at least as far from the Sun as the orbit of Saturn. Voyager 2 had sampled this region to 150 Rj, as had Galileo during one highly elliptical orbit of the planet. New Horizons established that volcanic material issued by Io was carried to large distances within the tail. In fact, it noted the passage of a series of magnetically confined 'plasmoids' filled with hydrogen and helium leaking from Jupiter's atmosphere and ionosphere, and also sulfur and oxygen from Io. As these observations were underway, in April the vehicle executed its first 5-day hibernation test. Solar wind and particle instruments observed the tail continuously until mid-May, by which time New Horizons was 1,655 Rj from Jupiter, and then they recorded the tail flapping back and forth in response to the solar wind, at some times being inside the magnetotail and at other times in the sheath and magnetopause which surrounded it. The Jovian encounter concluded on 21 June with the termination of these observations, at which time the spacecraft was 2,565 Rj (1.22 AU) from the planet. Remarkably,

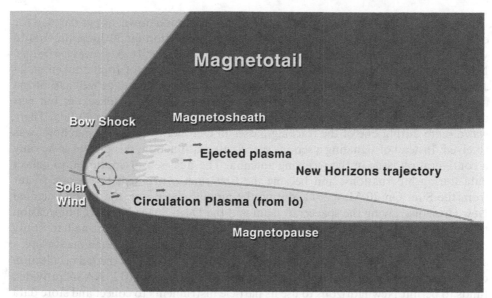

The unprecedented trajectory of New Horizons within the Jovian magnetosphere.

even at that distance the period of Jupiter's rotation was still modulating particle parameters.[333,334,335,336]

For most of its remaining cruise to Pluto, New Horizons was to spend 10 months per year in hibernation, with month-long wakeups in early summer and two 10-day wakeups each November and January. Once per week the Deep Space Network would check a simple beacon signal which indicated the spacecraft's status, and once per month download detailed telemetry. On one occasion, the vehicle suffered a computer reset that issued a 'red' beacon code for the first time. During an intense instrument calibration exercise, it took distant images of Uranus, Neptune and Pluto in order to characterize them at phase angles inaccessible from Earth. The two large Kuiper Belt 'dwarf planets' Makemake and Haumea were also imaged, as well as several other objects.

A 2.37-m/s course correction on 25 September refined the trajectory for Pluto. In mid-December the vehicle again went through solar conjunction. In February 2008 it was placed into a hibernation which was to run until September apart from a 2-month hiatus for data download and 'housekeeping' tasks starting in May that would include a reorientation designed to maintain Earth in the narrow beam of the high-gain antenna. In early June New Horizons crossed the orbit of Saturn – the first spacecraft to venture that far from the Sun since Voyager 2 in 1981, although on this occasion the planet was not nearby – and so became the third farthest operational spacecraft. During this hibernation, it obtained sparse data on dust in the outer solar system. When revived in September 2008 it turned its telescope toward Neptune, more than 2 billion km distant, and took a sequence of images that showed both the planet and its moon, Triton, proving the imager capable of detecting a faint object alongside a much brighter one.

For the 2010 summer check-up, New Horizons was spun down, as per the plan for every even-numbered year. It performed a course correction on 30 June, the first in years, and then the camera was used to image Jupiter (over 16 AU away), Uranus, Neptune and Pluto, as well as some of their satellites. Pictures of Jupiter taken on 24 June showed the bright, saturated planet, with a hint of a phase, as well as Europa and Ganymede. A main computer self-reset occurred during the checkout but was quickly corrected and the probe was able to resume its hibernation on 30 July. There was a scare during one of the tracking passes in October, when no 'semaphore' was received. Instead of signaling a serious spacecraft malfunction, this proved to be only a configuration issue in the receiving antenna. The dust sensor continued to collect data during hibernations, and became the first such instrument to operate so far from the Sun. On 20 May 2011, during the annual check out, the Moon occulted New Horizons, giving the spacecraft team and the Deep Space Network an occasion to rehearse radio measurements to be made at Pluto and Charon, and to study electrons near the lunar surface. Such an experiment had been attempted using a spacecraft in deep space only once before, by Pioneer 7 in 1967. A second occultation occurred during a short wake-up on 21 January 2012. From 2012, NASA provided funds to permit New Horizons to use its particle instruments to collect and store data on the deep interplanetary environment also during hibernation. In May 2012, it carried out a "stress test" of the most intense 22 hours centered on the closest approach to Pluto, during which it simulated all of the activities and observations that were planned for this encounter phase. A little over a month later, with all of the engineering telemetry and scientific data from the test transmitted to Earth, the vehicle was put back into hibernation. During the 2013 summer wake-up, a full rehearsal of the "core sequence" for the encounter was carried out from the 5th to the 14th of July. This was one year earlier than initially planned, in order to have more time to solve any potential problem that may show up. Moreover, earlier in July, New Horizons used the long-distance camera to take six pictures of the mission target, still about 6 AU away, that resolved for the first time Charon as a separate dot to one side of Pluto. It was then put back into hibernation in late August. It will be revived in June 2014 for 2 months for a trajectory correction, to take new images of Pluto and Charon, and to rehearse some encounter activities. In particular, an initial observation campaign will be carried out, lasting one week to cover a full revolution of Charon.

New Horizons will cross Neptune's orbit in late August 2014, 25 years after the Voyager 2 flyby. By coincidence it passed through that planet's trailing Trojan zone. At the time of writing only two trailing asteroids are known. Both of them were discovered by a survey for possible objects along New Horizon's trajectory. The spacecraft passed by one of them, designated 2011 HM105, in November 2013 at a range of 1.2 AU, but the team running the spacecraft chose to assign precedence to preparations for the Pluto encounter, and therefore the asteroid went unobserved.[337] In January 2015 it will be the turn of another small object, currently known only by its preliminary, non-official designation of VNH00004, which will come within 0.5 AU of the probe. It is possible that some observations might be made to gain data on the appearance of a small Kuiper Belt object at phase angles inaccessible from Earth.

Such data would not only yield scientific information on the surface texture, but would also assist in planning future distant encounters by revealing how bright an outer solar system object would appear to an approaching spacecraft.

The Pluto encounter will span about a year. Most of the preparatory work was done in the years immediately after launch, prior to the dispersal of the engineers and scientists with a detailed knowledge of the spacecraft. Moreover, because the budget for operations would be drastically reduced after 2009 for the remainder of the cruise, the planning had to be completed by then and the encounter sequences rehearsed. After awakening on 7 December 2014, the distant approach phase will start in January 2015 and consist mostly of imaging for optical navigation purposes. In fact, due to the uncertainty of the target's ephemeris, which still amounts to several thousand kilometers (and also to the fact that it has traveled less than one-third of an orbit since its discovery) New Horizons is more reliant on optical navigation than any previous mission. In the second approach phase, lasting from 4 April to 23 June, New Horizons' observations will also image the smaller satellites and refine their orbits. It is expected that by about 75 days before the encounter the telescopic camera will start to supply images of Pluto that surpass those of the Hubble Space Telescope. The third approach phase will last until several days prior to closest approach. During this phase, when New Horizons is half a Plutonian day from its target the telescopic camera will be able to document with a resolution of about 40 km the hemisphere that will be in darkness at the time of closest approach, thereby enabling the single New Horizons mission to achieve some of the scientific objectives expected of the two-probe Pluto Fast Flyby. Optical navigation and imaging of the satellites will last until 24 hours before closest approach.

New Horizons will reach its minimum distance from Pluto at 11:50 UTC on 14 July 2015. The flyby will resemble the Voyager 2 encounter with Uranus, owing to the tilt of the orbital plane of the satellites of Pluto. The spacecraft will approach the system nearly face-on, like a bull's-eye. The trajectory is required to yield solar occultations for both Pluto and Charon for ultraviolet atmospheric studies, and an Earth occultation of at least Pluto in order to sound its atmosphere using the radio signal. The plan is to fly 12,500 km above the surface of Pluto at a relative speed of 13.8 km/s. The narrow-angle camera will take high-resolution regional images while the multispectral imager obtains global low-resolution pictures, charts the surface composition and temperature field, and provides ultraviolet measurements to study the structure of the upper atmosphere and the rate at which gas molecules escape to space. The challenge in making these observations is that the viewing fields must span all of the uncertainty in Pluto's ephemeris, which is not expected to be less than 2,500 km – comparable to the diameter of Pluto itself. Ionized atmospheric particles that are 'picked up' by the solar wind will be detected and analyzed by the plasma sensors. New Horizons will also search for exchanges of atmospheric gases that simulations suggest occur between Pluto and Charon. It is possible that atmospheric gas molecules could reach the Lagrangian point between Pluto and Charon, some 15,000 km from the surface of the former, and spill into Charon's gravity sphere. This way of transferring mass between two celestial bodies is known to occur, for example, in certain close binary systems where a large, swollen star spills material to a white dwarf companion, until enough mass is transferred to trigger the nuclear fusion reactions of a type Ia supernova.

At the time of closest approach, Charon and Hydra will be on the same side of Pluto as the spacecraft (although Hydra will be much farther away) and Nix will be on the opposite side. Although the distances from these objects will limit imagery to a resolution of a few kilometers per pixel, this should be sufficient to determine their shapes and general morphologies. It should also be possible to obtain pictures of the night-side and polar region of Pluto illuminated by 'Charon-shine' 10,000 times fainter than sunlight at that heliocentric distance about 1 hour after closest approach using the wide-angle camera. The same observations will be repeated hours later by the long-range camera, and again a few days later on the night-side of Charon itself. These should give at least a rough map of the night hemisphere and of polar frost deposits and the large-scale features.[338,339,340] As New Horizons downloads its data after the encounter, it will test using both redundant transmitters simultaneously to double the data rate and return the entire Pluto dataset in less than the 9 months it would take using a single transmitter. Depending on how this experiment turns out, the encounter will be wrapped up some time in early 2016.

Of course, as New Horizons was cruising towards Pluto scientific knowledge of its target increased, notably with the discovery of fresh ice and ammonia hydrate deposits on the surface of Charon that might be evidence of the existence of water geysers. A re-analysis of telescopic pictures taken between the discovery of Pluto in 1930 through the early 1950s clearly indicated that frost deposits migrate around its surface with the changing seasons in response to the ellipticity of its orbit and the obliquity of its axis. Other observations showed fine structures in the atmosphere of Pluto and monitored the mixing ratio of methane, which declined, possibly as the atmosphere cooled after the 1989 perihelion. In contrast to the atmospheres of the planets in the inner solar system, Pluto's tenuous atmosphere seems to be warmer than the surface, on which layers or patches of methane frost may be present.[341,342]

The atmosphere of Pluto was expected to contract after perihelion in 1989, but in reality its pressure was observed to increase during the 2000s and within a decade it expanded dramatically to some 3,000 km from the surface. Carbon monoxide had been tentatively detected in 2000 and its concentration was shown to have significantly increased, probably as patches of carbon monoxide ice were exposed to sunlight. This suggested a strongly coupled atmosphere and surface, with carbon monoxide helping to regulate the temperature of the atmosphere as Pluto receded from the Sun after perihelion. Carbon monoxide was only the second gas detected after methane. Nitrogen was expected to be the most common gas, although it would be undetectable spectroscopically from Earth. Moreover, carbon monoxide spectra suggested that it could be streaming away from Pluto in a comet-like tail.[343]

Pluto and Charon were observed using the Hubble Space Telescope's Advanced Camera for Surveys in 2002 and 2003 and new, more accurate color maps than those obtained for Pluto in 1994 were constructed. A striking result was that, despite the fact that they were receiving ever more illumination, the northern polar regions seemed to have rapidly brightened. A few spots appeared to have moved, or to have changed their appearance. Amongst the features that had not changed (and actually seem to have been present since at least the 1950s) was a vast bright region which spectroscopic observations had shown to be rich in a frost of carbon monoxide and could be the source of gas in the atmosphere. This latter feature will be prominently visible to New Horizons around closest approach. Hubble also discovered a strong ultraviolet absorber on the surface of Pluto, suggesting complex hydrocarbons or nitriles produced by the interaction of ice with energetic solar particles and cosmic rays. The map of Charon, on the other hand, remained almost featureless as always; but this could in large part be due to its small size.[344] If the interior of Pluto possesses enough decaying potassium isotopes, there could even be a liquid-water ocean buried under the external icy crust.

Two more Hubble discoveries were announced in July 2011 and one year later. While searching for faint rings around Pluto, the space telescope discovered instead a fourth and a fifth moon. Named Kerberos and Styx, these are respectively only 14 to 40 km and 10 to 25 km across and travel in orbits between Nix and Hydra. The possibility has not been ruled out that Pluto has a faint ring system. In fact, the escape speeds from the smaller moons would be comparable to their orbital speeds, meaning that debris could end up in orbits of almost any inclination, and form a torus or a cloud of dust instead of a ring. As a precaution in case something is found, a "safe haven by other trajectories" will have to be established. Nine possible such trajectories of increasing distance from Pluto were considered, but only a few were studied in detail, down to the exact observation sequences. Even a collision with a 1-mm particle would be catastrophic at a relative speed of almost 14 km/s. Searches for additional satellites, debris and rings will be carried out by New Horizons from 64 to about 18 days out, and a bail-out can be executed as late as 10 days prior to closest approach or 10 million km away with little effect on the fuel budget.[345]

As a result of an 18-month study on bail-out trajectories it was decided to stay on the baseline, since Charon should have cleared the targeted region of all dangerous dust, reducing the probability of impact to just 0.3 per cent. However, if the baseline trajectory proved to present a serious risk of impact, the main backup strategy would be to perform the most critical part of the flyby with the high-gain antenna facing forward, as the Cassini spacecraft did in crossing the ring plane at Saturn. Tests in which projectiles were fired at hardware showed that this would provide protection from particles up to several millimeters in size. A second backup trajectory would be to fly New Horizons much closer to Pluto, to within 3,000 km of its surface, where the thin atmosphere should have cleared most dust.[346,347,348]

The final portion of the New Horizons mission should involve at least one flyby of a Kuiper Belt object. The vehicle's trajectory through the Pluto system is obliged to produce eclipses and occultations, and owing to the extremely small mass of Pluto the encounter will produce only a minimal gravitational deflection. Hence most (if not all) of the targeting for the Kuiper Belt phase of the mission will have to be achieved by propulsive maneuvers. The targeting burn could be carried out as soon as 60 to 90 days after the Pluto flyby, or be put off to as late as 2016 or 2017. About 34 kg of fuel is expected to remain on board for this part of the mission. Whereas the Pluto–Kuiper Express spacecraft was to have searched for a target by aiming its camera in the direction of travel and awaiting an opportunity, New Horizons must rely on terrestrial telescopes.

Some of the largest telescopes on Earth started to survey the cone of space accessible to the spacecraft in search of objects as far out as 55 AU only during the spring of 2011. In fact, this search could not start until Pluto and any nearby Kuiper Belt objects exited some of the most crowded star fields of the Milky Way in 2011. Algorithms and techniques for an automated search of this type were tested in the context of the Pluto–Kuiper Express proposal, and resulted in the discovery of two new Kuiper Belt objects. It is estimated that up to a dozen possible targets could be identified.[349,350] The telescopic survey relies on the help of amateurs, in a "citizen science" project called IceHunters that solicits members of the public to sift through pictures of the area of sky to suggest possible Kuiper Belt targets. Twenty-four objects were identified in archive data from 2004 and 2005, and 18 more in 2011. They include 2011 JW31, 2011 JY31 and 2011 HZ102, all of which will pass within 0.2 AU of the baseline trajectory of New Horizons in 2018, and 2011 JX31 passing within 0.4 AU in mid-2020. None of these, however, is deemed reachable by New Horizons.[351] To these must be added an object discovered years before the mission was launched: 150-km (15810) 1994 JR1. New Horizons would nominally pass more than 0.5 AU from it in June 2016, and a large maneuver right after the Pluto flyby using a substantial amount of fuel would be required to arrange a flyby. Since this would

An image of Neptune and Triton taken by New Horizons. Beginning in 2013 the camera was capable of resolving Charon from Pluto.

probably preclude other encounters, scientists and engineers would prefer to target a different, yet-to-be-discovered object. If a good candidate is found, New Horizons will take navigation pictures to improve its ephemeris as early as possible, in order to make the best possible use of the limited propellant remaining. Statistical analyses indicate that the mission ought to be able to reach several objects exceeding 35 km in size, and possibly make the first targeting maneuver several weeks after concluding the Pluto flyby. A number of distant flybys (of the order of millions to tens of millions of kilometers) will be made of other Kuiper Belt objects, which will be imaged to search for satellites in order to nail down precisely the rotation period, etc.

After encountering Kuiper Belt objects, New Horizons will depart the solar system trailed by the final stage of its launcher – which will pass Pluto at a range of several hundred million kilometers. But unlike its Pioneer and Voyager predecessors, neither New Horizons nor its spent stage carry messages for any aliens that may recover them in the future – evidently due to the scale of the bureaucracy involved in preparing such a message! Instead, the spacecraft has mementos such as US flags, the state quarters of Maryland (APL's home) and Florida (the launch site), a fragment of the first private manned spacecraft (SpaceShip 1), CD-ROMs with signatures and pictures of people who supported the mission, the 1990 American *Pluto: Not Yet Explored* stamp which inspired so many mission proposals, as well as three small memorial plaques and, poignantly, a vial holding several grams of the ashes of Clyde Tombaugh, the discoverer of Pluto who died in 1997. It should be possible to track New Horizons out to 55 AU in the direction of the constellation Aquila and plans are being made for a deep heliospheric mission using the solar wind and energetic particle instruments, as well as the dust detector, lasting to the late 2030s or 2040s, as long as the RTG will allow the vehicle to operate.

BATTLESTAR JIMO

The development of space nuclear applications has been an exercise in frustration. In the 1960s it was expected that nuclear rocket stages would facilitate manned missions beyond the Moon after the Apollo program concluded in the 1970s. In the 1980s the Strategic Defense Initiative, known as 'Star Wars', envisaged using nuclear reactors for power-hungry applications in space. In the early 2000s the prospect of nuclear applications was revived. The Los Alamos National Laboratory dusted off projects for 'Gas Core Nuclear Rockets' that could slash the duration of a manned mission to Mars to less than a year, and the Sandia National Laboratories, a wholly owned subsidiary of Lockheed Martin, worked on a trash-can-sized fission reactor capable of providing the electrical energy to power low-thrust ion engines for robotic deep-space missions. In fact, Sandia proposed using this nuclear electric propulsion (NEP) technology for a mission that would deliver a 1,000-kg payload to the Kuiper Belt after a cruise of 10 to 15 years. Another study envisaged a 12-tonne NEP spacecraft delivering Viking-class landers to one large and one small 'Kuiperoid'. Each lander would take a sample from a depth of at least 10 meters for return to Earth. Using NEP, the mission would take less than 20 years to complete: starting by spiraling out

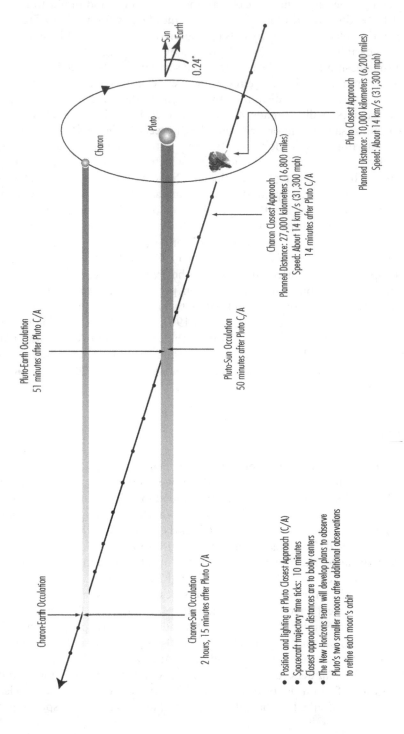

Sun
Earth
0.24°

Charon

Pluto

Pluto Closest Approach
Planned Distance: 10,000 kilometers (6,200 miles)
Speed: About 14 km/s (31,300 mph)

Charon Closest Approach
Planned Distance: 27,000 kilometers (16,800 miles)
Speed: About 14 km/s (31,300 mph)
14 minutes after Pluto C/A

Pluto-Earth Occultation
51 minutes after Pluto C/A

Pluto-Sun Occultation
50 minutes after Pluto C/A

Charon-Earth Occultation

Charon-Sun Occultation
2 hours, 15 minutes after Pluto C/A

- Position and lighting at Pluto Closest Approach (C/A)
- Spacecraft trajectory time ticks: 10 minutes
- Closest approach distances are to body centers
- The New Horizons team will develop plans to observe Pluto's two smaller moons after additional observations to refine each moon's orbit

The trajectory of New Horizons relative to Pluto and Charon during the July 2015 flyby. The actual distances may be slightly different. The smaller moonlets are not shown.

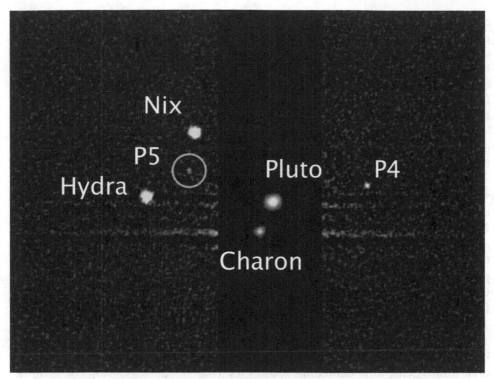

Pluto, Charon, Nix, Hydra, Kerberos (P4) and Styx (P5) seen by the Hubble Space Telescope on 7 July 2012. (NASA, ESA, and M. Showalter of the SETI Institute)

of Earth orbit, accelerating for a time, then coasting and finally braking to rendezvous with the target. Alternatively, it would be possible to reach Saturn in just 4 years or Neptune in 9 years. There was also a NEP proposal to reach Pluto sooner than New Horizons could do so, and NASA Administrator Sean O'Keefe said this would be more sensible than spending 15 years cruising to perform a fast-flyby mission.[352,353,354,355] Sandia and NASA jointly established the Nuclear Systems Initiative in order to develop nuclear power for use in deep-space exploration.

JPL promptly initiated studies of a Jupiter Icy Moons Tour, drawing on scientific priorities set by the National Academy of Sciences calling for focused exploration of Europa and its subcrustal ocean and, if possible, the same for Ganymede. These studies were aimed at comparing what could be done by drawing power from solar arrays, RTGs and fission reactors. The latter technique was selected as the most promising, and it was proposed that a technology demonstration mission be sent for approval by Congress in 2003. The project received early go-ahead, and detailed feasibility studies were initiated with the aim of demonstrating the safe and reliable use of a nuclear fission-powered and electric propulsion spacecraft. So the Nuclear Systems Initiative, renamed Project Prometheus after the giant in Greek mythology who taught men how to harness fire, expanded from a $950 million budget spread across 5 years into a $3 billion request starting from 2003. Of course, it had to compete for financial resources with other projects – which soon came to include returning the Space Shuttle to flight after the loss of Columbia in February 2003 and the imperative to finish the International Space Station within the decade. Prometheus was to be further developed by a team led by JPL that included engineers from the Department of Energy's Office of Naval Reactors whose normal focus was developing reactors for the aircraft carriers and submarines of the US Navy.

The Prometheus demonstration was to be the Jupiter Icy Moon Orbiter (JIMO), a giant spacecraft that would spend up to 6 years in the Jovian system, at different times orbiting Callisto, Ganymede and Europa. Io was not included because the shielding required to protect the spacecraft from the radiation that close to the planet would be prohibitive. For each moon, JIMO would investigate the composition and structure of the surface, decipher its geological history, and assess the scope for life. Moreover, it would wholly map Ganymede and Callisto at meter-scale resolution, and at least half of Europa. The interest in these three 'icy moons' derived from Galileo's discovery of liquid water beneath their icy crusts at depths ranging from possibly as little as several kilometers for Europa to hundreds of kilometers for Callisto. In the case of Europa, the likelihood of hydrothermal activity on the ocean floor made this the best candidate in the solar system for finding extraterrestrial life. The JIMO mission was not expected to launch before 2011, and probably sometime around 2015. It would be the most expensive planetary mission ever, with its multi-billion-dollar budget dwarfing even the Galileo and Cassini flagships and placing it on a similar scale to undertakings such as the Hubble Space Telescope or even the early development phase of the Space Shuttle.

After the Prometheus bus was demonstrated by JIMO, it would be used for other missions. In fact, by enabling deep-space missions to be flown in durations of at most 20 years, NEP would facilitate a comprehensive program of exploration that included Saturn and Titan orbiters, Neptune orbiters with atmospheric and Triton probes, Pluto orbiters, Kuiper Belt rendezvous missions, fast interstellar precursor flights 200 AU in the direction of the interstellar 'bow shock' of the heliosphere, heavy orbiters for Mars and Venus, comet and asteroid sample-returns, asteroid rendezvous-and-divert, lunar and Mars surface missions, and Mars cargo deliveries. This was certainly an enticing prospect.

As the US had flown only one nuclear reactor in space and that was in the 1960s, most of the technology required for Prometheus would be new. The objective was a reactor able to provide up to 200 kW reliably for about 20 years. There were several open issues. Could a reactor be reliably operated in space for that long? How should its heat be converted into electricity? And how should it run an ion engine? In fact, seven areas were identified as requiring technical advancements. These concerned the space-rated reactor, energy conversion systems, rejection of excess heat, electrical propulsion, high-power and high-data-rate communications, radiation hardening, and low-thrust trajectory optimization. Three types of reactor were studied: liquid metal-cooled, heat-pipe cooled, and gas-cooled. Liquid metal systems using liquid lithium would be the least massive, but the US had never tried this technology and it would be difficult to test on the ground. Furthermore, freeze and thaw cycles would make the system prone to single-point failures which could threaten the entire mission. A heat-pipe system using liquid sodium coolant would be more reliable and easier to test, but was perceived to be difficult to integrate into a converter and heat exchanger. Gas-cooled systems using helium or xenon would be simple and easy to test, but again posed integration difficulties and single-point-failure issues.[356] Of course, the project would have to address the inevitable public opposition to the use of nuclear power. The first step would be to 'nuclear rate' the launch vehicle in much the same manner as one is 'man-rated' by increasing its reliability to a level deemed satisfactory, and possibly adding an emergency escape rocket to carry the reactor away from a failing rocket. Also, the plan was to bring the reactor to a 'critical' state only after it had been inserted into a 'nuclear-safe orbit', as to launch it in that state would be too dangerous. Hence, unlike an RTG, the uranium fuel would be essentially non-radioactive prior to reactor startup.

About 80 per cent of the heat that the reactor generated would have to be dumped by radiators. But would a very large radiator be at risk of damage by impacts in the dusty environment of a planet such as Jupiter and Saturn? Two of the technologies being studied for power conversion did not involve moving parts: thermoelectric and thermophotovoltaic generators. Mechanical systems would be more efficient but had lower reliability due to their failure-prone moving parts. The latter included Stirling-cycle engines and Brayton-cycle turbines. In both cases, the higher efficiency would require less energy to be dissipated, which would translate into smaller radiators. A number of different types of ion and electric engines were assessed. They had to be capable of running for up to 120,000 hours, which was four times the longest duration of an ion engine in a laboratory, and almost eight times the total firing time of Deep Space 1. For JIMO, the engine would have to provide a total velocity change of up to 35 km/s, sufficient to escape Earth's gravity, reach Jupiter, enter orbit around that planet, and then maneuver between the three selected Galilean satellites, spending some time in a low polar orbit around each. The orbital design would probably exploit the science of chaotic celestial mechanics, employing such exotic topics as 'weak stability boundaries' and 'manifold captures'. Due to the number of celestial bodies involved it would be extremely sensitive to perturbations. An accidental loss of thrust for several hours during the approach to Europa could easily result in a major perturbation of the spacecraft's orbit that would be difficult to recover from, or even cause it to crash into its target.

The switch from using solar panels or RTGs to nuclear energy would increase the power available to an interplanetary spacecraft by several orders of magnitude, to at least 100 kW. Although this would facilitate unprecedented communication rates of 10 Mbits per second, expensive changes and upgrades to the Deep Space Network would probably be required to exploit this capability. Extensive tests of materials and their resistance to radiation from both the reactor and the Jovian environment were also required. Project Prometheus therefore represented a major challenge.[357]

NASA issued three study contracts to Boeing, Lockheed Martin and Northrop Grumman to define a preliminary spacecraft concept and determine the technological challenges. In September 2004 Northrop Grumman got a supplementary $400 million 5-year contract to initiate development of the non-nuclear portion of JIMO, with the requirement that the bus be capable of being adapted for other missions. NASA also gave study contracts to its own field centers and to industry to propose instruments that could be used on such missions. In fact, five NASA centers worked on aspects of Prometheus. The Glenn Research Center set up a 2-kW Brayton converter and ran it alongside an NSTAR ion engine left over from Deep Space 1 in the first combined test of this type. Also, two promising designs for a large thruster known as Heracles were tested. NASA centers investigated the technologies and materials for radiators, and studied the facilities which would be required at Cape Canaveral to handle and process nuclear reactors.

In parallel with these engineering studies, a workshop of scientists in 2003 sought to identify suitable scientific objectives. The actual payload suite was not defined in detail, but it was obvious that it would include a radar to map the thickness of the icy crusts of the moons as the primary objective, and a laser altimeter to measure their tidal deformations. There would also be a large camera, an infrared spectrometer, a magnetometer, particles and fields instruments on a rotating platform to sweep the space around the vehicle, and a monitor to characterize the dust environment of each moon. The spacecraft would probably be large enough to be able to carry a number of deployable payloads. One might be an entry probe to resume the investigation of Jupiter's atmosphere begun by Galileo, with the primary objective being to measure the water content. By penetrating an anomalously dry 'hot spot', Galileo was unable to measure this directly.[358] However, such a probe would require a relay subsatellite that would further increase the complexity of the mission. Other payloads could be small penetrators that (for example) would deliver to the surface of Europa a seismometer or instruments to undertake chemical analyses of the ice and non-ice components. Non-instrumented projectiles could instead be launched by an electromagnetic rail-gun to perform cratering experiments.[359,360,361,362] After the workshop, a 38-member science definition team was created and this issued its conclusions in February 2004. These included a recommended payload mass of 1,500 kg (the heaviest ever assigned to a planetary mission) with very-high-resolution optical and spectroscopic instruments, high-resolution radars and active laser spectrometers. It was further recommended that as much as 25 per cent of the payload be assigned to a lander that would undertake geophysical and biological investigations of Europa, in particular seeking evidence of biochemical processes. JPL made an internal study that considered not only a soft lander but also a hard lander that would use cushioning airbags and a rough lander that would use crushable shock absorbers. It was decided that whilst short-duration landers could use batteries, longer duration ones would require RTGs. The orbiter's scientific payload would address three main themes for each icy moon: its interaction with the Jovian environment, including tidal effects; the thickness of the crust which capped its ocean; and the prospects for biology. A radio sounder capable of measuring the depth of the interface between the ice and the liquid water would also perform plasma and magnetospheric observations of the environment around Jupiter and around each target moon.[363,364]

However, there was considerable opposition to JIMO in the scientific community. In their most recent decadal survey they had merely called for a radar-equipped orbiter for Europa. The enormous and expensive 'battleship' seemed concurrent to the views of the survey, rather than descended from them. In fact, the American Physical Society stated that the feasibility and cost of JIMO were "highly questionable", and doubted it was "the best way to carry out the highest priority science". Actually, the JIMO plan seemed premature because the existence of oceans beneath the icy surfaces of Callisto, Ganymede and Europa remained to be proved.[365,366] But NASA was undeterred by the criticism.

The result of the various studies was a common bus, referred to either as the Deep Space Vehicle or as the Prometheus Spaceship. With a nuclear reactor, radiators, ion engines and payload, it would be 58 meters long and its mass of 29–36 tonnes would include 12,000 kg of xenon fuel and a 1,500-kg allocation for the mission module and instruments. Because the spacecraft would be so heavy, it was to be launched in up to six parts and automatically docked and assembled in orbit. However, NASA had little practical experience in this field. In April 2005 its Demonstration of Autonomous Rendezvous Technology (DART) failed when the active spacecraft collided with its target. And a proposed robotic servicing of the Hubble Space Telescope was eventually rejected in favor of a manned Shuttle flight. America would not achieve its first autonomous docking until June 2007, with the Orbital Express mission of the Department of Defense's Space Test Program. Even the assembly of the JIMO spacecraft in orbit therefore posed a major challenge to NASA.

At the front of the Prometheus Spaceship was to be the liquid-metal-cooled reactor and the Brayton converter. The reactor would be enclosed in a heat shield that would prevent its destruction in the event of a launch accident which caused it to re-enter the atmosphere. As the heat shield was discarded, it would expose an inner shield against micrometeoroids. The radiator boom would account for most of the vehicle's length. This would either be deployed in segments or extended as a truss, and would hold a series of panels in a distinctive 'Christmas tree' arrangement. Set as far as possible from the reactor was the bus for mission avionics, scientific instruments, radar antennas and ion thrusters – the latter in a pair of pods, each of which carried three large Hall-effect electrical engines, four ion thrusters and six smaller engines for attitude control. Small solar arrays would supply power until the reactor was activated.

The 'launch campaign' would start in May 2015 by placing a number of stages in orbit loaded with propellant. They would dock and await the Prometheus Spaceship. Assembly of the stack would take place between October 2015 and January 2016, at which time the vehicle would fire and jettison the rocket stages to depart. Once clear of Earth, the spacecraft would extend its reactor boom and radiator, activate its core and start thrusting with its main engines. (An alternative scenario was to use the ion engines to depart but such a spiraling escape would take a long time.) Over the next few years, JIMO would thrust to match Jupiter's orbit around the Sun and gently slip into a distant orbit around that planet in May 2021. It would then initiate a ballet of thrusting phases and gravity-assists designed to enable it to spend 60 days at each of Callisto and Ganymede and then 30 days at Europa. During this tour, the spacecraft would undertake medium-resolution global mapping of Io and monitor

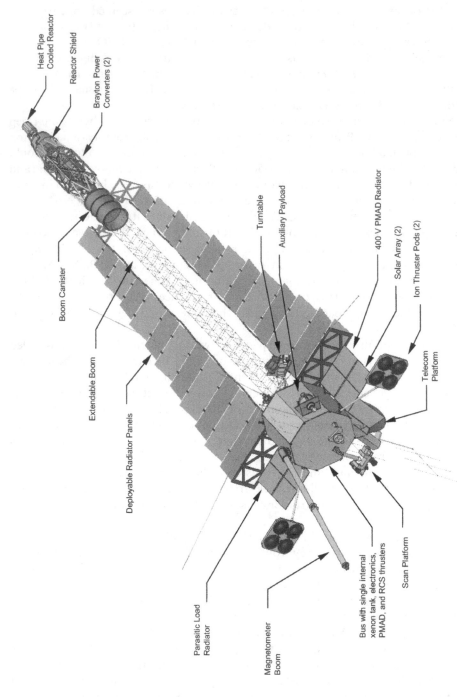

Heat Pipe Cooled Reactor

Reactor Shield

Brayton Power Converters (2)

Boom Canister

Extendable Boom

Deployable Radiator Panels

Parasitic Load Radiator

Magnetometer Boom

Bus with single internal xenon tank, electronics, PMAD, and RCS thrusters

Scan Platform

Turntable

Auxiliary Payload

400 V PMAD Radiator

Solar Array (2)

Ion Thruster Pods (2)

Telecom Platform

One possible configuration envisaged for the nuclear-powered Jupiter Icy Moon Orbiter.

its volcanic activity. However, to achieve the recommended Io imaging resolution of 1 km from a position orbiting Callisto would require a telescope with optics 3 meters in diameter! Relaxing this requirement to 1 km from a position orbiting Ganymede would reduce the optical system to a more realistic 1-meter diameter. While traveling from one icy moon to the next, the spacecraft would study Jupiter's dynamic atmosphere.

One possible follow-on Prometheus mission was the Neptune Icy Moons Orbiter (NIMO), using a Jovian slingshot to reach Neptune in about 15 years, entering orbit to conduct a 3-year study of that planet, and then entering orbit around its large moon Triton. As heavy radiation shielding would not be required at Neptune, the scientific payload could be increased to 3,000 kg; sufficient to carry a pair of landers. These RTG-powered long-duration landers would be deposited on the surface of Triton using a rocket system analogous to the Mars Science Laboratory's 'Skycrane' (of which more in the next chapter) and they would have a panoramic camera, spectrometers to analyze the surface, a gas chromatograph and mass spectrometer to analyze the atmosphere, a meteorological station to study the tenuous atmosphere in detail, and (because Voyager 2 observed geyser activity in 1989) a seismometer to measure microseisms.[367]

Meanwhile, in the aftermath of the Columbia accident in 2003 NASA was obliged to realign its priorities in response to President George W. Bush's instruction that the agency should resume human exploration of the Moon as a precursor to a mission to Mars. As a result, in February 2004 Prometheus was transferred from the Office of Space Science to the Exploration Systems Mission Directorate which NASA created to implement this new strategy, and the project was told to support a nuclear lunar surface power plant. For its part, JPL resumed its internal studies of a conventional Europa orbiter. A preliminary mission and systems review of JIMO was planned for 2005, but by then NASA had announced that it was considering alternatives and was eager to seek mission profiles that would involve considerably lower technical, schedule and operational risks. This implicitly recognized that the JIMO plan was unrealistic. Such alternatives might include nuclear-powered Mars communications satellites or Moon orbiters, Venus high-resolution radar mappers, missions to the asteroids, solar polar orbiters, and near-Earth sentinels against threatening asteroids or solar storms. These first Prometheus missions would have reactors delivering just 20 kW, shorter lifetimes and less stringent radiation protection requirements. The spacecraft would be capable of being launched whole, eliminating the need for autonomous docking and assembly. However, studies showed that these alternative missions would not be significantly cheaper than predicted for JIMO. In 2006 the budget for Prometheus was reduced, to transfer resources to the development of the successor to the Space Shuttle. This effectively killed Prometheus. With its annual budget cut from $270 million to just $10 million, NEP was unlikely to get off the ground anytime soon! By the time the Prometheus project was shut down, it had consumed a total of $464 million.[368] In an ironic twist, just as Prometheus was killed NASA drew up the specifications for the Ares V heavy-lift launcher to dispatch the new manned lunar lander. This rocket could have lifted JIMO in one piece and boosted it to Earth-escape velocity. In fact, it does not take much to see why JIMO was destined to fail: it was proposed for a political agenda rather than a scientific one, and it was a weak agenda at that; it could not be launched in one go by any existing rocket; and at up to $16 billion (not including the cost of developing the nuclear reactor and its associated systems, and the cost of the multiple launchers per mission) it was exceedingly expensive, ruling out a variety of other and more worthy investigations.[369],[370]

The United States was not the only country to study nuclear electric propulsion. In fact, the former Soviet Union had launched many fission reactors into space at the height of the Cold War to power its ocean reconnaissance satellites, so Russia was in principle better able to implement this revolutionary technology. Indeed, as early as 1989 the 'Generation' research center established by the Lavochkin Association and the Moscow Aviation Institute (MAI), with the assistance of the IKI (Institut Kosmicheskikh Isledovanii; the Russian Institute for Cosmic Research), started working on a NEP solar probe as an alternative to the conventional YuS 'Tsiolkovski' Jupiter and solar probe.[371] The NEP Tsiolkovski was to be powered by a 'Topaz' reactor with an output of 100 kW. Behind the reactor would be a telescoping conical radiator and the bus incorporating a cluster of electrical thrusters. In this configuration, Tsiolkovski would have a mass of at least 15 tonnes, of which 400 kg would be scientific instruments. On being inserted by a Proton rocket into orbit at an altitude of 800 km, the spacecraft would use its engine to spiral away from Earth and then slowly alter the eccentricity of its orbit to reduce its perihelion and turn its inclination perpendicular to the ecliptic. The conventional form of this mission would have used a Jupiter slingshot to shape its orbit.[372] Contemporary with JIMO, the Keldysh Research Center and the Lavochkin Association conducted a preliminary feasibility study for a Topaz-powered NEP orbiter for Europa. The 9.6-tonne spacecraft was to be launched either by a Proton or the proposed Angara. On arriving at Jupiter 4 to 5 years later, it would unfurl a huge antenna for a 30-kW radar capable of penetrating the icy shell of Europa to a depth of 100 km, possibly sufficient to map features on the floor of the subcrustal ocean.[373],[374]

THE STEREO SUN

The Solar Terrestrial Relations Observatory (STEREO) was the third project in an overlapping sequence of flexible and cost-capped missions to study the Sun and its interactions with Earth for the Solar Terrestrial Probes Program of NASA's Science Mission Directorate Heliophysics Division. The mission was first studied in the late 1990s, and development started in 2001 as a joint venture by APL and the Goddard Space Flight Center. The aim was to put two spacecraft into solar orbit, one ahead of Earth and the other trailing behind it to act as two separate 'eyes' in monitoring the photosphere (the visible 'surface' of the Sun) and corona to study how coronal mass ejections form, evolve and travel in the solar system. The capability of determining by 3-dimensional imaging where a coronal mass ejection was heading would provide forewarning of its arrival at Earth. This would be particularly valuable for operators of satellites in geostationary orbit, since these can be disabled by such eruptions; or for preventing damage to electric power lines on Earth. The mission could also contribute to one of the 'hot' scientific debates, namely the extent to which the Earth's 'global warming' was the result of human activities as against external sources such as solar activity.

Including propellants each STEREO spacecraft was 620 kg. The pair were to be launched together, one above the other, by a Delta II and then released into highly elliptical orbits with their apogees just beyond the orbit of the Moon. Several months later, one spacecraft would be steered into a lunar flyby that would deflect it into a solar orbit having a period slightly less than that of Earth, causing it to slowly draw ahead as STEREO-Ahead. One month later, the other spacecraft would make a

The two STEREO spacecraft mounted one on top of the other prior to launch.

lunar flyby that would deflect it into an Earth-lagging solar orbit as STEREO-Behind. As the mission progressed, the separation between the each spacecraft and Earth would slowly widen, giving an ever growing baseline for 3-dimensional observations of the inner solar system. After 2 years, this being the nominal duration of the mission, the angle between the Sun and the two spacecraft would be a 90-degree angle. Of course, sufficient propellant was carried to permit several additional years of operations.

The main difference between the two spacecraft was that the one on the bottom of the stack was reinforced to support the weight of the other at launch. The rectangular 1.1 × 2.0 × 1.2-meter bus supported solar arrays which spanned 6.5 meters and were rated for 596 W at the start of the mission. The bus had reaction wheels and 4.4-N hydrazine thrusters for 3-axis stability. The thrusters also targeted the lunar flyby. A 1.2-meter-diameter high-gain antenna provided a data rate to Earth of 720 kbits per second, but the arrangement of the payloads differed to enable the antennas to remain pointing at Earth and the instruments at the Sun as each spacecraft moved away from Earth with a different geometry.

Each spacecraft had a total of 13 instruments, with a solid-state memory to store up to 8 Gbytes of data when not in actual contact with Earth. The Sun–Earth Connection Coronal and Heliospheric Investigation (SECCHI, a tribute to the pioneering solar astronomer Angelo Secchi) package comprised extreme-ultraviolet and heliospheric imagers to observe the chromosphere (immediately above the photosphere) and the inner corona, and two coronagraphs, one having a field of view of the corona spanning 15 solar radii and the other only the innermost 3 solar radii. Its role was to identify coronal mass ejections at the moment of their eruption

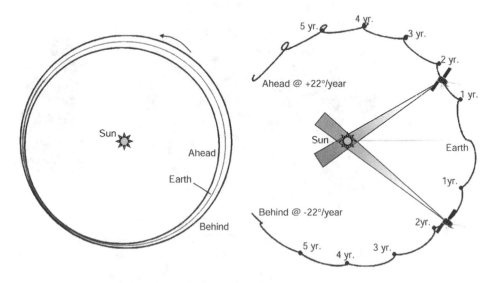

The orbits of STEREO-Ahead and STEREO-Behind relative to Earth, with (at right) their motions shown in a reference frame that is fixed on Earth.

and then follow their evolution as they moved out into interplanetary space. A suite of in-situ instruments consisted of four sensors on the body of the vehicle and three sensors at the end of a 60-meter boom to measure magnetic fields and electrons, ions, protons and energetic particles in the solar wind. A related instrument package measured the characteristics of plasma, protons, alpha particles and heavy ions. The final experiment (which followed French instruments on Soviet Mars missions in the 1970s and later on Ulysses) used a trio of monopole antennas to detect radio waves emitted by the Sun. Including the bus, instruments, launch and 2 years of operations and data analysis, the mission was budgeted at about $550 million.[375,376]

Launch was scheduled for April 2006 but a strike at Boeing (which supplied the rocket) caused it to be postponed, first to no earlier than June and then to October. As was usual for solar missions, the launch window was not particularly tight, and in this case the main constraints were the orbit-shaping lunar flybys – the injections of the two spacecraft into solar orbit would be refined to suit the actual date of launch. In the event, STEREO was launched at the very first attempt on 26 October 2006, and within half an hour the two spacecraft had been released into an orbit which had its apogee just beyond the radius of the Moon's orbit. On 15 December, following 2 months of testing, they had their first encounter with the Moon with STEREO-Behind passing at an altitude of 10,745 km and STEREO-Ahead at 5,937 km. As a result of this flyby, the apogee of STEREO-Ahead's orbit was extended to 1.75 million km to position it so that the Sun could draw it into the final 0.95×0.97-AU solar orbit with a period of 344 days. STEREO-Behind remained in Earth orbit until it encountered the Moon on 21 January 2007, when it made a flyby at an altitude of 16,029 km that injected it into a 0.99×1.09-AU solar orbit with a period of 389 days. Meanwhile, on 27 December 2006 the mission observed its first coronal mass ejection. Then, in January 2007 the heliospheric imager observed the bright comet McNaught passing through its field of view and detected what appeared to be a tail of iron atoms, which was the first time that such a thing had been found to accompany a comet.[377]

Over the next few years, STEREO provided some important contributions to solar physics, detecting and mapping a number of coronal mass ejections and cooperating with Earth-orbiting solar satellites and other probes. The mass ejection of 4 February 2008 was significant because by then the two STEREO spacecraft were separated by a sufficient angle to enable the full 3-dimensional kinematics of such an event to be reconstructed for the first time. To STEREO-Behind, the mass ejection occurred near the center of the solar disk and swept by both the spacecraft and Earth. From the point of view of STEREO-Ahead, however, the site of the event was nearer the limb of the Sun and the spacecraft was unaffected.[378] Observations of several dozen eruptions by STEREO led to the discovery that solar magnetic fields always constrain the plasma erupted in a similar manner, leading to what has been called the 'croissant model'. As expected, the mission was granted an extension.

In addition to the fortuitous observation of comet McNaught, STEREO has made some interesting and unprecedented studies of small bodies in the solar system. For 10 days in April 2007 Encke's comet was in view of STEREO-Ahead. Not only did the images reveal the presence of a surprisingly long tail of plasma extending in

The tail of comet McNaught crossing the field of view of STEREO.

excess of 10 million km from the small comet, they also recorded how the tail was swept away and then recreated by a coronal mass ejection. This 'tail disconnection' event seemed to be related to the magnetic field entrained in the mass ejection, rather than to the pressure of the plasma which it carried.[379] A remarkable event occurred in 2009, as the still fully operational spacecraft reached the stable Lagrangian points leading (L4) and trailing (L5) Earth by 60 degrees. Starting in March, the two vehicles were 'rolled' in order to obtain 'deep' images of these zones in a search of possible Trojan asteroids accompanying Earth around the Sun. It has been suggested that early in the history of the solar system a Mars-sized planet formed at one of these points and later impacted the proto-Earth and the ejected debris created the Moon. This hypothetical planet was informally named 'Theia', after the mother of Selene (the Moon) in Greek mythology. Any Trojans found today might be remnants of this event. In June 2009, STEREO-Ahead was used to image the near-

Earth asteroid Phaethon as it reached perihelion at 0.14 AU. Images of the asteroid, which, being so close to the Sun was impossible to see from Earth, revealed for the first time the release of a small cloud of dust. Phaethon was thus confirmed to be a rocky comet producing dust from the decomposition of hydrated minerals and from cracks in its surface around perihelion, when its temperature must exceed 700°C.[380] Images of the vicinity of the Sun taken by STEREO-Ahead during a 40-day period of low solar activity in 2008 and 2009 were carefully examined to search for Vulcanoids. A number of solar system objects were detected, including planets, dozens of comets and main belt asteroids, but nothing within the orbit of Mercury. This experiment further constrained the existence of Vulcanoids: no object larger than 6 km should exist, for it would have been detected, and no more than a dozen or so larger than 1 km, if any at all.[381] STEREO-Ahead has also captured the comet-like tail extending downstream of Mercury. Although a sodium tail had been known and observed from Earth for years, the tail seen by STEREO did not seem to be compatible with a sodium-only tail. Closer to Earth, images taken using the heliospheric imager showed slightly brighter areas of space corresponding to the orbit of Venus. This was proof of the existence of a ring of very fine dust orbiting the Sun and trapped by the gravity of Venus. A similar ring exists at Earth's orbit, and the ring for Venus had already been tentatively detected by instruments on the two Helios solar probes of the 1970s.[382] The mission also discovered a large number of 'sungrazing' comets, small objects probably at most several tens of meters in size that disintegrate as they approach within grazing distance of the Sun. Several longer-period comets were also discovered in the field of view of the coronagraph, which, along with a similar instrument on the Solar and Heliospheric Observatory (SOHO), witnessed the demise of the sungrazing "comet of the century" ISON as it made its perihelion passage in November 2013.

On 6 February 2011 the two STEREO spacecraft were exactly 180 degrees apart in their orbits, providing for the first time ever a view of the entire Sun. Moreover, as the two probes started to re-approach each other over the hemisphere of the Sun opposite from Earth, the entire surface continued to be monitored because NASA's Solar Dynamics Observatory in Earth orbit and ESA's SOHO at the Sun-Earth L1 Lagrangian point provided a near-side view. By early September 2012 they formed an equilateral triangle with Earth and Earth-orbiting solar telescopes.

In 2015 the two spacecraft will reach a point opposite to the Earth, on the far side of the Sun, pass one another, and start their return to the Earth's vicinity.

THE ORIGIN OF THE SOLAR SYSTEM

In 2000, one year after selecting Deep Impact and MESSENGER for the Discovery program, NASA invited another round of proposals, received 26 submissions and selected three for further study: the INSIDE Jupiter orbiter (one of the finalist in the 1998 round), the Kepler space telescope, and Dawn, which was a joint effort by JPL and the University of California at Los Angeles to rendezvous with and orbit around two asteroids in the main belt in order to conduct investigations relating to the origin of the solar system. Also under consideration was US participation in the French Mars Netlander mission. In December 2001 NASA announced the selection of both Dawn and Kepler as the ninth and tenth Discovery missions, respectively. Dawn has the distinction of being the first mission to attempt to enter orbit around more than one target.

The Dawn mission was made possible by proven ion propulsion technology and a fortunate alignment of two of the most interesting minor planets: Ceres and Vesta. It was to undertake morphological and gravity mapping, and determine the precise mass, shape and composition of each object. As the first and third largest members of the main belt, Ceres and Vesta are 'protoplanets', intermediate in size between small asteroids with the 'pristine' composition of the solar nebula and the planets whose material has been thermally evolved. Hence they ought to provide insight into the process of planetary formation. In addition, each had its peculiarities. Ceres seemed to be more of a 'wet' transition object with many of the same characteristics of the icy satellites of Jupiter, with indications of the action (if not the presence) of water ice. Vesta was more similar to the dry rocky bodies of the inner solar system, and the fact that it was the source of basaltic meteorites indicated that its material was differentiated and had been subjected to volcanic processes. One objective was to map 80 per cent of their surfaces at resolutions of 100 meters for Vesta and 200 meters for Ceres. Interestingly, their surfaces may preserve a history of cratering that may yield insight into the theory that Jupiter condensed in the outer solar system and migrated inward to its current orbit.[383]

Ceres was discovered on 1 January 1801 by Giuseppe Piazzi in Sicily. Orbiting at an average heliocentric distance of 2.77 AU it is only slightly less than 1,000 km in

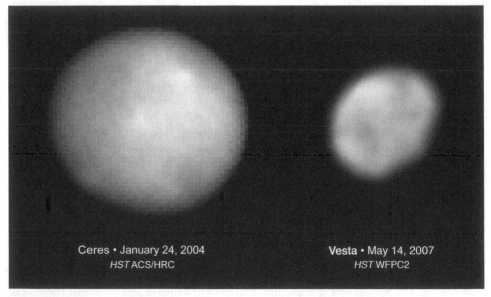

Ceres • January 24, 2004
HST ACS/HRC

Vesta • May 14, 2007
HST WFPC2

A composition of asteroids Ceres and Vesta (to scale) seen by the Hubble Space Telescope.

diameter, being 899 × 959 km (or 909 × 975 km according to another source). It was classified as a G-class object, a relatively rare type bearing similarities to the C-class. But little was discovered about it until the 1970s owing to the intrinsic difficulties of observing such a small body and the fact that its spectrum was fairly flat. In that decade photometry by both ground-based and space-based instruments measured the surface reflectance of both Ceres and Vesta, and they were investigated by radar. The most important breakthrough was ultraviolet spectroscopy of Ceres obtained in the early 1980s showing an absorption band believed to be caused by hydrated minerals. In addition to carbonates, it appeared to be covered with a dry clay-like material in a layer at least several centimeters thick. In fact, Ceres is similar in density to water-rich Callisto and Ganymede. Models calculated that water ice should be present at a depth no greater than several tens of meters from the surface, and spectral emissions by the water-related OH radical were seen from Earth. The discovery in the mid-1990s of a small number of 'main belt comets', seemingly ordinary asteroids that develop comas and tails at any point of their (almost circular) orbits, lent support to the theory that some asteroids can retain significant reservoirs of volatiles which can be exposed by impacts. More recently, observations by the 8-meter telescope of the Keck Observatory in Hawaii and by the Hubble Space Telescope resolved details on Ceres as fine as 50 km (giving an image with a resolution equivalent to a naked-eye view of the Moon). These observations refined estimates of its shape and measured its rotation period at just over 9 hours. Two roundish spots situated at opposite longitudes in the northern hemisphere could mark major impacts. One spot, 180 km wide, showed a bright center that could be a central peak. The other has been informally named Piazzi in honor of the asteroid's discoverer.[384] High-resolution observations also confirmed the lack of surface relief, which was consistent with a relatively fluidic ice-rich crust, possibly containing more water than all the terrestrial oceans. In 2006 Ceres was cataloged (together with such outer solar system bodies as Pluto and Eris) as a 'dwarf planet', an ill-defined class of objects that are neither planets since they have not gravitationally swept their vicinity clear of other objects, yet are sufficiently massive to be spheroidal, in contrast to the more irregular small asteroids.

Vesta was discovered by Heinrich Olbers on 29 March 1807. Although the fourth member of the main belt to be identified, it is the brightest and is actually visible to the naked eye against a very dark sky. It orbits at an average heliocentric distance of 2.34 AU. Its spectrum shows a heterogeneous surface and the presence of materials indicating that it must have differentiated to some degree during its history, which is remarkable. This discovery, made in the early 1970s, was very significant because it provided the first link between an asteroid and three different classes of basaltic or volcanogenic meteorites. Moreover, several other asteroids have been found to have a similar spectrum to Vesta, and together they constitute the taxonomic V-class. In the 1980s attempts were made to correlate the variability of Vesta's spectrum with its rotation in order to make crude maps of its surface composition in terms of classes of meteorites and types of minerals. Hubble Space Telescope observations showed Vesta to be ellipsoidal, 578 × 560 × 458 km in size, rotating in about 5.3 hours. The non-spherical shape was mainly the result of a crater 460 km wide and 13 km deep that removed more than half of its southern hemisphere in an impact with an 80-km object that must have come close to shattering the asteroid. The ejected debris probably accounts for the V-class of asteroids and the meteorites from Vesta. The geological context provided by the Dawn observations was expected to provide the first confirmation of the link between a class of meteorite and their parent body. Moreover, peering into the crater, Dawn would hopefully discern details of the differentiated interior of the asteroid, as rocks from the mantle were expected to be exposed on the slopes of the central peak. In a sense, Dawn would be lunar or Mars exploration in reverse, with global reconnaissance coming after sample return (via meteorites in this case).

The masses of Ceres and Vesta were fairly well determined by the perturbations they exert on each other's orbits and on those of Mars and of smaller objects. For example, as early as 1968 it was noted that every 18 years Vesta approached within 6 million km of the smaller asteroid (197) Arete, and measurements of their mutual perturbations gave the first reliable estimate of Vesta's mass. It was then far more precisely determined during the 1970s by tracking the Viking landers on Mars. Vesta was also involved in the first observation of an asteroidal occultation when it passed in front of a background star in 1958. By precisely timing the duration of an occultation as viewed by different observers, it is possible to directly measure the shape and size of the occulting object. In 1984 this technique was applied to Ceres.[385] Vesta's density is twice that of Ceres, consistent with it being of a rocky rather than an icy constitution.

As demonstrated by studies over more than two decades, a mission to orbit one main belt asteroid would be very difficult using conventional chemical propulsion due to the large amount of propellant required to rendezvous with and enter orbit around such a body. Even to do so for a single asteroid would require a more powerful (and so more expensive) rocket than the Delta II-Heavy earmarked for Dawn. To attempt two would have been out of the question. By sharing part of its scientific and engineering team, the ion propelled mission sought to capitalize on the experience of Deep Space 1. JPL was to supply the entire ion propulsion system and Orbital Sciences would build the spacecraft hardware and integrate its payload.

Apart from its propulsion system, Dawn, based on a communication satellite bus, was fairly conventional. It was an aluminum and carbon composite honeycomb box 1.64 × 1.27 × 1.77-meters in size, built around a carbon composite cylindrical thrust tube that enclosed tanks. The xenon fuel tank was 1 meter across, and consisted of a carbon composite shell with an internal titanium liner that was less than one-tenth of a millimeter thick. There was a fixed 1.52-meter-diameter high-gain antenna on one side of the bus, and a trio of low-gain antennas positioned to maintain communication at times when precise pointing was not possible. The system would provide a data rate of 124 kbits per second for the scientific phases of the mission. Attitude determination was to be achieved by a fully redundant system consisting of two star trackers, triple gyroscopic and accelerometer platforms, and no fewer than 16 Sun sensors. Attitude control would be by a combination of four redundant reaction wheels, ion engine thrust and a dozen 0.9-N hydrazine thrusters

The Dawn spacecraft with one solar panel deployed, during ground preparations. One of the three ion thrusters is visible at the bottom.

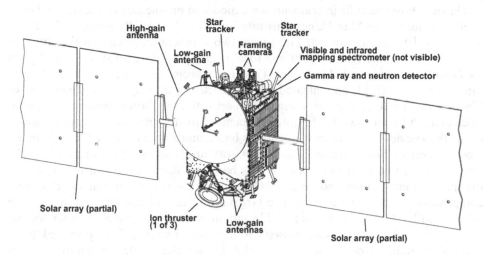

The Dawn asteroid orbiter.

in two groups of six. As with the pioneering Deep Space 1, Dawn required a lot of power to operate its ion propulsion system. A further complication was the need to continue firing even at large heliocentric distances. The bus mounted a pair of gimbaled solar wings, each with 5,740 solar cells. With an efficiency rating of about

28 per cent, the system was capable of delivering 10.3 kW of power at 1 AU, but just 12 per cent of this at Ceres. With its 10-segment accordion-like wings fully deployed, Dawn had a total span of 19.74 meters, making it one of the largest interplanetary vehicles ever launched by the US. A nickel-hydrogen buffer battery was available for launch and for transients in power consumption at large heliocentric distances.

The ion propulsion system would make all maneuvers during cruise, rendezvous, orbit insertion and shaping. To provide the required change in velocity, the engine would operate for two-thirds of the 8-year duration of the mission. To assure 2,000 days of thrusting (fully three times Deep Space 1's total firing time) and still have sufficient margin of capability, the ion propulsion system included three identical thrusters – but only one would be operated at any given time. These 30-cm-diameter xenon engines had 112 throttle levels with thrusts between 19 and 91 mN, although the upper part of this range would not be achievable beyond a heliocentric distance of about 2 AU due to power supply limitations. Like Deep Space 1, the engines were gimbaled on two axes to be able to accommodate shifts in the center of mass as the mission progressed and to provide attitude control. They were all mounted on the base, with one on the central axis and the others offset (one towards the instrument side and the other towards the high-gain antenna side) and angled in order to deliver their thrust through the center of mass. One of the technological issues was whether the propulsion system would be able to run reliably for years of continuous operation. In preparation for the mission, an ion engine left over from Deep Space 1 was run for over 30,000 hours to evaluate its performance and rate of degradation.

Dawn's three scientific instruments were mounted on one side of the bus, in body-fixed positions. The Max Planck Institute for Solar System Research in Germany developed the framing camera. There were two identical units, each with refracting optics of 19 mm aperture and 150 mm focal length providing a 1,024 × 1,024-pixel CCD with a 5.5-degree field of view. The filter wheels had one clear slot and seven filters across the visible and infrared ranges selected to investigate the mineralogy of Vesta. Only a few of the filters would be useful for the flatter spectrum of Ceres. Each camera had its own 8-Gbit internal memory to store raw data for transmission to Earth. The design of this instrument was based on those carried by Venus Express, Rosetta and its Philae lander. The gamma-ray and neutron spectrometer was the only US-built instrument onboard. It was based on hardware flown on the Lunar Prospector lunar orbiter and on Mars Odyssey, and was to determine the elemental composition of the surfaces of the two target asteroids and also detect water ice and other volatiles. Italy contributed a visible and infrared mapping spectrometer derived from instruments on Cassini, Rosetta and Venus Express. This provided good spectral resolution from the near-ultraviolet through the visible range into the near-infrared, and was to identify minerals on the surfaces of the asteroids with good spatial resolution. The combined observations of the spectrometers would facilitate identification of a wide variety of hydrated minerals and enable the role of water in the geological evolution of Ceres to be characterized. It might then be possible to deduce that certain meteorites originated from Ceres. Of course, the Doppler monitoring of the radio signal would chart the gravity fields of the asteroids to determine their masses, moments of inertia, localized mass concentrations, internal structure, etc. A magnetometer was to have been carried, but it had to be sacrificed early in the design phase to accommodate an increase in the cost of the mission. A laser altimeter was deleted even earlier.[386] An intrinsic magnetic field, possibly the fossilized remnant of the period when the asteroid had a molten metallic core, has been proposed to explain the fact that the surface of Vesta seems to show less evidence of the effects of 'space weathering' than the lunar maria, even after taking into account the lower density of the solar wind at Vesta's distance from the Sun. Moreover, some of the meteorites that originated from Vesta have been found to have retained some magnetization. Not having a magnetometer, however, Dawn was not able to evaluate this possibility.[387]

The spacecraft's launch mass was 1,218 kg, including 425 kg of supercritical xenon for the ion propulsion system and 46 kg of hydrazine for the attitude thrusters. This was sufficient for a total velocity change over the full duration of the nominal mission of about 11 km/s, which, remarkably, was comparable to that provided by the Delta launcher.[388,389,390] Although celestial mechanics and the flexibility of the low-thrust propulsion trajectory would allow a broad launch window opening before 2006 and closing in October 2007, the target was the second half of June 2006. The opening of the 20-day window reflected the readiness of the spacecraft team and budget constraints. Coincidentally, the broad launch window included the 200th anniversary of the discovery of Vesta. A launch in June would ensure an arrival at Vesta in July 2010 for an 11-month reconnaissance, after which the vehicle would cruise to reach Ceres in August 2014. A delay beyond late 2007 would rule out attempting a two-asteroid mission until the 2020s. Unfortunately, as with other recent Discovery missions, Dawn suffered several problems during its development that were so significant that it only narrowly escaped being canceled. Two specimen xenon tanks breached while being tested at pressures lower than that for qualification. Moreover, JPL was having difficulties converting an Earth-orbiting spacecraft design to a deep-space spacecraft design, and with the ion thrusters and associated power conversion and distribution units. In the case of Deep Space 1, the thrusters had been supplied by the Glenn Research Center, and by now the team which designed the NSTAR thrusters had dispersed and the vendors for some of the components had ceased trading.

In September 2005 the project asked NASA for a funding increment to complete the development, and headquarters demanded that work be halted while an independent review considered whether the mission was feasible as envisioned and calculated the cost implications. By then, the spacecraft was almost 90 per cent complete. The review reported in January 2006 that although Dawn had been poorly managed, there was no particular technical reason to prevent it from achieving its objectives so long as realistic funding was awarded. The mission could be launched one year late and be completed for an additional $73 million, inflating its total cost to $446 million. But on 2 March NASA canceled the mission. The international team appealed directly to NASA's Administrator, and was granted a thorough question-and-answer session between agency representatives and mission and JPL managers. On 27 March NASA reinstated the mission with an increased budget and a launch in the early summer of 2007.

The solution to the problem with the tank involved reducing the xenon load from 450 to 425 kg (which still enabled the full mission to be carried out) and reducing its preflight ground temperature from 40°C to 30°C. To make up for the delay, a Mars flyby was introduced. In fact, a flyby had been a considered early on as a way to increase the spacecraft's mass margins. The issue with the power units proved to have been a result of a poor test arrangement and an incorrect repair. Finally, in order to preclude another dispersal of knowledge on how to design an ion propulsion system, the Dawn team was asked to improve its archival documentation.[391]

Dawn arrived at Cape Canaveral on 11 April 2007 to undergo final preparations. Owing to delays in the production of the Delta launcher the first launch attempt was scheduled for 30 June. It then slipped to 7 July as a result of problems with hardware on the pad. On 11 June, whilst mounting Dawn on a spin table for final balancing, a wrench inadvertently slightly dented the back of a solar panel – fortunately without damaging the solar cells. Lift off was then delayed by thunderstorms, generally poor weather, and difficulties with an aircraft and a ship intended to relay telemetry. The potential effect of further slippage on preparations for the launch of the Phoenix Mars lander (which had a narrow window in August) prompted NASA to postpone Dawn to a window that opened on 26 September. If the mission did not get off by 15 October, it would no longer be able to reach both Vesta and Ceres. Dawn was

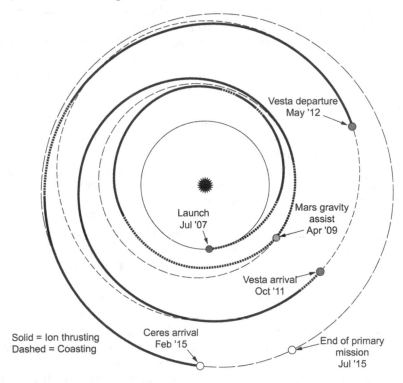

The trajectory of Dawn from Earth to Ceres via Vesta (and Mars). The dates are not definitive.

removed from the Delta and stored in a clean room while Phoenix was dispatched from an adjacent pad. When the effort switched back to Dawn, adverse weather precluded the first launch attempt. On 27 September 2007 a ship strayed into a restricted area downrange and caused a brief hold in the countdown, but once the way was clear the clock resumed and the vehicle lifted off without incident.

After a coast of some 40 minutes in a low orbit, the second and third stages fired in succession and Dawn was released into a 1.00×1.62-AU solar orbit. Several minutes after separation, the spacecraft fired its thrusters to cancel its inherited spin, allowed the xenon fuel to settle, and then opened its solar panels to face the Sun. The systems were tested during a 2-month calibration phase. The first 27-hour test firing of the central ion thruster on 6 October served to measure its performance and to monitor relevant parameters at several different throttle levels running up to the maximum. The angled thrusters were also tested. For these tests, the spacecraft thrusted towards Earth and the resultant Doppler shift was measured to calculate the acceleration imparted. While engineers checked the engines, scientists tested their instruments. The camera was first activated on 18 October and calibrated over the following months by taking pictures of star fields in the constellation of Cancer, standard reference stars, Saturn and miscellaneous star fields, during which it obtained some excellent pictures of the Eta Carinae nebula.[392] On 17 December, after all required tests had been successfully concluded, Dawn oriented itself to point its central engine in the optimal direction to commence its ion propelled mission. As in the case of Deep Space 1, the cruise would be a succession of long periods of thrusting interrupted on a weekly basis for telemetry dumps during which thrusting would be suspended in order to point the high-gain antenna at Earth, interleaved with longer periods of ballistic flight for housekeeping and miscellaneous calibrations. The propulsion system functioned perfectly, and the first thrusting period lasted over 10 months.

On 8 August 2008 Dawn reached aphelion at a point beyond the orbit of Mars. On 31 October the engine was turned off to initiate about 7 months of ballistic cruising which would include the Mars flyby. In thrusting for 82 per cent of the time since December 2007, it had consumed 72 kg of xenon and achieved a 1.81-km/s velocity change which placed it into a 1.22×1.68-AU orbit. A 2-hour correction using the ion propulsion system on 20 November refined the Mars flyby in February to an altitude of 542 km; slightly higher than the planned 500 km. As the error in the post-encounter trajectory would easily be within the capacity of the ion engine to correct, an inbound refinement was deemed unnecessary. The spacecraft spent the end of 2008 in solar conjunction. As Dawn headed in from aphelion, it crossed the orbit of Mars on 18 February 2009. It approached from the night-side, with the Sun occulted by the planet. It sped by at a relative speed of 5.31 km/s, with the closest point of approach above the volcanic region of Tharsis. On crossing onto the day-side, it was to take pictures of regions over which the Mars Express orbiter would pass an hour later. This would enable the performance of the cameras on both vehicles to be cross-checked. Other calibration imagery was to involve smearing the view so as to obtain uniformly illuminated frames. The gamma-ray and neutron spectrometer was to take calibration spectra of Mars that could be compared with those of other missions orbiting the planet. Dawn was to continue imaging Mars on the outbound leg for a week or so to calibrate optical navigation for the approach to Vesta. In all, the plan called for about 1,600 pictures.[393] Unfortunately, a hitherto unsuspected flaw in the software of the attitude control system caused the spacecraft to enter safe mode soon after closest approach, halting the instrument calibration sequences. When the spacecraft was restored to operation 2 days later, the data which had been recorded prior to the anomaly was downloaded and provided some high-resolution imagery and excellent spectra by the gamma-ray and neutron spectrometers, but nothing from the imaging spectrometer. However, the main objective of the flyby had been the gravity-assist, which successfully deflected Dawn into a 1.37 × 1.84-AU orbit and steepened its plane relative to the ecliptic by more than 4 degrees in order to match the 7.1 degrees of Vesta's orbit. The 2.6 km/s of velocity change from this flyby was equivalent to an extra 100 kg of xenon for the ion propulsion system.

Dawn reached perihelion on 16 April 2009, shortly after new flight software was uploaded from Earth. On 8 June (after a 7-month coast during which the engines had been operated for only about 10 hours) thrusting was resumed, and in November Dawn entered the asteroid belt. By early June 2010, Dawn had beaten the 4.3-km/s total velocity change record of Deep Space 1. At about the same time, one of the reaction wheels started to develop excessive friction and was shut off to facilitate troubleshooting. But attempts to revive it were ineffective. Controllers opted to transfer attitude control to the hydrazine thrusters in order to save the remaining wheels for orbit reconnaissance at Vesta and Ceres. In fact, the orientation of only one axis would have to be controlled by chemical thrusters, as the other two would be controlled by gimbaling the ion thrusters. Moreover, software was developed that would allow the spacecraft to control its attitude using only two reaction wheels and thrusters.

During 2010, thruster number 2 fired for 304 days, consuming less than 79 kg of xenon. Engine number 3 was restarted in early 2011 for the final leg of the flight to Vesta. At about the same time, orbital mapping operations at Vesta were rehearsed. In March, thrusting was stopped and the instruments were checked and exercised. The cameras were tested after a long period of inactivity by taking pictures of a star field near the border between Pisces and Cetus. The thrusters were also calibrated and their minute push measured. After a week of tests, Dawn resumed thrusting for the encounter. The interplanetary cruise ended on 3 May, and the approach phase began. About 189 kg of xenon remained on board. The first navigation images of the target were taken on that same day. Vesta, about 1.21 million km distant, was still a bright dot only a few pixels wide. One week later, the mapping spectrometer took its first spectra. The gamma-ray spectrometer was also activated in early May. Attitude control was switched back over from thrusters to reaction wheels. These far more precise actuators would be used until July 2012, when Dawn would leave Vesta heading for Ceres. During the first 6 weeks of the approach, one imaging sequence was performed every week. On 1 June, Vesta was observed during one full rotation for the first rotation-characterization campaign. By mid-June, the imagery began to rival the best from the Hubble Space Telescope, and the frequency of imaging was

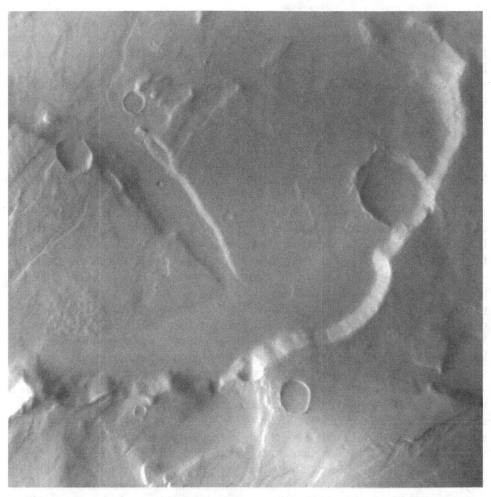

A 50-meter resolution view of the Tempe Terra region of Mars taken by Dawn. (NASA/
JPL/MPS/DLR/IDA, and the Dawn Flight Team)

increased to twice weekly. Dawn was then only as far from Vesta as the Moon is
from Earth.

Approach movies started to show hints of surface features and texture variation
on the asteroid, itself seen at an angle and phase relative to the Sun inaccessible from
Earth. The pictures were already sharp enough to show the central peak of the south
polar crater. Vesta appeared to be less spherical than other bodies in the solar system
of similar size. Unlike those, in fact, which included Enceladus and Miranda, Vesta
was known to be made of rigid rock instead of low-strength and 'plasmable' ice. It
was the largest completely unexplored object to be visited by a spacecraft since the
Voyager 2 flyby of the Neptunian system in 1989. These distant observations were
used to improve knowledge of Vesta's spin axis orientation, already broadly known
from terrestrial observations, in order to ensure that the spacecraft would be able to

enter a polar orbit. The asteroid was already known to complete a rotation in 5 hours and 20 minutes. The axial inclination was found to be significantly different than that determined from terrestrial and Hubble imagery, so much so that the northern spring equinox would occur later than expected, on 20 August 2012, when the spacecraft would have already departed Vesta if it stuck to the nominal plan. The visible and infrared spectrometer also started to obtain images and spectra to provide scientific context.

On 27 June Dawn experienced a sudden loss of thrust that was traced back to a reset in the ion propulsion control system that was attributed to a cosmic ray hit. This caused the xenon valves not to open properly, and the probe entered safe mode. The vehicle was back in operation by 30 June using the backup control and interface unit. Extensive tests carried out during July cleared the malfunctioning control unit. Lost were two imaging sessions, one on 28 June during the safe mode, and the other on 6 July in order to recover the lost thrusting and enter orbit when initially planned. In fact, the time of capture occurred about 15 hours earlier than planned because of this glitch. During the safe mode, moreover, the infrared spectrometer reset itself.

For the final leg of the journey, the thruster was switched from the number 3 back to the number 2. Just after Dawn was recovered on 1 July, the camera took more pictures of Vesta. The probe was then well south of the equator and had a direct line of sight of the central mound of the polar crater, centered at about 75°S, which, 180 km across its base, looked like a "gigantic belly button". The mound rose about 22 km above the surrounding terrain, making it the second tallest mountain in the solar system after Olympus Mons on Mars. It was unusually domed and much higher and wider than similar central peaks in craters on the Moon or Mercury. This appeared to be due to the curvature of the target body coming into play as the size of the crater approached its diameter. The surface of the southern hemisphere otherwise looked remarkably smooth, with only a few craters in the north, and smooth terrain around the equator. Apparently the giant crater was formed only in the last billion years, and showered the rest of the surface with debris that hid older features. On 9 and 10 July, during the second rotation-characterization campaign, Dawn also took three sequences of 72 images of the space around Vesta to search for any satellite. Although satellites were deemed likely for such a large, spheroidal body, none larger than several meters in size were detected. (A number of other main belt asteroids were detected moving in the background.)

In contrast to a conventionally propelled mission, there was no critical, fast-paced orbit-insertion maneuver. Instead, the ion thrusters gradually matched the vehicle's heliocentric speed with that of its target, so that it gently came under the asteroid's gravitational influence. Dawn smoothly went from orbiting the Sun to orbiting Vesta as it continued to thrust. This occurred at 04:48 UTC on 16 July. The precise time of the insertion could not be determined beforehand, since it depended on the relatively poorly known mass of the asteroid. When Dawn entered orbit around Vesta at an altitude of 16,000 km, it was at a relative speed of a mere 27 m/s (about 100 km/h). Vesta was 1.25 AU or 188 million km from Earth, and would be in opposition to Earth in late July. The spacecraft's ion thrusters had been on for 23,000 hours or almost 70 per cent of the mission so far, and had provided more than 6.6 km/s of

This image of *Vesta* from a distance of 5,200 km was taken by Dawn eight days after orbit insertion and shows the rough southern hemisphere and the equatorial grooves. (NASA/JPL-Caltech/UCLA/MPS/DLR/IDA)

A view of Vesta taken shortly after orbit insertion from a distance of 9,500 km. It looks down on the south pole and the central mountain of the Rheasilvia basin. (NASA/JPL-Caltech/UCLA/MPS/DLR/IDA)

velocity change, consuming 252 kg of xenon. Confirmation that Dawn had indeed entered orbit came more than 24 hours later when the probe suspended thrusting and turned to communicate with Earth. Along with the engineering data returned on this occasion were sequences of images taken earlier in the day.

The first image returned from orbit looked straight down at the polar crater and its central mound. The floor of the crater was crossed by parallel ridges and grooves, making it look quite like the Uranian satellite Miranda. Moreover, as with Miranda, there were kilometers-tall cliffs. Even more impressive grooves up to 15 km wide with steep walls and flat bottoms ran parallel to the equator, spanning two-thirds of the circumference of the dwarf planet. Individual troughs were up to 380 km long. These could have been formed by the south polar impact, which crushed the surface, making overlapping compressional wave-like tectonic features and ridges. However, the grooves looked more cratered than the southern crater, as if they were older. Or perhaps the floor of the giant crater had been resurfaced, making it appear younger. Conspicuously absent were smooth deposits of ejecta or pools of rock melted by the impact. There were also enigmatic fresh-looking small craters with dark streaks on their sides. The origin of the dark material was uncertain, and it could be either carbon-rich material deposited by low-speed impacts or excavated iron-rich volcanic material. Multispectral images taken at the same time implied compositional differences at the surface. The 475-km polar crater was officially named Rheasilvia after Rhea Silvia, the mythical Vestal virgin mother of Romulus and Remus, the founders of Rome. Other features would be named after Vestal virgins of Roman history and famous Roman women.

Three days were then dedicated to acquiring images, spectra and rotation movies. Dawn then flew over the night-side and one week after orbit insertion it was over the day-side on the northern hemisphere, where it was winter. The surface poleward of 52°N was in darkness. The visible portion of the northern hemisphere appeared to be more cratered than the southern terrains seen so far, and included three shallow, flat-floored craters in a pattern that resembled a snowman, but crater counts implied that the two hemispheres were about the same age. On 25 July thrusting was suspended at a height of 5,200 km in order to conduct scientific observations over the next four days. Full rotation movies were obtained, with the spacecraft flying over a range of different latitudes. On 2 August Dawn reached the survey orbit altitude of 2,700 km, in which it officially started scientific observations 9 days later. This activity would last seven 69-hour orbits, over about 20 days. Observations were mainly carried out using the camera and infrared spectrometer, while the gamma-ray and neutron detectors collected background data, the asteroid being still too distant to provide a significant signal. Data were collected and stored on board over the day-side and then relayed to Earth while over the night-side. This and subsequent orbits were designed to prevent the craft from entering the shadow of Vesta, so that the solar panels would continue to generate power. The two-way radio link between Earth and the spacecraft was used to precisely measure the Doppler shift and thereby map the gravitational field and internal structure of the asteroid. The mass in particular was established to be within a few percent of that determined from Earth. Gravity tracking confirmed Vesta to be a differentiated body, with a three-layered structure which included a high-density metallic core spanning 40 per cent of its diameter, a mantle, and a lower density, porous crust on average less than 20 km thick but reducing to zero in the deepest portion of the floor of Rheasilvia, where the mantle itself was exposed. Moreover, the gravity field appeared to correlate well with the topography. The metallic core was probably associated with a fossil magnetic field, but Dawn was not equipped to detect it. Although the infrared spectrometer had a glitch that prevented data collection on two out of seven orbits, it made in excess of 13,000 observations, or more than 3 million spectra, covering 63 per cent of the visible surface at 700-meter resolution. On the other hand, almost the entire illuminated surface was imaged by the camera at least five times at 260-meter resolution, producing a total of more than 2,800 images and 1,179 stereoscopic

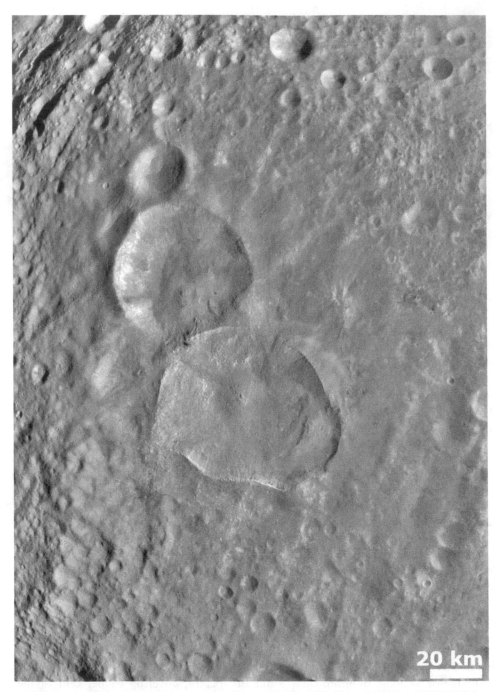

The 'snowman' cluster of craters on Vesta. (NASA/JPL-Caltech/UCLA/MPS/DLR/IDA)

images covering 80 per cent of the surface. Remote-sensing observations showed a geological variety unrivaled by any other minor planet visited by space probes, but despite the presence of pyroxene on the surface, a common volcanic mineral, failed to uncover any clear, unambiguous volcanic features such as cones, domes and flows.

In fact, the surface was entirely dominated and saturated with craters. Three-dimensional topographic reconstructions showed that the south polar crater had an '8' shape, and thus comprised two craters superimposed, with Rheasilvia being the more recent, sitting on an older, largely obliterated 400-km basin named Veneneia. Observations also showed the presence of a second, broader set of troughs north of the equator that had an older and more degraded appearance. Rheasilvia itself was mapped in detail, and found to lack many of the features of the large basins on the Moon or Mercury. It had a broadly hexagonal outer rim with steep scarps up to 20 km high, and its floor consisted of rolling plains with a roughly radial spiral pattern extending to the rim of the crater. Overall, it resembled craters on the icy satellites of Saturn, in particular Hyperion, Rhea and Iapetus. A clockwise spiral pattern of fractures was seen on the floor of the crater, consistent with a crater that formed under the influence of the rapid rotation of the asteroid. Other depressions were identified that could be ancient obliterated impact basins. There were also ponds of smooth material resembling those of Eros, although on a larger scale. The age of Rheasilvia agreed well with that of the asteroids sharing spectral features with Vesta, indicating that these are indeed ejecta from that huge impact, Spectra showed the surface to resemble all three known classes of Vestan meteorites. The northern hemisphere was enriched in one type and Rheasilvia in another, which was seemingly exposed from a greater depth by the impact. Such characteristics were evidence of a complex evolution of the magma layers that covered the asteroid's surface at different times, indicating a differentiation between the crust and the underlying mantle.[394,395,396,397,398,399,400]

The reconnaissance of Vesta by Dawn also triggered a mapping controversy between astronomers and planetary geologists. Whereas Vesta's prime meridian had been defined by the International Astronomical Union (IAU) on the basis of Hubble images as passing through the center of the dark Olbers Region, mission scientists established a different reference longitude system whose prime meridian crossed the prominent 700-meter-diameter crater Claudia, displaced some 155 degrees from the IAU prime meridian.[401]

On 31 August Dawn resumed thrusting to deliver a velocity change of 65 m/s and descend to the 680-km orbit with a period of about 12.3 hours in which most of the imaging would be undertaken, including color and stereo imaging by orienting the vehicle to point its camera sideward instead of straight down. It thrusted through 18 revolutions, pausing frequently for orbit determination because Vesta's irregular gravity field was not yet precisely known, until it reached this high-altitude mapping orbit on 18 September. After two short corrections and imaging sequences, Dawn restarted scientific observations on 1 October. It completed six 10-revolution mapping cycles, acquiring an almost complete stereoscopic topographic map of Vesta, as well as more than 7,000 pictures with pixel scales of 65 meters and 15,000 spectral frames covering all but the northern polar regions. From spectra acquired on the survey orbit and images collected during this phase, it was possible to determine that Vesta did not exhibit a lunar-like weathering and discoloration as a result of the formation of a microscopic coating of iron on the exterior of regolith grains exposed to the solar wind. The main evolution process of the surface seemed instead to be the deposition of layers of carbon-rich and water-rich material by impacts, and their subsequent spreading. This material constituted the dark areas of the surface, while pyroxene made the bright terrains. Hydrated minerals had already been reported on the surface of Vesta by terrestrial spectroscopic observations, but the measurements had been considered controversial because they contrasted with the composition of Vestan meteorites. Hundreds of dark, carbon-rich asteroids and comets would have crashed onto Vesta during the last 3.5 billion years and made a dark blanket several meters thick. Moreover, impacts in the asteroid belt would occur at relatively low speeds, several times lower than on Earth or on the Moon, implying that mechanical destruction processes would be at work in the creation of the regolith, rather than vaporization and melting. Olivine was expected within Rheasilvia, as it is found in Vestan meteorites whose composition is a match to the mantle material excavated by that impact. However, no olivine-rich areas were detected on the floor of the gigantic crater, implying that Vesta was never a fully differentiated and layered protoplanet. Instead, the infrared spectrometer detected olivine on the walls of the northern hemisphere craters Arruntia and Bellicia and in impact ejecta. This discovery could allow scientists to rule out some models of the formation and evolution of Vesta and its interior.[402,403,404,405,406]

The transition to the low-altitude mapping orbit began on 1 November, and would take more than five weeks to complete, Dawn pausing every 3 days in order to perform orbit measurements. During the descent, Dawn was briefly a synchronous satellite of Vesta, when its orbit period matched the rotation period of the asteroid. It also briefly entered safe mode on 4 December while turning to the attitude to fire the engine, and finally restarted observations on 12 December. The low-orbit phase was the longest one, with the vehicle on average 210 km above the irregular surface of Vesta. The low orbit had a period of 4 hours 21 minutes on average and was tailored to obtain high signal-to-noise gamma-ray and neutron spectra and gravitational field data using one of the low-gain antennas. Of course, remote-sensing instruments also returned images and data at high resolution. In fact, 13,000 images with pixel scales as fine as 20 meters and over 2.6 million spectra were obtained covering the whole of the illuminated surface of the asteroid. Weekly maneuvers ensured that Dawn never strayed into the shadow cone of the asteroid, for which it was not designed. In this low orbit, the gamma-ray spectrometer measured a composition consistent with Vestan meteorites and confirmed that the interior of Rheasilvia had a different composition from the remainder of the surface, indicating a differentiation of the crust prior to its being excavated by that impact. The most amazing result from this instrument, however, was the detection of an excess of hydrogen in the equatorial regions, indicating the presence of large quantities of water mixed with the regolith. Moreover, the higher resolution at pixel scales better than 20 meters facilitated the identification of hundreds of pits and irregular depressions on the floors and around craters up to hundreds of meters deep that could represent sites of water sublimation after impacts had exposed the deposits. Pits were most common in the region of the 70 km crater Marcia, one of the youngest large impacts on Vesta, and ranged in size from the resolution of the camera to 1 km. As in the case of dark terrains, volatiles were assumed to be exogenous and carried by hydrated, carbon-rich meteorites to form part of the regolith layer. Subsequent impacts would exhume the deposits and enable them to evaporate. To corroborate this hypothesis, the terrain in Rheasilvia, that had been excavated relatively recently by the impact, was mostly hydrogen-poor.[407,408,409]

After completing about 800 low orbits over 141 days, on 1 May 2012 the spacecraft restarted thrusting to progressively widen its orbit for a second high-altitude orbit phase in a sun-synchronous, almost polar orbit 680 km above the surface, which it reached on 6 June. Observations restarted on 15 June and ran until late July. During this second phase, the seasons had changed on the asteroid: more of the northern hemisphere was illuminated, and the overall lighting conditions had changed. Six 10-orbit mapping cycles with different viewing angles were made to derive 3-dimensional views of the surface and some 4,700 images were obtained. By the end of this phase, more than 80 per cent of the surface had been mapped under favorable illumination. No obvious structure antipodal to Rheasilvia was observed.

By 25 July the scientific part of the orbital mission at Vesta was concluded, having significantly exceeded the requirements. Over 31,000 pictures had been collected, as well as more than 20 million spectra. The ion thrusters were fired and the spacecraft restarted spiraling out. It was planned to stop the engine in mid-August for 4.5 days in order to map the entire surface of Vesta, including most of the northern polar regions illuminated at a grazing angle, during four rotations. Dawn would then escape on 26 August, several weeks later than planned. Part of the motivation for this delay was the need to have a longer low-altitude phase in order to collect sufficient high-quality gamma-ray spectra. Moreover, a less conservative reassessment of the power available on the spacecraft had showed that Dawn could remain 40 days more at Vesta and still reach Ceres in 2015. On 8 August, however, the spacecraft entered safe mode due to excessive friction in a reaction wheel, the second malfunctioning wheel of four. As a result, thrusting had to be interrupted while the problem was diagnosed and the departure from Vesta was further delayed. Dawn was then at an orbital distance of more than 2,100 km. Since the reaction wheels would not be used during the 30-month cruise to Ceres, it was decided to leave them off and use the thrusters for the rest of the departure. Moreover, the hybrid attitude control strategy using two wheels and thrusters would probably be used at Ceres. The engine was finally reactivated on 17 August. Departure observations were curtailed, except for some final distant parting shots from 6,000 km over two full rotations on 25 and 26 August, several days after the northern spring equinox. Due to an issue with attitude control tolerances, some of the pictures showed the limb of Vesta or empty space adjacent to it, but the data collected were deemed satisfactory overall. Dawn finally escaped Vesta's gravitational grip at 06:26 UTC on 5 September 2012, 11 days later than planned.[410,411]

Dawn was then to pursue about three-quarters of a solar orbit, most of it spent thrusting in order to reach Ceres in early 2015. Communication sessions would be

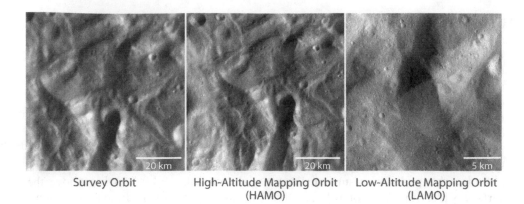

Survey Orbit   High-Altitude Mapping Orbit   Low-Altitude Mapping Orbit
                         (HAMO)   (LAMO)

The three different orbital altitudes reached by Dawn during its mission in orbit around Vesta allowed scientists to map its surface at a wide range of resolutions. (NASA/JPL-Caltech/UCLA/MPS/DLR/IDA)

carried out every 4 weeks instead of once a week to save hydrazine for attitude control. However, twice per week the throttle would be reduced to redirect power to the low-gain antenna and give some basic information on the health of the spacecraft and confirm, in particular, that it was still thrusting. These and other changes should ensure that sufficient hydrazine will be available for the exploration of Ceres, even after taking into account possible anomalies. By late September 2013, Dawn had thrust for 1,410 days, or 64 per cent of the time it had spent in space. Its thrusters had used 318 kg of xenon to yield a total velocity change of 8.7 km/s. The cruise from Vesta to Ceres will require 21,000 hours of thrusting, and a velocity change of 3.4 km/s, in particular to increase the inclination of its orbit from 7.1 to 10.6 degrees. The spacecraft will pass relatively close to a number of asteroids and if funds are available to support an encounter it might detour to inspect one or more of these objects. One flyby candidate is Arete, already mentioned as the object that first enabled Vesta's mass to be determined.

The Ceres rendezvous phase will begin in February 2015. A total of six imaging sessions will be carried out during the entire approach in order to characterize Ceres, its rotational state, and the possible presence of small satellites. Early during these sessions, the resolution of the cameras will best that of the Hubble Space Telescope. On approaching Ceres, Dawn will undertake a similar ballet to that which it carried out at Vesta: entering an orbit at an altitude of 5,900 km with a period of about 10 days sometime in early April, reducing to 1,300 km and 17 hours, and finally to 700 km and 9 hours. The observation strategies have been revised in order to minimize the use of reaction wheels, which will now be used only during the lowest orbital phases. However, contingency plans are in place to ensure that the mission can be carried out using only thrusters if that proves necessary.[412,413] The primary mission is scheduled (by funding) to conclude in July 2016. If Ceres is indeed found to have an environment potentially favorable to life, the main end-of-mission option is to leave the spacecraft in a 'quarantine' orbit around it that should not crash on Ceres for at least half a century. Unlike NEAR at Eros, it will not be possible to soft-land Dawn on Ceres owing to the asteroid's stronger gravity.[414] It had been rumored that if sufficient fuel remains the spacecraft might attempt a flyby of the large asteroid (2) Pallas as it crosses the ecliptic in December 2018, but the JPL team denies having investigated such a possibility.[415]

Approved along with Dawn was a Discovery mission that was not to investigate any of the bodies of the solar system. However, it is briefly described here because it would require a spacecraft flying in solar orbit. It was intended to use a telescope in space to detect planets around other stars, in particular Earth-sized planets. JPL and the Ames Research Center teamed with Ball Aerospace as the main industrial partner for this mission, which was named in honor of Johannes Kepler, who gave support to the Copernican heliocentric hypothesis by formulating the laws of planetary motion. Fittingly, Kepler published his first two laws in *Astronomia Nova* in 1609, exactly four centuries before the launch of the spacecraft.

Since 1995 hundreds of planets have been discovered to orbit other stars, but most are large ones, probably made of gas just like Jupiter, orbiting around their stars in only a few days and hence so close-in that their temperatures must be extremely high and unsuitable for any form of life. The next step is to discover small rocky planets in the 'habitable zone', the annular region around a star in which liquid water can exist on the surface of a planet. The discovery of such planets would be a significant step in the search for extraterrestrial life.

The Kepler spacecraft was meant to search for tiny dips of brightness as a planet transits in front of its parent star. As a transit can occur only if the orbit of a planet is more or less edge-on to Earth, only a small fraction of all the stars accompanied by planets will produce transits. Nevertheless, several large planets and smaller 'super-Earths' had already been discovered by ground-based and space-spaced telescopes using this technique. Kepler would make far more precise brightness measurements than were possible from below Earth's turbulent atmosphere. It would also stare at a selected star field to provide uninterrupted observations lasting several years so as to see planets making repeated transits. The chances of success were estimated from the fact that the COROT (COnvection, ROtation and planetary Transits) satellite launched by France in December 2006 discovered several medium and large transiting planets despite its using a smaller telescope with a narrower field of view and its being able to dedicate only a few months to a single star field. The Kepler spacecraft resembled the Spitzer Space Telescope in that it had a telescope module (without the helium dewar), a mission module and flush solar panels. Overall, it was 2.7 meters in diameter and 4.7 meters high. It carried a Schmidt telescope with a 0.95-meter aperture and a 1.4-meter-diameter primary mirror capable of providing a 15-degree field of view. The telescope had a photometer fitted with a 95-megapixel CCD array cooled by a mixture of liquid propane and ammonia. The spacecraft was 3-axis stabilized by reaction wheels and (when needed) by thrusters. It had four fine-guidance star sensors mounted at the focal plane beside the photometer to provide the required very high pointing accuracy. The launch mass of 1,053 kg included 12 kg of hydrazine for the eight 1-N thrusters of the attitude control system. Four roof-shaped solar panels would provide over 1,100 W of power. There was a high-gain antenna for scientific data transmission and two smaller omnidirectional antennas for housekeeping data and routine commanding. In fact, the spacecraft was to store its photometric data in a 16-Gbyte memory and turn once every month to aim the high-gain antenna at Earth to download its observations at 10 Mbits per second, which is the highest rate of any NASA deep-space mission to-date. Like several other recent Discovery missions, Kepler came close to being canceled in 2007 owing to its escalating cost.

Kepler was to be launched by a Delta II during a window running from 5 March to early June 2009, but was delayed until 7 March. After spending about 45 minutes

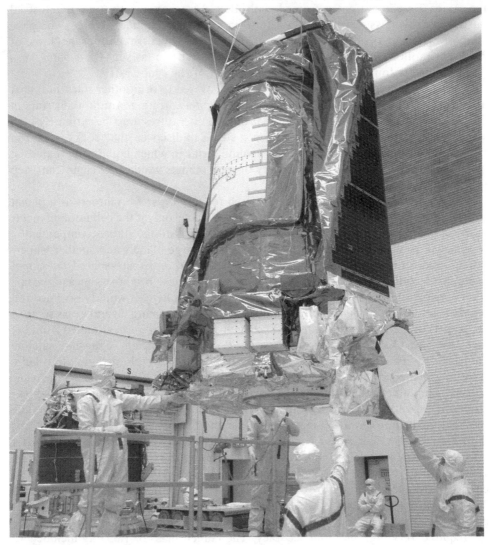

The Kepler space telescope being prepared for launch. It is currently the only mission of the Discovery program not targeted at a solar system object.

in a 185-km parking orbit, the second and third stages fired to put the spacecraft into a 0.967 × 1.041-AU heliocentric orbit. Although slowly migrating away from Earth, the vehicle would still be within 0.5 AU of us at the end of its primary mission. After a preliminary 'dark' calibration of the photometer, a month after launch the cover of the telescope was jettisoned to initiate observations. As the system was being tested, it observed a known giant planet transiting its star and the light curve was so 'clean' that it was possible to infer the presence of an atmosphere around the planet.[416]

The mission was to insert Kepler into a heliocentric orbit with a period of 371 days which would trail Earth, and then spend 3.5 years continuously staring at 21 star fields in the Milky Way centered on the constellations of Cygnus and Lyra to accurately record the brightness of 100,000 stars of a wide variety of spectral types every 30 minutes. In addition to 'hot Jupiters', the data would reveal smaller objects orbiting further out. To detect and confirm an Earth-sized planet in the habitable zone of a star would need repeated transits over several years. Although the dip in brightness for such a planet orbiting around a solar-type star would be fewer than 100 parts per million, it would display a distinctive profile. The science team hoped to find about 50 planets of the same radius as Earth, 185 of about 1.3 Earth radii and 640 of about 2.2 Earth radii. They also expected about 12 per cent of the stars to possess a number of planets, most of which would be gas giants.[417,418] Subsequent missions could use the Kepler data in more focused searches for indicators of life, and to directly image the planets and glimpse their surface. The total cost of the Kepler mission was estimated at $600 million including operations.

Although a discussion of the discoveries of Kepler during its first years in orbit would be beyond the subject of this book, it must be noted that the mission had been complicated by the discovery that solar-type stars have more vigorous gas churning motions than expected, making it more difficult to identify the rare transits of Earth-like planets in this background 'noise'; so much so, in fact, that an extended mission was deemed essential, with an 8-year mission being the minimum for a statistically-significant signal.[419]

After a senior review, the Kepler mission was awarded $60 million to continue operating until 30 September 2016, and possibly longer, unless the craft was incapacitated by a hardware failure. In fact, in July 2012 one of the four reaction wheels failed, leaving the telescope without backups. Kepler meanwhile completed its primary mission in November 2012, having discovered in excess of 2,300 planet candidates, hundreds of which had been confirmed as such, including hundreds of Earth-sized ones.

During the communication session of 14 May 2013, the probe was found to have entered safe mode 2 days previously, and to be slowly spinning with its solar panels facing the Sun. Troubleshooting indicated a second reaction wheel failure, leaving it with only two functioning wheels and thrusters to control its attitude. In any case, scientific operations that required the pointing accuracy which only reaction wheels could provide seemed compromised. Tests were made to determine whether any of the failed wheels could be run backwards, or by applying more torque, but NASA had to concede that it would not be able to revive any of them. It asked the scientific community for ideas on how best to use the spacecraft's remaining capabilities.

Proposals were also put forward to use it to detect near-Earth asteroids, although the telescope had been optimized for photometry rather than for imaging. In the end, a stabilization system was devised to fly Kepler parallel to its orbit around the Sun and use solar radiation pressure to maintain the orientation stable. Tests in October took pictures of an area in Sagittarius to check the stability of the spacecraft in this attitude. In this orientation, Kepler could be used for a number of astronomical observations concerning not only transiting planets around other stars but also active galaxies and supernovas. It would be able to observe areas of the sky for approximately 90 days before it would need to be reoriented. A decision on this "Kepler 2" mission was expected at the end of 2013.

A VENUSIAN 'METEOSAT'

The European Venus Express signaled a resurgence of interest in Venus, a planet recently neglected by solar system exploration. This interest may possibly continue into the 2010s, with increasingly complex missions. For a while, however, Venus Express was to cooperate with Japan's Akatsuki (Daybreak; also known as Planet-C and the Venus Climate Orbiter, VCO). As a 'meteorological satellite' for Venus, Akatsuki was to apply well-established techniques to study the planet's environment in order to derive a complete model of its weather patterns and climate, and facilitate a comparison between Earth and its 'twin'. In addition to surveying meteorological phenomena globally and regionally and observing the phases of cloud formation, the mission was to investigate the mechanics of the super-rotation and vertical motions of the atmosphere. Also, it would seek evidence of lightning and volcanic activity.

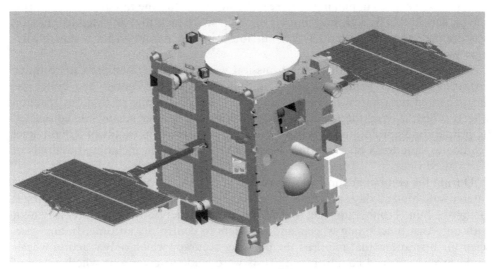

A CAD rendering of the Akatsuki Venus Climate Orbiter (also known as Planet-C). The orbital maneuvering engine, visible at the bottom, would be badly damaged during the first attempt to enter orbit around Venus. (JAXA)

A model of the H-IIA rocket used to launch Akatsuki.

Planet-C was the third of a series of missions for planetary exploration managed by the Japanese Institute of Space and Astronautical Sciences (ISAS), after the Suisei Halley probe (Planet-A) and the failed Mars orbiter Nozomi (Planet-B). Other deep-space missions such as Hayabusa were mainly engineering tests for the MUSES (MU rocket Space Engineering Satellite) series. The Venus orbiter emerged from tradeoff studies for the Planet-B mission of targeting Mars versus Venus. After the launch of Nozomi, Japanese scientists and engineers turned their attention to Venus, and first proposed the Venus Climate Orbiter project to ISAS in 2001, envisaging a launch in 2007 or 2008 on an M-V, already used for Nozomi and Hayabusa. As in the case of Nozomi, the spacecraft would start off in an eccentric Earth orbit. It would then enter a 1-year solar orbit and use a flyby of Earth to lower its perihelion so as to reach Venus some 18 months later.[420] The mission was finally approved by the new Japanese Aerospace Exploration Agency (JAXA) in 2004. In a rationalization following the creation of JAXA, the M-V was phased out in favor of the H-IIA that had been developed by the former National Space Development Agency (NASDA), a more powerful launcher that could send a slightly heavier orbiter directly to Venus from the Tanegashima Space Center. As a result, although its launch slipped to May 2010, the spacecraft would enter orbit around Venus in December of the same year to initiate a 24-month orbital mission.

The spacecraft had an orthodox appearance, with a boxy 1.04 × 1.45 × 1.40-meter structure which supported a 1.6-meter high-gain antenna and, opposite it, the main 500-N bipropellant engine for major course corrections and maneuvers such as orbit insertion. It had a pair of solar wings, each with 1 degree of freedom and an area of 1.4 square meters. They would provide a total of at least 700 W at Venus's distance from the Sun. Much of its hardware was derived from existing Japanese spacecraft, including other interplanetary missions, but several new technologies were adopted. These included two flat, slot-array high-gain antennas, a small one being to receive commands and a large one connected to a 20-W X-band transmitter to return data. In addition, the ceramic thruster of the main engine was the first one ever to use silicon nitride. At times when the spacecraft could not be oriented to aim its main antennas at Earth, there were two low-gain receiving antennas sticking out from the bus and two steerable medium-gain transmission antennas. The data rate to Earth would vary between 4 and 32 kbits per second depending on the distance to our planet. The craft was to maintain 3-axis stability using four reaction wheels and a dozen hydrazine monopropellant thrusters mounted at the corners of the bus; eight with 23 N of thrust and the others with 3 N of thrust for roll attitude control only. The mass at launch was about 501 kg, including 33 kg allocated to the payload and 189 kg of fuel in a single hydrazine tank. The mission was estimated to cost about $275 million.[421,422,423,424,425,426]

In its operating orbit, Akatsuki was to maintain one side facing Venus, and all instruments were mounted on this face. It carried four different cameras to be used for observations at wavelengths ranging from the near-infrared to the ultraviolet, in addition to a high-speed camera for lightning and airglow studies. The four cameras included two infrared CCD instruments, one centered at 1 micrometer, the other at 2 micrometers, with which to observe the atmosphere of Venus, measuring particle sizes, tracking water vapor and carbon monoxide, and searching for traces of recent volcanism. A lower resolution long-wave infrared camera would measure cloud-top temperatures. The ultraviolet camera would track the motions of the atmosphere by monitoring sulfur dioxide as well as the still mysterious "ultraviolet absorber". Just like Venus Express, by exploiting a variety of wavelengths the cameras on Akatsuki would be able to observe cloud structures in the altitude range 50 to 90 km. Later analysis of sequences of images using algorithms used with meteorological satellites would provide a comprehensive picture of wind speed and direction over the entire disk of the planet. Furthermore, radio-occultation experiments using an ultrastable oscillator would obtain information about the vertical dynamics of the atmosphere. Atmospheric observations by Akatsuki were to be complemented by the Japanese EXCEED (EXtreme ultraviolet spectrosCope for ExosphEric Dynamics) telescope that was scheduled to be flown on a small satellite in 2013 and to observe the plasma environment of the planet from Earth orbit.

Thanks to the switch from the M-V to the H-IIA, Akatsuki's launch would also drop off three small satellites in Earth orbit and two others in solar orbit. The most interesting of the latter pair was the $1.3 million IKAROS (Interplanetary Kite-craft Accelerated by Radiation Of the Sun) technology demonstrator, the first solar sail ever to test propulsion by radiation pressure in space. Although sails had been under study by all the major space agencies since the Halley rendezvous proposal of the 1970s, actual flight experience had been unremarkable.[427] The US Planetary Society and the Russian Lavochkin Association jointly funded the Cosmos 1 prototype with eight 15-meter triangular blades which had a total area of 600 square meters, but it was lost in 2005 owing to the failure of the Russian converted submarine-launched ballistic missile on which it was riding. NanoSat-D, a 10-square-meter technology demonstrator by the NASA Marshall center, was similarly lost in 2008 on an early flight of the US privately developed Falcon 1 launcher. Contact with the backup was lost after it was deployed from a 'mother satellite' during December 2010. The only successful tests up to 2010, involving only the deployment, and not the 'sailing' phase, were suborbital deployments of 10-square-meter 'clover-leaf' sails by JAXA.

Although NASA has been studying solar sail propulsion for more than 30 years, it would need a demonstration flight on the scale of the Deep Space 1 mission for ion propulsion to prove it. This may be achieved by the 1,200-square-meter Sunjammer due to fly in late 2014 as the secondary, NASA-sponsored, payload on a Department of Defense launch to the L1 Lagrangian point. On the other hand, JAXA is well advanced in its tests, and is even planning a mission to Jupiter which, like Hayabusa, will perform a demonstration of various technologies whilst also making scientific observations. Its technologies will include a novel hybrid solar sail/ion propulsion using thin-film solar cells as part of a sail 50 meters in diameter that will also power a high-specific-impulse ion engine. This spacecraft could release a small atmospheric probe for the planet and a small Jupiter orbiter for magnetospheric studies, and then be maneuvered to fly by at least one Trojan asteroid; (588) Achilles is the nominal primary objective for this phase of the mission. The mission could be launched around 2020 and reach Jupiter in the second half of the 2020s after one Earth flyby, and the Trojan asteroids around 2030. A Trojan reconnaissance may well be able to sample the material from which Jupiter formed.[428],[429] The next step in solar sail tests in preparation for the Jupiter and Trojan mission was IKAROS, a small craft to deploy a square sail with 20-meter diagonals that would demonstrate orbital maneuvering over an interval of about 6 months. It was initially proposed as a standalone mission, but was then earmarked to piggyback with Akatsuki. Otherwise, the H-IIA rocket would have flown with ballast attached! Development started in early 2008. The cylindrical spacecraft was about 80 cm tall and 1.6 meters in diameter and had a mass of 308 kg, including the 16-kg solar sail membrane. It was to release and unfold the X-shaped main sail masts, each with a 500-gram mass on its tip, rotating at 25 rpm. After the booms were deployed, stoppers would be released to allow four triangular membrane segments to deploy dynamically. The sail itself was made of a 0.0075-mm-thick film of aluminum deposited on a polyimide resin substrate, with insets of 0.025-mm thin-film solar cells. The attitude of the central spacecraft would be controlled by a combination of a new type of gas-liquid thruster 'burning' non-toxic fuel, and liquid-crystal patches on the sail that would change their reflection characteristics so that, by switching them on and off in synchrony with the spin, they would create a minuscule imbalance in solar radiation pressure

The IKAROS probe before being integrated with the launcher. The sail is wrapped around the cylindrical body of the spacecraft. (JAXA)

The minuscule UNITEC 1 was the first spacecraft built by radio amateurs to be put into solar orbit. Unfortunately, contact with it was lost shortly after launch. (JAXA)

and generate a torque to cause the entire spacecraft to slowly bank. The attitude, spin, and axial orientation were determined by a Sun sensor and by the Doppler-shift of radio waves transmitted by the low-gain antennas, mounted offset from the spin axis.

Two scientific instruments, a dust counter covering 3 per cent of the sail surface with piezo-electric sensors of differing thickness and sensitivity, and a gamma-ray burst detector would be carried.[430] The spacecraft would communicate with Earth by a 7-W transmitter connected to two low-gain antennas mounted on the drum's periphery, one on the top and the other on the bottom side to be used in different phases of the flight. As there would be periods when Earth was more or less in the plane of the sail, communications would be difficult. It also had a medium-gain antenna and an antenna for precise tracking. A spring-actuated system was to ensure separation and spin-stabilization of a pair of cylindrical pods, each of which was 5.5 cm wide and 5 cm tall and contained a wide-angle color camera, a radio system and a 15-minute-duration battery. These were to provide pictures of the deployed sail. With IKAROS, JAXA would become the only space agency in the world to have flown experiments with both of the main low-thrust orbit change technologies: electrical propulsion (with Hayabusa) and solar sails.

Another secondary payload to fly with Akatsuki was the small satellite UNITEC 1 (University space engineering consortium Technology Experiment Carrier). It had a mass of just 26 kg, and was built by a consortium of some 20 Japanese universities and colleges to test long-duration technologies for electronics, communications and tracking. It was the first privately funded university or amateur satellite ever to be inserted into solar orbit. It was a box 39 × 39 × 45 cm across and almost completely covered with solar cells. It was unstabilized, and was to make a flyby of Venus with a miniature camera and a radiation counter. It had omnidirectional antennas and a transmitter of less than 10 W of power.[431]

Assembly of the Venus Climate Orbiter started in June 2009. It was delivered to Tanegashima in March 2010 for integration with the launch vehicle and the other payloads. The launch window extended from 17 May to 2 June. In an unprecedented move, JAXA announced the name of the spacecraft, Akatsuki, in advance. A launch on the first day of the window was scrubbed because of bad weather and rain, but all went well on the second try, on 20 May. On reaching parking orbit, the launcher coasted for 12 minutes, releasing the three subsatellites. It then restarted to insert its interplanetary payload into a 0.72 × 1.07-AU orbit. Some 28 minutes after launch, Akatsuki was released southeast of Hawaii, followed 15 and 20 minutes later by the IKAROS and UNITEC 1 spacecraft respectively. The small university satellite was then renamed Shin'en (Abyss). Unfortunately, contact with it was lost after less than 24 hours. Meanwhile, Akatsuki calibrated its cameras by taking images of the night-side of Earth from a range of 250,000 km. During the interplanetary cruise, it was to undertake observations of the zodiacal light from a variety of perspectives.

With Akatsuki safely on its way, all eyes turned to the solar sail demonstrator. At the end of May IKAROS started going through the steps leading to sail deployment: it was spun up from 2 rpm to 25 rpm, and on 26 May the sail mast tip masses were released under the eye of the monitoring cameras. However, after the sail masts had deployed about 5 meters, the deployment was paused in order to give the engineers time to investigate why the semi-deployed sail was slowly spinning up. The masts were completely deployed by 8 June, then the stops blocking the sails were released and the four parts of the sail dynamically unfolded to achieve the square shape. At that time, IKAROS was 7.4 million km from Earth. Images returned by the onboard cameras showed the sail membrane properly deployed. Moreover, telemetry showed that the thin-film solar cells embedded in the sail were generating power. It was the first time that a solar sail had been correctly deployed in space. It took the spacecraft almost 5 hours to dampen out the vibrations and imbalances, but the behavior nicely matched simulations. On 14 June the first of the two camera pods was ejected at a speed of 65 cm/s and took pictures of the fully deployed sail. The second pod was ejected on 19 June at a speed of 35 cm/s and provided pictures from much closer in. Moreover, in the meantime the sail was spun down from 2.5 to 1.3 rpm. Together, the two camera pods took more than 80 pictures which clearly showed the sail with the liquid crystal panels switched on and switched off.[432] Accurate tracking revealed that the sail was being pushed by a solar radiation pressure thrust of 1.12 mN. The next step was to demonstrate attitude control using the liquid crystal panels, and this was accomplished on 13 July, finally meeting all the criteria for mission success.

Meanwhile, Akatsuki made its first midcourse correction on 28 June, with a 13-second thruster-firing for a change in velocity of 12 m/s. The engine slightly over-

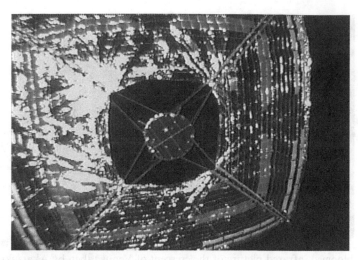

The fully deployed IKAROS square solar sail, photographed in space by one of the jettisonable cameras. It is the first solar sail to have demonstrated deployment in space, as well as propulsion and attitude control using solar radiation pressure. (JAXA)

performed. For 24 hours on 22 and 23 October one of the infrared cameras scanned almost the whole ecliptic plane to observe zodiacal light, but the images were mostly spoiled by sunlight reflecting in the camera optics. Two days later, Akatsuki was turned to Earth, some 30 million km away, to calibrate its instruments. The infrared cameras had no problems in imaging both our planet and the Moon beside it, but the ultraviolet camera could not detect the latter.[433] Three more fine-tuning corrections were performed using the 23-N thrusters on 8 and 22 November and on 1 December.

Upon reaching Venus, the spacecraft was to fire its main engine to enter a 4-day elliptical, almost equatorial orbit with a periapsis of about 550 km that matched the 'inverted' direction of the planet's axial rotation. It was then to refine the parameters of its orbit in three steps for a periapsis at 300 km, an apoapsis at 79,000 km (about 13 planetary radii) and a period of 30 hours. Its speed at apoapsis would be roughly synchronized with the 60-m/s super-rotation winds at an altitude of about 50 km, so that it would be able to track atmospheric features continuously for up to 20 hours at a time. It was to take moderate-resolution pictures every 2 hours covering most of the disk. In fact, unlike most planetary orbiters, Akatsuki was to obtain most of its observations at apoapsis. In this way, its scientific studies would complement rather than repeat those of Venus Express, which was dedicated to high-resolution imaging of the northern hemisphere and wide-area observations at high southern latitudes. Nevertheless, at periapsis Akatsuki would undertake close-up and limb observations. During its primary mission lasting two terrestrial years, Akatsuki, working together with Venus Express, was to provide almost uninterrupted monitoring of winds in the southern hemisphere and of volcanic activity, if it indeed exists.

Akatsuki arrived at Venus late on 6 December (early on 7 December in Japan). At 23:49 UTC the 12-minute orbit insertion burn began. Two minutes later, the vehicle passed behind Venus, as viewed from Earth, starting what was expected to be a 22-

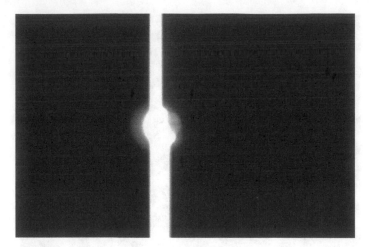

A heavily bloomed infrared picture of the crescent of Venus taken by Akatsuki shortly after the missed orbit insertion. The darker patch on the night-side of the planet corresponds to the continent of Aphrodite Terra. (JAXA)

minute radio blackout. It would then make an hour-long pass through the planet's shadow. But contact was not regained as planned, and when it was, near the end of the solar occultation, it was evident that something had gone awry. The spacecraft had placed itself into safe mode, oriented itself to face its solar panels to the Sun and was slowly spinning on that axis for stability. This made communication difficult, because the high-gain antenna could not be aimed at Earth to establish a high-speed data link. Contact was limited to a 40-second window per 10-minute rotation, as the medium-gain antenna beam swept by Earth. In these conditions it took hours to determine what had gone wrong, and whether Akatsuki had achieved Venus orbit. Engineers estimated that if the engine burn had lasted for a little over 9 of the planned 12 minutes, then the vehicle should have entered a highly eccentric planetary orbit that would enable the intended mission to be recovered by further maneuvers. But as tracking data slowly accumulated, it became apparent that Akatsuki had not been captured. In fact, the aborted burn and the gravity assist from the unintended flyby of Venus had placed it into a solar orbit that was interior to the planet. Attitude control was rapidly regained and the high-gain antenna was finally aimed back to Earth 23 hours after the aborted burn in order that 28 Mbits of recorded telemetry could be downloaded. To verify that they were still working properly, images of the thin crescent and night-side of Venus were taken two days after the flyby using three of the five cameras. They showed the thin crescent of the planet from a distance of 600,000 km and again the next day from 890,000 km. These included the first mid-infrared views of the atmosphere of Venus. Long-wave infrared observations imaged cold belts along the parallels as well as warmer streaks extending along the meridians.[434] Infrared images of the night-side even recorded details of the surface in the Aphrodite Terra continent. This was the first failed Venus orbit insertion maneuver in the entire history of solar system exploration.

The telemetry suggested that Akatsuki had suffered a large attitude disturbance around an axis perpendicular to the thrust line of the engine 2 minutes 32 seconds into the burn, which caused safeguards to switch off the engine and enter safe mode. Bit by bit, it was determined that there had been a problem in the lines that supplied helium to pressurize the fuel tank. Nitrogen tetroxide vapor had apparently seeped into a non-return valve and reacted with hydrazine to form crystals of ammonium nitrate salt, clogging it sufficiently to prevent it from opening. During the burn, the tank ended up delivering a pressure of the gas remaining in its ullage, i.e. operating essentially in the 'blow down' mode. This was consistent with the fact that the acceleration declined from the time of ignition to when the burn was curtailed. But it was not clear what had caused the attitude perturbation that invoked the safeguards. Possibly it was an imbalance in the center of mass as fuel and oxidizer were consumed at different rates, but there was a suspicion that the non-optimal propellant mixing ratio and fuel starvation might have disrupted the thin film of fuel that adhered to the walls of the nozzle to keep it cooled, which could have damaged or even 'burned through' the main thruster, with a jet of hot gas inducing a sideways thrust to initiate the attitude disturbance. Six seconds later, after closing the main engine fuel valves, the attitude control system tried to regain control by switching from thruster stabilization to reaction wheels. Six minutes after initiating the burn, the spacecraft entered safe mode and adopted an attitude which would enable it to spin around the Sun line and generate electrical power. Meanwhile, pressure in the fuel tank returned very slowly to normal levels. The aborted burn had still managed to give a velocity change of about 135 m/s and, together with the unplanned gravity-assist from Venus, it had placed Akatsuki into an orbit ranging between 90 and 110 million km of the Sun with a period of 203 days.

This scenario was disconcertingly similar to the problem that marred Nozomi as it attempted to leave Earth orbit in 1998 to head for Mars. As then, Japanese project officials announced a new profile that could save the mission at the cost of delaying it to the mid-2010s. It calls for Akatsuki to complete eight orbits around the Sun in order to make a second approach to the planet in November 2015. Hopefully, during this long cruise the spacecraft will withstand solar radiation better than Nozomi and Hayabusa. It may be placed into hibernation for long periods, but some scientific observations are likely, including a survey of zodiacal dust. Celestial navigators looked at possible asteroid encounters and discovered that, provided Akatsuki's engine could be used to target the return to Venus, two small objects could be flown by with only minor adjustments. In fact the spacecraft would pass several million kilometers from Venus in 2017 unless its orbital period were reduced during either the second or third perihelion in November 2011 or June 2012. A relatively large burn would therefore be required, as well as a new orbit insertion burn. While fuel did not seem to be an issue, because 80 per cent of it remained, the condition of the main ceramic thruster was not yet known and could still make the recovery impossible. Tests were made on Earth to replicate the known conditions of the burn and showed that the nozzle was likely to have cracked. The actual engine in space would be tested later in 2011. If it proved to work, then it would be used for the second attempt. Otherwise, oxidizer would be dumped to reduce the mass of the spacecraft and attitude control thrusters would be used instead, although this would obviously result in a severely shortened orbital mission because Akatsuki would not have the fuel to reach the originally intended orbit.

Akatsuki reached perihelion on 17 April. Controllers were busy monitoring its temperature to see how it behaved in an environment different from that for which it was designed, receiving a solar heat flux some 40 per cent greater than if it had gone into orbit. Meanwhile, taking advantage of the position of the spacecraft at the time, in the vicinity of Venus and sunward of it, a campaign was executed to characterize the atmosphere of the fully illuminated planet at angles inaccessible from Earth. The planet, 10 million km away, spanned only a few pixels.[435] The data suggested the presence of large particles in the upper cloud layers, possibly related to an increase in sulfur dioxide reported by Venus Express. Moreover, the 4-day super-rotation period of the atmosphere was clearly detected in ultraviolet reflectivity data. Shortly before perihelion, Akatsuki had to be reoriented to place some hardware in the shadow, because it had exceeded its maximum permitted temperature. Moreover, it was believed that the multi-layer insulation was degrading and could not possibly withstand a number of perihelion passes. Akatsuki went through its first solar conjunction in late June, and on 2 September it fired its main engine for 2 seconds for the first integrity check; the first time it had been used since the aborted orbit insertion burn. The crippled engine unfortunately provided only one-eighth of the expected thrust. A 5-second burn one week later only confirmed the shortcoming. If the first burn had been successful, the plan was to make a 20-second burn to validate attitude control strategies. The tests however confirmed that the engine produced only about 40 N, and proved that the nozzle had probably broken off and gas was being spewed in all directions. It was then decided to perform orbit insertion using attitude control thrusters, and 64 kg of nitrogen tetroxide was dumped through the damaged main engine injectors during October. In order to avoid the risk of the injectors freezing and clogging, oxidizer was dumped in three sessions, at most 9 minutes long, on 6, 12 and 13 October. By the end of the third session, the engine was clearly jettisoning only helium gas used for pressurization. The first 600-second, 90-m/s maneuver was then carried out on 1 November, followed by another nine days later. Finally, on 21 November, a 70-m/s burn reduced the aphelion distance by 3 million km and the orbital period to 199 days, adjusting the course to return to Venus. Moreover, it moved the point of encounter to near the aphelion of Akatsuki, reducing the relative speed. Several mission recovery strategies are being studied. If the spacecraft is found to be in good shape, it could make a flyby in November 2015 and be retargeted for orbital insertion in June 2016, when it should be able to reach an orbit better suited for atmospheric science observations. An even later insertion would further improve the orbit, but any decision will have to take into account the status of the hardware and instruments. Achieving the planned retrograde orbit in November 2015 would normally place the probe in a zone where solar perturbations would rapidly lower its periapsis and cause it to enter the atmosphere. Two scenarios are under study. One is a fairly conventional strategy of entering orbit at aphelion, but it would require a maneuver of 80 m/s prior to arrival in order to move the encounter to the first half of December 2015 and establish the correct trajectory. In

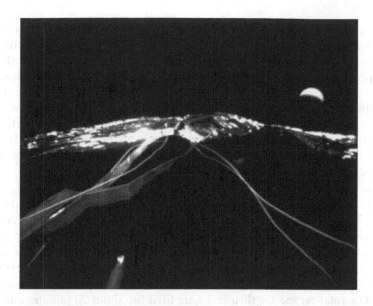

The crescent of Venus with the stretched sail in the foreground seen by one of the engineering cameras on IKAROS. (JAXA)

the second, the spacecraft would first be inserted into a direct orbit with its apoapsis almost 1 million km from the planet. Once at apoapsis, solar gravity perturbations would serve as a brake and reverse the motion, effectively inserting the vehicle into a retrograde orbit; i.e. in the same direction as Venus spins. Over time, moreover, solar perturbations would move the apoapsis onto the day side of the planet. But no decision on the orbit insertion strategy will be taken before 2014.[436,437] Ideally, it would be put into a one-week elliptical orbit and spend 6 days taking global images of the atmosphere and surface, as well as close-up images of the atmosphere, surface and planet limb near periapsis. This orbit would also give multiple opportunities for periodic radio-occultations in order to profile the atmosphere.[438]

While control of Akatsuki was being regained and its status evaluated, IKAROS continued its mission. During 5 months of operations to November 2010, the solar sail had changed its velocity by 100 m/s by the use of solar radiation pressure alone. Traditional attitude control laws for spin-stabilization were attempted, but failed due to the difficulty in controlling a craft that had a large flexible appendage like a sail. On the other hand, the liquid crystals had allowed IKAROS to rotate 180 degrees to remain face-on to the Sun over six months, or about half an orbit. Thrusters were used for quick maneuvers or to maintain the spin rate, but otherwise they were rarely used. Other engineering tests concerned the degradation of the solar sail material on being exposed to the Sun, as well as a radio astronomy experiment to determine the position and velocity of IKAROS with a precision 20 times better than previous JAXA missions. Dust fluxes were determined to be larger than recorded by Nozomi one decade earlier and a number of gamma-ray bursts were detected, assisting in the effort with Earth-orbit satellites to triangulate their sources.

The velocity change imparted by the sail was sufficient to ensure that IKAROS made it to Venus one day after Akatsuki. Moreover, while both had been dispatched at launch to fly over the day-side of the planet, solar sailing had moved the closest approach point 80,000 km over the night-side. One of the wide-angle cameras that monitored the sail at deployment was used to take a low-resolution image of the sail with the little crescent of Venus in the background. The sail, at the time, was too far from Earth for its non-directive antennas to return the picture to Earth at high speed, and communication opportunities were relatively rare, so much so in fact that it took two weeks to transmit the single picture.[439] The long-silent Shin'en also flew by the planet sometime in December.

The solar sail test of IKAROS was declared complete in January 2011, when the mission was extended to March 2012 in order to test less conservative navigation techniques. Most of these tests would be completed by late June 2011. After that, depending on their results, more 'aggressive' techniques of solar sailing would be attempted, changing the spin rate and Sun angle over wider ranges. At one point, the spin rate was reduced to just 0.2 rpm. By early September, only 3.4 kg of the original 20 kg supply of gas for spin control remained. On 18 October, after gaining experience with very slow rotation speeds, thrusters were fired for about 20 minutes and the spin direction changed. There were fears that the sail membrane, kept in shape by centrifugal forces, would get entangled but this did not happen. After a communication on 24 December, IKAROS should have put itself into hibernation due to the scarce power available as it approached aphelion and also the increasing distance from Earth, which made communications difficult. And the gas supply was nearing depletion, making attitude control ineffective. JAXA, however, remained confident that it would be able to return the spacecraft to normal operations in the summer. After 8 months of silence, a carrier wave from IKAROS was received on 6 September 2012 and again two days later. Soon after that, however, it went through solar conjunction, meaning that contact would not be able to be re-established until mid-2013. On 20 June 2013 IKAROS was contacted, and the telemetry confirmed the solar sail was in low-power hibernation. It should return to the vicinity of Earth after a total of 5 years.

With the Nozomi Mars orbiter lost in interplanetary space, Hayabusa marginally successful after an incredible series of mishaps, Lunar-A canceled after its surface penetrators were judged to be technologically immature, and the Kaguya (SELENE) lunar orbiter spectacularly successful, it would seem to be time for Japan to review its management of scientific missions. It is worth noting that Kaguya was the only mission of the group to have been started by the former NASDA space agency, all the others began as ISAS projects. The impression, at least from the outside, is that, as with NASA's 'faster, cheaper, better' missions, the Japanese engineers, operating under tightly constrained budgets, are cutting corners to save on cost.[440,441]

A CHINESE CLEMENTINE

The People's Republic of China apparently first began to study the theory behind deep-space missions in the 1960s and 1970s. Concepts for lunar missions are known to have been investigated in the mid-1960s, and in 1978 the science and technology minister announced that "space surveyors" were being planned.[442] However, these projects did not mature, and the Chinese program was redirected to applications satellites. Nevertheless, starting in the early 1990s a variety of ambitious plans were formulated for human flights (the first of which occurred in 2003), lunar probes, and deep-space probes. Following a number of joint cosmic radiation experiments flown on balloons, China also wished to participate in the Russian missions to Mars that were scheduled for the 1990s. However, planetary exploration was not a priority, and this was deferred until after the first lunar missions had been flown. Then in 2000 the director of the Chinese National Space Administration (CNSA) announced that his country intended to play an active role.

After much speculation, the Chinese lunar exploration program was officially unveiled in 2003. It was named Chang'e after a mythical Chinese lunar goddess, and was to consist of three incremental phases. In the first phase, lunar orbiters would be sent to map the surface of the Moon in detail and thereby test a number of different techniques. Next, lunar landers would carry instrumented autonomous rovers to the surface. In the third phase, robotic landers would collect samples of lunar rock and dust and return them to Earth, a feat that had previously been accomplished only by the former Soviet Union. Human missions could follow at a later date if political and economic support was favorable.[443] In parallel with these studies, Chinese scientists and engineers of the Harbin Institute of Technology Deep Space Center showed some interest in flying a NEAR-class asteroid mission. The 1-tonne orbiter would be equipped with cameras and spectrometers to study the topography, morphology and mineralogy of the target body, and would perform a single Earth flyby and possibly also an asteroid encounter along the way prior to entering orbit around either near-Earth asteroid (4660) Nereus, or preferably (1627) Ivar which, at 6.2 km instead of 1 km across would provide a more varied scientific environment. The concept evolved to a multi-target mission. Engineers would have liked to launch it around 2010, but it apparently was deferred until after the first lunar missions, as well as the Martian Yinghuo mini-orbiter (of which more in the next chapter) had been flown. Asteroid landers have also been the subject of theoretical studies, along with Deep Impact-like experiments and a mission to soft-land on a comet or asteroid and test an engine capable of pushing it off collision course with Earth.[444,445,446]

The first lunar orbiter, Chang'e 1 (or CE1) was launched in October 2007 using a Long March 3A rocket, the Chinese three-stage equivalent of the US Atlas–Centaur from the southern range of Xichang, where geostationary communications satellites are usually launched. The Chang'e lunar orbiter was built by the Chinese Academy of Space Technology (CAST) and was based on the Dong Feng Hong 3 (DFH-3, "East is Red") communication satellite bus, consisting of a central box 2.2 × 1.72 × 2.2 meters in size to which were attached two wings of three solar panels each with a total span of 18.1 meters and an area of 22.7 square meters providing up to 1,450 W of power. The spacecraft was 3-axis stabilized to ensure that the cameras and other instruments were always pointed at the Moon. The orientation was determined by solar sensors, star cameras and by an inertial gyroscopic platform and controlled by two redundant clusters of six 10-N thrusters as well as by reaction wheels. Also mounted on the spacecraft for large maneuvers was a bipropellant (hydrazine and nitrogen tetroxide) 490-N main engine. Chang'e 1 had a launch mass of 2,350 kg, of which 1,200 kg was propellant. A 60-cm parabolic high-gain antenna provided high speed communications and there were omnidirectional antennas operating at slower rates. Deep-space antennas had been constructed in Beijing and Kunming for the lunar projects. A 65-meter antenna was under construction near Shanghai, and a 35-meter dish was being built at Kashgar in the northwestern region of Xinjiang for planetary missions. In addition, a 64-meter antenna at Jiamusi in the northeast was to be completed in 2012 in time to be used for Yinghuo at Mars. A third antenna will be completed in Patagonia at the tip of South America in 2016. The European deep-space antennas were often used to upload commands to Chang'e 1.[447] At the end of its mission, on 1 March 2009, the spacecraft was crashed onto the lunar surface.

The second Chang'e mission made use of the spare Chang'e 1 orbiter to obtain higher resolution, close-up views of the lunar surface and of candidate landing sites, and to perform a number of technological tests for the lunar landing as well as deep-space planetary missions such as the use of X-band tracking and precise localization of the vehicle in space using interferometric networks of radio-telescopes. It was announced from the start, moreover, that the probe would probably be flown to other targets in deep space after orbiting the Moon. Studies had been completed on possible extended missions to near-Earth asteroids that could be encountered within 300 days of leaving the Moon. Several candidates were found among the Apollo and Aten objects, proving the mission to be feasible.[448]

Reportedly massing 2,480 kg, Chang'e 2 was heavier than Chang'e 1, and it had a modified payload including a much improved 3-dimensional 6,144-pixel dual push-broom CCD camera, a laser altimeter, gamma-ray and X-ray spectrometers, high-energy solar particle and solar wind ion detectors, and a microwave radiometer to measure the temperature of the lunar surface, and 128 gigabits of solid-state data storage. It also had four CMOS engineering micro-cameras to monitor critical events such as the firings of the main engine and the deployment of the solar panels and high-gain antenna, and demonstrate an obstacle-detection landing camera for future landers. Unlike Chang'e 1, it was to be launched by a Long March 3C, a more powerful form of the Long March 3A using two liquid-fueled boosters. The second Chinese lunar orbiter reportedly cost the equivalent of $134 million.[449,450]

Chang'e 2 lifted off from Xichang on 1 October 2010 and reached lunar orbit 112 hours later. Unlike Chang'e 1, which flew in a 200 km orbit, Chang'e 2 orbited at 100 km, although early in the mission the periapsis was reduced to as low as 15 km in order to rehearse landing maneuvers and to obtain images with a resolution as fine as 1.05 meters. A mere 8 months after launch, the lunar mission was concluded. The Moon had been fully mapped at a resolution of 7 meters from an altitude of 100 km and part of it, including candidate landing sites, at higher resolution, but several hundred kilograms of fuel remained on board, equivalent to an estimated total delta-V of about 1,100 m/s. A number of alternative extended missions were studied, including a proposal calling for a circuitous route that would involve phases at the L1 and L2 Lagrangian points of both the Earth-Moon and the Sun-Earth systems, a near-Earth asteroid flyby, or a return to Earth and final high speed lunar impact.

The importance of selecting a target asteroid with an accurate ephemeris was recognized from the very start by celestial navigators at the State Laboratory of Astronautical Dynamics. As an alternative, only individual parts of this highly ambitious plan could be flown.[451] Accordingly, on 9 June 2011, at the end of a maneuver which involved two engine burns and used more than 75 per cent of the remaining fuel, Chang'e 2 escaped the Moon heading for the L2 Lagrangian point in deep space more than 1.5 million km beyond the Earth's orbit, in order to test the Chinese deep-space network over greater distances than lunar orbit, test control of a distant spacecraft, investigate charged particles at the Lagrangian point, and monitor X-ray and gamma-rays from the Sun. It entered a halo orbit around the L2 point on 25 August, becoming the first spacecraft of any nation to travel from lunar orbit to a Lagrangian point.[452] During the next 8 months it conducted engineering tests as well as observing the solar wind downstream of Earth. In particular, it tested solar sailing and "solar windmilling" techniques for attitude control. Its data was returned at a rate of 750 kilobits per second, down from 3 megabits per second in lunar orbit because the range had increased almost four-fold.[453] Meanwhile Yinghuo was launched piggybacked on the Russian Fobos-Grunt spacecraft, but that failed to leave Earth orbit. Chinese sources indicated that following the L2 excursion, and depending on the conditions of the spacecraft, Chang'e 2 could be sent back to the Moon, to the L1 Lagrangian point 'upstream' of Earth, or to undertake a flyby with a near-Earth asteroid or comet. Some 115 kg of fuel remained on board, equivalent to a delta-V of about 120 m/s. At about the same time, there were contacts between the engineers "flying" Chang'e 2 and astronomers of the Purple Mountain Observatory in Nanjing on the possibility of an extended mission to an asteroid.

Although it had been announced that Chang'e 2 would remain in a halo orbit until at least the end of 2012, in January of that year the Beijing Aerospace Control Center called for proposals to use the probe beyond L2, with the objective of leaving the halo orbit as soon as possible. Various alternatives were studied, including some to repeatedly fly past the Earth and Moon, to visit the L1 and L2 Lagrangian points of the Sun-Earth system, to fly past a hundred-meter-sized asteroid, and to reach the L4 stable Lagrangian point of the Sun-Earth system in 2017 in a STEREO-like orbit. In March, however, the simpler solution proposed by the Chinese Academy of Space Technology was adopted, calling for a flyby of a near-Earth asteroid. No fewer than 38 potential targets had been identified for an encounter in late 2012 or 2013, but only three could be realistically targeted. While two targets would require relatively little fuel to be reached, leaving the option open for further asteroid encounters in a further-extended mission, they were all smallish objects less than 1 km across and their orbits were not very precisely known. The eventual choice was one of the best known and best observed near-Earth asteroids: (4179) Toutatis.[454,455] This was to have been a target for the US military Clementine 2 probe. It was first imaged by radar in 1992 and revealed to be a binary object involving a pair of kilometer-sized bodies in contact.[456] Observed in the infrared by ground-based telescopes, Toutatis showed an undifferentiated composition similar to some chondritic meteorites and to NEAR's target (433) Eros.[457] After discovery by French astronomers in 1989, it was realized that Toutatis would make six close passes of Earth at 4-year intervals, with the next one, in December 2012, being the last. The choice of Toutatis was a clever one, as the flyby would occur 7 million km from Earth, enabling Chang'e 2 to return data despite not having been designed to do so over such a distance. Moreover, thanks to the frequent close passes and radar observations, the orbit of the asteroid was known with an uncertainty of just a few kilometers, eliminating the need for an autonomous system to recognize and track it in imagery taken by the spacecraft. Chang'e 2 would merely need to be steered to the exact position that the asteroid was predicted to occupy at the time of the encounter. Nevertheless, in May Chinese astronomers at the Purple Mountain Observatory began an observation and orbit determination campaign on Toutatis in cooperation with large telescopes in Hawaii and Chile. Although such an effort was not strictly needed for such a well-known object, it was a demonstration of what would normally be required for an asteroid encounter.

On 15 April a 6.2-m/s burn marked the departure of Chang'e 2 from its halo orbit. Optimization of the trajectory was unfortunately only carried out afterwards, but it still allowed the team to save a sizeable amount of fuel that could be used to ensure a successful flyby. A large, 105-m/s burn on 31 May was the first Toutatis targeting maneuver. The second maneuver was on 9 October. Chang'e 2 was by then in an almost Earth-like solar orbit ranging between 1.022 and 1.035 AU, with a period of 381 days. The asteroid mission was revealed at a meeting of the Chinese Academy of Sciences in June, when it was hinted that the flyby would occur on 6 January 2013, at about 29 million km from Earth. However, the vehicle was recovered by Western telescopes monitoring small asteroids and space junk and it was possible, from the orbit thus determined, to calculate the encounter date as 13 December 2012, only a day after Toutatis flew by Earth at a range of 6.9 million km. A number of corrections were planned in order to arrange a flyby with a geometry suitable for communicating with Earth and solar illumination conditions for imaging – although the event was seen as a technology demonstration for a future deep-space mission rather than an opportunity for science.[458,459,460] It was initially suggested that the spacecraft would pass by several hundred to a thousand kilometers from Toutatis, yielding images with a resolution of 70 meters per pixel or better. However, the push-broom science camera would not be able to provide more than a handful of pictures because it would have to wait for the asteroid to cross the field of view of the two linear CCDs, or would require the vehicle to rock back and forth to acquire each image. Other observations could use the laser altimeter and the microwave radiometer. Due to these limitations, engineers controlling Chang'e 2 decided to target the encounter at a point just several kilometers from the asteroid and to use one of the engineering cameras, the one employed to monitor the solar array deployment, instead of the scientific camera proper. This was the only narrow-angle camera of the four on board. It had a mass of just 358 grams and was designed to take color movies using a 1,024 × 1,024-pixel CMOS detector mated to 54-mm focal length optics with a field of view of 7.2 degrees.[461]

While the encounter would yield only webcam-like uncalibrated images from a camera not designed for science, significant science could be extracted besides basic shape and topography. In particular, it would be possible to reconstruct the geology and cratering history, and gain insight into surface processes and even the internal structure. In November, the 65-meter radio-telescope in Shanghai was inaugurated and rapidly integrated in the Chinese deep-space network, and from the end of that month to just prior to the encounter it was used to pinpoint Chang'e 2's position relative to distant quasars and to refine its trajectory. A trajectory correction on the 10th was postponed, and finally performed on 30 November. Instructions based on the latest orbit determination were uploaded to the probe the day before the encounter, when the final course correction was also carried out. The probe would miss Toutatis by over 100 km, so a 3.3-m/s correction moved the closest approach distance much closer. It is important to stress that Chang'e 2 was steered without reverting to optical navigation. Imaging on the inbound leg would have been impractical because Chang'e 2 was approaching the night-side of Toutatis. Instead, the probe was rotated so that the camera axis was parallel to the direction of the relative motion with the asteroid. Some 65 minutes out, the solar panel was oriented in a position that would not interfere with the camera, with the non-illuminated side visible in the field of view. The camera was switched on 10 minutes before the flyby, and would remain on for 25 minutes. Around closest approach, the diminutive camera was to take five pictures per second for more than 100 seconds. At 08:29:55 UTC (16:29 in Beijing) on 13 December, Chang'e 2 flew by Toutatis at a relative speed of 10.73 km/s. About 4.5 gigabits of imaging data were returned over the ensuing days at some 20 kilobits per second. The closest approach distance was initially announced as an unexpectedly narrow 3.2 km. It was later recalculated by analysis of the relative motions of the vehicle and asteroid in images. One team derived a minimum distance of 1.32 km from the center of mass of the asteroid, equivalent to just 770 meters from its surface. Another team fixed it at 1,564 meters, with an uncertainty of just 10 meters. In any case, this was the closest flyby ever of a small solar system body. Four seconds after the flyby, Toutatis had entered the field of view of the camera. The first frames were spoiled by motion blurring. The first non-blurry image was taken at a distance of 22 km, showing a portion of the asteroid partially occulted by the shadowed solar panel. The first photo non-obstructed by the solar panel was taken 6 seconds after Toutatis had entered the field of view, and at a distance of about 66 km.

After Itokawa and comet Hartley 2, Toutatis was only the third object previously imaged in high resolution by radar to be visited by a spacecraft. It was confirmed to be a binary and to be shaped "like a ginger root", with a squarish primary body and a roundish secondary. The best resolution on the larger lobe was about 3 meters, while the smaller lobe, which was covered by the solar panel at closest approach, was three times poorer. Its reddish surface appeared mostly smooth, and covered with regolith and clusters of boulders. A prominent scarp marked the edge of the large lobe. All in all, it was rather similar to Itokawa. Some circular depressions had sharper rims that could be interpreted as muted craters. Smaller and better defined craters were seen in higher resolution images obtained at closer range. A total of nineteen craters larger than 100 meters were seen. Moreover, the larger lobe seemed to have a higher crater density than the smaller one. There was a big boulder, almost 100 meters across, sitting on the "neck" between the two lobes. Interestingly, the

A model of the Chinese Chang'e 2 lunar orbiter that went on to perform a flyby of asteroid (4179) Toutatis.

The solar panel deployment monitoring camera carried by Chang'e 2. It was this 'webcam-like' engineering instrument that was used to image asteroid Toutatis.

neck appeared to be devoid of craters, hinting at the presence of regolith pooling in the area. The extremely close flyby raised the possibility that the tracking data could be used to determine the mass and, from a shape model, the average density of the asteroid. At the time of writing, however, no results for such a study seem to have been published in the literature.[462,463,464,465,466,467]

About 5 kg of fuel remained after the flyby. Since this was capable of imparting a delta-V of less than 10 m/s, it effectively negated any further meaningful extended mission. Tracking by the Chinese deep-space network continued as long as possible for experience in controlling a distant spacecraft in preparation for future deep-space and Mars missions. By mid-2013, the spacecraft was being prepared for reduced long-distance operations, configuring the propulsion system into blow-down mode and reconfiguring the communication system. In this phase of its mission, Chang'e 2 would assess the long term survival of hardware, the capability of autonomous flight and test long-distance tracking and telemetry. By mid-July, the probe was 50 million kilometers from Earth. By a pure coincidence, the trajectory of Chang'e 2 after flying by Toutatis would take it to the vicinity of the L5 Lagrangian point of the Sun-Earth system in mid-2016.

Based on the success of this first deep-space mission, more probes could be sent to Mars, Venus, and near-Earth objects. Apparently a meeting of Chinese scientists was

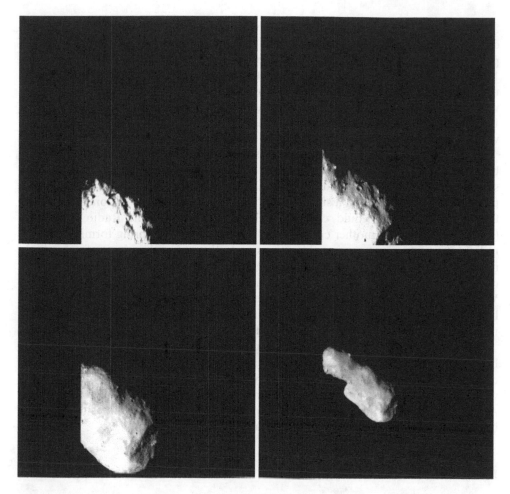

A sequence of images of Toutatis taken during Chang'e 2 extremely close flyby. The left portion of the closest images is covered by the non-illuminated side of the solar panels. Compare these images to radar images of the same asteroid in Part 2, page 358. (Chinese Academy of Sciences)

held in May 2012 to start the design of a proper asteroid mission. Launched by a Long March 3-class rocket in the second half of the 2010s, the mission would use xenon ion thrusters, first tested on an experimental Chinese satellite as recently as November 2012, to fly by (12711) Tukmit in August 2018, to briefly orbit (99942) Apophis between April and September 2020, and finally to land on (175706) 1996 FG3 in late 2023. The latter was also to be the target of ESA's Marco Polo-R sample return mission. Other multi-asteroid missions have been described in the Chinese technical literature.[468] After that, a sample-return from a main belt asteroid is on the cards.

A mission has been proposed by Chinese and American researchers that would exploit a CONTOUR-like profile to make a flyby of comet Wirtanen in December

2018 and of Schwassmann–Wachmann 3, with a return to Earth between these two encounters. It has been suggested that "some far-sighted US scientists" might propose instruments to fly on the Chinese probe and/or that NASA could retarget its 1970s-vintage, revived International Cometary Explorer (ICE) to comet Wirtanen at the same time. The Chinese probe could then explore the inner coma and nucleus while the US spacecraft, lacking a camera, would explore the tail.[469] However, the US Congress has a resolution preventing NASA from any form of space cooperation with China; an unprecedented move, even at the time of the Cold War.[470]

Possible Jupiter and solar polar missions are also known to be under study: one a ballistic flight to Jupiter that would be followed by insertion into a high inclination solar orbit, like Ulysses, and the other using a low thrust (probably electric) propulsion system to reach a similar orbit. The mission itself, provisionally planned for launch in 2016, would consist of the Solar Polar Orbit Radio Telescope (SPORT): a spinning mother craft and up to ten tethered offspring forming a VHF radio antenna to observe coronal ejections and other solar phenomena from a high inclination orbit with a perihelion around 0.5 AU.[471],[472] Moreover, the orbital design of a proper Galileo-like Jupiter mission reaching the giant planet through Earth and Venus gravity assists as well as several asteroid flybys has appeared in the technical literature.[473] Some of these expeditions could use the heavy Long March 5 launcher that is expected to enter service in the mid-2010s.

However, despite asteroid and Mars missions, and these future projects, China has for a long time looked destined to remain marginalized in planetary exploration due to the lack of openness of its military-run space program. This could change with the recent establishment of a National Space Science Center of the Chinese Academy of Sciences to centralize, select and plan programs, as well as to ease and increment international cooperation in space science.[474]

BACK TO THE KING

In the 1990s, studies of low-cost Jupiter missions by NASA's Outer Planet Science Working Group produced the MEASURE-Jupiter concept of multiple small missions launched at intervals of 2 to 3 years, similar to the Mars Surveyor program. These solar-powered spacecraft would be launched by the relatively inexpensive Delta II rocket. Missions could include miniature atmospheric probes and Galilean satellite penetrators, but the two favored ideas were the Io Skimmer and the Jupiter Auroral Observer. The former would make a single very close flyby of Io, passing within 45 km to sample its neutral atmosphere and take 1-meter-resolution pictures. The latter would be placed into a 100-day eccentric polar orbit to observe Jupiter and its polar regions and auroral emissions around periapsis and then download data and recharge its batteries during the rest of time.[475] Another low-cost Jupiter mission, the already mentioned INSIDE Jupiter, was proposed in 1998 as a Discovery-class mission and selected for further study, but not implemented.

Building on these proposals, and the Jupiter Polar Orbiter with Probes concept of the 2003 'decadal survey', a Jupiter mission called Juno (named after Jupiter's wife in

Roman mythology) was devised by scientists at the Southwest Research Institute in Texas and by JPL as the second New Frontiers flight after the New Horizons Pluto flyby. It was one of seven proposals received after an announcement of opportunity issued by NASA in 2003. Juno and Moonrise, a lander with two rovers to collect and return to Earth 2 kg of samples from the lunar Aitken basin, were selected for further study. Selection of one mission was expected for 2005 with launch early in the next decade. Moonrise appeared more politically viable, nicely dovetailing on President Bush's recently announced drive for human exploration of the Moon in the wake of the Columbia Shuttle accident. Juno's use of solar panels appeared to clash with the administration's emphasis on space nuclear power, embodied in the JIMO proposal. In spite of this, in June 2005 NASA announced the selection of Juno, cost capped at $842 million including launcher. Launch was initially scheduled for 2009 but it was soon slipped to a more realistic 2011 date, with arrival at Jupiter 5 years later after an Earth gravity-assist. The spacecraft itself would be built by Lockheed Martin with instruments from JPL, the Southwest Research Institute, APL, the Italian Space Agency ASI and other NASA and foreign centers and universities.

Juno would address the interests of atmospheric scientists, as well as investigate the auroras, the magnetosphere, and the internal structure, composition and evolution of the planet, with imaging ranked as a low priority. The primary scientific objective was to precisely determine the abundance of water, oxygen and ammonia. It will be recalled that the measurement of water had been one of the objectives of the Galileo atmospheric probe, but that had apparently entered the atmosphere within a rare 'hot spot', or the Jovian equivalent of a desert, and so it detected very little of it.[476] These measurements would enable scientists to discriminate between different, and often conflicting theories as to the formation of the giant planet. In particular, they could prove or disprove a suggestion that Jupiter formed at a different position in the solar system and then 'migrated' to its present position, as has been proposed to explain a number of phenomena and characteristics observed in the solar system, including the 'late heavy bombardment' of asteroids on the inner planets 3.9 billion years ago, the formation of the Kuiper Belt, the different asteroid populations within the main belt, and the small mass of Mars.

The Juno mission would also map the magnetic and gravitational fields in order to determine the interior structure of the planet and discern whether it had a solid core. A detailed measurement of the gravitational field, moreover, would enable scientists to infer whether the interior was in equilibrium or was undergoing deep, churning convective motions that could affect the atmosphere. It would also study the polar magnetosphere, auroras and their interactions with the atmosphere. To address these objectives, Juno would employ a spin-stabilized, solar-powered spacecraft – making it the first mission to the outer solar system not to be powered by an RTG – and it would become only the second Jovian orbiter after Galileo.

The Juno spacecraft used hardware largely based on the Mars Reconnaissance Orbiter, and consisted of a central bus connected to three large solar panels arranged in the shape of a windmill. The central structure comprised a propulsion module 3.5 meters in diameter, an electronics module (also known as the 'vault'), and the fixed high-gain antenna mounted on top of the stack, giving it a total height of 3.5 meters.

The hexagonal composite propulsion module contained four tanks for 1,280 kg of hydrazine and 752 kg of nitrogen tetroxide, which fed a 645-N engine of the same type as used by the Mars Global Surveyor orbiter. As in the case of Cassini, a shield would protect the nozzle against impacts by micrometeorites and dust in the Jovian system. Twelve additional monopropellant hydrazine thrusters mounted in 4 clusters would be used for small trajectory corrections and for changing the orientation of the spacecraft. Also mounted on the module were the sensors of most of the instruments, as well as a toroidal antenna for use when the high-gain antenna could not point at Earth and in particular during the Jupiter orbit insertion maneuver. On the top deck of the propulsion module was the thermally insulated vault, a $0.8 \times 0.8 \times 0.6$-meter titanium box with 1-cm-thick walls that held the command and data handling computer, power and data units, a pair of inertial measurement units, transmission systems and other electronic subassemblies. The thick metal walls were to protect the solid-state electronics from most of Jupiter's energetic particles in the radiation belts. The original intention was to make the vault of an exotic lightweight honeycomb tantalum structure, but this turned out to be overly complex compared to aerospace-standard machined titanium. The downside was that each wall weighed 18 kg. Inside the box were also the electronics and in some case even the detectors of a few instruments whose sensors were mounted on the outside. The titanium structure alone was more than 80 kg, and over 200 kg when the electronics boxes were fitted. Mounted on top of the vault was the 2.5-meter high-gain antenna for the Ka- and X-bands. This doubled as a sunshield while the vehicle was in the inner solar system. The radio system was partly provided by the Italian Space Agency, and it

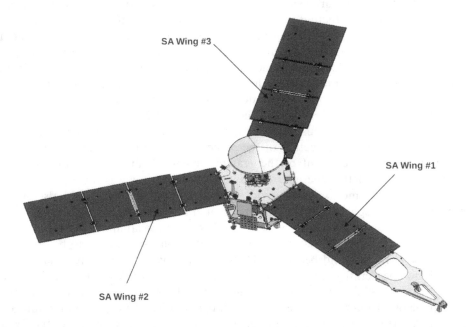

The NASA Juno Jupiter orbiter.

The Juno titanium 'radiation vault' being mounted on the propulsion module.

could be operated at a maximum data rate of 15 kbits per second. This slow rate was feasible because there were no 'bandwidth-hungry' instruments. To preclude electric charges from building up in Jupiter's energetic particle environment, the outer surface of the vehicle was covered with, or built from, conductive materials.

Attached to the sides of the propulsion module were the three solar arrays, giving Juno an overall span of about 22 meters. Smaller, 2.02 × 2.36-meter panels were mounted nearest the body and their size then increased to a total size of 8.86 × 2.65-meters, for a total surface area of 60.3 square meters. Two of the solar panel 'wings' consisted of four hinged panels, and the third had three panels and a composite boom to position magnetometers away from the central bus. The panels were covered with high-efficiency solar cells which had a thick glass cover to shield them from Jovian radiation. Although they could generate over 14 kW at Earth, their output would be just 400 W in Jupiter orbit. About half of this scarce power would be consumed by electrical heaters simply to keep the electronics warm. Augmenting the solar panels were two rechargeable batteries. The vehicle was to spin for stability, rotating at 1 rpm for cruise, 2 rpm for science sequences, and 5 rpm for increased stability during major engine burns. The spin axis, which was also the high-gain antenna axis, would normally point at Earth, but for science sequences it could be reoriented for remote-sensing by the instruments.

Nine instruments for a total of 25 sensors were onboard Juno. The mission made use of well-known technologies, so none of the instruments required any particular development effort. A microwave instrument consisted of six separate radiometers operating at different wavelengths, feeding detectors which were protected inside the vault. The instrument was based on the experience gained using the Cassini radar as a passive radiometer during the Jupiter flyby, and was to measure the absorption of electromagnetic waves by water, ammonia and other molecules in the atmosphere. The six wavelengths would enable Juno to probe to depths as great as 500 km below the outer ammonia cloud deck, much deeper than had been directly sampled by the Galileo atmospheric probe, where the ambient pressure would be 200,000 hPa, some 200 times sea-level pressure on Earth. The instrument was a key orbit-design driver, as microwave measurements of water in the atmosphere would be totally obscured by the radiation belts if they were taken at large distances from the planet. Hence, a periapsis at an altitude between the radiation belts and the planet's cloud tops was a necessity.

Two fluxgate magnetometers were mounted on a dedicated boom, one 10 meters from the center of the vehicle and the other at 12 meters. They were to map the polar magnetosphere in three dimensions. High-precision star trackers would provide very accurate data on the orientation of the craft while the magnetometer collected data. These cameras also had a special object-tracking mode that could be used to search for asteroids during the cruise to Jupiter, and to study the zodiacal light, Jovian ring and very small satellites. This instrument was a collaboration by NASA's Goddard Space Flight Center and the Danish Technical University. An auroral distribution experiment would map the distribution of low-energy electrons and the composition and velocity of ions carried by Jupiter's magnetic field to the auroral regions, including sulfur ions that originate from Io. An infrared auroral mapper built by the Italian Space Agency would take images and spectra of the upper layers of the atmosphere in order to study the chemistry and dynamics of the auroral regions and the interactions between the atmosphere, aurora and magnetic field to a depth of several tens of kilometers. This instrument would be able to penetrate to a depth where the ambient pressure would reach 7,000 hPa, just short of the depth at which the Galileo atmospheric probe had succumbed to the increasing pressure and temperature. A plasma-wave experiment would record Jupiter's radio waves using two mutually perpendicular 4-meter-long antennas, projecting from the lower deck of the propulsion module. An energetic-particle detector would measure ions of hydrogen, oxygen and sulfur from Io in the polar magnetosphere. And an ultraviolet spectrograph would provide spectra and images of the auroral emissions. The gravitational field of Jupiter was to be studied by the Doppler effect imposed on the carrier signal of the spacecraft's radio system.

The final instrument was a lightweight color camera for public outreach purposes and to provide scientific context to the mission. Its optics provided a 58-degree field of view designed to fit the entire disk of the planet in a single frame when over the polar regions, and gave a resolution of about 50 km at the poles and up to 3.5 km at the equator. It had a carousel of three colored filters and a methane filter. Imaging of the satellites could theoretically be carried out, but the range to the Galileans would be such that they would span only a handful of pixels. This was done on purpose, to minimize perturbations and orbit control maneuvers. Moreover, imaging would only be possible on the rare occasions when one of the moons happened to pass through the plane of the vehicle's orbit. The camera was initially to be a radiation-hardened version of the Mars Science Laboratory descent camera, but it was redesigned with better electronics partly based on the Mars Reconnaissance Orbiter context camera. Like that instrument, it was a push-broom imager, in this case using the spacecraft's spin to build the second dimension of the image. It was designed to last at least eight orbits, during which it was expected to take at most a hundred images, with software to reduce radiation-induced 'noise'. In keeping with its educational role, the raw data would be immediately released to the public as it arrived.

The Juno spacecraft had a total mass of 3,625 kg at launch. There had been a controversial proposal to carry a small sliver of a bone of Galileo Galilei, but in the end it carried instead a plaque and three aluminum figurines provided by the Lego toy company of Galileo and the mythological Jupiter and Juno.[477] The mission had a 22-day window that opened on 5 August 2011. The rocket was to be an Atlas V 551, the same configuration, with five strap-on boosters, as was used by New Horizons. Even this rocket, however, was not powerful enough to send the massive spacecraft directly to Jupiter. The launch vehicle was to place the spacecraft into an orbit whose aphelion was in the asteroid belt, and a two-part deep-space maneuver executed in late August and September 2012 would lower the perihelion to 0.88 AU in order to set up an Earth flyby. The gravity-assist of a 500-km Earth flyby on 9 October 2013 would increase the spacecraft's heliocentric velocity by 7.3 km/s and stretch its aphelion to Jupiter. During this flyby, Juno would spend 20 minutes in eclipse for the first and final time since launch. Most of the instruments were to be tested and calibrated during the encounter, if analyses showed that they would not be put at risk of overheating by operating at a distance from the Sun much less than that for which they had been designed. Although Juno would make two passes through the asteroid belt, the fact that it was a spinning vehicle with a wide-angle camera made it unlikely that any targets of opportunity would be investigated.

Juno was to arrive at Jupiter on 5 July 2016 on a trajectory which would take it so near the planet that at periapsis their relative velocity would exceed 71 km/s. A 30-minute insertion burn would place the spacecraft into a preliminary 'capture' orbit with a period of 107 days in which the instruments would be powered on, calibrated and begin to collect data. A 37-minute maneuver at the first periapsis would reduce the period to 11 days and establish the final science orbit, which would be roughly polar with its periapsis just 5,000 km above Jupiter's cloud tops (equivalent to 0.07 planetary radii) and an apoapsis of 2.7 million km out, well beyond the radius of the orbit of the outermost Galilean satellite, Callisto. This elliptical, polar orbit avoided the bulk of the radiation belt, since the vehicle would only cross it at apsides, and in particular at periapsis when it was moving the fastest, crossing the equator at some 60 km/s. This meant that the solar arrays and electronics would receive only minimal radiation damage on each pass. Moreover, the low periapsis, set halfway between the upper atmosphere and the innermost halo ring, would reduce the risk of damage by dust particle strikes. A clever choice of the orbital orientation ensured that the craft would be in continuous sunlight for generating power using solar panels. At first, the orbital plane would be almost perpendicular to the Earth-to-Jupiter line. This orbit would be in constant sight of the Earth and the Sun, and would have a periapsis over

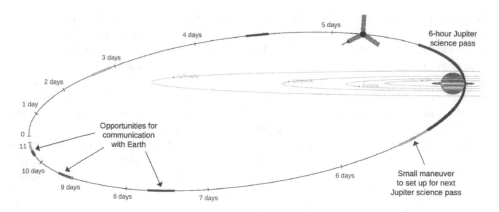

The planned orbit of Juno around Jupiter.

the dusk side of Jupiter, with Juno passing roughly along the terminator. Over the duration of the mission, the precession of the orbital plane would see the longitude of periapsis slowly move towards noon. By the end of the mission, Juno would pass periapsis over the Jovian local mid-afternoon.

Juno would operate its instruments only around periapsis, and spend most of the remaining time recharging its batteries and transmitting data to Earth. Each north-to-south periapsis pass would provide a 6-hour window for data collection, followed by a small trim burn 4 hours after closest approach in order to target the required orbit orientation for the next pass. Each periapsis would need to be 192° of longitude from the previous one to give an even grid for measurements. Remote-sensing instruments were designed to return all of their data in the first half of the mission, in order not to make them too heavy with long-term radiation protection. The Doppler effect would be essentially zero in an orbit oriented perpendicular to the Earth-to-Jupiter line, but as the orbital plane precessed toward noon this would become easier to measure, and at that time the gravity measurements would start. During gravity science passes the remote-sensing instruments would not be used, because the geometry would prevent Jupiter from crossing their field of view. The precession of the orbital plane from the dusk side towards noon would not be the only perturbation of the orbit. The latitude of the periapsis would also move north during the mission, so that by about the 30th revolution the outbound leg would start to pass deep into the radiation belts. Hence, Juno would accumulate most of its radiation exposure during the last seven orbits of its nominal 33-orbit mission. After repeatedly dipping deep into Jupiter's radiation belts, a significant extended mission seemed unlikely. In any case, by the end of the primary mission, as the latitude of periapsis moved north, the orbit of Juno would pass through the equatorial plane of the system close to the orbit of Europa and because the vehicle was not sterilized prior to launch it would need to be disposed of in order to prevent it from striking Europa and contaminating it with bacteria or other biological matter. As a result, on reaching apoapsis after its 33rd periapsis, a deorbit maneuver would cause it to dive into the Jovian atmosphere near 34°N on 16 October 2017, in a similar manner to which Galileo was disposed of 15 years previously.

Compared to the Mars Science Laboratory and the James Webb Space Telescope, in development at the same time and whose costs increased out of control, Juno was a rare example of a mission coming in on cost and on schedule, in the end costing an estimated $1.107 billion, including the Atlas launcher and operations up to its destruction. In contrast, NASA would probably further delay its Europa orbiter, planned for the 2020s, in response to budget reductions. Hence, occasional imaging of the Galilean satellites assumed some urgency, in order to fill a coverage gap which could otherwise last for decades.[478,479,480,481,482,483,484,485,486]

Assembly of Juno was finished in early 2011 at Lockheed Martin Space Systems in Denver, and it was flown to Cape Canaveral on 8 April to be integrated with its solar panels and prepared for launch. It was installed on the Atlas V rocket in late July. The launch of Juno assumed a symbolic value, by being NASA's first after the retirement of the Space Shuttle, whose final mission ended on 21 July. The launch of Juno was delayed some 50 minutes by a helium pressurization problem, but it got off within the 69-minute window of 5 August. After about half an hour of coasting in a low parking orbit, the Centaur stage restarted and, 53 minutes after launch, the spacecraft was released. Some 39 seconds later it established contact with the Canberra Deep Space Network station, then deployed the solar panels and spun itself up to 1 rpm for cruising. The Centaur stage had successfully achieved the planned heliocentric orbit of 1.0×2.26 AU, inclined just 0.1 degree to the ecliptic, and with a period of about 25 months. In fact, the launch was so accurate that, as in many previous cases, there was no requirement for the spacecraft to perform a clean-up maneuver. However, the propellant spared would remain unused, since there were no asteroid flybys planned and little prospect of a mission extension. Four days after launch, the antennas of the plasma-wave instrument were deployed. Moreover, the engine shield was exercised and all of the thrusters (with the exception of the main engine) were successfully test fired. Starting in late August, the other instruments were checked out. When tested on 26 August, the camera took images of Earth and the Moon, as a bright and a dim 'star' about 10 million km distant.

With Juno's launch, Dawn's arrival at Vesta, MESSENGER's arrival at Mercury, the Stardust flyby of Tempel 1, and the launches of the Mars Science Laboratory, Fobos-Grunt, and the insertion into lunar orbit of the GRAIL (Gravity Recovery and Interior Laboratory) for the Discovery program, 2011 was another landmark year for solar system exploration. This was even more remarkable for the fact that, except for the loss of Fobos-Grunt, all of these missions were run by NASA, an agency which had been criticized as having "a great future behind it" as a result of the absence of a clear vision by politicians for manned spaceflight after the retirement of the Space Shuttle.[487]

The Juno spacecraft settled in for a routine cruise, during which the plasma-wave sensor and magnetometer would collect data on the deep-space environment. It made its first 25-minute, 1.2-m/s maneuver on 1 February 2012. On 14 March the camera took 21 images of the sky to test how the instruments interacted with each other. The image covered a 360-degree strip of the sky and included a number of recognizable constellations, including Ursa Major, as well as extensive glare from the Sun. The first part of the deep-space maneuver was made on 30 August. It lasted 21 seconds short of 30 minutes, delivered a velocity change of 344 m/s and used 376 kg of fuel. The second part was scheduled for four days later. However, when the valves were closed at the end of the burn there were anomalously high readings of the pressure in one of the tanks, and engineers requested a few days more in order to better assess the situation. Meanwhile, the probe passed aphelion. For the second burn, the engine start sequence was modified to lower the temperature of the misbehaving line and thereby avoid an excessive pressure. The 388-m/s maneuver on 14 September went exactly as planned, and no anomalies were encountered this time. Another shorter burn on 3 October fine-tuned the trajectory for the Earth flyby. In 2013, Juno made another maneuver on 7 August and then passed perihelion on 31 August at a heliocentric distance of 131.2 million km, the closest it would ever get to the Sun.

A number of activities were planned for the Earth flyby. The spacecraft was to demonstrate data collection as a rehearsal for operations in Jovian orbit, and use the well-characterized terrestrial magnetosphere to test and calibrate instruments. Most of the payload, which had been off for several months, with the exception of the magnetometer, was turned on 4.5 days out. Only the microwave radiometer and low-energy particle detector would remain off during the whole of the encounter. The particles and fields instruments were to make coordinated observations with three THEMIS (Time History of Events and Macroscale Interactions during Substorms) magnetospheric satellites and two Radiation Belt Storm Probes in Earth orbit. The plasma-wave experiment was to try to detect lightning at closest approach, while the radio experiment would try to detect artificial transmissions. The remote sensing instruments would also be making observations. The ultraviolet spectrometer would be trained on the Earth before and after closest approach to observe the airglow on the day-side and the aurora on the night-side. It was also to collect 4 hours of lunar ultraviolet data on the inbound leg for comparison with that collected by the Lunar Reconnaissance Orbiter. In the event, the camera obtained seven four-color images of the Earth, in addition to one lunar image and one of the distant mission target, Jupiter. Moreover, tracking data was being collected when possible to investigate the 'flyby anomaly'. The star camera attached to one of the magnetometers produced an amazing video of the Blue Planet approaching with the Moon zooming by. A similar video is planned for the Jupiter approach. And the radio wave experiment clearly detected Morse signals sent to the spacecraft by radio amateurs worldwide in unison.

Juno hit a minimum distance from Earth of 559 km at 19:21 UTC on 9 October, somewhere over the South Atlantic, off the coast of South Africa. Due to the lack of NASA facilities in this region, ESA stations in South America and Australia were called in to support the flyby. Even so there was a 27-minute gap in data around the time of closest approach. About 2 minutes before closest approach, Juno entered the shadow of the Earth, where it would remain for 20 minutes, in the only eclipse of its entire mission. Just 4 minutes later, as it was receding from the dawn side of Earth, over the Bay of Bengal, the ESA ground station in Perth acquired signals showing it to be in safe mode, but transmitting. Telemetry showed it to have entered safe mode some 10 minutes after closest approach, when its self-protection logic was triggered by exceeding conservative power-consumption limits whilst in eclipse. At that point

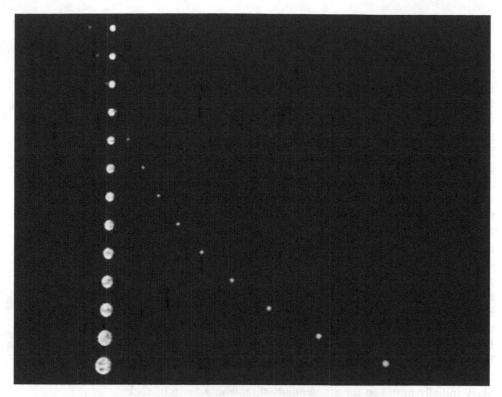

This amazing sequence of the Earth approaching and the Moon zooming by was captured by the star tracker of the magnetometer experiment on Juno during the October 2013 flyby.

it turned off the instruments and non-critical systems, and oriented the solar array to face the Sun. The craft was retrieved from safe mode within 2 days, by which time it was in a 0.98 × 5.44-AU solar orbit heading for Jupiter. The microwave radiometer was turned on 10 days after the flyby, and observations of comet ISON or what remained of it were scheduled for January 2014.[488] During the cruise to Jupiter, the star tracker of the magnetometer would collect scientific data, in particular providing a survey of the little-known small asteroid population of the main belt.

Only weeks after Juno was launched, scientists published new models of the Jovian core suggesting that, in the prevailing conditions of pressure and temperature, solid rock as well as ice can dissolve into the liquid metallic hydrogen layers. In this case Jupiter, like gas giants in other star systems, may not have a solid core after all. When it arrives at Jupiter, Juno should be able to determine whether or not this is true.[489]

AND FINALLY...

One by one, as the spacecraft stationed at the L1 and L2 Lagrangian points of Earth's heliocentric orbit exhaust their trajectory control propellant reserves, they will drift off to join those deliberately put into solar orbits. They include the Solar and Heliospheric Observatory SOHO and Advanced Composition Explorer ACE at the L1 point, and the Wilkinson Microwave Anisotropy Probe, Planck, Herschel, Chang'e 2, Gaia, and the James Webb Space Telescope at the L2 point. The first of these, the Wilkinson Microwave Anisotropy Probe, was 'de-orbited' by a 20-minute burn on 8 September 2010. Launched in June 2001, this 840-kg spacecraft spent 9 years observing the cosmic microwave background that is taken to be a signature of the Big Bang. After further maneuvers it was abandoned in a 1.00×1.09-AU solar 'graveyard' orbit. Chang'e 2 then left L2 in April 2012 bound for Toutatis. In 2013, Herschel and Planck, launched together in 2009, were also abandoned in solar orbit. In the case of Herschel, the infrared space telescope had radiation monitors like the ones carried by Rosetta, and scientists had expressed interest in keeping contact with the spacecraft as long as possible to provide radiation data. ESA unfortunately was not able to fund this mission extension. After technological tests and several burns to put Herschel on course for an Earth-trailing 1.04×1.06-AU orbit that would prevent it from impacting Earth for the next several centuries, it was commanded to shut off the transmitter on 17 June. Likewise, Planck left L2 on 14 August and entered a 1.00×1.10-AU orbit prior to being switched off on 23 October. As once happened for a spent Saturn V stage, from time to time one of these spacecraft will approach Earth, be spotted, and more than likely initially mistaken for a potentially threatening lump of rock.

REFERENCES

1 NRC-2002
2 NRC-2003
3 For the CNSR Rosetta mission see Part 2, pages 114–117
4 Schwehm-1994
5 Muirhead-1999
6 Weissman-1999
7 Woerner-1998
8 Tan-Wang-1998
9 Flight-1999
10 Warhaut-2003
11 Claros-2004
12 Keller-2007
13 Stern-2007a
14 Coradini-2007
15 Balsiger-2007
16 Gulkis-2007
17 Kissel-2007
18 Colangeli-2007
19 Riedler-2007
20 Carr-2007
21 Glassmeier-2007a
22 Nilsson-2007
23 Burch-2007
24 Trotignon-2007
25 Eriksson-2007
26 Kofman-2007
27 Pätzold-2007a
28 Berner-2002
29 Nielsen-2001
30 Glassmeier-2007b
31 Koschny-2007
32 Bibring-2007a
33 Di Pippo-1999
34 Ercoli Finzi-2007
35 Goesmann-2007
36 Wright-2007
37 Klingelhöfer-2007
38 Spohn-2007
39 Auster-2007
40 Seidensticker-2007
41 Mottola-2007
42 Bibring-2007b
43 Biele-2002
44 Ulamec-2002
45 Ulamec-2003
46 Ball-1999
47 Kronk-1984a
48 Lamy-1998
49 Flight-2002
50 Furniss-2002
51 Furniss-2003
52 Flight-2003
53 Schilling-2003
54 Barthelemy-2003
55 Clery-2003
56 Kronk-1984b
57 Lamy-2007
58 Agarwal-2007
59 Hansen-2007

60 Elwood-2004
61 Jäger-2004
62 Barucci-2007a
63 Keller-2007
64 Coradini-2007
65 Ferri-2004
66 Coradini-2007
67 Lämmerzahl-2006
68 Keller-2005
69 Küppers-2005
70 Ferri-2005
71 Montagnon-2005
72 Biele-2005
73 Edberg-2006
74 MPEC-2007a
75 MPEC-2007b
76 Barucci-2007a
77 IAUC-8315
78 Weissman-2007
79 Küppers-2007a
80 Carvano-2008
81 Barucci-2008
82 Nedelcu-2007
83 Fornasier-2007
84 Jorda-2008
85 Lamy-2008a
86 Lamy-2008b
87 Accomazzo-2008
88 Lodiot-2009
89 Keller-2008
90 Morley-2009
91 Keller-2010
92 Jutzi-2010
93 Marchi-2010
94 Küppers-2007b
95 Barucci-2007a
96 Carvano-2008
97 Barucci-2008
98 Lazzarin-2009
99 Weaver-2009
100 Belskaya-2010
101 Carry-2010
102 Weaver-2009
103 Snodgrass-2010
104 Fornasier-2011
105 Andrews-2010
106 Siersk-2011
107 Pätzold-2011
108 Coradini-2011

109 Weiss-2011
110 Morley-2012
111 Lamy-2008c
112 Tubiana-2008
113 Kelley-2009
114 Lowry-2012
115 Glassmeier-2007b
116 Verdant-1998
117 Weissman-2012
118 Muñoz-2012
119 Jansen-2013
120 For Mariner 10 see Part 1, pages 171–196
121 Potter-1985
122 Potter-1990
123 Slade-1992
124 Harmon-1992
125 Paige-1992
126 Harmon-2001
127 Harmon-2002
128 Ksanfomality-2003
129 Cecil-2007
130 For early Discovery Mercury missions see Part 2, pages 350–351
131 Jonaitis-2003
132 Yen-1989
133 McAdams-2006
134 NASA-2008a
135 Covault-2004a
136 van de Haar-2004
137 For the Galileo laser link experiment see Part 2, page 230
138 Smith-2006
139 Neumann-2006
140 Adler-2009
141 McAdams-2006
142 Rengel-2008
143 Izenberg-2007
144 Solomon-2007
145 Margot-2007
146 NASA-2008a
147 Prockter-2009
148 Murchie-2008
149 Head-2008
150 McClintock-2008a
151 Strom-2008
152 Robinson-2008

153 Slavin-2008
154 Zurbuchen-2008
155 McClintock-2008b
156 Lawrence-2009
157 Zuber-2008
158 Anderson-2008
159 For Planet Vulcan see Part 1, pages lv–lvi
160 Schumacher-2001
161 Evans-2002
162 Stern-2000
163 Vokrouhlick-2000
164 Izenberg-2009a
165 Watters-2009
166 Denevi-2009
167 Zuber-2009
168 McClintock-2009
169 Izenberg-2009b
170 Slavin-2009
171 Glassmeier-2009
172 Blewett-2009
173 Smith-2009a
174 Kelly Beatty-2009
175 NASA-2009a
176 Prockter-2010
177 Slavin-2010
178 Vervack-2010
179 Nittler-2011
180 Peplowski-2011
181 Head-2011
182 Blewett-2011
183 Anderson-2011
184 Zurbuchen-2011
185 Ho-2011
186 Kerr-2011a
187 Zuber-2012
188 Smith-2012
189 Lawrence-2012
190 Neumann-2012
191 Paige-2012
192 Kelly Beatty-2012
193 McNutt-2012
194 McAdams-2012
195 McAdams-2013
196 Belton-1996
197 Kronk-1984c
198 Belton-2005
199 Blume-2005
200 Warner-2005a

201 Blume-2005
202 Warner-2005a
203 NASA-2005a
204 Williamsen-2004
205 Blume-2005
206 Warner-2005a
207 Bryant-2005
208 Warner-2005b
209 Frauenholz-2008
210 Frauenholz-2008
211 Sunshine-2006
212 Dornheim-2005
213 Tytell-2005a
214 Yamamoto-2007
215 For the 1991 outburst
     of comet Halley see Part
     2, page 89
216 A'Hearn-2005a
217 Meech-2005
218 Tytell-2005b
219 Feldman-2006a
220 Feldman-2006b
221 Feldman-2006c
222 Bensch-2006
223 Milani-2006
224 Lisse-2006
225 Cochran-2006
226 Lara-2007
227 Blume-2005
228 AWST-2005
229 Kronk-1984d
230 Kronk-2005
231 Benest-1990
232 Carusi-1985
233 A'Hearn-2005b
234 Encrenaz-2005
235 Crovisier-2005
236 Christiansen-2008
237 Ballard-2008
238 Bennett-2004
239 Bortle-1986
240 Christiansen-2008
241 Ballard-2008
242 Rieber-2009
243 Cowan-2009
244 Sunshine-2009a
245 A'Hearn-2008
246 Crovisier-1999
247 Colangeli-1999

248 Groussin-2004
249 Lowry-2001
250 Lisse-2009
251 Wellnitz-2009
252 Hartogh-2011
253 Ferrin-2010
254 A'Hearn-2011a
255 A'Hearn-2011b
256 Bodewits-2011
257 Hermalyn-2011
258 Meech-2011
259 Schultz-2011
260 Thomas-2011a
261 Grebow-2012
262 Larson-2013
263 Farnham-2013
264 Gimenez-2002
265 For ESRO's Venus
     Orbiter see Part 1, page
     262
266 Taverna-2002
267 Winton-2005
268 Fabrega-2003
269 Fabrega-2004
270 Svedhem-2005
271 ESA-2001
272 McCoy-2005
273 Warhaut-2005
274 Maddé-2006
275 Fabrega-2007
276 Accomazzo-2006
277 Zarka-2008
278 Markiewicz-2007
279 Piccioni-2007
280 Drossart-2007
281 Bertaux-2007
282 Barabash-2007a
283 Zhang-2007
284 Pätzold-2007b
285 Russell-2007
286 Titov-2008
287 Piccioni-2008
288 Hoofs-2005
289 Mueller-2008
290 Svedhem-2007a
291 Ingersoll-2007
292 Svedhem-2008a
293 Robertson-2008
294 Smrekar-2010

295 Luz-2011
296 Mahieux-2012
297 Vaubaillon-2006
298 Lai-2012
299 For Pioneer Venus and
     asteroid Oljato see Part
     1, page 281
300 Svedhem-2007b
301 Zhang-2012
302 Damiani-2012
303 Svedhem-2008b
304 Cowen-2012
305 Svedhem-2013
306 For Pluto missions of
     the 1990s see Part 2,
     pages 374–379
307 Guo-2002
308 Guo-2004a
309 Kusnierkiewicz-2005
310 APL-2006a
311 Stern-2002
312 Stern-2004
313 Stern-2007b
314 For the Oort Cloud see
     Part 1, page lvi
315 Noll-2005
316 Buie-2005
317 Stern-2005a
318 van de Haar-2006
319 Tubiana-2007
320 Olkin-2006
321 Guo-2002
322 Guo-2004a
323 For the spot seen by
     Pioneer 10 see Part 1,
     pages 136 and 139
324 Schenk-2008
325 Grundy-2007
326 Spencer-2007
327 Retherford-2007
328 Showalter-2011
329 Reuter-2007
330 Baines-2007
331 Gladstone-2007
332 Showalter-2007
333 McComas-2007
334 McNutt-2007
335 Krupp-2007
336 Morring-2007

337 Sheppard-2010
338 APL-2006a
339 Stern-2002
340 Stern-2004
341 Schaefer-2008
342 Lellouch-2009
343 Greaves-2011
344 Buie-2010
345 Weaver-2012
346 Weaver-2013
347 Grundy-2013
348 Morring-2013a
349 Trujillo-1998
350 Spencer-2003
351 Buie-2012a
352 Lipinski-1999
353 Scott-2000
354 Lenard-2000
355 Morring-2002
356 Balint-2007
357 Morring-2003a
358 For results of the Galileo atmospheric probe see Part 2, pages 243–247
359 Young-2003
360 Spilker-2003a
361 Shirley-2003
362 Dissly-2003
363 Prockter-2004
364 Green-2004
365 Morring-2004a
366 NAS-2006
367 Balint-2005
368 Morring-2005a
369 JPL-2005
370 GAO-2005
371 For YuS and Tsiolkovski see Part 2, pages 130–132
372 Pichkhadze-1996
373 Gafarov-2004
374 Gafarov-2005
375 APL-2006b
376 Morring-2005b
377 Fulle-2007
378 Wood-2008
379 Vourlidas-2007
380 Jewitt-2010

381 Steffl-2013
382 Jones-2013
383 Turrini-2009
384 Carry-2007
385 Cunningham-1988a
386 Rayman-2004a
387 Fu-2012
388 Rayman-2004b
389 Rayman-2005
390 Russell-2004
391 Iannotta-2006
392 Schröder-2008
393 Maue-2008
394 Russell-2012
395 Jaumann-2012
396 Marchi-2012
397 Schenk-2012
398 De Sanctis-2012
399 Reddy-2012a
400 Bell-2012
401 Hand-2012a
402 Pieters-2012
403 McCord-2012
404 Clark-2012
405 Jutzi-2013
406 Ammannito-2013
407 Binzel-2012
408 Denevi-2012
409 Prettyman-2012
410 Rayman-2012a
411 Rayman-2012b
412 Rayman-2013
413 Raymond-2013
414 Rayman-2004b
415 Schmidt-2009
416 Borucki-2009
417 NASA-2009b
418 Carlisle-2009
419 Gilliland-2011
420 Yamakawa-2001
421 ISAS-2001
422 Imamura-2006
423 Imamura-2007
424 Nakamura-2007
425 Nakamura-2008
426 Ishii-2009
427 For Solar Sails see Part 2, pages 20–23 and 347–348

428 Kawaguchi-2004
429 Kawaguchi-2010
430 JAXA-2009
431 UNITEC-2009
432 Sawada-2010
433 Satoh-2011
434 Taguchi-2012
435 Satoh-2011
436 Nakamura-2012
437 Hirose-2012
438 Imamura-2013
439 JAXA-2011
440 Hand-2007
441 Berner-2005
442 AWST-1978
443 Ulivi-2004
444 Cui-2004
445 Cui-2005
446 Coué-2007a
447 CLEP-2009
448 Chen-2011a
449 Zhao-2011
450 Huang-2012a
451 Yang-2012
452 Wu-2012
453 Huang-2012b
454 Gao-2012a
455 Gao-2012b
456 For Clementine 2 and Toutatis see Part 2, pages 356–358
457 Rcddy-2012b
458 Liu-2012
459 Li-2012a
460 Li-2012b
461 Yue-2011
462 Li-2013
463 Huang-2013a
464 Hu-2013
465 Tang-2013
466 Zou-2014
467 Huang-2013b
468 Chen-2011b
469 Farquhar-2013
470 See Part 1, pages 65 and 117 for examples of early US-USSR cooperation
471 Wu-2011

472 Li-2011
473 Chen-2013
474 Cyranoski-2011
475 Wallace-1995
476 For the Galileo
     atmospheric probe
     results see Part 2, pages

243–247
477 Nature-2010
478 Matousek-2005
479 NASA-2011
480 Kayali-2008
481 Kayali-2010
482 Zander-2011

483 Morring-2011a
484 Morring-2011b
485 David-2011
486 Werner-2011a
487 April-2011
488 Hansen-2013
489 Wilson-2011

11

Red Planet blues

A MARTIAN 'SPY-SAT'

Several months after initiating the development of the twin Mars Exploration Rovers, NASA drew up its future Mars Exploration program architecture. The 2001 orbiter and the 2003 rovers were to be followed by the launch of Mars Reconnaissance Orbiter (MRO) in 2005. This would image the surface at a resolution of several tens of centimeters and recover some of the science of the lost Mars Climate Orbiter. After that, the agency planned to launch the first of the low-cost Mars Scout missions and a long-range roving science laboratory in 2007. The program envisaged the first sample-return mission in 2011 and the second in 2013.

Mars Reconnaissance Orbiter was to advance knowledge of the Martian climate and of the processes which shaped the surface of the planet; to identify sites where water may be present, or may have once existed, and assess their suitability to life; and to characterize future landing sites. To do this it would employ 'spy satellite'-resolution visible and near-infrared imagery, monitor the weather and climate, and search for additional evidence of water ice. It would target water-modified rocks and ice deposits discovered by Mars Odyssey, the Mars Exploration Rovers and Mars Express, and in particular enable the results of the rovers to be extrapolated to a broader context. And it would make high-resolution observations of the sites where Mars Global Surveyor saw 'gullies' suggestive of recent flows of fluid, and use a radar to look for subsurface ice. It was also to demonstrate high-accuracy navigation at arrival to prevent the repetition of another Mars Climate Orbiter-like catastrophe, by being capable of determining its orbit independently of tracking. And it would serve as an orbital relay for future landers and demonstrate high-data-rate telemetry using the Ka-Band. Lockheed Martin was selected in October 2001 to develop the spacecraft and to provide operational support. Because the company had expertise and software inherited from decades of operating US spy satellites, it would control the spacecraft from its facility in Colorado.

Owing to its planned very low orbit around the planet, the spacecraft was built to stricter sterilization requirements than previous orbiters, more resembling those of landers. At 1,031 kg, its dry mass was greater than previous Surveyor orbiters. With a full load of propellant it would have about twice the mass of Mars Global Surveyor. In fact, when the spacecraft proved to be lighter than the expected target liftoff weight, it was possible to add 51 kg of additional hydrazine fuel, taking the overall load to 1,196 kg. The main angular body was made of carbon-composite and aluminum honeycomb. This contained the propulsion system and electronics. Most of the instruments were on the deck that would continuously face the planet. On one side perpendicular to the Mars-facing deck was a gimbaled 3-meter high-gain antenna for a 100-W amplifier in the X-Band and a 35-W amplifier in the Ka-Band. The large size of this antenna (almost as large as the rest of the spacecraft) was required to provide data rates of between 300 kbits per second and 1.5 Mbits per second depending upon the relative positions of Mars and Earth and the size of the Deep Space Network antenna. It was one of the most capable communications systems ever launched on a planetary mission. In view of the lessons learned from Mars Global Surveyor, this antenna was to be deployed early in the mission, prior to aerobraking. Two low-gain antennas were also carried attached to the parabolic dish for use just after launch and during critical events such as orbit insertion when the main antenna could not be pointed Earthward. Hinged on the same face as the antenna were two 5.35 × 2.53-meter solar panels with 2 degrees of freedom, with a total output of 2 kW at Mars. While in the planet's shadow, the spacecraft would draw upon nickel-hydrogen batteries. The panels gave the orbiter a maximum span of 13 meters. Given the experience of Mars Global Surveyor during aerobraking, the panels on this new orbiter were designed to accommodate any realistic change in thermal loads without failure. Furthermore, because the solar panels and high-gain antenna would be on the trailing side this would bestow passive 'shuttlecock stability' during aerobraking.

Rather than a single bipropellant engine, Mars Reconnaissance Orbiter had the novel orbit insertion engine architecture of a cluster of six 170-N monopropellant engines delivering a total thrust of about 1 kN which had been cannibalized from the canceled 2001 Mars Surveyor lander. The reuse of existing thrusters not only reduced mission costs, but also increased reliability because orbit insertion could still be achieved with five operative thrusters by just lengthening the burn. For this reason the burn itself would have to be controlled by accelerometers which would sense when the spacecraft had attained the intended speed and shut the engines off. Six 22-N thrusters augmented the primary engines during the orbit insertion burn, and would be used for smaller corrections. The propulsion system was capable of providing a total velocity change of about 1,545 m/s, two-thirds of which would be required for orbit insertion. Attitude would be determined by an inertial platform which operated with star trackers and Sun sensors. Four reaction wheels and eight 0.9-N thrusters would control the spacecraft's attitude. It had an innovative dual-redundant 'cross-strap' attitude control system in which the primary system could use sensors of the backup system and vice versa, with relatively fast switching and reaction times.

Six instruments made Mars Reconnaissance Orbiter the best-instrumented US orbiter since the ill-fated Mars Observer.

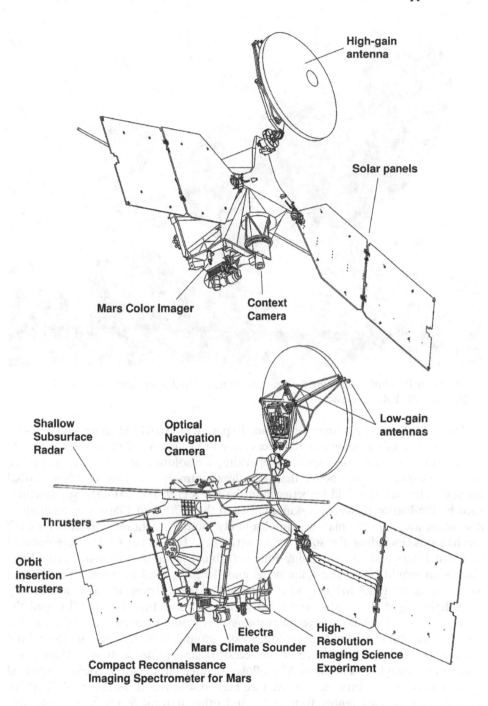

A diagram of Mars Reconnaissance Orbiter, which was equipped with the highest resolution camera ever sent to Mars.

The High Resolution Imaging Science Experiment (HiRISE) camera flown on MRO. (NASA/JPL/Ball Aerospace)

The High-Resolution Imaging Science Experiment (HiRISE) comprised a 50-cm telescope with a focal length of 12 meters mated to fourteen 2,048 × 128-pixel CCD arrays. It was theoretically capable of providing a resolution of 25 to 32 cm per pixel, which was several times better than the already amazing imager of Mars Global Surveyor. In fact, it would be better than the US KH-9 'HEXAGON' spy satellites used by the National Reconnaissance Office in the 1970s and 1980s. For the highest resolution imaging, the plan was to absolutely minimize vibration by switching off mechanisms, including the solar arrays actuators. Ten of the CCDs were covered with red filters and the remaining four with blue, green and near-infrared filters. Each scan would be a mere 6 km wide, and the individual high-resolution images would range up to 28 Mbits in size. The mass of the instrument was minimized by using lightweight glass optics and a graphite-epoxy structure, but it still tipped the scales at 65 kg. The landing sites of past missions would be of special interest to the team. Successful static landers would provide 'ground truth' to evaluate the orbital camera. It would seek evidence of unsuccessful landers, in particular Mars Polar Lander and Beagle 2. The entire Mars Polar Lander target ellipse could be covered by a mosaic of ten images, but even at high resolution it would be difficult to distinguish a crashed lander from rocks and other natural features. It would also have been interesting to locate the Soviet probes Mars 2, 3 and 6 but their landing sites were too poorly defined. In addition to identifying possible targets of interest for ongoing rover missions, HiRISE would be able to examine candidate sites for future missions, most immediately for the 2007 Phoenix Mars lander and the Mars Science Laboratory rover. The ability to identify individual rocks would be vital in selecting safe sites.[1] Context for the high-resolution pictures would be provided by another camera that used a 10.8-cm telescope and a 5,000-pixel linear CCD to provide 30-km-wide images with a resolution of 6 meters.

A visible and infrared imaging spectrometer with 544 channels was to build on the results of the Mars Express mineralogy instrument by seeking mineralogical evidence of water and hydrothermal sources in areas as small as 20 meters across. Although it was ideal for matching deposits of a particular mineral with a specific geological feature, the tradeoff was the narrower field of view of just several tens of kilometers. Due to this limitation, the initial targets would be chosen from areas already identified by the European spectrometer. The instrument consisted of a 10-cm Ritchey-Chrétien telescope mated to a pair of spectrometers, and was mounted on a gimbal to enable it to track specific targets independently of the orientation of the spacecraft. One of the team's objectives was to identify sites for high-precision landings.

When operating together, the high-resolution camera, the context camera and the spectrometer would consume up to 95 per cent of the available data link. They would be capable of acquiring up to four targets per orbit, or as many as 20 targets per sol, generating between 20 and 90 Gbits per sol; as compared, for example, to 0.7 Gbits for Mars Global Surveyor and 1.0 Gbits for Mars Odyssey. To store this data, the spacecraft had two solid-state recorders with a total capacity of 100 Gbits and dedicated lossless image compression hardware.

A climate sounder was to map the thermal structure of the atmosphere up to an altitude of about 80 km and monitor seasonal and spatial variability. It would also collect data for use in planning the arrival of landers. Consisting of a pair of 4-cm telescopes on an articulated pedestal, this instrument was a reflight of one that was lost on both Mars Observer and Mars Climate Orbiter. A color camera with a field of view which would enable it to view the planet from limb to limb was to provide daily images of the weather with a resolution of several kilometers. This instrument was a spare from Mars Climate Orbiter. The subsurface radar was provided by the Italian Space Agency, ASI, and was a derivative of the Mars Express instrument. With a 10-meter antenna operating in frequency bands between 15 and 25 MHz, the radar had a vertical resolution of between 10 and 20 meters, and a horizontal resolution of 300 meters along the orbital track and 7 km perpendicular to it. The goal was to 'sound' to a depth of about 1 km at higher resolution than was possible using the instrument on Mars Express. It would be mostly operated over the night-side, where the ionosphere was weakest and when the imaging instruments would not be in use.

In the final days of the interplanetary cruise a 6-cm navigation camera would acquire images of the planet and its moonlets to enable controllers to identify the position of the spacecraft in a demonstration of the kind of navigation that would be required to achieve a precision landing. The UHF package would provide not only a command and data link at up to 1 Mbits per second to/from the surface of the planet, but also a navigational signal for arriving landers. Scientific data would be obtained by radio tracking and by onboard accelerometers during aerobraking passes. The overall scientific data from the primary mission was expected to be of the order of 34 Tbits, marking a 10-fold increase over previous missions.

The cost of the Mars Reconnaissance Orbiter mission was estimated at about $720 million. Whilst relatively inexpensive by space exploration standards, it was clear that the Mars Exploration program had abandoned the 'cheaper' aspect of the 'faster, cheaper, better' mantra.[2,3]

The launch was initially assigned to a Delta III or Atlas III-class rocket, but this was changed to the newly developed Atlas V Evolved Expendable Launch Vehicle with a Centaur stage that was a single-engined derivative of the workhorse of solar system exploration. Remarkably, the first stage of this distant relative of the old Atlas ballistic missile was powered by a Russian engine. Although the sixth launch for this type of vehicle, it would be the first for NASA and the first on a deep-space mission, so the agency took the occasion to carry out detailed performance tests, in particular because the Atlas V was to launch the New Horizons mission to Pluto later in the year. The window for Mars would open on 10 August 2005 and last for 21 days. Mars Reconnaissance Orbiter would cruise for 7 months and reach its target in mid-March 2006, entering a 35-hour capture orbit with a periapsis at 300 km. Aerobraking would begin one week after orbit insertion. An estimated 500 atmospheric passes would be needed to lower the orbit to one that was almost circular with its periapsis over the south pole. The 92.7-degree Sun-synchronized orbit would cross the day-side equator at 3 p.m. local time. This would provide similar lighting conditions to Mars Global Surveyor. The orbit would be much lower than previous Mars orbiters to deliver a higher resolution and the ground track would repeat every 17 days. There would be an Earth occultation on almost every revolution. The primary science mission was to run to December 2008. After that, Mars Reconnaissance Orbiter would double as a relay satellite and end its nominal mission on 31 December 2010. However, it was expected that the propellant would be more than adequate to sustain a mission extension to 2014.[4]

The spacecraft was flown to Cape Canaveral in April 2005 for integration, final testing and mating with its launcher. The Atlas launcher itself experienced several glitches which required repeating the countdown rehearsal and propellant-loading, but otherwise had little impact on the mission.[5] A software problem resulted in a postponement but the mission got underway on 12 August. The Centaur coasted in low parking orbit for several minutes until it was over the southern Indian Ocean, then it reignited and achieved escape velocity. West of Australia 58 minutes after launch, Mars Reconnaissance Orbiter separated and immediately deployed its solar panels and high-gain antenna. Three minutes later, it established contact with the Japanese Uchinoura station. It was in a 1.013×1.680-AU orbit. On 15 August the color imager was aimed at Earth for calibration purposes. A 7.8-m/s correction on 27 August canceled the deliberate launcher injection bias and placed the spacecraft itself on course for Mars. On 8 September, with the spacecraft slowly slewing, the high-resolution camera obtained its first scans of the Moon and of the globular star cluster Omega Centauri. More scans of deep-space objects were taken in October and December. A second scheduled course correction was made on 18 November 2005, but the third and fourth refinements nominally scheduled for 40 and 10 days prior to arrival were deleted as unnecessary.

The approach phase began in early February 2006. Starting on 10 February the navigation camera began to snap pictures of Phobos and Deimos against the stars in order to evaluate the possibility of using the two Martian moonlets to precisely navigate future missions. Navigating a spacecraft in this way could yield its exact position with an error as low as 100 meters on the day prior to arrival, as compared to about 750 meters on the basis of radio tracking. The flight plan had options for two refinement maneuvers in the final 24 hours of the approach, but these were not needed. As a result, a mere 10 kg of hydrazine was spent during the cruise, leaving plenty for mission extensions. With only 35 minutes remaining, the propellant tank pressure was increased in readiness for orbit insertion. This was always a moment of concern for those who remembered the loss of Mars Observer. Some 16 minutes prior to the burn the spacecraft slewed to the correct firing attitude and, at 21:12 UTC on 10 March, the engine cluster ignited to fire against the velocity vector and the 22-N thrusters held its attitude. Twenty-two minutes into what was expected to be a 27-minute burn, the spacecraft disappeared behind the south polar limb of the planet. Within minutes of it reappearing over the northern hemisphere half an hour later, radio tracking confirmed that it had slowed by 1,015 km/s and was in a 426 × 45,000-km capture orbit with a period of 35.5 hours, essentially as planned. The engines had actually underperformed due to a slightly lower pressure in the tanks, but the burn had been extended by 33 seconds. The maneuver had consumed over 770 kg of hydrazine.[6]

As the orbiter passed its 10th periapsis on 24 March, the first four full-color high-resolution images were taken showing a 3.5-km-long portion of the southern highland Bosporus Planum. Although taken at an altitude of some 2,500 km, these strikingly demonstrated the camera's ability by showing details only a few meters across, including wrinkle ridges. Four more pictures were taken on the next pass. The climate sounder took low-resolution images of the north pole and also a blurry self-portrait image of the orbiter's payload deck that showed the 'forest' of other instruments. No other pictures were planned until after the spacecraft had achieved its final orbit. The radar antennas would also remain stowed until after aerobraking was complete.[7,8] The walk-in for aerobraking began several days later, with the

Shortly after orbit insertion, the Mars Climate Sounder experiment onboard MRO took this 'self-portrait' of the instrument deck. The high-resolution camera is the instrument on the right.

periapsis first being cut to 333 km and then gradually lowered to about 107 km during orbits 24 and 34. Accelerometer data was used in near-real-time to monitor density, density gradients, orbit-to-orbit variability, upper atmosphere wind speeds and other atmospheric phenomena. The periapsis latitude happened to be over the southern polar vortex and profiles gained during passes that were well within the vortex were quite smooth and matched predictions almost perfectly, but profiles at lower latitudes showed substantial variations in density over intervals as brief as several seconds.[9] By 30 August the apoapsis was down to 486 km and the time it would take for the orbit to decay sufficiently for the vehicle to crash to the surface had reduced to 2 days, so the aerobraking was halted by initiating the walk-out that would lift the periapsis out of the atmosphere. The altitude during aerobraking had dipped as low as 98 km. A total of 426 atmospheric passes over 147 days had shed 1,200 m/s. No fewer than 27 maneuvers had been performed, including collision-avoidance maneuvers for Mars Global Surveyor, Mars Odyssey and Mars Express. The final orbit was close to the required orientation. Three maneuvers and a period of drifting then set up the 3 p.m. equator crossing with the spacecraft in its planned 250 × 316-km 'primary science orbit' with its periapsis in the south. A variety of engineering tasks were undertaken during the 2 months spent drifting, including deploying the radar antennas and checking its transmitter, opening the door of the infrared spectrometer and calibrating the high-gain antenna. On 24 September the climate sounder started to collect data, and 5 days later the first high-resolution images to be obtained in the operating orbit showed a portion of Valles Marineris.

One of the first tasks for Mars Reconnaissance Orbiter was to image candidate sites in the north polar region for the Phoenix Mars lander, due to arrive in less than 2 years' time. This campaign was carried out with some urgency in order to ensure that the sites were observed before clouds could mask them in late summer or early fall. As the spacecraft passed through solar conjunction in October all but two of its instruments were turned off, the exceptions being the climate sounder and the wide-angle weather camera.[10] The vehicle resumed full operation in early November as Mars exited from solar conjunction, with all the instruments finally ready and commissioned. Also in early November, it exercised its radio relay link with the Spirit rover. In its first week of operations, Mars Reconnaissance Orbiter returned more data than both Spirit and Opportunity during 3 years on the surface, and by February 2007 it had returned more data than any other mission to the Red Planet. One of the first imaging targets was Mawrth Vallis, a long outflow channel in the ancient cratered terrain near the boundary with the northern lowlands. Mars Express data implied the presence of clays. The high spatial and spectral resolution of the new instruments allowed precise identification of the type of mineral, of the extent of the deposits and their stratigraphic relationships. The walls of the valley were revealed by the imaging spectrometer to expose a number of different clay-like minerals, indicating a wide and complex range of water-related activity. Many locations showed evidence of hydrothermal sediments. Such data would improve understanding of the early geological history of the planet.[11]

The increase in resolution from Mars Global Surveyor to Mars Reconnaissance Orbiter revolutionized our scientific appreciation of the planet in much the same manner as had Mars Global Surveyor after the Vikings. As program scientist Steve Saunders reflected, "every time we go to Mars with a new set of instruments, we see a different planet".[12] Although a total of 9,137 images were taken by the high-resolution camera during the primary mission, they covered less than 1 per cent of the planet's surface. This total included 960 stereoscopic pairs to facilitate precise 3-dimensional reconstruction of the topography of interesting features such as the landing sites of the Spirit and Opportunity rovers. The scientific themes included: composition and photometry, impacts, volcanoes, hydrothermal processes, fluvial and tectonic activity, stratigraphy, layering processes, landscape evolution, wind-driven aeolian processes, glacial processes, rocks and regolith, mass wasting, polar geology, seasonal processes, climate change, landing sites and future exploration. In addition, a significant number of observations could not be classified as any of the above.

To cite a few of the latter ... On 26 November 2006 the camera acquired the longest image ever attempted, a 120,000-line scan of the portion of sky where the just-failed Mars Global Surveyor was calculated to be in an attempt to assess its

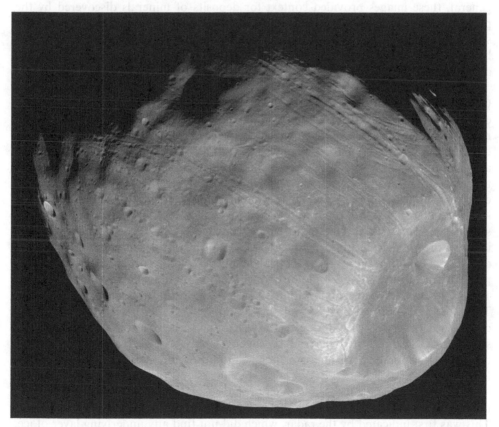

Phobos as viewed by MRO on 23 March 2008 from a distance of 5,800 km. Each pixel is less than 6 meters on the surface. (NASA/JPL/University of Arizona)

state; unfortunately it was not captured in the image. On 11 January 2007 it was aimed at Jupiter, 3.88 AU distant, and returned some very-high-resolution views of that planet and its major moons. In a similar experiment on 3 October 2007 the Earth and Moon system was imaged, but our planet was overexposed in all but one channel. On 23 March 2008 it imaged Phobos in color, showing it to be distinctly reddish, probably as a result of being covered by a veneer of material blasted off Mars by impacts. This meant that a mission to sample Phobos would also collect samples from the planet in a much easier way (but without helpful geological context) than by flying a Mars sample-return mission. There was also an attempt to image Deimos in November 2006. The diminutive moonlet was missed on that occasion but was caught in February 2009. Most remarkably, in May 2008 a picture was obtained of the Mars Phoenix lander descending on its parachute.

To address the composition and photometry theme, the camera and infrared spectrometer teams worked jointly to characterize sites identified by Mars Global Surveyor, Mars Odyssey and Mars Express (in particular clays in Nili Fossae and Mawrth Vallis) as candidate landing sites for the Mars Science Laboratory rover. In general, these images provided context for deposits of minerals discovered by the previous spacecraft. For example, chlorides detected by Mars Odyssey in the southern highlands were found to form bluish deposits in low-lying areas such as basins, and they most probably represented places where surface water had pooled and subsequently evaporated to leave behind salty crystals.

Impact cratering was one of the most popular themes, as Mars Reconnaissance Orbiter not only imaged craters for themselves but also to study gullies and strata exposed on their walls, bedrock uplifted in central peaks, etc. Fresh craters several meters in size which had been produced by meteorite impacts only months earlier, discovered by Mars Global Surveyor, were revisited by the high-resolution camera and an additional 75 were discovered by the context camera. One example formed in the northern hemisphere sometime in the summer of 2008 actually comprised a cluster of smaller craterlets, probably resulting from the atmospheric break-up of the incoming meteoroid. A larger pit appeared between January and September in that same year, spectroscopic data indicated its white ejecta blanket to be made of water ice, and the blanket disappeared over several months just as sublimating ice should do. Remarkably, if the shallow ice slab implied by these observations were also present beneath the Viking 2 site in Utopia Planitia, at a similar latitude, then that 1970s lander would have come within 10 cm of discovering water ice on the planet! The effects of seasonal dust storms on the appearance of these extremely young craters was observed, with dark ejecta rays tending to be removed or buried during such events. Scientists were particularly eager to study how fresh craters in the polar regions were modified, to improve understanding of the active processes at these latitudes.

A variety of volcanic terrains and features were imaged. One result was proof that the 'equatorial ice sea' discovered by Mars Express was actually a single lava flow. This was first indicated by the radar, which did not find any underlying layer of ice, and subsequently confirmed by the high-resolution camera. Images showed that whatever resurfaced the area, it must have occurred within the last 200 million years,

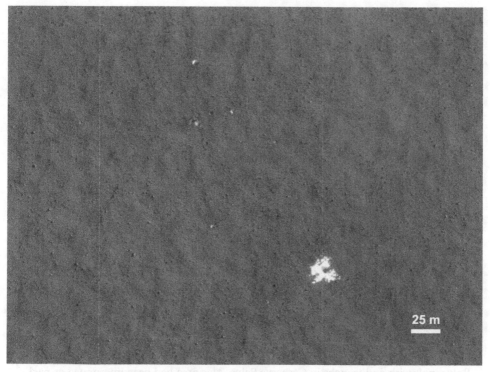

25 m

A small, fresh impact has unearthed ice immediately below the surface of Mars. Dozens of such craters were seen by MRO. (NASA/JPL/University of Arizona)

judging from crater counts. They showed, moreover, a patterned, polygonal and spiral terrain resembling in morphology those which form over terrestrial lava flows and lakes. Many of these patterns have no equivalent on Earth's pack ice. It was therefore concluded that the plain was a well-preserved, and probably young volcanic structure.[13] Wrinkle ridges in Hesperia and Solis Planum, and faults and grabens in Tharsis were the subject of studies pertaining to tectonic processes such as folding and faulting.

Fluvial and hydrothermal processes were a relatively popular theme in terms of the number of pictures taken. In particular, the gullies that seemed to have formed during the Mars Global Surveyor mission were targeted. The fact that their bright appearance had not changed in the months since their discovery implied that they were not bright as a result of being ice-rich, but for some other reason. Four more young-looking gullies were discovered. Moreover, infrared spectra did not show traces of the hydrated minerals that would be expected if salty groundwater were released. Laboratory studies demonstrated that in the weak gravity of Mars it was possible for flows of dry, unconsolidated granular debris to create such features, suggesting that they might not have been created by liquid water after all.[14] Long-duration studies of new gullies showed that these were more likely to form during the winter, further indicating that water was not involved, as this was the season in

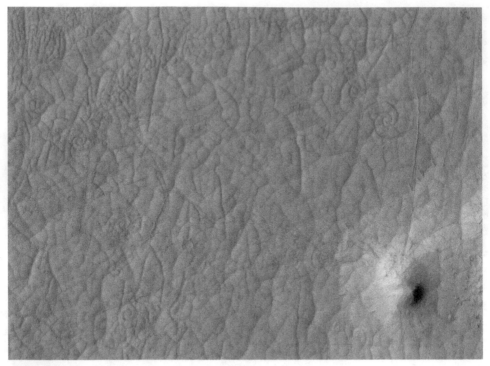

Polygonal swirls and troughs in Cerberus Palus. This area had been interpreted as pack ice rafts from Mars Express images (see Part 3, page 353) but MRO's higher resolution images proved it to be of volcanic origin, and similar to 'lava coils' in Hawaii. (NASA/JPL/University of Arizona)

which water was least likely to melt. Researchers instead suggested that the gullies were due to the winter accumulation of carbon dioxide frost triggering avalanches of dry sand. Some of the gullies even had bright objects on their aprons in early spring. These might well be slabs of carbon dioxide ice traveling downslope on a cushion of sublimating gas and plowing furrows along the way. Nevertheless, older gullies still appeared to be due to the release of liquid water. Clear signs of the presence of water ice in the upper crust were found in several recent craters where channels, branched fans and ponds were formed as the impact melted ice. Scientists had hypothesized that impacts on ice would trigger local rainfall, but no evidence was apparent to substantiate this. On the other hand, long-term monitoring by the orbiter showed that dark rivulets formed and grew, before disappearing. Unlike the controversial gullies, these occurred only in midsummer and on equator-facing slopes in seven places at intermediate southern latitudes. The rivulets began at rocky outcrops during the warm season, when the temperature probably averaged around 0°C, at times growing by tens of meters per sol, until early autumn, and then faded during the cold season. They were usually a few meters wide and up to hundreds of meters long. The simplest explanation was that they formed as a mix of water and salt containing enough of the latter to act as an efficient antifreeze, dampening and hence darkening the soil. Unfortunately, these 'returning slope lineae', as they were designated, were smaller than the pixel size of the infrared spectrometer, so it was not possible to analyze their composition to confirm the presence of salty water. Another puzzle was the process by which such a water reserve could have formed. It could be either groundwater or atmospheric water vapor adsorbed by hygroscopic salts, but both explanations had substantial difficulties. Neither was it clear why rivulets would occur only at intermediate southern latitudes.[15]

The young Athabasca Valles was a target of particular interest because it was believed to be the youngest example of a flood- or glacier-carved outflow channel. But high-resolution images showed it to have been formed by a geologically recent lava river that erupted from a fissure in Cerberus Fossae. The glacier-like features were produced by lava flows. This was confirmed by Mars Odyssey's gamma-ray spectrometer, which showed Athabasca Valles to be one of the driest areas on the planet.[16] This showed once again that orbital imagery can be deceptive, and that, as in the case Spirit at Gusev, surface missions sent to investigate putative water-related features can all too easily find themselves at lava-covered sites.

Vastitas Borealis, the lowest portion of the northern plains, had been interpreted as a deposit of very fine sediments on the floor of an ancient ocean. Seen at very high-resolution, however, no materials consistent with this process were observed and, even more remarkably, meter-sized boulders were omnipresent. A long-lived ocean ought to have ground all rocks down to pieces no larger than a grain of sand. Either there had been no ocean, or the sediments that it laid down were masked by volcanism.[17] Outflow channels were imaged in the hope that the much improved resolution would reveal flood-transported boulders. (In fact, this had been one of the scientific objectives of the original mid-1980s proposal for the Mars Observer camera.) Boulders were seen associated with dunes within Holden crater. Related to fluvial research were observations of the stratigraphy of layered deposits which might preserve the history of the changes in environmental conditions. Horizontal layers deriving from lake-like settings were imaged in craters Holden, Gale, Jezero and Columbus, where standing water could have persisted for thousands of years, as well as in Xante Terra. Unsurprisingly, some of these settings made the shortlist of Mars Science Laboratory candidate landing sites. Traces of localized alteration by water were discovered along fractures in the layered terrain of western Candor Chasma, where sulfates had been detected by Mars Express. From high-resolution imagery, it was inferred that this alteration occurred entirely underground and was subsequently exposed by erosion.[18]

It was possible to use stereoscopic observations to determine the 3-dimensional structure of the strata and measure their true thicknesses. This was done for stair-stepped layers in four craters in the near-equatorial Arabia Terra that are believed to record more ancient surface conditions than, for example, the polar terrain. This reconstruction showed that although the thickness of the layers appeared irregular from an overhead perspective, there was actually a pattern and statistical analyses revealed significant periodicities which suggested that the layering had occurred in rhythmic climate cycles. For example, in the crater Becquerel a 10-layer sequence repeated 10 times. Although this could be linked to periodicities in the eccentricity of

Extensive layering in Becquerel crater. (NASA/JPL/University of Arizona)

the planet's orbit of the Sun, axial variations were probably more significant. In particular, small axial wobbles with a period of 1.2 million years are able to mimic the 10-to-1 pattern seen in Becquerel crater.[19]

The landscape evolution theme was particularly diverse, and actually imaged sites all across the planet. The wind processes theme also included joint studies with the Opportunity rover and showed that basaltic sand was continuously being blown out of Victoria crater to create the characteristic dark 'tails'. Studies of the predominant wind direction at Columbia Hills were carried out in cooperation with the Spirit rover. In other observations bright streaks downwind of small ice-filled craters and ice-rich dunes were imaged, and a number of dust devils were caught in the act. The context camera provided a record of dust devil activity over a wide range of latitudes. Studies in ice-related processes targeted a number of middle-to-high-latitude sites. Polygonal patterns 5 meters across were found to be present in almost every image of the northern polar region, a pattern that seemed to imply the effect of thawing cycles on ice at shallow depth. Mid-latitude terrains that looked ice-rich and were presumed to have been deposited in the most recent era in which the spin axis was highly inclined, proved to be of a greater age. Dust-covered ice mounds were imaged at mid-latitudes. These could be related to gullies and to the presence of shallow groundwater. Traces of ancient glaciers were found in sites such as the Argyre basin. Imaging of dune fields in Nili Patera hundreds of days apart detected,

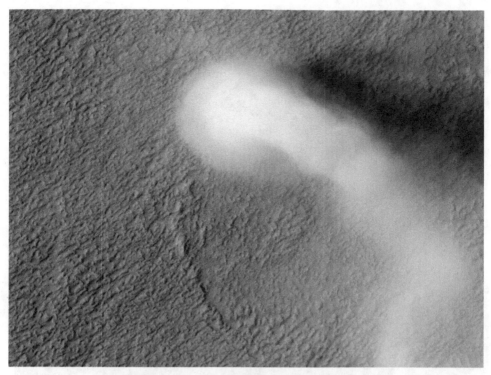

A 'monster' dust devil caught in the act by the MRO high-resolution camera in the late afternoon in Amazonis Planitia. (NASA/JPL/University of Arizona)

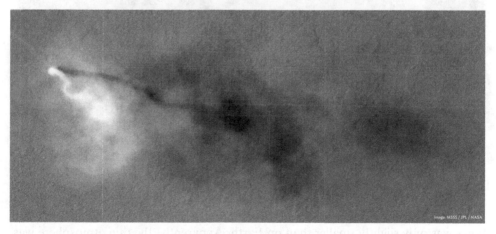

A context camera image of the same dust devil. The plume is about 140 meters across. (NASA/JPL/University of Arizona)

An avalanche caught in the act by MRO at the edge of a northern polar cliff on 19 February 2008. (NASA/JPL/University of Arizona)

for the first time, dune ripple migration on Mars. Dune fields had long been believed to be more or less static, given the thin atmosphere, and to have been formed when the planet had a thicker atmosphere. Remarkably, the computed displacement of dunes was only slightly smaller than on Earth. Apparently, the thin atmosphere was compensated in some manner by the lower gravity, making the transport of sand easier. Moreover, dune migration may be helped by peculiar topographies creating unusual wind circulation patterns.[20,21]

The mass wasting theme targeted slopes, landslides and possible landslide sources, as well as gullies and mesas. The most amazing image of mass wasting was actually

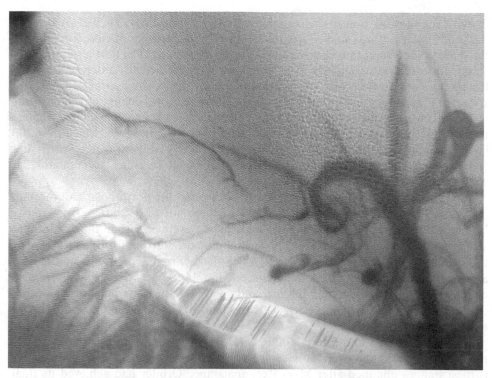

Delicate, tattoo-like dark trails left by dust devils on a light-toned terrain on Mars. (NASA/JPL/University of Arizona)

obtained for the polar geology theme. It was taken on 19 February 2008 and showed a 700-meter-tall 60-degree layered cliff uncovered by the retreating early-spring northern polar cap at a latitude of 84°. There were at least four avalanches in the act of raising clouds of dust hundreds of meters across. The avalanches appeared to have been caused by the thaw reducing the cohesivity of the mix of rocks and carbon dioxide ice. In a related observation, images of a portion of Cerberus Fossae revealed a field of boulders which had recently rolled downslope; so recently, in fact, that the wind had not yet obliterated the tracks they created. For some scientists, this was proof of recent seismic activity, all the more so as Cerberus Fossae is located near the volcano Elysium Mons.

Polar geology was another productive theme. The thickness of layered deposits was reconstructed by 3-dimensional data, and the stratigraphy studied in detail to sub-meter resolution. Images resolved the thickness of the individual layers of the northern terrain, showing them to be as thin as 10 cm but often covered with dust that masked their real thickness.[22] Layering at the south pole appeared to be more degraded, complicating the analysis. Moreover, the southern polar terrain showed peculiar structures such as networks of rectilinear cracks. Radio waves scattering within these cracks could explain why the radar appeared unable to penetrate these terrains to their base. Over 500 images targeted seasonal patterns and processes, in many cases showing dramatic changes in very short timescales. The sublimation of the south polar cap was studied from December 2006, paying particular attention to the dark 'spiders' and fans discovered by previous missions. The northern polar cap sublimation started in January 2008. Scientists were particularly interested in how frost sublimation interacted with polar dune fields. Defrosting patterns similar to southern fans were seen on some dune fields. Morphological changes from one year to the next were investigated for the non-aeolian surface changes and climate change theme. Terrains targeted included the south polar 'Swiss cheese' (of which over 100 images were taken), mid-latitude sites that appeared to giving off water, scarps and troughs, and indeed any location anywhere that could be either accumulating or sublimating ice.

A campaign to repeatedly image dune fields around the north polar cap through 2008 and 2010, spanning a local year, showed them not to be static, as previously believed, but to be shaped in dramatic ways by the cycle of carbon dioxide ice and strong wind gusts. The sublimation of carbon dioxide, starting from the ice base at the brink of the dune, created gas jets which destabilized sand grains and caused avalanches, alcoves and gullies that reshaped the dunes. Moreover, winds were so effective in eroding and modifying the landscape that the scars of past avalanches could be obliterated within a year. This mechanism of dune modification has no terrestrial counterpart.[23]

Investigating past and future landing sites was a primary objective of the high-resolution camera, and over 400 images were taken for this theme. One of the very earliest images obtained after Mars Reconnaissance Orbiter had achieved its final mapping orbit in October 2006 was a stunning view of the Opportunity rover and its tracks on the rim of Victoria crater. Even although the image lacked some of its color channels, within hours of its acquisition it was being used by the rover team to select traverse targets. As related in previous volumes of the series, the landing sites of the Vikings and Mars Pathfinder were photographed and hardware from these missions was finally precisely located.[24,25] The first image of the Viking 2 lander was irrecoverably smeared by vibrations imparted by the high-gain antenna actuator inadvertently 'unparking'. The entire landing ellipse of Mars Polar Lander was photographed but was found to be strewn with bright and dark features on the meter-scale that could mimic the lost lander, making its identification difficult if not actually impossible. Part of the landing ellipse of Beagle 2 was also imaged, as were the 'general areas' of the Soviet Mars landers, but a rigorous search would require hundreds if not thousands of images. The best method seemed to be to seek the parachutes, but it appears that the parachute of Mars 2 never opened and it is not known whether Mars Polar Lander and Beagle 2 deployed their parachutes.[26] In spite of this, while scrutinizing high resolution images of the predicted point of landing for Mars 3, Russian enthusiasts identified objects on the surface which, in follow-on images as well, looked like the parachute, retrorocket pack, heat shield, foam shock absorber, and long-silent lander of the old Soviet mission. In addition to candidate sites for the Phoenix Mars lander, more than 50 sites were imaged for the Mars Science Laboratory rover. A number of public participation programs were also introduced, including using the Internet to enable members of the public to

The bright circular object at the center of this image could be the parachute of the Soviet Mars 3 lander, found 41 years after its arrival. Other smaller features in the environs have been tentatively identified as the retrorocket package, heat shield, foam shock absorber and the lander itself. (NASA/JPL/University of Arizona)

propose targets and to participate in the search for missing missions, in particular Mars Polar Lander.[27],[28]

Whilst the context camera did not produce such stunning scientific results, as of 2010 it had mapped 50 per cent of the planet at 6-meter resolution. The imaging infrared spectrometer mapped the composition of about 66 per cent of the surface at 200-meter resolution. It detected evidence of hydrated silica, or opal, indicating that liquid water existed in some places as recently as 2 billion years ago. One of the opal deposits was in Valles Marineris. In Nili Fossae and two adjacent regions the imaging spectrometer found a relatively large deposit of a mineral that proved to be magnesium carbonate. High-resolution pictures revealed these deposits to be in bright eroded mesas. The carbonate deposits appeared always closely associated with clays and olivine, suggesting that carbonates and clays formed when olivine, which readily erodes in the presence of water, interacted with water at some time early in the history of Mars. This proximity suggested that the waters that formed clays in Nili Fossae were not as acidic as those which made hematite at Meridiani, as acidic water would have dissolved the carbonates.[29] The instrument detected carbonates in several craters where the impact had penetrated kilometers below the surface. It is possible, in fact, that the missing carbonates were simply buried deep beneath younger volcanic rocks. Calcium or iron carbonates were found on the rim of the giant 500-km crater Huygens, amid clays. Other carbonates were detected in the vicinity of a small crater within the crater Leighton. Due to its high resolution, the spectrometer detected 10,000 deposits of clays, many of which were too small to have been detected by Mars Express. The deposits were confirmed to be located primarily in the ancient southern highlands. Clays were often associated with the walls, central peaks and ejecta of craters, indicating that the clays formed ancient, deeply buried stratigraphic layers. Moreover, they were often buried under olivine layers that formed at times when water was no longer present in quantity on the surface, for otherwise the olivine would have weathered away. The instrument also detected the mineral kaolinite, which (on Earth) is associated with hydrological systems. But the fact that these deposits were rare and only a few hundred meters across argued against widespread water.[30]

An analysis of all northern hemisphere craters with diameters exceeding 30 km was undertaken in cooperation with the mineralogy instrument on Mars Express. These craters were deemed large enough to have dug through the layer of volcanic rock that blankets the entire hemisphere to depths ranging from hundreds of meters to thousands of meters. Nine such craters showed spectral signs of clays and other hydrated minerals, evidently unheated by the impact, that belonged to the planet's ancient surface. This confirmed that the wetter environment that had left so many traces in the southern highlands also influenced the northern hemisphere, although all traces of it there have been buried by lava, dust and sediments.[31] The existence of buried hydrated minerals did not in itself mean that there was once an ocean in this low-lying region, but it was encouraging for that hypothesis.

Infrared observations of Phobos and Deimos were also carried out, resolving in particular the disk of the latter for the first time in this part of the spectrum.

The shallow radar investigated the layered terrains centered on both poles, with a

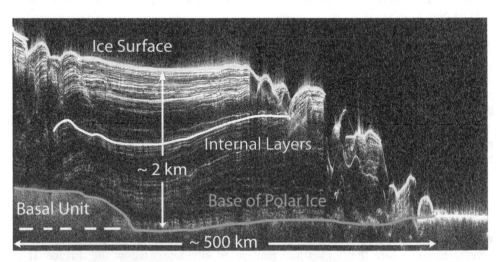

A cross-section of the northern polar cap from the shallow radar instrument on MRO. (NASA/JPL-Caltech/ASI/UT)

resolution sufficient to study their stratigraphy in detail. The data for the northern polar terrains confirmed the periodicity, with finely spaced layers of reflecting ice clustering into packets alternating with layers of mainly pure ice of lower radio-wave reflection. These regions could have formed at times when the orbit of Mars was more circular and the climate was more stable, with only the occasional storm to deposit dust; or they could have been formed at times of small and stable spin-axis tilt. Moreover, the measurement of the deformation of the underlying rocky surface implied a thickness for the Martian northern polar lithosphere of more than 300 km. A layer several hundred meters thick of almost pure ice was discovered to underlie the rugged Deuteronilus Mensae on the edge of the northern hemisphere lowlands. Chasma Boreale, also on the fringe of the northern polar cap, is a 500-km-long canyon resembling the Grand Canyon on Earth. Its origin had long been disputed, with one hypothesis being a catastrophic flood triggered by volcanoes which melted the base of the polar cap. However, radar slices through the canyon revealed it to be relatively ancient, with a stratigraphy which ruled out the melting origin. Instead, Chasma Boreale appeared to derive from the way in which the pre-existing terrain and ice deflected winds and controlled where and how the layered terrain accumulated. A similar process appeared to have been responsible for the spiraling troughs which radiate outwards from the pole, in that the troughs seem to have been carved by winds of the anticyclonic atmospheric structure located over the northern pole. Wind-carved canyons in Promethei Lingula and the region of Australe Mensa in the southern hemisphere were also studied. Layers in Australe Sulci appeared to be recently exposed at the surface. Lobate flows at southern mid-latitudes along the rims of the Hellas and Argyre basins had long been suspected to mask water ice deposits at shallow depths. This was confirmed by the radar, which probed the internal structures of many such features and their contexts. Deposits appeared to consist of massive ice glaciers covered by relatively thin layers of dust and debris.

No fewer than 58 radar passes by MRO were dedicated to mapping channels buried under the surface of Elysium Planitia, believed to have been covered by volcanism within the last few hundred million years. These allowed scientists to reconstruct the history of Marte Vallis, the longest young outflow channel of the region, in detail. It had been a mystery whether Marte Vallis, over 1,000 km long and 100 km wide, was carved by a single, long event or by a number of catastrophic floods that occurred in the last half billion years. Such studies had been hindered by the fact that the area had been subsequently partially filled with lava flows that masked most of its geological history. Radar observations revealed that smaller branches of the channel were carved during the first phase, leaving behind four streamlined islands that are still visible above the volcanic floor. In the second phase, a deep and wide single channel was etched. The origin of the floods was traced back to an underground water reservoir in the Cerberus Fossae fractures. What is most interesting of Marte Vallis is that, given the uncertainty of the dating of Martian geological features, it could be as young as 10 million years.[32,33,34,35,36,37,38] The radar detected 'reflection free' zones, areas at the base of the residual south polar cap scattering back extremely faint signals. These were interpreted as deposits of low-porosity carbon dioxide ice, in places as much as 700 meters thick, holding the equivalent of up to 80 per cent of Mars' atmosphere. The area matched a terrain marked by collapse pits and other signs of the release of volatiles. These deposits would probably melt and enter the atmosphere when the axial tilt of the planet increased, in turn increasing its density. These deposits held orders of magnitude less carbon dioxide than was required for an increased greenhouse effect and a warmer, wetter Mars, but if they completely evaporated they would lead to more frequent and violent dust storms. On the other hand, their evaporation could create a few 'oases' where liquid water could exist.[39,40]

While the radars on Mars Reconnaissance Orbiter and Mars Express have been successful in finding subsurface water ice deposits, thereby showing ice to be very common on the planet, they have not succeeded in locating any of the liquid water presumed to have survived from the early era – if such 'aquifers' remain then they must be at depths beyond the reach of orbital radars.[41]

A remarkable observation, in addition to routine monitoring of the weather, was carried out by the climate sounder. Operating during its first southern winter, the instrument identified, for the first time, polar carbon dioxide clouds discharging snow, thereby proving that snow, as opposed to the atmosphere freezing onto the ground, was the primary means of forming the polar cap during winter.

Having a low periapsis over the south pole, the orbit of Mars Reconnaissance Orbiter was particularly suited for gravity studies of this polar cap. Radio tracking provided a measurement of its mass which, when combined with volumetric data derived from altimetry obtained by Mars Global Surveyor, proved its density to be only slightly greater than that of water ice, meaning that water ice must be its main ingredient. This made the southern polar terrains of Mars one of the largest deposit of water in the inner solar system.[42] Tracking of Mars Reconnaissance Orbiter for gravitational studies also provided interesting insight into the dichotomy between the planet's extremely flat northern lowlands and the rugged southern highlands. One theory proposed to explain this involved a giant impact. Another theory involved a hemisphere-scale mantle circulation that either diminished the crust in the northern hemisphere or reinforced it in the southern hemisphere. Although the 'Borealis basin' was a poor fit to the boundary of the dichotomy, the boundary was deformed by the development of the Tharsis region. Using gravity data from Mars Global Surveyor and Mars Reconnaissance Orbiter to 'subtract' Tharsis and thereby measure the thickness of the underlying crust, it was determined that the dichotomy boundary could be fitted by a $8{,}500 \times 10{,}600$-km ellipse that could mark the impact of a planetesimal approximately 2,000 km in diameter. Altimetry of adjacent regions, such as Arabia Terra, showed the possible presence of multiple rings. Although this basin would be the largest such structure yet identified in the solar system, the impact would have released only about 1 per cent of the energy of the catastrophic strike on our planet which is believed to have formed the Moon. One objection to the giant basin theory of the Martian lowlands was that even if it did not disrupt the entire planet, the evidence of such an event would have been masked by its own impact melt. But simulations have shown that low impact velocities and oblique angles can produce basins such as Borealis. One way to further investigate the matter would be to land seismometers to measure the thickness of the crust and determine its structure and composition, to compare the bedrock in the highlands with that of the lowlands.[43],[44],[45],[46]

In early 2008 Mars Reconnaissance Orbiter synchronized its orbit with the landing site of the incoming Phoenix Mars lander in order to record telemetry from that craft during its entry, descent and landing. After serving as the primary relay for Phoenix, the orbiter completed its primary science mission in December 2008. On 26 August 2009 it rebooted its computer and entered safe mode for the fourth time since the start of the year. In debugging the anomaly, engineers discovered a potential, albeit unlikely, mission-ending condition that involved multiple reboots. To eliminate this scenario, software patches were written to amend data files in the onboard flash memory and these were uplinked in late November and thoroughly exercised. All science and data-relay operations were suspended during this hiatus. In early December the orbiter was released from safe mode, but science observations could not be resumed until the instruments had been checked and recalibrated.

A 2-year mission extension to 2012 was granted in order to investigate surface changes such as dune field migration and Swiss cheese pits. It was to characterize the candidate landing sites for the Mars Science Laboratory and ExoMars rovers and then relay for either or both of these missions. As of April 2010 it had 290 kg of propellant remaining, of which 120 kg was intended for supporting the Mars Science Laboratory. It was consuming propellant at a rate of about 15 kg per annum, and so ought to be able to continue throughout the 2010s and possibly beyond.[47] This was particularly important, as no space agency currently had plans for an orbiter with a sub-meter imaging resolution. Unfortunately, during the summer of 2011 a problem with one of the CCDs of the high-resolution camera prompted two precautionary shutdowns in a fortnight. While the behavior of this sensor was being investigated, the camera resumed operations with 13 CCDs and with a ground track image width limited to 5.4 km instead of the usual 6 km. The most limiting factor on the mission was the hardware failures in the telecommunication system, which had eliminated some of the redundancies.

By early 2012 the orbiter had returned in excess of 155 Tbits of data, including over 22,000 high-resolution images covering 1.4 per cent of the surface, and only 0.25 per cent of it in stereo. On the other hand, the context camera had mapped over 75 per cent of the surface of the planet at 6-m resolution, and 6 per cent of it in stereo. The radar had taken in excess of 10,000 swaths.[48] There had also been a proposal to revive the optical navigation camera used at arrival, in order to use it to locate the failed Mars Global Surveyor. This exercise would rehearse the technique of locating a sample canister whose orbit was not precisely known.[49]

During the summer and fall of 2013, the high-resolution camera on the orbiter was used to observe the bright comet ISON. The first images were taken in late August, when the comet was still some 148 million km from Mars, but it was not detected. Then ISON passed within 10 million km of the planet in early October and in late September the spacecraft took images of it as a bright spot possessing a faint coma. Observations were also attempted from the ground by the Opportunity and Curiosity rovers. These observations revealed that the nucleus was probably less than 1 km in size, which could have been one of the reasons why it did not survive its close perihelion pass in November.

On or about 19 October 2014 Mars Reconnaissance Orbiter, as well as other Mars orbiters, might be called upon to make observations of the long-period comet C/2013 A1 Siding Spring, which may pass about 135,000 km from the planet. This pass is so close, in fact, that the planet (and probes on the surface and in orbit) will probably pass through the comet's coma at a relative speed of some 56 km/s, raising concern about large cometary particles impacting the orbiters at high velocity. The comet's nucleus itself, believed to be several kilometers across, could be resolved by the high-resolution camera, although the instrument is not optimized to image such a faint object. If this is done, it would be the first Oort Cloud comet to be observed at close range by a spacecraft. There is also the possibility that ionized gases accompanying the comet might trigger short-lived auroras interacting with the weak fossil magnetic field that could be observed by the orbiters.[50]

The accumulated results of the recent orbital and surface missions support the hypothesis that abundant water existed in the past on a warmer Mars. But scientists remain divided about the times and durations of these wetter periods, and they are coming to the view that there were only brief spells during which precipitation would not have been sufficient to form extensive water drainage systems. In fact, all of the typical water-related Martian geological features and phenomena appear to be explicable by brief bursts of water remaining liquid for mere decades, or centuries at most. If this is so, then these brief periods may have been triggered in the first billion years of the planet's history by the occasional impact of asteroids or comets injecting hot vapor into the atmosphere. This was confirmed by the interpretation of spectral observations collected over a period of several years by Mars Express and Mars Reconnaissance Orbiter of thousands of clay-bearing outcrops. Aluminum-rich clays that would have formed on the surface were relatively rich. However, iron- and magnesium-rich clays that formed at shallow depths seemed to be scattered all over the planet. Furthermore, minerals were found within clay deposits compatible only with their having been formed in warm subsurface hydrothermal environments. The implications of this for finding evidence of ancient life are not clear.[51,52]

POLAR LANDER STRIKES BACK

After losing Mars Climate Orbiter and Mars Polar Lander, NASA transformed its recently initiated Mars Surveyor program into the Mars Exploration program, one part of which called for a series of 'low-cost' Mars Scout missions to complement the systematic approach by providing a means of rapidly responding to scientific discoveries. As with the Discovery program, each such mission would be led by a Principal Investigator. The first Mars Scout was to be launched during the August 2007 launch window, in a gap between the 2005 Mars Reconnaissance Orbiter and the 2009 Mars Smart Lander (later renamed the Mars Science Laboratory), with an initial cost cap of $325 million. NASA received no fewer than 24 proposals from universities and research centers. These suggested orbiters, rovers, polar landers, aircraft and gliders, lander networks and penetrators. Four were selected for a 6-month detailed study, one of which would fly. These included the Phoenix lander, the Aerial Regional-scale Environmental Survey (ARES) airplane, SCIM (Sample Collection for Investigation of Mars) and the MARVEL (Mars Volcanic Emissions and Life scout) orbiter which would specifically seek signs of volcanic or organic activity.

Proposed by the Arizona State University, JPL and Lockheed Martin, SCIM would reach Mars in May 2008 and use a 1,200-km flyby to arrange a second pass in April 2009, when the atmosphere was most likely to be dusty. On this second flyby the closest point of approach would be as low as 40 km, at which time an inlet at the extreme front end of the spacecraft would let in an atmospheric sample and a port on its side would open to expose an aerogel sample collector to capture dust particles. This flyby would lower the spacecraft's perihelion to the vicinity of the Earth's orbit. A course correction maneuver would deliver the samples to Earth in January 2011. The spacecraft would be enclosed in a conical thermal shield for protection and passive aerodynamic stabilization while intersecting the Martian atmosphere. It would carry two dish-shaped solar panels which, along with the high-gain antenna, would be stowed during this time. Inside the shield were a large propellant tank, avionics and a return capsule of Genesis and Stardust heritage. SCIM was expected to collect at least a liter of atmospheric gas and 1,000 large dust particles, along with millions of smaller particulates. The mission would have less stringent planetary protection requirements than a surface sample-return because the samples would have been sterilized by solar ultraviolet in the upper Martian atmosphere.[53,54,55]

The ARES proposal sought to revive the Micro-mission Mars Airplane. The objective was to make measurements of the atmosphere, surface and interior using cameras, magnetometers and spectrometers mounted on a small aircraft that would fly a 500-km route at an altitude of 1 or 2 km over a duration of about 1 hour. The mission was to be launched on a Delta II and arrive at Mars one year later. The entry would be similar to the Mars Exploration Rovers, except that instead of impacting the planet the cruise stage would perform a deflection maneuver in order to make a flyby during which it would collect data from the aircraft and relay it to Earth. The target site would be selected to enable Mars Reconnaissance Orbiter to serve as a backup for data relay. The airplane would consist of a blended fuselage and wings that spanned 6.25 meters, had a total area of 7 square meters, and supported a pair of folding tail booms for an inverted-vee tailplane. After being lowered from the backshell on a bridle, the aircraft would be released attached to a parachute which would hold it stable whilst it deployed its appendages, then the parachute would be jettisoned. In its operating configuration, ARES would have a length of 4.45 meters. On reaching level flight at the end of a dive, the propulsion system would be activated. After studies indicated that very large propellers would be needed in the thin air, it was decided to use a bipropellant rocket. The fully fueled mass of the aircraft was 175 kg. After it had flown an essentially straight path, possibly incorporating 180-degree turns, the vehicle would be commanded to impact on the surface in a way which would minimize the breakup of the airframe and thereby

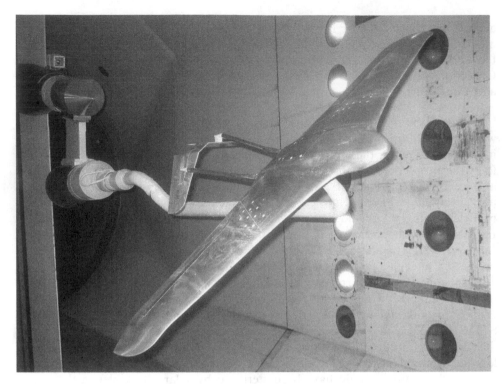

A wind tunnel model of the ARES Martian airplane.

reduce the risk of biological contamination. The Reynolds numbers and flight speed of Mach 0.62 to 0.71 were selected in order to enable designers to test the aerodynamics in realistic conditions in terrestrial wind tunnels and high-altitude balloon drops. After a series of tests in simulated Martian conditions using NASA's Langley transonic tunnel, a half-scale model was flown in the summer of 2002, culminating in the successful demonstration of deployment and flight leveling at an altitude of over 30 km. Later full-scale tests confirmed the soundness and readiness of hardware.[56] The fact that ARES was not selected was hardly surprising because, whilst it would raise public interest in the program, scientists believed that an aircraft flight on Mars would not mark a significant improvement over orbiters equipped with cameras that provided centimeter-scale resolution. Despite the extremely brief mission of an aircraft, it is possible that they could perform 'niche' missions such as 'sniffing' for the sources of methane.

MARVEL was an orbiter carrying instruments to study the Martian atmosphere and locate concentrations of water vapor leading to the identification of reservoirs of subsurface water. These places, of course, could be benign environments for life to develop.

In August 2003 NASA selected the most conventional Scout proposal, and the one with the firmest scientific case.

The Phoenix Mars mission proposed by the University of Arizona and JPL was designed to capitalize on Mars Odyssey's discovery of water ice located near the surface in the north polar regions by sending a lander to address three questions: Can the polar regions support life? What is the history of water and water ice at the landing site? How is the climate affected by the dynamics of the atmosphere above the polar region? As the name connoted, the spacecraft was to reuse much of the hardware which had been designed for the old Mars Surveyor program, including instruments from Mars Polar Lander and the entire 2001 lander which, upon its cancellation, had been put into secure and controlled storage. In fact, because its availability was stated in the announcement of opportunity this was not the only Mars Scout proposal to reuse the lander. For example, Urey was a JPL and US Geological Survey mission to perform an absolute dating of Martian rocks by reusing not only the 2001 lander but also the Marie Curie rover and instruments developed for the 2003 sample-return lander.[57] The winning proposal included a detailed plan to retrieve the spacecraft from storage, test it, analyze and solve all of the known failure scenarios, and upgrade it. Of course, the fact that a spacecraft was already available made more funds available for analyses and modifications, as opposed to developing new systems. The total cost of the mission was estimated at $417 million.

In addition to correcting known shortcomings, the team conducted a thorough study which identified another dozen issues that had escaped scrutiny, and these were overcome. The critical landing radar was subjected to more tests than all the previous Mars landers combined, culminating in towing it on a cable slung from a helicopter for over 60 hours across a variety of terrains with different geology and texture and at various vertical and horizontal speeds. All the problems which this revealed were resolved. This accounted for much of the $30 million overrun in the cost of the mission. It was discovered, for example, that echoes from the discarded heat shield could have caused the radar to switch off the engines much too soon. A system to use the radar data to assess horizontal winds and, if necessary, perform a separation maneuver was added to prevent the lander from becoming fouled with the spent canopy. Because the original thrusters of the mothballed 2001 lander had been cannibalized for use by Mars Reconnaissance Orbiter, the final 26 seconds of the descent would be braked by a dozen new 293-N hydrazine engines. To prevent the engines from getting too cold (as perhaps occurred in the case of Mars Polar Lander) heaters were mounted on the thrusters themselves, which were thoroughly tested for hammering effects, etc.[58] Some modifications were mandated by the launch window being less favorable in 2007 than in 2001; specifically Mars being some 0.2 AU further from the Sun at the time of the mission's arrival. Moreover, the unfavorable window imposed a much higher energy requirement at launch, obliging the use of a more powerful version of the Delta II rocket. On the plus side, an arrival nearer to aphelion would reduce the spacecraft's atmospheric entry speed by over 1 km/s to 5.7 km/s.[59] The final configuration strongly resembled Mars Polar Lander, consisting of a more or less circular deck 1.5 meters across that would be held 53 cm above the surface by three strutted legs. All of the instruments were mounted above the deck, while the systems and electronics boxes were below it. To the sides were deployed two 1.8-meter-diameter decagonal solar arrays, each with an area of 4.2 square

A rendering of the Phoenix lander sitting on the Martian arctic. (NASA/JPL-Caltech)

meters, that gave the lander a total span of 5.52 meters. When necessary, the arrays would be supplemented by a pair of redundant batteries. The mass of the lander itself was approximately 410 kg, including 67 kg of hydrazine. At 59 kg versus a total launch mass of 664 kg, the instrument package had the highest payload mass ratio of any Mars lander.

Directly inherited from the 2001 lander was a 2.35-meter backhoe robotic arm with four degrees of freedom for collecting samples and exposing the ice that was believed to be present only a few centimeters below the surface. A motorized rasp was added to the back of the scoop late in the development to ensure that it would be able to collect a sample of scraped concrete-hard frozen soil. The rasp was built by the same New York company that supplied the rock abrasion tool for the Spirit and Opportunity rovers. In 2003 the Committee on Space Research (COSPAR) of the United Nations had revised spacecraft sterilization requirements to include the notion of 'special regions' in which terrestrial bacteria might be able to proliferate. One such region was the subsurface of Mars, so the robotic arm was required to be heat-sterilized, and its sterilization was protected during ground-handling by a roll-off tent-like shield. Moreover, for the first time even the internal insulation of the Delta II rocket's aerodynamic shroud was heat-sterilized.

The Thermal and Evolved Gas Analyzer (TEGA) was essentially the same as the instrument carried by Mars Polar Lander. It comprised eight mini-ovens, each capable of holding between 20 and 30 milligrams of soil. A vibrating sieve with a 1-mm pitch was added in front of the inlet to an oven in order to admit only the smallest particles. Like the instrument from which it was derived, TEGA was to seek volatile compounds by heating the sample, analyzing the gases released, and recording the absorbed heat as a function of the applied temperature to identify the phase changes. The instrument would first apply a mild warming to 35°C to melt ice, then to 175°C to release gases, and finally to 1,000°C to decompose organics, salts and water-altered minerals. The temperature at which any water was released would assist in identifying the kinds of hydrated minerals present.

The other instruments included a Canadian-supplied light detection and ranging instrument that would fire a laser beam upward and record the light backscattered by dust, clouds, ice crystals, etc, and so determine the vertical structure of clouds passing overhead, detect snow precipitation, and profile turbulent and convective mixing in the lower atmosphere. A microscopy, electrochemistry and conductivity analyzer was added specifically for the mission. This had a low-power microscope for imaging targets just several millimeters across at a resolution of 4 micrometers (about 8 times better than the microscopes on the Spirit and Opportunity rovers) and an atomic force microscope, the first on a planetary mission, that would use a silicon needle to image individual dust particles at a resolution of 10 nanometers. A wheel with 69 different substrates ranging from sticky surfaces to magnets was to supply samples to this microscope. The package also included a thermal and electrical conductivity probe that consisted of four metallic spikes attached to the robotic arm, and would be driven into the soil for measurements. Within the wet chemistry experiment were four cells containing pure water. After a sample of soil had been soaked and stirred, it would be tested for salts, acidity (pH), alkalinity, oxidation potential, etc, in order to determine whether ice immediately beneath the surface would be conducive to life. It would also be able to identify positive ions, and test for carbonates by adding acid to the cell to prompt any carbonates present to release carbon dioxide. Three sample runs were planned, one from the surface, one from below it, and one from any ice present. The fourth cell was to serve as a spare. A block of very-low-carbon ceramic was placed within reach of the scoop and rasp in order to assess the quantity of 'stowaway carbon' carried from Earth in the analytic instruments prior to undertaking a test for Martian organics.

The camera was similar to that on Mars Polar Lander, which was itself derived from Mars Pathfinder hardware, but it used much larger 1,024 × 1,024-pixel CCDs and wheels carrying 12 filters ranging from the visible to the infrared. As on Mars Pathfinder, the camera was mounted on a coiled mast that would position its optics 2 meters above the ground. A second camera was carried just above the scoop on the robotic arm to provide close-ups of the samples. Although it was not equipped with filters, this mobile camera could take 'color' images by illuminating the scene using blue, green and red LEDs. A third, downward-looking camera was to take a sequence of color images during the final three minutes of the descent in order to assist in characterizing the landing site. A 1.12-meter meteorology mast carried a trio of temperature sensors at different heights, and was topped by a wind gauge in the field of view of the main camera. The meteorology experiment also included pressure sensors. The original Scout proposal had a neutron spectrometer to detect water, but this was deleted early on.

The arm-mounted camera was used to image areas inaccessible to the main camera. LED arrays were used to obtain color images. (NASA/JPL-Caltech)

The lander was to fly to Mars enclosed in a 2.65-meter aeroshell similar to that of previous missions, with a small solar-powered cruise stage on top which would be discarded shortly prior to arrival at Mars. In addition to the engines for the final descent to the Martian surface, the lander had thrusters which fired through cutouts in the backshell. These included four 15.6-N engines for course corrections, and four 4.4-N attitude control thrusters which were to maintain 3-axis stability during the interplanetary cruise. All of the engines shared the same monopropellant hydrazine supply. In order to improve the accuracy of landing, three course corrections were scheduled for the last 18 days of the approach and trajectory determination would be done twice daily instead of twice weekly. And throughout the cruise NASA would make extensive use of the two ESA tracking stations in Australia and Spain in order to improve trajectory determination. It was initially intended to retain the guided entry profile of the Mars Surveyor lander which was intended to achieve a landing within a 10-km ellipse, but this was soon judged unnecessarily complex and deleted. Moreover, it would have required opening the parachute at too low an altitude. After considering a lifting but unpowered trajectory, the team decided to employ a simple ballistic trajectory. After entry, the lander was to be slowed using a single 11.8-meter parachute as in the case of Mars Polar Lander, but modified to increase its drag characteristics and with a nylon canopy instead of a polyester one. It would then release the backshell with the attached parachute and continue to the surface using its engines. It was to land at a vertical velocity of about 2.5 m/s, but could tolerate a horizontal rate of no greater than 1.4 m/s if it were to avoid tipping over.[60] Having learned from the failure of Mars Polar Lander, engineers wrapped a UHF antenna around the backshell in order to provide a wide primary lobe for a radio link to an orbiting relay satellite during the descent. In contrast to most missions, the Phoenix lander would rely solely on a helical UHF antenna and orbiters to receive commands and return data to Earth. This was made possible by the high latitude of the landing site, which made orbiter passes more frequent. Mars Reconnaissance Orbiter and Mars Odyssey would be the primary relays, with Mars Express serving as a backup. Despite the shortcomings of this communications strategy, the Mars Phoenix lander would be able to return hundreds of megabits per day.[61,62,63,64]

Whereas the 2001 Mars Surveyor lander was to have been targeted in a broad equatorial belt, Phoenix was to be capable of landing anywhere between latitudes 65° and 72° in the northern hemisphere because this was where Mars Odyssey had detected ice-rich soil. Compared to the southern target assigned to the Mars Polar Lander mission, the northern plains were generally flatter and at a lower elevation, meaning that there would be more atmosphere available for parachute braking. The specific site was to have as few rocks exceeding 50 cm in size as possible. A larger rock would not only pose a landing hazard, it could also interfere with the deployment of the delicate solar panels. A preliminary selection was made using images and altimetry from Mars Global Surveyor and visible and infrared images from Mars Odyssey. Four regions 20 × 7 degrees in size were selected. 'Region B', near 130°E, became preferred after it was determined that the ice should be found within 10 cm of the surface. This area was imaged by Mars Reconnaissance Orbiter at 30-cm resolution as soon as possible after its orbit circularization in late 2006 and prior to entering solar conjunction. It revealed the region to be strewn with large boulders. Apparently, the thawing cycles that created polygonal cracks in the frosty terrain also drove buried boulders to the surface. It was therefore discarded and a new candidate sought. A broad valley was found in 'Region D' just to the west of a 10-km crater informally named Heimdal. This was hastily inspected before January 2007, when clouds and hazes were expected to impair imaging. The valley looked relatively free of rocks, had benign slopes, and could accommodate a 100 × 20-km landing ellipse safely outside Heimdal's ejecta blanket, nestled between the extensive Vastitas Borealis plain and the hills of Scandia Colles. An automatic rock-counting algorithm was used to ensure a 98 per cent chance of making a safe landing.[65,66] Mars Express and Mars Reconnaissance Orbiter both provided near-infrared data which confirmed the presence of water ice at shallow depth.

A 22-day launch window with dual opportunities 40 minutes apart on each day would yield arrival at Mars between 25 May and 5 June 2008. The surface mission would last just 3 months. Unlike other missions, only a modest extension could be granted because the ensuing winter would encrust the lander in a thick blanket of carbon dioxide ice.[67] After a delay of 1 day, Phoenix lifted off on 4 August 2007. Despite communications problems between the spacecraft and Earth and between telemetry aircraft and the ground, it was successfully dispatched on a trajectory for Mars. In fact, the launch was so precise that 10 kg of fuel was saved from the first course correction and made available for more precise final targeting.[68] Six days into the mission, Phoenix fired its thrusters for 197 seconds and the 18-m/s change in velocity converted the 95,000-km miss distance into a collision course. An issue that was identified late during development was that when the descent imager sent images to the data handling system this could interfere with the operation of the gyroscopic platform, jeopardizing the landing. Lacking time and money to remedy this fault, it was decided early in the mission to switch off the descent imager. This action was made less painful by the fact that Mars Reconnaissance Orbiter could easily obtain very-high-resolution images to contextualize the touchdown point.[69] The spacecraft refined its course again on 24 October 2007 and on 10 April 2008. And then a maneuver performed 4 days out from Mars targeted an area at the far end of the landing ellipse. Seven minutes prior to entry, the cruise stage was discarded at an altitude of about 650 km and left to burn up in the atmosphere. Phoenix then turned to face its heat shield forward and established the critical contact with Mars Odyssey, the primary landing relay. The radio carrier was also monitored using the radio-telescope at Green Bank in Virginia. Mars Express flew over the area during the entry and descent phases, with its relay antenna providing a backup recording of telemetry and using its camera and other instruments to monitor the atmosphere.

Phoenix penetrated the Martian atmosphere at an altitude of 125 km, traveling at about 5.7 km/s. Soon thereafter, at the peak deceleration of 9.2 g, the heat shield reached a temperature in excess of 1,400°C. At an altitude of about 13 km, having slowed to Mach 1.68, the parachute was deployed and 15 seconds later the forward heat shield was released. Several weeks earlier an engineer at Lockheed Martin had noticed that at this point in its descent Phoenix would pass through the field of view of the high-resolution camera on Mars Reconnaissance Orbiter, and asked if a picture could be taken to provide an additional source of data for an investigation in the event of a loss like Mars Polar Lander. Despite some concern that operating the camera might interfere with the all-important radio relay link, it was decided to go ahead. Using the latest navigation data available, the orbiter was able to score a 'first' by snapping a dramatic image of Phoenix some 47 seconds after it deployed its parachute, at a range of almost 800 km and viewed against the backdrop of the crater Heimdal. Owing to the oblique angle of 26 degrees at which the picture was obtained, Phoenix, which was at the time 9.2 km above the surface, appeared to be plunging right into the crater. As a bonus, the picture included the free-falling heat shield as a dark spot. At an altitude of 960 meters and descending at 55 m/s, the backshell was jettisoned. The lander fell freely for a few seconds to align its center of gravity with the velocity vector, then gave its engines five short pulses to warm them up. The maneuver to avoid the backshell and parachute was not required, as the wind was minimal. Finally, Phoenix rolled into an attitude that placed its back to the Sun so that the area within reach of the robotic arm would be in shadow and so minimize ice sublimation during sampling. Phoenix had 37 seconds of powered flight, with three thrusters firing continuously and the remaining nine pulsing at up to ten times per second in response to inputs from the radar and computer. By the time it was at a height of 50 meters it had practically zeroed its horizontal velocity and was maintaining a constant rate of descent of 2.38 m/s. It landed at 23:38 UTC on 25 May, coming to rest tilted by just 0.25 degrees. For the first time there were three active missions on the surface of Mars.[70,71]

One minute after landing, Phoenix ceased transmitting when Mars Odyssey set below its horizon. Valves of the helium pressurization system were then opened to prevent the warming and expanding gas from damaging or even rupturing the lines later in the mission. This, however, would prevent Phoenix from taking off for a short hop as Surveyor 6 did on the Moon in 1967 to provide a new perspective of the landing site. After waiting 15 minutes for the dust to settle, the lander used its battery power to deploy its solar panels, camera and weather station masts. Contact was established about 2 hours later, as Mars Odyssey again passed overhead. The first indication that events were going to plan was telemetry showing positive solar power. If the panels had failed to deploy, the batteries would have been able to sustain the lander for little more than one sol of nominal activity.

As expected from orbital imagery, the landing site was much less scenic than the five areas already inspected by landers. It was remarkably flat with a few pebbles scattered across the pattern of polygonal 'pillows' several meters across created by the seasonal expansion and contraction of ice, similar to the Canadian or Siberian

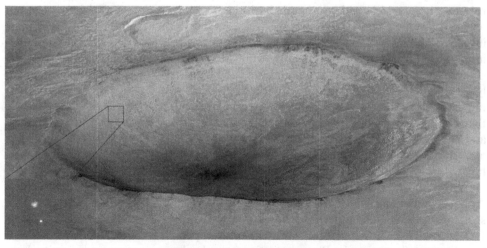

Phoenix on its parachute, photographed by MRO. From the oblique geometry of the picture it would appear the vehicle was about to land in the large crater Heimdal, but that was far in the background. (NASA/JPL/University of Arizona)

An early panorama of the Martian arctic, taken shortly after landing by the Phoenix lander. (NASA/JPL-Caltech)

permafrost. Only a distant range of hills disturbed the otherwise level horizon. The polygons were separated by narrow troughs 20 to 50 cm deep. Larger rocks were remarkably scarce and there were no dunes or ripples. Fortuitously, the lander's robotic arm would be able to sample both the frozen ice-rich soil of the polygon on which the vehicle had landed and the ice-poor soil in one of the troughs. On first inspection no ice was exposed at the surface, but this was expected since it would be in the permafrost. An image taken by Mars Reconnaissance Orbiter 22 hours later showed the lander, its heat shield, and the backshell with its parachute still attached lying 300 meters away; confirming that a bright spot seen by the lander was indeed the backshell. Phoenix was standing at the center of a darker ellipse 10 meters across that could mark where the retrorockets had blown away dust. From these images and the sparse landmarks visible, the site was identified at 68.219°N, 234.248°E, almost 20 km west of Heimdal, about 1,200 km from the north pole and 4.1 km below 'sea level'.[72]

With Phoenix safely on the surface, control was handed over to the University of Arizona in Tucson, which would command further operations. On the first sol the sequences to unlatch and move the arm were issued via Mars Reconnaissance Orbiter, but were not enacted because the relay appeared to be in standby mode. In the absence of commands, Phoenix performed a backup sequence and sent its data later that same sol. A different glitch prevented full communication on Sol 2. The next sol the arm was unfolded using a seven-step maneuver involving rotating the

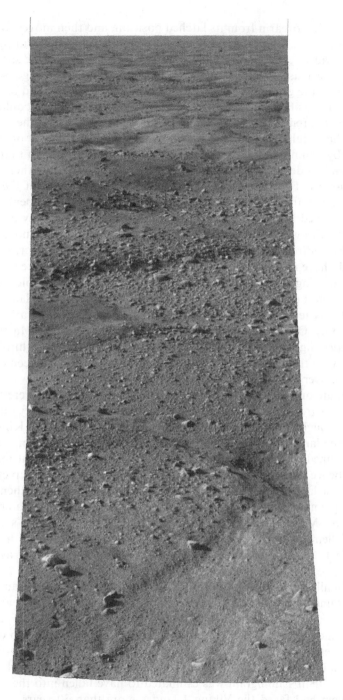

Polygonal terrain at the Phoenix landing site. (NASA/JPL-Caltech)

wrist to release the forearm from its latched position, and then moves to release the elbow and extract it from the roll-off tent-like biological shield. During this time, the lander provided its preliminary low-resolution panorama of the site and tested the meteorological laser. On Sol 4 the camera on the arm took images of a landing pad and some of the ground beneath the lander. In the area exposed to the blast of the retrorockets was Holy Cow, a bright hard-looking feature that could have been a rock but looked remarkably like a slab of ice. Images taken on the following sol showed another bright slab dubbed the Snow Queen, possibly part of the same ice table buried by a thin blanket of loose dirt. Simulations had shown that the pulsed thrusters would erode more dust than the Viking landers did using their continuous jets. The shadows of the legs provided an estimate of the depth of the ice slab of about 5 cm. It was an excellent start to the scientific mission. Temperatures during the first sols ranged from $-80°C$ in the early morning to a relatively mild $-30°C$ in the afternoon, with winds blowing from the northeast at 20 km/h and a pressure of 8.55 hPa. This indirectly confirmed that the ice was made of water, since carbon dioxide would have rapidly sublimated at such temperatures. The team drew upon fairy tales in naming features and targets near Phoenix. For example, Snow White resided on the crest of the polygon Wonderland, the nearest to the lander. Another polygon that was accessible to the arm, albeit somewhat farther away, was dubbed Humpty Dumpty. On Sol 5 the arm made its first contact with the Martian surface. Several practice digs and dumps were carried out, unearthing a whitish streak of material a few centimeters beneath the surface which could have been either ice or salt – imaging over the ensuing sols would reveal which, since ice was expected to slowly sublimate on being exposed to the Sun. The first samples were to be taken from the Dodo trench on one side of the Humpty Dumpty polygon, near the Rabbit Hole trough. Over several sols the scoop reached out to the King of Hearts region near the lander and the arm-mounted camera inspected samples that were lifted and dumped on the ground. Pictures showed a glint of bright material that again could be ice or salt. The camera was also used to obtain more pictures of the underside of the lander, including the Snow Queen ice table. It was then decided to rehearse sample manipulation in readiness for precisely releasing material into the analyzer ovens. Meanwhile, the microscope imaged small dust and sand particles which had fallen onto an exposed sticky surface – the sample tray had been left open to collect dust particles that were blown on top of the lander as it touched down.[73,74]

Phoenix finally performed its first sampling for the gas analyzer during Sol 11. This was obtained from the top centimeters of the location Baby Bear within the Dodo-Goldilocks trench at the interface of two polygons. This material was to be dumped into the fourth oven on the next sol. However, one of the two doors of the inlet to the oven did not open fully. And this was only the start of difficulties; for when the sample was dumped onto the instrument it failed to trigger the 'oven full' sensor and less than 1 milligram was estimated to have fallen into the crucible.[75] A similar problem had faced the Viking 1 lander more than 30 years earlier. It was decided to vibrate the sieve to shake some of the smaller particles loose. However, several sols of such activity filled the oven to less than 10 per cent of the desired level.

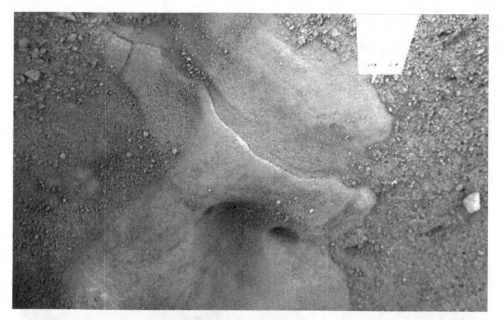

A close-up by the arm-mounted camera showing the 'Snow Queen' slab of ice unearthed by the retrorockets of the Phoenix lander. (NASA/JPL-Caltech)

Apparently the arctic soil was more cohesive than expected. In what was to be the seventh and final effort, on Sol 18 the sieve was automatically stopped after several seconds by the tripping of the 'oven full' sensor. By Sol 22 the first two heating stages were completed. When the sample was tested at 35°C no water was detected. This was not unexpected, as it seemed likely that the soil had cleared the sieve only by losing cohesiveness after the ice that it contained had sublimated in the polar summer sunshine. Moreover, the sample was from the exposed surface in an area which was expected to be mostly ice-free. Carbon dioxide was released at higher temperatures, then oxygen and eventually some rock-bound water. Two gas release peaks were recorded during the heating run. The lower temperature peak could indicate minerals that formed in the presence of water, such as goethite or kaolinite, or various sulfates, and the higher temperature peak could be explained by water-bearing rocks such as talc and serpentinites.

The mass spectrometer within the gas analyzer was also used to measure the isotopic ratios of carbon in the atmosphere, providing much more accurate results than the Vikings. If Mars had been an inactive planet, the atmosphere would be enriched in heavier isotopes, the lighter ones having being more readily 'eroded' by the solar wind. Surprisingly, lighter isotopes were found to be present in similar quantities to Earth's atmosphere. This implied that something was replenishing the lighter isotopes in the atmosphere, most probably volcanic activity. On this basis therefore, Mars must have been active during the last few hundred million years. However, oxygen isotopic ratios did not show the signature of any such activity, as they should if carbon dioxide was injected into the atmosphere by volcanoes. This

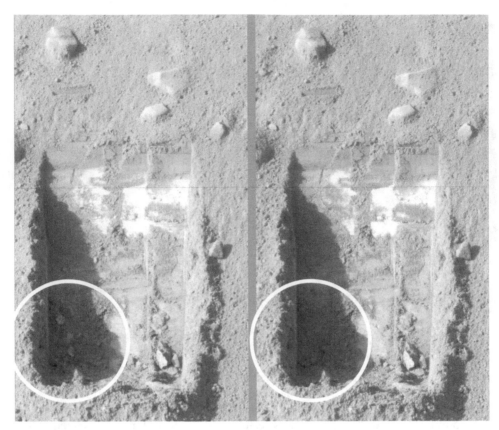

Two images of the Dodo-Goldilocks trench taken on Sol 20 (left) and 24 (right) showing the disappearance of some bright objects, probably small pieces of water ice. (NASA/ JPL-Caltech)

paradox could be solved if oxygen reacted with liquid water or brines in the recent past to create carbonates.[76],[77]

In these early operations, the gas analyzer developed a glitch that produced an intermittent short-circuit in a backup filament. Diagnosing this and developing a workaround took several sols. The problem was probably caused by the shaking to which the instrument had just been subjected. In fact, a similar short-circuit had been found several days before the instrument was shipped to Lockheed Martin for integration into the spacecraft. As the instrument was performing its first analyses, a means of sprinkling a sample by vibrating the arm using the scoop-mounted rasp was tested. This looked promising when tested on a clean surface, so much so that on Sol 17 it was used to deliver the first soil sample to the optical microscope, which observed thousands of particles, including small reddish ones mixed with larger glassy black ones that might have been of volcanic origin but could just as well have been hematite.[78]

Meanwhile, Phoenix continued to dig into the Dodo and Goldilocks trenches to a depth of about 7 or 8 cm, unearthing more white material. On Sol 22 it began to scrape the Cheshire Cat polygon but failed to reveal white material within the first few centimeters, probably because the blanket of soil was thicker there. The lander also suffered a software problem that generated so much housekeeping telemetry as to prevent scientific data from being stored in the flash memory. Although this resulted in the loss of most of the data collected on Sol 23, it was mostly weather monitoring and imagery that could easily be replaced.

Small clumps of bright material which were exposed on the floor of the Dodo-Goldilocks trench on Sol 20 were found to have disappeared when re-imaged four sols later, providing convincing proof that the material was not salt but ice that had sublimated away.[79]

A second microscope sample was collected on Sol 26. Before delivering this to the instrument, the arm-mounted camera took a 30-micrometer-resolution image of clods of fluffy red soil on the central part of the scoop blade.

On Sol 29 the first sample was obtained for the 'wet chemistry' package. It was taken from several centimeters depth at the small divot Rosy Red near the center of the polygon. On being mixed with water the sample showed a slightly alkaline pH of about 8.3 that contrasted with the analyses of the Viking landers. They had suggested the soil was acidic and permeated with a strongly oxidizing compound which was tentatively identified as hydrogen peroxide. The addition of acid to the solution should have lowered its pH but it was seen to remain almost stable, as if something was 'buffering' it. The most likely candidate for the buffering molecule was some form of carbonate. In fact, this characteristic of carbonates would make the Martian surface rather less inimical to life. Salts included magnesium, sodium, potassium and calcium as well as chlorine types but little sulfates and no nitrates. The most interesting result, and possibly the most important of the entire mission, was finding a concentration of ions of perchloric acid salts of around several per cent; i.e. relatively high. These perchlorate salts on Earth are used as a supply of energy by peculiar types of plants and bacteria. Runs by the gas analyzer would be used to confirm the presence of salts. What was more remarkable was that a mix of water with silicon, iron, calcium, chlorine and other elements would remain liquid even at sub-freezing temperatures, and such a fluid could even be responsible for the mysterious gullies. Despite the conclusions drawn from the Viking results, the soil at the Phoenix landing site proved to be "surprisingly hospitable to life", being no more hostile than the dry valleys of Antarctica.

On Sol 41 a sample from the crest of the polygon, at the interface between dust and ice, was taken by scraping Sorceress, just to the right of Rosy Red, and fed to the second wet chemistry cell.[80,81] The first electrical conductivity measurement was made on Sol 43 by driving the arm-mounted spikes into an undisturbed patch of soil, with the outcome indicating that electric charges were not carried between electrodes, in turn implying that the soil near the surface was mostly dry. Later in the mission, small amounts of water were observed to accumulate during the night. This instrument was also used regularly to measure atmospheric humidity. A series of pictures were taken of the lander's legs between Sol 8 and 44 to check whether specks seen just after landing were mud splashes, and the results suggested these had merged and moved. It was possible the combination of the legs being warmed by spacecraft heaters and the melting point of water being lowered by perchlorate had caused specks of ice to melt and flow. On the other hand, it may have been an illusion caused by the changing illumination. In any case, skeptics insisted that the resolution of the camera was insufficient to establish that there really were liquid droplets of water on the lander's legs.

By early July NASA managers were starting to criticize the apparent lack of progress with the mission, and reportedly asked the scientists to collect a hard-ice sample and, fearful that the short-circuit problem would recur and compromise the gas analysis instrument for good, urged that they should treat it as the final sample. This involved first scraping the bottom of the trench in order to expose frozen soil and then using the rasp to collect a pile consisting of a mix of soil and ice shavings containing at most several teaspoons of water. After several sols of scraping and rasping in sixteen different places in Wicked Witch, situated on the floor of Snow White, the result was a mere 3 cubic centimeters of ice mixed with dirt. Not even the tungsten carbide blade of the scoop could scrape the rock-hard ice.[82] Imaging using color filters confirmed that the material had the spectral behavior of water ice mixed in with a small amount of dust. The sample was to be delivered to oven 'number 0' of the analyzer. This time the inlet doors opened as intended but when the sample was sprinkled it again failed to trigger the 'oven full' sensor. It seemed that the frozen soil was so sticky that it adhered to the walls of the scoop "like ice cubes refreezing in a drink glass". Different methods of delivering the sample by vibrating the scoop and arm were tried, and it was eventually decided to obtain the next samples by looking for icy soil instead of hard-ice. In any case after two sols of exposure to the air during which some of the ice had sublimated and made the sample easier to handle, a small portion of it finally reached the oven and could be used for analysis. The soil was undergoing a phase change between $-2°$ and $+6°C$, indicating the presence of about 1 milligram of melting water ice. Unfortunately, too little water was present to determine the isotopic hydrogen-to-deuterium ratio, which could have provided insight into its history and in particular confirmation of the existence of an ancient ocean in the northern hemisphere. Chlorine was sought in the spectra but was not found at any temperature. This probably indicated that perchlorates were releasing oxygen (recorded on this run, as well as Baby Bear) but not chlorine itself. A peak near $400°C$ was possibly attributable to magnesium carbonate, or to the oxidation of carbon-rich material by perchlorate. Perchlorates would release oxygen at these temperatures, which would react with any organic molecules to form carbon dioxide and water, completely destroying them. In fact, this could have occurred in the Viking gas chromatographs, which failed to detect any trace of organics. The molecule which best fitted the results was magnesium perchlorate, which was consistent with the wet chemistry results. The sample then showed two phase changes at high temperatures: the first, at $725°C$, was probably carbon dioxide released by the decomposition of calcium-rich carbonates; but the second, at $860°C$, was not identified.

Collected on Sol 74, the third gas analysis sample was another surficial one from Rosy Red – with the exception of the first, all such samples were obtained in the

Wonderland polygon. The oven to be used on this occasion was number 5, but the doors only barely opened. Lamentably, the problem, a minuscule interference between the doors and the brackets at their bottom, had been discovered on Earth prior to the mission. Replacement parts were made and installed but apparently to the old flawed specifications and design! On Mars only oven number 0 properly opened its doors. Fortunately, the robotic arm managed to deliver samples in this configuration. In mid-August the arm enlarged the Cupboard trench and deepened another one named Burn Alive where samples could be collected for gas analysis. The Stone Soup trench, located on a trough between polygons, was deepened to almost 20 cm for the chemistry laboratory.[83] Images in the visible and infrared range of some of these trenches seemed to indicate that the perchlorate salts had been concentrated in small patches by water, perhaps thin films of it, which may have been liquid as recently as the summer or spring.[84] A fourth sample, the final one for the primary mission, was delivered to the gas analyzer on Sol 85. This was collected at intermediate depth in Burn Alive and delivered to oven number 7. The third and fourth gas analyses confirmed the results of the first two, in particular the presence of carbonates. However, it was not clear whether the carbonates had formed in-situ in the presence of water or, as suspected from data by Mars Global Surveyor, had been transported as dust by the Martian winds.

As the lander appeared healthy and there were ample power margins available, the mission was extended from late August to the end of September in order to use the remaining gas analyzer ovens and wet chemistry cells. With the Sun transiting at diminishing elevations and the first patches of frost beginning to appear, by late August there was some urgency to complete the primary objectives of the mission. By mid-September the shorter days caused the power level to drop to 65 per cent of that in late May, and the meteorological instruments were showing a decrease in atmospheric pressure.[85] The laser instrument mostly detected dust in the lower atmosphere, and its presence at altitudes as great as 4 km was an indication that turbulent motion was mixing the lower layers of the atmosphere and redistributing dust on a global basis. This dust loading of the atmosphere was seen to decrease dramatically after the summer equinox. Clouds were mostly seen at an altitude of about 10 km near summer solstice (just after landing) and then water ice crystals began to form closer to the ground in darkness. These low clouds persisted through the morning and dissipated as the atmosphere warmed, but they disappeared later and later with the advancing season. On Sol 99 this instrument made one of the most visually striking observations of the mission. Atmospheric profiles showed almost vertical streaks projecting from high altitude clouds toward the ground. The streaks were evidently snow falls, as often occur on Earth from cirrus clouds, with the ice crystals sublimating before they could reach the ground. Then, in the early morning of Sol 109, it observed snow falls extending down to the surface.

The camera also made some atmospheric observations: a dust devil was seen scouring the flat plain just before noon on Sol 104. Others were detected as sudden decreases of atmospheric pressure. Dust devils appeared only late in the mission, probably owing to some seasonal effect. Images taken on Sol 126 showed clouds

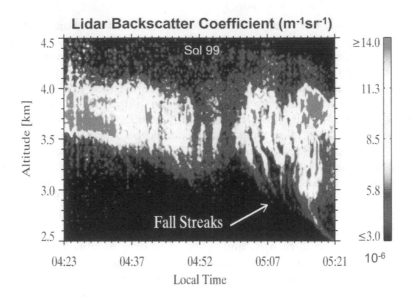

An 'image' obtained by the lidar instrument on Phoenix showing snow falling from the clouds. (NASA/JPL-Caltech)

forming, growing and dissipating, and these were presumed to be water ice clouds producing sublimating snow falls like those observed by the laser instrument.

Meanwhile, the analytic instruments were continuing to provide results. A final wet chemistry sample was taken with some difficulty from deep in the Stone Soup trench in the hope that perchlorates had collected there, but their concentration did not vary with depth. On Sol 120 a sample from Sam McGee was provided to oven number 1 and two sols later a sample from the organics-free ceramic tile was given to oven number 2. The doors of oven number 3 barely opened, so it was not used. The remaining oven, number 6, was filled with another sample from Rosy Red on Sol 131. Unfortunately, by then the valve to supply nitrogen to transport evolved gases from the ovens to the mass spectrometer had malfunctioned, making the data of dubious merit. One final experiment was prepared in which oven number 3 was to perform an analysis of trace gases in an atmospheric sample, thereby making the best use of a poor situation.[86,87,88,89,90]

Phoenix was to start powering off heaters one after another from 28 October in an effort to save its dwindling power to continue to operate at least the camera and the meteorology instruments in order to monitor the onset of winter and study the formation of the first carbon dioxide ice deposits. Since the first of the heaters to be switched off was the one that kept the arm warm, no further digging would be done. The arm itself was parked with the electrical conductivity spikes inserted in the ground to continue measurements. The weather was rapidly deteriorating, with temperatures declining to $-100°C$ during the 7-hour night and ice clouds and a mild dust storm reducing the solar power available for recharging the batteries.

To preclude the batteries draining overnight, the lander was commanded to switch off all non-essential equipment. On 27 October, the final day of 'high-power' activity, a dust storm unexpectedly engulfed the site and prevented sunlight from reaching the solar panels, tripping the lander into safe mode. It also made communications difficult for several sols. (It was this storm that almost killed off the Spirit rover, whose power generation hit a 5-year minimum.) No further signals were received from Phoenix after 2 November (Sol 152), thereby concluding more than 5 months of activity on the surface of the planet. Orbital relays continued to send commands to tell the lander to switch on its transmitter, but after a month of silence, and solar conjunction imminent, this effort ceased; the last attempt was by Mars Odyssey on 30 November. By then Phoenix had probably slipped to 'Lazarus mode' in which its computer ran basic housekeeping tasks but no longer accepted new commands.

Phoenix provided a total of about 25,000 pictures from its panoramic camera, the camera mounted on the arm and the optical microscope. The first image from the atomic force microscope was returned on Sol 43, and several dozen such scans showed the particles of dust to be usually more or less flat. A total of 31 samples were obtained by the arm from twelve trenches. These were fed to six of the eight ovens available and to three of the four wet chemistry laboratory cells. Two of the microscope slots remained unused.[91] However, not all of the scientific objectives were met. In particular, the study of the history of water fell short of expectations. As the scientists put it, this was "largely because Mars failed to cooperate". For example, it proved impractical to dig deeper than about 5 cm through the concrete-hard ice, making it impossible to measure profiles of salt distribution with depth. These could have revealed much about the ebb and flow of water during the brief spells of warmer climate. Such studies will evidently require some sort of drill or subsurface probe. Nevertheless, the Phoenix mission presented a scenario of Mars less hostile to life, at least at high latitudes, than the oxidizing or acidic soils found by the Viking landers and the more recent rovers.[92]

The silent Phoenix lander was imaged several times by Mars Reconnaissance Orbiter before the Sun finally set. The low winter temperatures would have fatally damaged its electronics and built up a thick layer of carbon dioxide ice. Images of the ice-shrouded lander were taken again in the summer of 2009 as the northern spring began and the Sun crept above the horizon. It was barely visible as a dark spot amid brighter wisps of frost and atmospheric haze. The team decided to try to resuscitate the lander to resume a minimal mission.[93] Attempts to revive Phoenix started on 18 January 2010 with Mars Odyssey listening in vain for transmissions. Another attempt was made in late February by which time the Sun was above the horizon for 90 per cent of a sol, but again there was no response. By early April the Sun was continuously above the horizon, reproducing the conditions of more or less one local year earlier when the spacecraft landed. Although Mars Odyssey made 60 passes over the site, no transmissions from the lander were detected. A fourth campaign was carried out in mid-May, just after summer solstice, but again there was no response. Images taken by Mars Reconnaissance Orbiter at about this same time showed no shadow from one of the solar panels, indicating that it must have been snapped off at

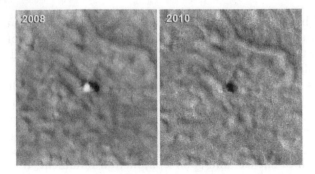

Phoenix on the ground, imaged by MRO before (left) and after winter (right). The solar panels apparently snapped off during the winter. (NASA/JPL/University of Arizona)

some time during the winter by the weight of the carbon dioxide ice encrustation. Moreover, the winter ice cap had almost obliterated any trace of the parachute. If the polar environment could so readily obliterate traces of a successful landing, this did not bode well for locating Mars Polar Lander, lost over a decade earlier at a southerly latitude.

BACK TO PHOBOS

In the early 1990s scientists at IKI and engineers at the Lavochkin Association in Russia began to consider obtaining samples from the surface of the major Martian moonlet Phobos. In terms of energy, it would be one of the simplest sample-return missions from beyond the Moon. The earliest ideas for this Fobos-Grunt (Phobos-Soil) mission envisaged a large vehicle based on the UMVL (Universalnyi Mars, Venera, Luna; Universal for Mars, Venus and the Moon) bus that had already been used for the Fobos missions in 1988 and was to be used for Mars 96, with an initial mass of 7,700 kg. This could be launched in 2003 on a Proton or an Energiya-M, a lightened form of the launcher developed in the 1980s for the Buran (Snowstorm) space shuttle and possibly also to establish a human outpost on the Moon. It was envisaged that the mission would land on Phobos in 2004 and return to Earth with at most a kilogram of samples after 2.7 years of flight.[94] There was also a proposal for the Russians to fly a US-supplied return capsule and instruments funded by the Discovery program.[95] But the already critical finances of the Russian planetary exploration program worsened after the failed dispatch of Mars 8 in 1996 and the Fobos-Grunt proposal had to be completely re-formulated. In 1997 the 'Planets and Small Bodies' division of the Academy of Sciences tried to develop a realistic national near-term planetary exploration policy. To achieve the desired science, three missions would be required: Luna Glob to investigate the internal structure of the Moon, Fobos-Grunt and Mars-Aster.[96,97] It was decided to fund Fobos-Grunt first.

By this time the original design had been replaced by one which would use an electric propulsion stage for the interplanetary cruise, and the spacecraft could be accommodated by a smaller and less expensive Molniya or Soyuz launcher.[98] This design included a Fregat-based stage to accelerate from low Earth orbit; an electric-powered cruise stage with nine proven Fakel SPT-100V ion thrusters and a load of 425 kg of xenon as fuel; an orbital module which itself included a propulsion module for Mars orbit insertion and for Phobos landing; an Earth injection module and a return module with the atmospheric capsule that would deliver the sample to Earth. The launch mass of the spacecraft (including the Fregat stage) would still be 7,250 kg. After dispatch in December 2004, Fobos-Grunt would slowly accelerate in heliocentric orbit and reach Mars within 800 days. Prior to Mars orbit insertion, the electric propulsion module would be discarded. Perhaps a small station would be delivered to the surface of Mars. From its preliminary Mars orbit, Fobos-Grunt would make navigational observations of Phobos and undertake remote-sensing of both Phobos and Mars. The landing phase would be conducted using low-thrust engines and would conclude with the firing of harpoons to hold the craft firmly on the surface in the minuscule gravity field.[99] In a sample collection operation over a period of several days, 175 grams of regolith would be retrieved and stowed in the return capsule. The injection module would then be fired to begin the 280-day cruise to Earth.[100] In the summer of 2008 the spacecraft would release the capsule to soft land in Kazakhstan. One proposal was for the spacecraft to pursue an extended mission by maneuvering to conduct flybys of several asteroids.[101]

Facing continuing financial problems, Fobos-Grunt was completely redesigned yet again, deleting the electric propulsion system. Interestingly from a technical point of view, this version of the mission would be the first Russian planetary craft to have electronics designed to operate in vacuum, thus enabling the large, heavy pressurized hulls of Soviet heritage to be discarded. In August 2003 the Russian Space Agency confirmed that Fobos-Grunt would be launched on a Soyuz-Fregat in October 2009, with the sample being delivered to Earth in July 2012. In this version, Fobos-Grunt would have an initial mass of 11,100 kg, most of which was propellant in the 'Flagman' stage, a modified Fregat with an additional jettisonable toroidal tank. The spacecraft itself was 1,400 kg. It would reach Mars in July or August 2010 for a 9-month orbital phase, starting off in a 3-day orbit which would be progressively adjusted to match to that of Phobos. Landing on Phobos would be performed in April 2011 using the 'Unified Transport Module' bus, which in this case was fitted with legs. Phobos always shows the same face to its parent, and all the proposed landing sites were on the side of the moonlet which faced away from Mars. The surface mission would be complicated by the need to generate power by solar panels and to communicate with Earth, because for much of the time the bulk of either Mars or Phobos would be in the way. Depending on the phase of the mission, the data rate could be as high as 16 kbits per second. A contingency sample was to be collected immediately after landing, with emergency procedures ensuring the ascent vehicle would lift off in the event of a communications failure. Otherwise, the collection phase would last one week. Originally, it was envisaged that the mission would use a coring drill to sample layering within the regolith as in the case of the 1970s Luna landers, but then it was realized that in such a low gravity field the act of driving the drill into hard soil could easily overturn the vehicle. It was therefore decided to use a scoop mechanism and take panoramic images to provide context. A piston inside the scoop would push the sample into the return capsule, capable of accommodating up to 20 scoopfuls adding up to a total of 400 grams.[102] The ascent stage of the 159-kg return vehicle would require to accelerate to a mere 1 to 10 m/s to escape the moonlet. To prevent its exhaust damaging the lander, a spring was to lob the ascent stage to a safe height before it lit its engine. The departure from Mars would be essentially the reverse of the earlier maneuver sequence, first with insertion into a circular orbit around Mars approximating that of Phobos, then into an eccentric orbit from which it would escape for an 11-month cruise to Earth. On approaching Earth the 10.9-kg, egg-shaped capsule, which would have topped the stack at launch like a cherry on a pie, would be released for atmospheric entry. It would not employ a parachute, but crushable shock absorbers would cushion the impact of touchdown.

The prospects of resuming planetary exploration improved with the increasing fortunes of the Russian economy boosted by the country's reserves of fossil fuels. By 2007, after a decade of austerity, IKI was receiving healthy funding and it was possible to allocate a total of 2 to 2.5 billion rubles to Fobos-Grunt, equivalent to about $160 million. Owing to the generation gap caused by the 10-year hiatus it was necessary to recruit ranks of young researchers, most of whom, ironically, had cut their teeth in analyzing data from US missions. As with Fobos and Mars 96, the Russians adopted a 'Christmas tree' approach to the payload, cramming 20 instruments with a total mass of 50 kg into the spacecraft to guarantee that scientific results would be obtained even if the sampling itself were to fail. The payload changed over time, but it was to include wide-field and a narrow-field CCD cameras for navigation and surface imaging, in addition to a stereoscopic microscope on a robotic arm. Several instruments would analyze the surface of Phobos. These included a complex mass-chromatograph and spectrometer to detect water and other volatiles and to measure the isotopic ratios of oxygen, hydrogen and carbon; a Mossbauer spectrometer to study iron-bearing minerals; a neutron spectrometer to study the composition of the regolith and look for hydrated minerals; a gamma-ray spectrometer to detect radioactive elements; a laser mass spectrometer to investigate the composition of the regolith to a depth of about 50 micrometers; and a mass spectrometer for secondary ions liberated from the surface by the solar wind. In addition, a Fourier spectrometer was to study the mineralogy and thermal characteristics of the surface of Phobos in daylight and in darkness, both while in space and on the surface; a radar sounder would probe the regolith with a vertical resolution of 2 meters; a thermal probe would measure the thermal conductivity of the regolith; and a seismometer would record seismicity. The orbital environment would be studied using a plasma-wave system to measure electromagnetic fields, a micrometeoroid detector and a dust counter. And, finally, precise tracking using an ultrastable oscillator would investigate the orbital and libration motions of Phobos, its mass and inertial characteristics, as well as performing fundamental physics experiments. It was to be an international mission, including contributions from the former Soviet republics of Ukraine and Belarus and inputs by ESA, Germany,

Cui Pingyuan, the director of the Deep Space Exploration Center of the Harbin Institute of Technology answers questions from Chinese television during the space exhibition held in Beijing in November 2003. Behind him is a model of a Chinese Mars orbiter. (From the Deep Space Exploration Research Center website)

France, Switzerland, Sweden, Holland, Austria and Hungary. Italy had planned to provide a dust counter and a thermal infrared mapper but this was not possible owing to financing constraints.[103,104] Two other instruments were to be flown despite not being part of the official payload suite. One sponsored by the US Planetary Society was to transport a small titanium canister containing ten types of micro-organism to Phobos and back to ascertain whether life is really able to survive 'hitching a ride' on a meteorite from one planet to another and thus seed the solar system with life. A Bulgarian instrument was to record radiation hazards during the outbound interplanetary cruise and in Mars orbit. Provision was made from the start of the project for a small deployable payload which could consist of a surface meteorological station, penetrator or balloon for Mars, or perhaps some kind of Deimos probe.

After years of rumors, in March 2007 the directors of the Russian and Chinese space agencies signed an agreement on the joint exploration of Mars with China exploiting the remaining capacity on Fobos-Grunt. Following a number of joint cosmic radiation experiments flown on balloons, China also wished to participate in the Russian missions to Mars scheduled for the 1990s. However, planetary exploration was not a priority and this was deferred until after the first Chang'e lunar mission was flown. In November 2003 the recently-created Deep Space Exploration Center of the Harbin Institute of Technology presented a model of a small Mars orbiter at an exhibition in Beijing. At about the same time, preliminary studies were announced for surface vehicles similar to the US Mars Exploration Rovers and landing systems similar to NASA's skycrane (of which more in the next section). In 2005 Russia invited China to consider participating in the Fobos-Grunt mission.[105] For China, the 2007 agreement meant that it would have the opportunity to piggyback a small (115 kg) payload on Fobos-Grunt and have it released into the 800 × 80,000-km capture orbit inclined at just 3 degrees to the planet's equator. Developed in a mere 23 months by the Shanghai academy of spaceflight technology, this spacecraft was named Yinghuo (literally Firefly, but also the ancient Chinese name for Mars, 'unpredictable fire').

The project called for Yinghuo to operate a payload of six instruments for one local year. A wide-angle color CMOS camera was to image Fobos-Grunt during the separation sequence. With a resolution of 500 meters at periapsis, this camera would be able to image Mars to monitor sandstorms and image Phobos for public outreach. Moreover, both Yinghuo and Fobos-Grunt would have the possibility of flying by Deimos and viewing its poorly mapped farside at the start of the orbital mission. An onboard memory would be able to store up to 10 images per orbit. It would also carry a Chinese-Swedish plasma package consisting of an electron and a pair of ion monitors and mass spectrometers, a radio-occultation sounder and a fluxgate magnetometer. The extremely eccentric orbit would facilitate exploring almost all areas of interaction between the solar wind and the planet, including the ionospheric bow shock, the magnetosheath, the pile-up region, and the tail and the plasma sheet. The lower ionosphere would be out of reach, but it would be able to be sounded by a double-frequency mutual radio-occultation experiment between Yinghuo and Fobos-Grunt, with the Russian vehicle transmitting and the Chinese vehicle receiving. Of course, the vertical resolution of this experiment would vary with the position of Yinghuo in its elliptical orbit. This would be the first orbiter-to-orbiter radio-occultation experiment. It would be able to 'sound' the ionosphere in the equatorial plane at altitudes between 50 and 500 km at local times that were inaccessible to orbiter-to-Earth radio-occultations. In particular, it would provide the first good data on the night-time ionosphere.[106] In addition, precise tracking at periapsis would chart the equatorial gravitational field in detail. Furthermore, the Chinese-Russian agreement called for the Hong Kong Polytechnic University to provide the Russian lander with a sample-grinding and preparation system derived from the 'dentist drill' which it had developed for the ill-fated Beagle 2 lander.

Not surprisingly, the Yinghuo spacecraft was similar to the model displayed by the Harbin Institute of Technology. It had a small 0.75 × 0.75 × 0.6-meter central bus, two solar panels with a 6.85-meter span, one of which had the magnetometer on its tip. The panels were to generate between 110 and 190 W of power. On one side of the bus would be a 0.95-meter high-gain antenna capable of 16 kbits per second of data return as well as localizing the spacecraft to within 1 km. A low-gain antenna would face in the same direction as the high-gain antenna. A 3-axis-stabilization system using solar sensors, an inertial platform, reaction wheels and ammonia gas jets would orient the spacecraft for solar power generation, operating the camera, the dual-satellite occultation system, and communicating by high-gain antenna. The mission profile imposed significant design constraints, in that when the spacecraft was at the apoapsis of its eccentric and almost equatorial orbit at the same time as Mars was eclipsing the Sun, it would be in darkness for periods of up to 9 hours, during which time the solar panels would be unable to generate power and it would chill down to temperatures approaching −180°C. However, laboratory tests showed

A model of the Russian Fobos-Grunt sample-return spacecraft. The tube on top was designed to deliver the samples to the return capsule. The Chinese Yinghuo orbiter was carried inside the truss between Fobos-Grunt and the modified Fregat stage.

that it would be able to survive temperatures as low as −260°C.[107] As mentioned, after the launch of the Chang'e lunar orbiter in 2007 this would be the second step in China's deep-space exploration, and it would exploit facilities built for the lunar mission, including communications antennas which were refurbished and increased to a diameter of 50 meters. The use of European and Russian deep-space facilities was also envisaged.[108,109,110]

The interface between Fobos-Grunt and Yinghuo evidently posed a challenge to the Lavochkin engineers. The first design placed the piggyback satellite above the sample-return capsule. On realizing that a failure of the release mechanism would jeopardize the primary objective of the mission, it was decided to 'cage' it between Fobos-Grunt and the Flagman stage. But this too created a few difficulties. It had been intended to use the propulsive stage only for the Earth-escape maneuver, and use the main engine on Fobos-Grunt for orbit insertion at Mars. With Yinghuo beneath the spacecraft, obstructing its engine, this would be impossible. In the revised plan the Flagman would be retained during the interplanetary cruise in order to perform orbit insertion, then it would be jettisoned, the subsatellite would be released, and all subsequent maneuvers would be made by Fobos-Grunt. Also as a result of the inclusion of Yinghuo, attitude control thrusters mounted at the tips of

A computer graphics rendering of the Chinese Yinghuo 1 small Mars orbiter.

the solar panels of the lander would no longer be effective at controlling the orientation of the whole stack. The system was redesigned by adding power-hungry reaction wheels, which in turn meant that the size of the solar panels had to be increased. The extensive redesign of the mission profile to accommodate the Chinese subsatellite required switching from the Soyuz-Fregat to the more powerful Zenit, which had never been used for an interplanetary mission. Ironically China had no need to rely on the Russians, as its own launchers had sufficient power. Indeed, Chinese engineers indicated that if a joint mission with Fobos-Grunt were too complex, then they might as well 'go it alone'. In fact, the basic bus for Yinghuo could be used for mini-satellites ranging from 80 to 150 kg, or 300 kg if a propulsion module for autonomous orbit insertion were to be added, and Chinese sources have clearly stated that it could be used for Mars, Venus or Moon orbiters.[111],[112]

Yinghuo arrived at Lavochkin in mid-2009 for integration with Fobos-Grunt. It was scheduled for launch in a window running from 6 to 26 October 2009, with an extension into November being possible if the payload were lightened. However, there were a number of issues. One was the deterioration of the Soviet-era deep-space communications facilities. Only the antenna at Ussurisk in the Far East was usable. Medvezkye Ozyora was in the process of being upgraded and brought back online. An agreement was reached to use ESA facilities, but whilst the European stations could receive downlink from Fobos-Grunt they would not be able to uplink commands.[113],[114],[115] And spacecraft hardware that was originally to have arrived at Baikonur in the summer was not delivered until late September. Finally, with the launch window imminent, IKI (but not Lavochkin or the Russian Space Agency) announced that Fobos-Grunt would have to be delayed to 2011. Russian scientists insisted that for the first time money was not a factor; there was simply not enough time to complete the necessary tests – the testing of some crucial items such as the landing radar had not even begun until August. Other hardware issues concerned the main flight computer and its software, which was being developed in-house by Lavochkin and required systematic debugging. Another reason for the delay was the decision to have the Polish Academy of Sciences supply a backup sampler for the robotic arm in case the surface of Phobos was too hard for the scoop. This new sampler, called 'Chomik' (Hamster), consisted of a small self-propelled penetrator that would operate in a similar manner to the percussive 'mole' of Beagle 2 which was designed to dig into either soft or hard materials. In addition to Chomik, the spacecraft would have two manipulators with which to sample soil of any hardness ranging between soft regolith and hard rock, one provided by Lavochkin and the other by IKI. The delay allowed more time to upgrade the deep-space antenna at Medvezkye Ozyora. Russia sought an agreement with Ukraine to use the facilities at Yevpatoria in Crimea with which the Soviet Union had communicated with its first deep-space missions. Unfortunately, none was reached. It was also agreed that the sample would return to the Sary Shagan missile test range in Kazakhstan, which was already equipped with all the necessary radar and optical tracking systems, thereby eliminating the need to equip the capsule with beacons and other active localization systems. Finally, the delay would allow the inclusion of some previously deleted instruments such as a secondary ion spectrometer, an infrared spectrometer and a low-speed dust detector.[116] Two additional bacterial colonies would be provided by the Russian Institute of Medical-Biological Problems and by the Faculty of Agrobiology of Moscow's Lomonosov State University. After several modifications Fobos-Grunt had a mass of about 13,535 kg fully fueled at launch, of which 1,560 kg were accounted for by the fueled lander. The return vehicle had a mass of 285 kg and was topped by a redesigned 11-kg re-entry capsule, conical instead of spherical. Power for the lander was provided by two wings of solar panels for a total area of 10 square meters. The return module was powered by a single 1.62-square-meter panel.[117] Tests were carried out during the whole of 2010 and, with the delivery of Yinghuo in December, the final assembly, integration and testing of the spacecraft could finally begin. Thermal vacuum tests were completed in June, in preparation for the shipment to the launch base at the end of September. The flight software was still

The Zenit launcher carrying Fobos-Grunt and Yinghuo being moved by train to the launch pad. Although the Zenit was introduced in the 1980s, this was the first time that it was used for an interplanetary mission. (Roskosmos)

running late, raising the possibility that, like Cassini, Fobos-Grunt would be launched with a basic software release on board. Full capabilities would be restored at a later stage by uploading the relevant code.

The Flagman stage arrived at the launch range in late September, followed by the Fobos-Grunt spacecraft proper and Yinghuo on 17 October. The two spacecraft were launched on 8 November, on the first day of the window; the first Russian planetary launch in 15 years. Fobos-Grunt and the Flagman stage separated from the Zenit launcher 11 minutes after liftoff, in a 207 × 348-km parking orbit. After coasting for 11.5 minutes, they were then to perform two burns, the first to raise the apogee to 4,162 km and the second, 2.4 hours later, on the perigee pass, to head to Mars. If all went well, they would reach the planet in October 2012, land on Phobos in February or March 2013 and return samples to Earth in August 2014. The lander would continue to operate on the surface of Phobos for one year. Telemetry received as the stack completed its first orbit confirmed that the solar arrays had deployed and the probe was oriented to the Sun. However, communications suddenly ceased during the second orbit, when the spacecraft apparently lost orientation as it passed from the day-side to the night-side of Earth. The two burns would occur as Fobos-Grunt headed northward, passing over South America and out of contact with Russian ground stations. The Russian Space Agency had not asked foreign tracking stations along the ground track to monitor the telemetry, and instead the crew of the International Space Station and amateur astronomers were called upon to optically observe the burn. It was amateur astronomers, in fact, who first noticed that the maneuvers had not taken place. The spacecraft had aborted the two firings, without wasting fuel or jettisoning the auxiliary propellant tanks, so that it would still be able to attempt its mission. Russian engineers would have days to recover the probe, reprogram it and perform the trans-Mars injection burns before the launch window closed or the orbit decayed. But they would first have to wait until the spacecraft again passed over Baikonur, which was the only tracking station equipped to upload commands. Even if communications were re-established after the launch window closed. Fobos-Grunt could still be put into a high parking orbit between the Earth and the Moon, where it would wait for the 2013 launch window while engineers overcame the problem.[118] Unfortunately, the designers evidently had not envisaged the spacecraft establishing a two-way link with the ground prior to making the first burn, as the toroidal Flagman tank masked the antenna. ESA offered the use of its network of ground stations, but no telemetry was received at first. Meanwhile, the silent spacecraft was slowly modifying its orbit. It seemed that the small firings that it made to stabilize its orientation were slightly changing the altitude of its orbit.

Finally, after a fortnight of silence, on 22 November an ESA station in Perth, Australia, managed to command a transmitter on Fobos-Grunt to switch on using a 15-meter-diameter antenna modified with the addition of a low-power 3-W transmitter to simulate a weak signal received in deep space. On another pass 24 hours later, telemetry was received, indicating that communications gear and solar panels were working nominally. Baikonur then also made contact, and a few days later another ESA ground station in the Canaries was also modified to receive

Fobos-Grunt. But the low and shrinking orbit offered communications windows lasting only several minutes before the probe set below the horizon. Commands were sent by Perth on 28 and 29 November to fire the engine and raise the periapsis, in order to provide longer communications windows. However, the engine did not fire and the vehicle remained in its decaying parking orbit. At about the same time, Fobos-Grunt was seen by space-tracking radars to shed a few small items of debris. By then, the rechargeable batteries as well as the emergency batteries were probably exhausted and it ceased to communicate. High-resolution imaging from the ground showed Fobos-Grunt to be structurally intact, but no longer pointing its solar panels at the Sun.

By December, the orbit had precessed to the point that it spent most of its time in sunlight, raising the faint hope that the batteries would recharge and enable contact to be renewed. But this did not occur and, with the launch window closing, ESA ceased attempting to communicate on 9 December.

With the spacecraft stuck in Earth orbit and incapable of communicating, all eyes turned to its eventual re-entry and the environmental effects of many tonnes of toxic propellants reaching the ground. In a similar situation, a 'dead' American military satellite whose abundant supply of hydrazine appeared to have frozen was destroyed in an opportunistic test of an anti-satellite system. Moreover, very small quantities of radioactive cobalt-57 were carried by Fobos-Grunt as an alpha-ray source for its spectrometers. However, it finally crashed, apparently harmlessly, on 15 January, in the Pacific west of Chile (remarkably, not far from where Mars 96 also reportedly re-entered in 1996).[119]

In December, meanwhile, Roskosmos announced the creation of a commission to investigate the causes of the failure, amid warnings by politicians that if obvious negligence were identified, those reponsible would be punished. The investigation delivered its conclusions in late January. The fault was attributed, perhaps in an excessively self-absolving manner, to electronic components that could have been latched-up by cosmic-ray strikes, knocking out both flight computers and thereby preventing the burn. If that really was the case, then one wonders what would have happened to the spacecraft if it had managed to depart the relative safety of Earth's magnetosphere, where such events are fairly rare. In any case, in the absence of a communications link, there was no way the problem could have been diagnosed in time and corrected. What is more troubling is that the investigation found several design faults and cases of negligence: the main flight control system had evidently been poorly tested on the ground, with little or no oversight or quality control. As a result of slack quality control, it was reported that counterfeit electronic parts were used, as well as non-radiation-hardened chips like the ones which apparently latched up. The suspect type of chip was evidently well known for its sensitivity to radiation, and to be permanently damaged by extreme latch-up events. It was also reported by independent sources that the flight control computer that was developed in-house by Lavochkin did not manage to pass a single ground test without fault, and lots of bugs, major and minor, were uncovered during the final preparation, to the point that the electrical cable routing had to be redesigned late in the development. These problems could also be due to Lavochkin employing young engineers with no experience on how to run a deep-space mission, let alone one as complex as Fobos-Grunt, while the older engineers hadn't flown a fully successful mission for 25 years.

Other theories were proposed to explain the failure, including a fanciful one (and carelessly endorsed by Russian space officials) that it was caused by interference from a US space-surveillance radar. A more realistic possibility was that the flight control computer became overloaded as components were switched on shortly after launch, causing it to crash and reboot. In any case, the loss of Fobos-Grunt underscored the fact that Russian planners had not 'learned the lesson' of Nozomi, CONTOUR and, worse still, of their Mars 96 probe: namely the poor tracking coverage at such a critical event as the departure from parking orbit. The failure, and the embarrassing Soviet-style attempts to 'cover it up', capped a string of failures by Russian spacecraft and launchers which included for the first time in years a launch in the manned spaceflight program. The new director of the Russian Space Agency had only a few months earlier deemed the 48 per cent of his budget dedicated to human spaceflight and the International Space Station to be excessive, and had

A MetNet penetrator prototype hanging on its inflatable decelerator. (Courtesy of Patrick Roger-Ravily)

The Russian MetNet probe.

announced that more funds would need to be redirected to application and scientific spacecraft. Fobos-Grunt demonstrated just how underfunded and poorly planned such an important scientific mission could be.

At one time the payload of Fobos-Grunt included a meteorological surface station for Mars as the demonstrator for a joint mission with Finland and Spain in which a number of small stations would be established for MetNet (Meteorological Network). It was initially funded by Russia to offset the Soviet-era debt to Finland. With an entry mass of 17 kg, the station would consist of a penetrator using an inflatable toroidal heat shield at atmospheric entry and a large inflatable braking and stabilization device. A 1-meter heat shield would be inflated to provide protection at hypersonic speeds. Once subsonic, an outer inflatable 1.8-meter stabilizer would open to slow it for an impact at about 50 m/s. The definitive MetNet penetrator would carry an entry accelerometer, a panoramic and possibly also a descent camera, barometer, thermometer, magnetometer, seismometer, soil temperature and humidity sensors and other instruments drawing power from a small RTG to collect data for several local years. Inflatable heat shields and decelerators have been extensively studied in Russia. They were assigned to the Mars 96 penetrators, but that mission was lost early on. They were first tested in flight in 2000. Extensive aerodynamic, thermal and penetration tests of MetNet have already been carried out, as well as drop tests from aircraft. In 2005 a decommissioned Volna submarine-launched missile was to have dispatched a MetNet station on a suborbital trajectory to conduct a re-entry stability test, but the rocket failed. There were several options for sending MetNet to Mars. The original demonstration in which a single penetrator would piggyback on Fobos-Grunt was discarded in favor of the Chinese orbiter. Another option was to deliver a single penetrator on a small ion-propelled cruise stage which would be launched by a Volna missile, or on a conventional cruise stage launched by a small Rokot missile derived from another decommissioned ballistic missile. In the early 2000s Lavochkin studied a $12 million Mars flyby and mini-lander mission that would be launched by a Shtil submarine-based missile equipped with a small kick stage derived from the Fregat. Of course, a Mars Express-class orbiter launched by a Soyuz-Fregat could deliver several penetrators. After the demonstration mission, a single Fobos-Grunt-like spacecraft would deliver at least a dozen MetNet landers. The first salvo would be released during the approach to the planet, with the others being targeted from orbit. The concept was designed to enable landers to be dispatched to almost any latitude. One penetrator could be landed near either Mars Pathfinder or one of the Vikings in order to compare atmospheric measurements over a long period of time.[120,121,122,123,124,125,126,127,128]

After Fobos-Grunt and MetNet, Lavochkin intended to use the 500-kg 'Unified Transport Module' bus for a number of ambitious planetary missions in much the same manner as it used the UMVL in the late 1980s and 1990s.[129] One such proposal was Mars-Grunt, in which the spacecraft would leave an Earth-return vehicle in orbit around Mars before landing to collect a sample. The lander would comprise a 2,750-kg descent module, a 750-kg landing platform with a robotic arm for sample collection, and a 1,700-kg 3-stage ascent rocket broadly based on the Fobos-Grunt return module. Exploiting studies dating back to the 1970s, the Mars entry profile would combine aerodynamic lift with a parachute. The conical heat shield of the entry module was to double as a high-gain antenna for use during the cruise. A dozen hinged panels on its periphery would provide a controllable geometry for the lifting entry profile and a surface for solar panels. The entire landing platform and ascent rocket would be enclosed by a 'bio-barrier' so as not to contaminate the landing site with terrestrial micro-organisms. Another concept for the entry vehicle was an enlarged form of the inflatable heat shield of the MetNet. After surface sampling, the ascent vehicle would rendezvous with the orbiter (to be derived from Fobos-Grunt) and this would return the samples to Earth. With a mass at launch of about 5,200 kg, the Mars-Grunt mission could be launched either by a Proton or by the Zenit-3SL rocket. A 6,000-kg version that would employ an ion propulsion stage for the outbound cruise has recently been unveiled which could be launched using the new Angara-Breeze in 2022. Like the 5M proposal of the 1970s, this could be preceded by a mission to validate the entry and landing techniques by delivering a rover.[130] In spite of the loss of Fobos-Grunt, Russia is still pushing for flying the MetNet and Mars-Grunt missions early in the 2020s.

Other possible missions included the Europa-Penetrator for orbital observations and subsurface investigation of Europa. This would consist of the standard module of Fobos-Grunt and a daughter spacecraft derived from a module developed for the Soviet-French Vesta asteroid proposal, carrying two 27-kg penetrators.[131] The spacecraft would have a large variable-geometry heat shield and backshell so that it could use the aerocapture method to decelerate into Jovian orbit. Thereafter the vehicle would approach Europa and fire its engines to slip into orbit around it. The penetrators were to strike the surface of the ice-covered moon at a speed of about 120 m/s. The launch mass of the spacecraft would be 6,800 kg. An alternative idea was a Europa lander. It has been reported that a Europa mission could be launched as early as 2020 or 2021 and reach its destination 7 years later. Other RTG-powered missions could explore Jupiter and Ganymede, Saturn and Hyperion (this being the Gipersat mission because Giperion is the Russian form of Hyperion) and Iapetus, Uranus, Oberon and Titania (the Obertur mission), Neptune and Triton (the Neptrit mission) and the Kuiper Belt.

Simpler derivatives of Fobos-Grunt included resuming the exploration of Venus with Venera-D (of which more later). Preliminary definition studies have also been carried out for small and medium-scale missions to near-Earth asteroids using the Soyuz-Fregat or even smaller launchers. These could include the Asteroid-Tur (for 'tour'), a mission to 'tag' the potentially hazardous near-Earth asteroid Apophis, and the Asteroid-P lander. The Asteroid-Grunt probe would return samples from a main belt asteroid. Lavochkin had hoped to be able to launch Asteroid-Grunt in the early 2010s, but the 2020s now seem more realistic. Its launch mass could exceed 5,000 kg, but most of this would be propellant.[132,133,134,135,136] When equipped with dust shields it would become the Kometa-Grunt sample-return spacecraft. Finally, in 2008 Lavochkin and IKI carried out a preliminary study of a Fobos-Grunt-based Mercury orbiter and lander mission.[137] By freezing development until the second half of the current decade, the failure of Fobos-Grunt has put all these projects in doubt. Perhaps, rather than chasing ambitious projects, the Russian Space Agency ought to concentrate on simpler, more focused 'Discovery-class' missions in order to rebuild the competence it lost during the 1990s and 2000s.

The Russian Space Agency is not alone in studying and planning sample returns from the moons of Mars. It is believed that a significant portion of the upper regolith of Deimos consists of material ejected by Martian impacts and swept up by the moonlet. This, and the fact that its smooth surface should be easier to land on than Phobos, makes it an attractive target for a sample-return. In the mid-2000s ESA conducted a number of Technology Reference Studies to identify technologies of relevance to scientific missions which could be attempted in the 2010s. One study was to collect 1 kg of regolith from Deimos for return to Earth. The probe would employ a touch-and-go strategy akin to that used by Hayabusa in which it would briefly touch the surface of the moonlet, obviating the need for a complex landing, anchoring and sampling device. It would comprise a propulsion stage responsible for all maneuvers leading up to sampling, and the return stage on which would be mounted solar panels, the communications system and the sample capsule. The pneumatic sampling system would consist of a penetrator to inject compressed gas into the soil. Of course, the communications lag across interplanetary space would require the entire operation to be performed autonomously. It was considered unlikely that a full sample could be collected in just one such rendezvous, so the plan envisaged making as many as five sampling runs. To build upon this study, ESA investigated a mission to return a sizeable sample from a near-Earth asteroid.[138,139]

THE MARTIAN SUV

As redesigned in 2000, the US Mars Exploration program envisaged launching the Mars Smart Lander (MSL) in 2007 to demonstrate technologies required for a sample-return mission. As initially conceived, this was to perform a precision landing at a site of scientific interest using hazard avoidance capabilities, and deliver a large rover with a range sufficient to investigate sites outside the target ellipse. The rover's payload would be designed to enable scientists on Earth to assess which samples would be worthy of collection and analysis. It was to address geological and geophysical objectives, including the elemental, chemical, mineralogical and isotopic composition of the surface; as well as atmospheric processes, the presence and past history of water, and the radiation environment on the surface. For the first time since the Viking landers, biology would be directly addressed by seeking traces of the chemistry associated with it, making an inventory of the organic carbon compounds on the site and identifying possible biology-related features. In order to provide access to any latitude and to sustain very long traverses without concern for dust deteriorating the output from solar panels, the rover would be powered by RTGs. In effect, this 'mobile Viking' would have more in common with the 1970s philosophy than with the 'faster, cheaper, better' mantra.

A major part of the technology demonstration would be the 'second-generation' lander whose target ellipse was to be at least an order of magnitude smaller than those of the Mars Pathfinder and Mars Exploration Rover missions. This would be achieved by a number of techniques. First, during the final days of its approach to the planet the spacecraft would use Phobos and Deimos as positional references to refine its atmospheric corridor; the technique evaluated by Mars Reconnaissance Orbiter. It would then perform hypersonic trajectory control and descend under a large parachute which could be steered. As it descended, a laser radar would scan the target for possible hazards and send commands to thrusters through a guidance system using algorithms derived from those which enabled the Apollo missions to make pinpoint lunar landings. After the 3-axis propulsion system had brought the lander to a complete halt at a height of several meters, it would free-fall and touch down with a vertical rate of 4 m/s. The propulsion and guidance systems were in a low-slung crushable pod with aluminum sheets and honeycomb to absorb some of the energy at impact in a controlled manner. Six spider-like outriggers and shock absorbers would enable the lander to remain upright even on a 30-degree slope or in contact with a rock about 1 meter tall. This configuration was extensively tested by JPL. Also investigated was a 'heavyweight' airbag system with a separate self-righting system.[140,141]

A computer rendering of the Mars Smart Lander, from which the Mars Science Laboratory was derived. (NASA/JPL-Caltech)

If Mars Smart Lander were successful, NASA hoped to fly the sample-return as early as 2011 involving a NASA lander and rover modeled on the 2007 mission, a lightweight ascent vehicle and a French sample-return orbiter based on PREMIER (Programme de Retour d'Echantillons Martiens et Installation d'Expériences en Reseau; Mars sample-return and network experiment establishment program).[142]

Also under consideration were Martian communications satellites. An Italian Space Agency proposal was named after the radio pioneer Guglielmo Marconi. It would operate in a quasi-polar orbit at an altitude of 4,000 km to provide multiple daily passes over any given site. By using a UHF communications link to send and receive commands and data for landers and an X-Band link to communicate with Earth it would be able to support data rates in excess of 200 kbits per second. This spacecraft would be derived from the bus developed for the Italian Cosmo Skymed

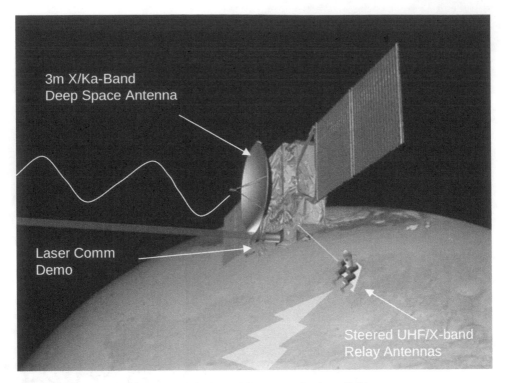

The proposed Mars Telecommunications Orbiter.

Earth observation satellite and mass up to 700 kg. It would have a nominal lifetime of 6 years, but would carry propellant and other consumables for 10 years. It could even be accompanied by a smaller Spanish relay satellite in a more equatorial orbit at a lower level to better support low-latitude landing sites. The satellite bus could then be reused in 2009 or 2011 for an Italian mission to provide full radar mapping of the subsurface of the planet.[143],[144],[145]

In parallel, NASA was seeking funding for the 2009 Mars Telecommunications Orbiter. It would be equipped with steerable UHF and X-Band antennas to receive transmissions from landers, and a 3-meter antenna to communicate with Earth in the X/Ka-Bands. The agency also desired this spacecraft to carry an experimental optical communications package which would use a 5-W far-infrared laser to send data to Earth at rates of 1 to 10 Mbits per second depending on how far apart the planets were in their separate orbits around the Sun. The system would consist of a 50-kg package and would be aimed at Earth either by reference to star trackers or by sighting on a beacon. The terrestrial receiver would consist of a large telescope on a dry site at high elevation, so that atmospheric turbulence would not 'destroy' the signal. Alternatively, the receiver could be installed in a geostationary satellite. Such a system would be able to return to Earth at least an order of magnitude more data than was possible using a radio system and, amongst other things, it would allow almost-real-time imaging by rovers and landers. A secondary function for the Mars

Telecommunications Orbiter was to provide navigational fixes for rovers operating on the surface and to serve as a positional reference for spacecraft approaching the planet. Furthermore, if it were to release a small subsatellite, track it and perform a rendezvous, it would usefully demonstrate a capability needed for a sample-return mission. The cost of the basic Mars Telecommunications Orbiter was estimated at $550 million. A further $170 million would be required for the laser demonstrator and $80 million for the rendezvous experiment. The spacecraft was to be launched by either a Delta IV or an Atlas V in September 2009, orbit Mars at an altitude of 4,500 km (much higher than scientific orbiters) to provide longer relay passes, and operate for at least 10 years.[146,147]

By 2005 NASA's annual budget devoted about $650 million to the exploration of Mars and this was projected to double by 2010. However, as a result of the new goals set for the agency by President Bush, namely to replace the Space Shuttle by the end of the decade and start a program designed to return humans to the Moon, the budget for the robotic exploration of Mars would eventually remain fairly flat. Planning for the sample-return was again halted and Mars Telecommunications Orbiter was canceled. Nevertheless, the program remained solid, with missions up to the end of the decade approved and securely financed. There was the possibility of recovering the Mars Telecommunications Orbiter's optical relay system by adding it to an orbiter scheduled for launch in 2013, but beyond that the future once again became nebulous.[148]

Meanwhile, the objective of the Mars Smart Lander was changed from being a technology demonstration into a proper scientific mission and it was renamed the Mars Science Laboratory (conveniently retaining the acronym MSL). However, its launch was postponed to 2009. It was intended to be a 'moderate' project costing about $650 million to develop and operate. But by the end of 2003 its overall cost, including development, RTGs, launch and operations, was projected at just under $1.5 billion – which was almost twice that of the combined cost of the Spirit and Opportunity rovers.[149] Science advisory groups recommended launching a pair of identical spacecraft for redundancy, as in the case of the Mars Exploration Rovers, but the funding ruled this out and so once again a mission would be reliant upon a single spacecraft. In 2005 the mission narrowly escaped being slipped by another 2 years in order to cover the cost of returning the Shuttle to flight after the Columbia accident and trimming the interval between retiring the Shuttle and introducing its successor, but these funding problems were partially alleviated by the cancellation of Mars Telecommunications Orbiter.[150,151]

The 'smart' landing system was deleted for the mission and it was decided quite early in the design to replace it with an innovative and complex 'skycrane' to deliver the MSL rover to the surface. In effect, it was a development of the concept in which an airbag-lander slung from a bridle was brought to a halt by braking rockets and then cut loose. The rocket-propelled skycrane, officially named the Powered Descent Vehicle, would hover several meters above the surface of Mars, reel down the rover, and then fly clear and make a crash landing. This would enable heavier payloads to be delivered to Mars with slower landing speeds than was possible using airbags. Moreover, it could be easily adapted to landing on other, airless bodies like Europa.

In fact, while airbags had scored a '3-for-3' success in delivering landers, the Mars Exploration Rovers stretched the system to its limit. Without a precursor technology demonstration, the challenge would be to make a skycrane work first time. The launch mass of the Mars Science Laboratory complete with skycrane, aeroshell, cruise stage and propellant exceeded 3,000 kg, which was three times that of a Mars Exploration Rover. It would not be possible to use a cheap Delta II as a launcher, so a larger and more expensive Delta IV or Atlas V would be required. In mid-2006 Lockheed Martin was awarded the $195 million contract to launch the mission on an Atlas V 541 rocket with four solid-fuel boosters. Meanwhile costs continued to escalate, reaching $1.7 billion by 2007. Although this was still much less in real terms than the Viking missions, it made the Mars Science Laboratory the agency's first flagship mission of the 21st century. Consideration was given to deleting the descent camera and the laser chemical analyzer to reduce costs. By the summer of 2008 the projected cost was $1.9 billion and further overruns were expected.

The spacecraft was to be spin-stabilized for the interplanetary cruise, with eight thrusters providing spin and trajectory control. The rover and its skycrane would be protected by the largest aeroshell employed on any non-reusable spacecraft, being fully 4.5 meters in diameter. For comparison, the Viking aeroshell was just over 3.5 meters and the main spacecraft of an Apollo mission was a mere 2.9 meters across. It would use two clusters of 4.35-N thrusters for attitude and trajectory control. These would be carried on opposite sides beneath the solar panel, and draw upon hydrazine from two 36-kg tanks. On arriving at Mars, the solar-powered cruise stage would be discarded. The aeroshell would then use eight 300-N thrusters to cancel its 2-rpm spin and adopt and maintain the attitude for entry. About 5 minutes before entry, a pair of 75-kg L-shaped balance masses would be jettisoned to offset the center of mass and create the conditions for a lifting entry profile. The ablative coating of the heat shield was changed early on from the cork-based material used since the Vikings to the same phenolic material used on the Stardust sample-return capsule, as aerodynamic heating was expected to be much higher than on earlier Mars landers. The heat shield itself would be heavily instrumented to collect aerodynamic data for use by future manned missions. A very accurate landing would be achieved by a combination of refining the trajectory during the final approach to the planet and active steering in the entry phase. The design of the aeroshell and the distribution of mass provided a lift-to-drag ratio at hypersonic speeds greater than that of the Vikings, bestowing significant aerodynamic lift. Also, thrusters would adjust the trajectory. Just before parachute deployment, six 25-kg tungsten ballast masses would be released to straighten the trajectory and correct the center of mass. Some 225 seconds after entering the atmosphere, the supersonic parachute would open to rapidly slow the capsule to subsonic speed. The 21.5-meter parachute would be a scaled-up version of the same design used by every US lander since the Vikings. In fact, it would be the largest, fastest, and highest-opening parachute yet for a Mars mission. Extensive testing had to be performed because the design was not necessarily validated by the smaller Viking parachute. Like the parachute developed for the Mars Exploration Rovers, it was tested in late 2007 in the world's largest wind tunnel at NASA's Ames Research Center in California.

After releasing the backshell, the rover would approach the surface connected to the skycrane, equipped with eight pressure-fed engines. These MR-80B engines were derived from those used by the Viking landers, albeit each with only one nozzle instead of a cluster of 18 nozzles, had a higher maximum thrust, and were throttleable between 400 and 3,000 N. They would be fed by three tanks of high-purity hydrazine. The engines were to be mounted in four pairs at the corners of the skycrane's truss structure and angled outwards in order to prevent their plumes from impinging on the rover. The skycrane was to zero its horizontal rates and slow its vertical speed to about 0.75 m/s before lowering the rover on a triple-stranded 7.5-meter nylon bridle. An electrical cable would also connect them. At the same time, the wheels of the rover would be deployed using a hydraulic system similar to that by which an aircraft deploys its landing gear. When touchdown was detected, the skycrane would stop the descent, sever the bridle and electrical cable, then ascend vertically for some time, pitch over at 45 degrees and continue thrusting at full throttle to depletion. It would then crash at a safe distance from the rover. Within 6 minutes of entering the atmosphere the rover would be on the surface and essentially ready to drive.[152,153,154] Telemetry from this first use of such an innovative means of landing would be imperative. Both Mars Odyssey and Mars Reconnaissance Orbiter would be phased to be overhead and record data at a high rate. A direct-to-Earth link was also envisaged, consisting as usual of simple 'semaphores' indicating the status of the spacecraft.

A full-scale model of the Mars Science Laboratory 'Curiosity'.

The 'end effector' of the robot arm on Curiosity, pictured after landing on Mars. (NASA/JPL-Caltech)

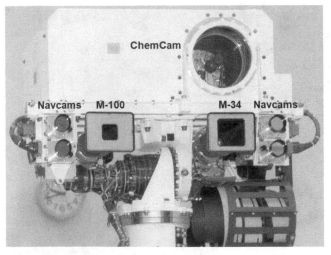

The mast 'head' of Curiosity showing the location of the navigation cameras, the two scientific color cameras (with 34-mm and 100-mm optics) and the laser chemistry 'camera'. (NASA/JPL-Caltech)

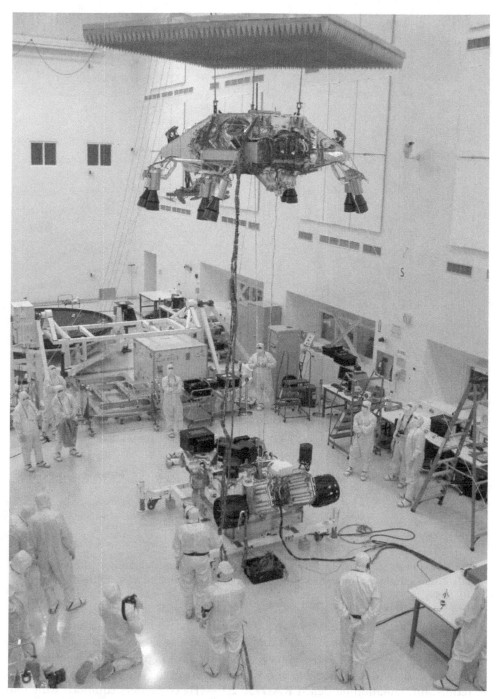

The Curiosity rover during integration tests with its 'skycrane' (above).

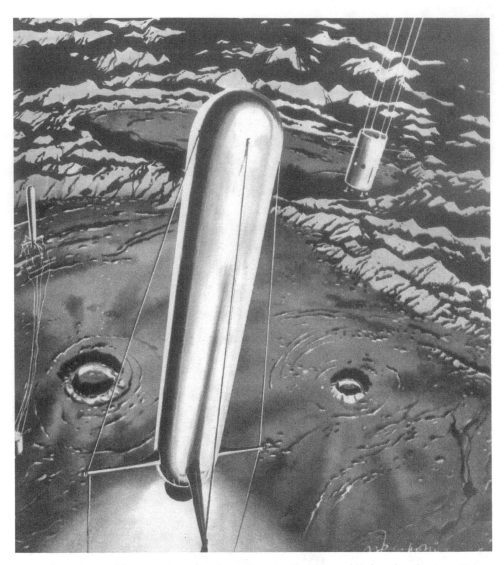

A 1950s vintage 'skycrane' to deliver humans onto the Moon while keeping the crew at a safe distance from the nuclear rocket. (Drawing by M. Jacoponi, from 'Oltre il Cielo' n. 40, 1959)

The Mars Science Laboratory rover itself would be about the size of a compact car, equipped with six 50-cm wheels and an enlarged version of the 'rocker-bogie' mechanism of Sojourner and the Mars Exploration Rovers providing a wheelbase of 1.1 meters and a ground clearance of 66 cm. To control the vehicle, it carried a pair of redundant, radiation-hardened RAD750 200-MHz computers equipped with 256 megabytes of RAM and 2 gigabytes of flash memory. Almost all of the electronics and part of the instruments would be housed within a boxy body 2.7 meters in length with its top deck 1.1 meters above the ground. Just like Spirit and Opportunity, the new rover would have a mast at the front carrying cameras and other instruments to provide a line of sight at a height of 2.1 meters. On the back would be the RTG in its case, somewhat resembling the stinger of a scorpion and giving the vehicle a predatory appearance. It would be able to scramble over 60-cm-sized rocks, drive about 200 meters per sol and travel a distance of at least 20 km during a period of a local year. Half of the driving time would be devoted to identifying and avoiding hazards and path planning, and its top speed on firm flat terrain would be 4 cm/s. A 'scarecrow' engineering model was extensively tested in drops and traverses in the Mars Yard at JPL. On Mars it would have a direct-to-Earth radio link using a steerable high-gain antenna provided by Spain for data rates of 500 to 32,000 bits per second. However, as in the case of Spirit and Opportunity, the preferred means of communication was to be by a UHF link to an orbiter, with the speed to Earth depending on which orbiter was transmitting.

The Mars Science Laboratory spacecraft would have a launch mass of 3,839 kg; comprising the 539-kg fully fueled cruise stage, the 385-kg heat shield, the 349-kg backshield and parachutes, 300 kg of jettisonable ballast, the 1,370-kg skycrane fully fueled with 387 kg of propellant, and, of course, the 899-kg rover.

The scientific payload was selected in late 2004. Since the Vikings, US landers had addressed the issue of biology only obliquely by the strategy of 'following the water'. The Mars Science Laboratory was to 'follow the carbon' by seeking traces of ancient life in addition to extant organics and 'biogenic' gases such as methane. It was to assess the site as a potential environment for life, either past or present. To do this it would inventory carbon-bearing molecules, hydrogen, nitrogen, oxygen, phosphorus and sulfur; establish the nature of organic compounds; and identify chemicals which might have a biological origin. It would investigate the chemical, isotopic, and mineralogical composition of the surface at the site, and study the processes which formed and modified the rocks and soil. It would investigate atmospheric evolution processes and determine the present state and distribution of water vapor and carbon dioxide. Finally, it would characterize the radiation environment. To accomplish these tasks it had ten instruments.

A stereo camera mounted on the mast would consist of two 1,200 × 1,200-pixel CCDs. It would be the first camera on the planet to be capable of providing high-definition full-color videos at up to 8 frames per second. Images and videos would be stored in solid-state internal memories with a capacity of 8 Gbytes. Initially, the two cameras were to have zoomable optics that, for example, would have facilitated a detailed reconnaissance of Burns Cliff from a safe distance had they been mounted on Opportunity. To save on cost, these were replaced by a pair of fixed-focal-length cameras, with one at each extreme of the zooming range. A medium-angle camera would allow preliminary reconnaissance, and a narrow-angle camera would provide detailed images. The development of the zoomable camera continued in parallel until it became evident that this could not be readied in time. The right-hand camera was equipped with a telephoto lens with a 100-mm focal length and a field of view 6 degrees. The left-hand camera was equipped with a medium-angle lens with 34-mm focal length and a field of view three time larger.

The two could produce 3-dimensional views only of the portion of the field of view common to both. These cameras would not employ filter wheels for color imaging, they would instead use microscopic Bayer filters mounted directly on the CCD pixels, just like consumer digital cameras. A color calibration target with a magnetic dust experiment was mounted on the rover deck, as on the Mars Exploration Rovers and Phoenix Mars lander. Filters, on the other hand, were carried for narrow-band and Sun imaging.

The chemistry and micro-imaging camera developed jointly by the US and France would be mounted on top of the mast and would consist of a small, 11-cm telescope mated to a laser and to fiber optics leading to three spectrometers, plus a CCD camera. Concentrating up to about 0.4 joules of energy on a pinhole-sized surface, the laser would be capable of stripping off dust and vaporizing rocks at ranges out to 7 meters, in order that the composition of the underlying rock could be studied in 6,144 separate channels by three spectrometers jointly spanning from the ultraviolet through the visible to the infrared that could identify single elements as well as hydrated minerals, ices, organics, etc. It was envisaged that this would zap as many as 20 targets per sol and perform at least 14,000 analyses during a 700-sol mission. From the same standoff distance, the camera would be able to resolve details as fine as 1 mm in order to provide high-resolution context for its analyses. A calibration target with a number of different rock samples that the laser could shoot was mounted on the back of the rover, close to the RTG. Two instruments were to be mounted on a 2.25-meter robotic arm with five degrees of freedom. These comprised an alpha-particle X-ray spectrometer to determine the elemental composition of material and a 'hand-lens imager' which was a $1,600 \times 1,200$-pixel microscope camera with a wider field of view and a resolution 2.4 times better than its counterpart on the Mars Exploration Rovers. Placed 2.1 cm from a target, the camera would have a spatial resolution of 0.014 mm per pixel. Moreover, it was equipped with light emitting diodes to operate at night and to obtain color images, as well as with adjustable-focus optics enabling it to image over a wide range of distances. On the other hand, compared to the similar instrument on Spirit and Opportunity, the spectrometer would be able to integrate spectra faster, and thus possibly analyze more rocks, and obtain them at any time of the day thanks to a cooling system. The uncooled spectrometer on Spirit and Opportunity was limited to taking spectra at night. In addition, a body-mounted X-ray diffraction and fluorescence analyzer would assay the minerals in samples. A composite gas chromatograph/mass spectrometer and tunable laser spectrometer equipped with 59 reusable quartz ovens and a common sample processing and distribution system would be the first US instrument since the Vikings dedicated to detecting organics such as simple amino acids and carboxylic acids which might provide insight into biological or pre-biological chemistry. The laser spectrometer consisted of two lasers traversing a 20-cm gas-filled cell no less than 81 times in order to perform an accurate analysis of it. It would take periodic measurements of methane, collected through a heated inlet located on a side of the rover, to measure its abundance and seasonal variation. This would be able to detect as little as 100 parts per trillion of methane, far less than reported by orbiters or ground-based telescopes. On the other hand, if methane was present in several tens of parts per billion, the instrument would be able to distinguish the isotopic ratios of carbon in it, which on Earth would be a proxy for its organic or chemical origin. It was also to extend the study made by Phoenix of oxidants in the soil. A Russian instrument would irradiate the soil with neutrons and measure how they were backscattered in order to infer the presence of hydrogen in the form of ice or hydrated minerals in the near-subsurface. It would be able to detect as little as 0.1 per cent of water down to a depth of 2 meters. A number of Spanish sensors housed in part inside the rover body and on two small booms would characterize the Martian environment around the rover. These would include thermometers, anemometers, pressure and humidity sensors mounted on the camera mast to return the first continuous meteorology data since Viking 1 failed. Another sensor would record the ultraviolet radiation. And a detector would characterize cosmic rays and solar wind particles and radiation reaching the surface of the planet. This instrument, like an analogous one initially planned for the Mars Surveyor 2001 lander and another flown on Mars Odyssey, was primarily to obtain data relevant to a human expedition to the planet. A similar instrument was to be delivered to the International Space Station in order to provide a direct comparison of the radiation environment in low Earth orbit. Unlike the Mars Odyssey detector, which was mounted on the outside of the probe, it would fly to Mars buried inside the heat shield, reproducing the position and shielding of an astronaut flying to Mars.

A descent camera was to provide geological context for the landing site, as well as pinpointing it for study by the high-resolution camera of Mars Reconnaissance Orbiter. Unlike the 'rudimentary' descent imager on the Mars Exploration Rovers, this 1,600 × 1,200-pixel color camera was to provide about 100 seconds' worth of high-definition video-style data at up to 5 frames per second. As in the case of the Mars Exploration Rovers, there would also be four hazard-avoidance cameras in pairs at the front and rear of the chassis and two black-and-white mast-mounted navigation cameras.

The rover was to feature a sample acquisition, preparation and handling system capable of brushing and abrading rocks, acquiring core samples from as deep as 5 cm in a rock, grinding small rocks, rock cores and gravel to grains small enough to be fed into the analytical instruments, and acquiring as many as 70 samples of rock or dust. As a backup means of collecting samples the arm would be equipped with a spoon that was 4.5 cm wide and 7 cm long. Prior to passing a sample to the chromatograph, the preparation system would be able to treat it by a chemical process to increase the concentration of the organic compound of interest.[155,156] However, after launch it was discovered that drill samples would probably be contaminated with Teflon (i.e. with carbon and fluorine) from the drill seals. Even so, scientists were confident they would be able to distinguish carbon from Teflon from Martian carbon.

A 40-kg Multi-Mission RTG was developed from the SNAP-19 power source used by the Viking landers and deep-space Pioneers. It had eight plutonium oxide-fueled heat source modules. The radioactive material was supplied by Russia. This was the first RTG since the Vikings designed for convective cooling in a planetary atmosphere. The thermocouples would have an efficiency of just over 6 per cent and a power output of about 125 W at the start of the mission.[157]

The Mars Science Laboratory was to be subjected to a sterilization process of a thoroughness comparable only to that of the Vikings. The RTGs would present an unusual contamination issue in the event of a failed landing. Scientists were not so much concerned about radiation as the releasing of a heat source that could cause melting of subsurface ice and spark chemical or even biological reactions which involved 'stowaway' micro-organisms from Earth. For this reason, the candidate landing sites were required not to have reserves of water ice within 1 meter of the surface.

Free of solar power restrictions, the planners were able to consider landing sites anywhere within 60 degrees of the equator and as high as 1 km above 'sea level'. The ability to use aerodynamic lift to steer during atmospheric entry meant that the target itself could be reduced to several tens of kilometers in diameter, enabling an individual feature to be selected. The fact that a skycrane could accommodate slopes steeper than an airbag system enabled rougher sites to be considered, and it was decided that MSL should be able to deal with slopes of 15 degrees and obstacles ranging up to 50 cm in size. Given the specifications and constraints, the question of where to land was even more complex than for previous missions. Landing site workshops started in mid-2006, three years in advance of the planned launch. No fewer than 35 sites were initially offered. These included suggestions to return to Meridiani Planum and, in fact, a spot just outside the northern margin of this plateau even made the 'top ten'. In a certain sense, the workshops underscored the existence of a 'rover gap'. Given the discoveries made by Mars Express and Mars Reconnaissance Orbiter, there were so many interesting sites and so few surface missions in the budget to make follow-up investigations. By the time of the second workshop, in 2007, observation by the Mars Reconnaissance Orbiter spectrometer boosted the prospects of the sites at which clays had been observed. Proposals included Gale, a 154-km crater just south of the equator that had a mountain near its center to provide access to a 5-km stratigraphic column. Images and spectra of mounds in the crater showed layers near the bottom of this stack to include clays, followed higher up by sulfates mixed with clays and then by younger, unidentified material. This showed how the environment inside Gale had changed over the aeons as water became more and more acidic. Gale had already been considered for the Mars Exploration Rovers, but it proved impractical to squeeze a 100-km landing ellipse within it whilst avoiding the walls and mound. Moreover, the rover primary mission traverse of just 600 meters would not give access to any interesting site. The 67-km crater Eberswalde had a well-preserved delta where an ancient river discharged into a lake and left clay sediments which could preserve traces of ancient life. Another excellent candidate was the layered material in the crater Holden. But engineering concerns meant the mission could not be sent to any of the scientifically interesting canyons, as these were likely to be too windy for a safe descent. This meant that a landing in Melas Chasma, one of the side canyons of Valles Marineris, had to be deleted early on.[158]

The third workshop, in September 2008, reduced the shortlist to five sites. This

The landing sites considered for MSL, as well as the landing sites of previous NASA missions. It was eventually sent to Gale crater.

meeting highlighted a difference between "the spectroscopists" and "the geologists". The former argued for a site which had been shown by orbital reconnaissance to harbor water-bearing minerals, whilst the latter argued for where the geology showed the past presence and action of water. The spectroscopists favored the highlands near Mawrth Vallis where more than half of the surface was made of clays, or for the clay deposits in Nili Fossae from where the rover could safely drive into a canyon. The geologists argued that uncertainty over the origin of hydrated minerals could complicate interpreting the data. They preferred sites such as Gale, Eberswalde and Holden where the geological history was better understood. But this argument had its weakness, in that Spirit was sent to a site where the geology was thought to be well understood and it proved to be different. Gusev crater was initially thought to be a lakebed, but turned out to be mostly covered with volcanic basalt. Eventually Eberswalde became the preferred target, although it was not yet a definitive decision since engineering issues could force a rethink, in particular because the southerly latitude might pose thermal and insolation problems.[159]

The Mars Science Laboratory was to have been launched in a window running from mid-September to early October 2009, with arrival at Mars in the summer of 2010. By late 2008 it was clear that the mission would not be ready in time. This translated into a slippage of 26 months. One of the main issues was technical and concerned the actuators of the rover, more than 30 of which were needed for tasks including driving and steering the wheels, moving the arm, acquiring and handling samples. The reduction gears of the actuators were initially to have used a 'wet' lubricant but it was decided to use a 'dry' lubricant instead which would enable

A mosaic of Mars Odyssey images of Gale crater, the landing site chosen for MSL. The actual target site lies near the 12 o'clock position in this image. (NASA/JPL-Caltech/ Arizona State University/MSSS)

heaters to be eliminated. Meanwhile, the gears were changed from steel to titanium, which was lighter, but no means was found to make dry lubricant work well with titanium and the design reverted to steel and wet lubrication. This delayed development by about 9 months and drove up the cost. There were also issues with finalizing the mission software, in part because the hardware itself had not yet been fixed. Finally, a serious problem was discovered with propulsion line welding which required some time to overcome. In December 2008 NASA announced that the mission had been postponed to a launch in 2011 with arrival at Mars the following year. For the first time since 1996, the US intended to skip a Mars launch window. Another 2-year delay was threatened by a little-reported contamination issue on the drill bit and wheels as well as on the heat shield insulation blankets. Go-ahead was given only because the landing site had to be water-poor and therefore unlikely to give any chance of survival to stowaway terrestrial bacteria and microbes.[160] The actuators finally cleared testing in February 2010. Helicopter trials of the landing radar were performed several months later, culminating with high-altitude high-speed performance checks using a NASA supersonic aircraft.[161,162,163]

The MSL mission was already one of the most expensive US planetary missions, and this delay increased the cost to a total of some $2.3 billion – the projected overrun of $353 million was almost the total cost of a Discovery or Mars Scout-class mission. To pay for this, NASA was forced to scavenge other planetary programs, including future missions to Mars, studies for a post-2020 sample-return, the development of technologies for future planetary missions, and a joint NASA/ESA flagship mission intended for the outer solar system. This was just another episode in a budget crisis in which a host of missions, including the James Webb Space Telescope, the successor to the Hubble Space Telescope, were running over budget while NASA was trying to complete the International Space Station, retire the Space Shuttle, develop a new man-rated spacecraft and launch vehicle, and set up a program to return humans to the Moon.

On the positive side, the postponement of the Mars Science Laboratory would enable engineers to learn lessons from the difficulties that the Phoenix lander faced in handling its samples, and modify their instruments accordingly. However, other than such tweaks the spacecraft would not be changed, since revising the scientific payload would only drive up the cost. Meanwhile, hardware testing continued. In a significant milestone 2 months before the postponement, JPL engineers mated for the first time the cruise stage, aeroshell, skycrane and rover. After a competition in which members of the public offered names, it was announced that the new rover would be named 'Curiosity'.

The postponement meant that the final selection of the landing site would be delayed to 2011. With the list having been reduced to a handful of contenders, a discovery cast doubt on the process. Scientists announced that using terrestrial telescopes they had measured the distribution and seasonal behavior of atmospheric water and methane over the northern hemisphere of Mars and found that methane occurred in extended plumes which originated in discrete regions, reminiscent of some of the most active terrestrial hydrocarbon seeps. Sites of enhanced methane emissions included an area east of Arabia Terra where water vapor was also present, Nili Fossae, and an area southeast of Syrtis Major where the surface showed indications of interacting with a volatile-rich substrate.[164] Nili Fossae had only recently been deleted from the target list, and this finding made its proponents argue for it to be reconsidered. In fact, a new call for suggestions was made and seven more sites were added to the shortlist. Scientists met in December 2009 to discuss whether any of these candidates were stronger than those already shortlisted. In the end, two sites (a diverse assemblage of minerals in northeastern Syrtis Major and chloride and clay deposits in Margaritifer Terra) were chosen for high-resolution imaging by Mars Reconnaissance Orbiter and thermal analysis by Mars Odyssey so that they could be better assessed as possible landing sites. But significant safety concerns were found in both cases, and so neither site made it to the main candidate list.

By the fifth workshop, in 2010, engineers felt confident that they would be able to land on any of the candidate sites, and so it was entirely up to scientists to select one. Only in 2011 were some choices made. Scientists could not convincingly prove that Holden had held a lake since there was no river or canyon discharging into it, so it was deleted. Also deleted was Mawrth, in part because its formation was judged to be poorly understood and also because, by analogy with the flat expanse of Meridiani Planum explored by Opportunity, it lacked geological diversity. Eventually, only Eberswalde and Gale remained in the running. Given that the workshops could not reach a consensus on a single candidate, the target site was selected in the end by a small group of managers and mission scientists, and the choice of crater Gale was announced during July. Named after the Australian amateur astronomer Walter Frederick Gale, the crater is located just 5 degrees south of the equator, straddling the Martian north–south dichotomy. It is also one of the lowest points on the planet after Hellas and the northern plains. As a point of comparison, it is slightly smaller than the Chicxulub crater in Yucatan whose formation is believed to have caused the extinction of the dinosaurs. Gale was selected for its geology, morphology and mineralogical diversity, which apparently provided a long and varied record of environmental conditions on Mars as it changed from wet to cold and ice-covered to arid.

The 21 × 14-km landing ellipse would be located north of the central mound, and close to a prominent fan of debris coming from the northern wall of the crater. Viewed by Mars Odyssey's infrared camera, the fan, later officially named Peace Vallis, showed a higher thermal inertia than its surroundings, a sign that it could include consolidated material, possibly cemented together by water. After possibly visiting this debris, the rover was to cross a small dune field to arrive at the base of the mound, where a canyon would provide direct access to layered terrain, the lowermost layers of which are iron-rich clays that could provide mineralogical and morphological clues of the past existence of water. In particular, scientists were seeking to determine whether the stratification of the mound represented sediment deposited on the floor of a lake or the accumulation of wind-blown material from impacts and volcanic eruptions. After finishing with clays, the rover would likely climb to the sulfate layers. Finally, if granted a mission extension, it might try to reach the summit of the mound – informally named Mount Sharp after the late Caltech geologist Robert P. Sharp, who participated in many of the early NASA Mars missions; officially it is named Aeolis Mons.[165,166,167] A minor factor in the selection of Gale was apparently also the fact that the site was almost at the antipodes of the Mars Exploration Rover Opportunity, so that the two rovers could carry out their missions without the transmissions from the first interfering with those from the second.

The first components of the spacecraft, the heat shield, backshell and cruise stage, arrived in Florida on 12 May 2011, followed by Curiosity and the skycrane on 23 June. Budgetary problems were not over yet, however, and many issues remained open as late as June 2011, according to an internal NASA 'watchdog' report. The issues were difficulties in completing the avionics, radar, drill, mobility system and the sample analysis instrument suite. The solution suggested was to increase the budget by several tens of millions of dollars. Moreover, like previous probes, the rover would be launched without software for landing and surface operations because that was still under development and would be uploaded in flight. Funding woes were not the only difficulties faced at this late phase of the development. The possibility of another delay occurred when the composite backshell was mishandled, by inadvertently lifting it for a few seconds while attached to a 1-tonne table. Fortunately, preliminary assessments, later confirmed by non-destructive tests, verified that it had not been damaged. The launch window opened on 25 November 2011 and ran to 18 December. The launch was delayed a day to allow time to replace a battery of the "flight termination system" (i.e. autodestruct) of the launcher, but on 26 November no problems were encountered and Curiosity – "the Hubble Space Telescope of Mars exploration" as NASA associate administrator for space science John Grunsfeld called it – lifted off as planned. Some 44 minutes later, the spacecraft separated from the last stage of the launcher, which executed a maneuver to preclude it accidentally hitting Mars. Images taken by cameras on the Centaur showed the solar panel-side of the probe leisurely moving clear. A little more than 10 minutes after that, the probe made contact with Deep Space Network antennas. It was then in a 0.98 × 1.54-AU orbit that would pass by Mars at a range of 61,200 km; so close to nominal that the first course correction scheduled for 15 days into the mission was postponed for several weeks.

During the check-out phase, a star tracker glitch was encountered which put the probe, spinning at 2.05 rpm, into safe mode. On 6 December the radiation detector was turned on. This would be the only instrument to collect data during the cruise, monitoring energetic particles and recording secondary particles created as they hit the walls of the cruise stage. It noted the solar storm of January 2012, the largest in years, which also hit Earth.

A small burn was executed on 22 December to assess the status of the propulsion system after concern about a possibly defective propellant valve, and then the first of six course corrections was made on 11 January, removing the intentional launch bias. A second maneuver on 26 March targeted Gale crater for the first time. During the flight, engineers were able to reduce the landing ellipse to just 7 × 20 km as a result of better predictions of surface winds and atmospheric conditions. Curiosity would therefore be able to land nearer to the slopes of the mound that it was to investigate, and halve the driving distance.

On 14 July, the radiation detector was turned off in preparation for landing. It had collected data on solar particles and galactic cosmic ray fluxes for most of the cruise, recording in particular five solar flares whose particles passed all the way through the heat shield. Data on the galactic cosmic rays were the most eagerly awaited, for they would be useful in assessing the health of a human crew traveling to Mars. It was thus confirmed that humans on a low-energy ballistic flight to Mars would receive a radiation dose three times that which would result from 6 months in low Earth orbit and several hundred times that of an abdominal X-ray computed tomography scan. This would yield a statistically significant increase of the risk of contracting a fatal cancer. Solar particles on the other hand, looked relatively benign, although this was probably due to the fact that the flight happened to take place in a period of relatively quiet Sun.[168],[169] Tracking data was collected almost continuously for the last 45 days in order to determine the atmospheric entry point with the required accuracy, and for the last 28 days even more precise data was collected twice per day by using distant quasars and orbiters around Mars as references.

After pursuing a type-1 trajectory, the spacecraft arrived at Mars on 6 August 2012. It could have undertaken a type-2 trajectory for a slower arrival velocity but this would have conflicted with the scheduled departure of the Juno orbiter bound for Jupiter. However, only 2 months before landing, the Mars Odyssey orbiter, the prime relay for the descent and landing phases went into safe mode over a reaction wheel issue. Engineers soon brought it back online using a backup wheel that had not been used since launch in 2001. It then re-entered safe mode for 21 hours less than a month before arrival. This latest glitch put the use of Mars Odyssey as a relay during the descent at risk, but finally, with less than a fortnight to spare, the orbiter was able to fire its thrusters to advance its overflight of Gale by 6 minutes. Two further maneuvers were planned for Curiosity, but were not needed. Six days prior to arrival, the latest predictions of the entry point were uploaded to the navigation system. The rover's computer would then use this data and the landing site coordinates to guide itself during the hypersonic flight in the atmosphere. All three operational orbiters would be overhead at the time of landing because in addition to Mars Odyssey, Mars Express and Mars Reconnaissance Orbiter would participate – with the latter trying to obtain pictures similar to the one that it took of Phoenix descending on its parachute. Moreover, during the first minutes of entry the rover would broadcast status 'tones' toward Earth. The Opportunity rover was still operational, exploring crater Endeavour, and would be returning data over its high-gain antenna in order to free the orbital relays for its larger sister.

At 05:00 UTC on arrival day, the cruise stage was released and left to burn up in the Martian atmosphere, followed 5 minutes later by the first two balance weights. Ten minutes after the cruise stage was released, the rover, still cocooned inside the heat shield entered the atmosphere at an altitude of about 125 km and at a speed of 6.1 km/s. It was later determined that Curiosity had hit the atmosphere 200 meters off the aim point, at a speed 11 cm/s off, and with an angle 0.013 degrees shallower than the 15.5 degree target. Four minutes into the descent, the remaining six balance weights were released in order to return the center of mass of the capsule close to its geometrical center. And some 34 seconds after that, the parachute was deployed. Curiosity was then 10 km above the surface and traveling at some 500 m/s. At about this time, the Earth passed below the local horizon as expected, cutting the direct telemetry link. All of the remaining phases of the landing were still followed in near 'real time' at JPL (actually, with a 14-minute delay owing to the distance between Mars and Earth) using the 'bent pipe' relay of Mars Odyssey. The front heat shield was let go some 20 seconds later, turning on the descent radar. Meanwhile, the descent camera had already started taking pictures and it imaged the spent heat shield falling off and finally hitting the ground. The dark dune field located between the landing site and the base of Aeolis Mons was clearly visible, as were many small craters. Mars Reconnaissance Orbiter passed almost overhead and repeated its feat with Phoenix by capturing an image of Curiosity nested inside its backshell and dangling under its large white parachute, only 1 minute and 3 km before landing. The orbiter was looking almost straight down from a distance of 340 km. The long picture swath also happened to include the freely falling spent heat shield. The descent under the parachute lasted for 100 seconds, during which the speed of Curiosity relative to Mars was cut from 500 to 100 m/s. At an altitude of about 1.6 km the backshell was released, carrying the parachute with it, and the skycrane, clutching the rover, was left to fall freely for 1 second prior to activating its eight engines. These thrusters were first operated at a throttle setting in the range 50 to 70 per cent in order to maneuver laterally and to avoid contact with the backshell. They then cut the vertical speed and zeroed the horizontal speed. On approaching the surface, four of the thrusters were throttled back to just 1 per cent, while the others thrusted at 50 per cent. At a height of 20 meters and some 12 seconds before touchdown, the skycrane started to unreel the rover.[170,171]

At 05:17:57 UTC Curiosity was delicately deposited on the Martian surface by the skycrane at a vertical speed of 0.75 m/s and a horizontal speed of 4 cm/s. The site was at 4.5895°S, 137.4417°E, only about 2.4 km east and 400 meters north of the intended target, and some 6 km from the base of Aeolis Mons, 28 km from the rim of Gale, and close to the edge of the alluvial fan. It was later dedicated to the late American science fiction writer Ray Bradbury. The rover came to rest tilted a mere 4 degrees and facing east-southeast. At that time it was late afternoon in Gale crater. A number of 'thumbnail' images were taken through the transparent dust covers that protected the hazard cameras. These images were fuzzy and disturbed by dust that had been sprayed onto the covers, but they were able to show what looked to be a dark dust plume from the impact of the skycrane. By the time of the second, shorter

A few seconds after it was released, the descent camera on Curiosity took this picture of the jettisoned heat shield falling away. Dark dunes are visible on the floor of Gale crater. (NASA/JPL-Caltech/MSSS)

This picture taken by MRO shows Curiosity on its parachute. (NASA/JPL/University of Arizona)

A hazard-avoidance camera picture taken by Curiosity shortly after landing on Mars, showing Aeolis Mons (also known as Mount Sharp) in the background. (NASA/JPL-Caltech)

contact with Odyssey one orbit later, the hazard camera covers had been opened, and yielded clearer pictures showing Curiosity's long shadow projecting on the dusty crater floor with a few small rocks and Aeolis Mons in the distance. Also returned were some low-resolution versions of the descent camera frames.

Mars Reconnaissance Orbiter images taken later on the day of landing showed Curiosity on the surface at the center of a butterfly-shaped area disturbed by the skycrane's exhaust, as well as the crater dug by the heat shield, some 1,500 meters away, the backshell and parachute 615 meters away, and a dark patch where the skycrane impacted and exploded some 650 meters to the west. (About 140 kg of fuel remained in the tanks when the skycrane released Curiosity, much more than initially expected.) Images taken by the context camera pinpointed the impact sites of the six ballast weights, some 12 km distant. Craters formed by the cruise stage and the two tungsten ballast weights were found months later 80 km northwest of the landing site and actually outside Gale crater. The cruise stage had apparently broken up in the atmosphere and dug two separate craters. Images taken over the next several months showed the parachute canopy flapping in the wind.

As the Sun rose the following day, Sol 1 of the surface mission, Curiosity was commanded to unlatch and deploy the high-gain antenna. In a rare glitch, it could not at first be correctly aimed at Earth. Several instruments were also exercised,

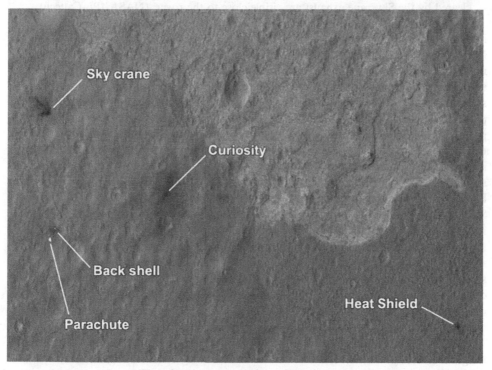

The landing site of Curiosity imaged by MRO about 24 hours after landing. (NASA/JPL/University of Arizona)

An early panorama of the landing site taken by the navigation camera on Curiosity shortly after landing, showing the distant rim of Gale crater. Two of the areas that were cleared by the Skycrane exhaust are visible close to the rover. (NASA/JPL-Caltech)

including the arm-mounted camera, which took its first picture of the surface of Mars through a transparent cover. On Sol 2, Curiosity finally deployed the camera mast and took the first images with the navigation and science cameras. The wide-angle camera was used first, and provided a 130-picture color panorama of the surroundings which was transmitted over many sols, engineering data still having priority. The azimuth and elevation of every picture in the panorama had been pre-programmed before landing without knowing precisely where the rover would be, so the sequence omitted the summit of Aeolis Mons. Higher resolution black-and-white images were returned by the navigation camera. In the vicinity of the rover the skycrane's engines had removed dust in four spots, revealing what was under the layer of dirt. The shallow scar marks were named Burnside, Goulburn, Heburn and Sleepy Dragon. Goulburn in particular exposed what appeared to be individual rocks cemented together. Also near the rover, a number of enigmatic rocks were seen that consisted of light-colored material embedded within an otherwise dark matrix. A portion of the northern wall of Gale was visible, as well as a network of erosion-smoothened valleys and alluvial streams. The field of dark dunes seen during the descent was visible to the south, between 2 and 4 km off, beyond which were layered mesas, hills up to tens of meters high and canyons at the base of the mountain. In front of the dunes was a rock strewn, gravel-like surface.

After Sol 4, Curiosity had to stand down while its software was updated. The no longer needed functions used for descent and landing were erased, and new ones to drive the vehicle and operate the robot arm were uploaded. The software update was completed on Sol 8, and the two computers were rebooted to enable Curiosity to begin to calibrate and exercise its instruments. Meanwhile, scientists were busy scrutinizing orbital images and designing the first scientific traverses. It was decided to forsake the putative alluvial fan to the north and instead head for a spot 400 meters to the east-southeast where three geological units intersected. This was named Glenelg, after a geological site in northern Canada. These geological units included layered bedrock, a heavily cratered and possibly old terrain, and the dusty terrain unit on which Curiosity landed.[172,173,174,175]

Instruments meanwhile continued to be exercised, calibrated and tested. The meteorology station returned its first data. Typically, air temperature ranged from near zero to −75°C. The only major glitch discovered was that one of the two sets of wind sensors, mounted on the camera mast, was not providing meaningful data. Some exposed circuit boards of the rear-facing sensor may have been permanently damaged, possibly by small pebbles kicked up during the landing. The position of the sensor, moreover, was such that it would be difficult to image it with the arm-mounted camera, which could not be positioned closer than half a meter from it. The ability to determine the direction of winds was partially compromised, but the forward-facing sensor was working perfectly well.

Next up was a test of the chemistry laser. Coronation, an angular rock with flat faces that was some 8 cm across and about 2.5 meters from Curiosity was selected as the first target. On Sol 13, thirty 14 millijoule laser pulses were fired at it during a 10-second sequence, and the spectrum of the vaporized rock was recorded. It appeared to be a rather typical Martian basalt. The spectrometers detected oxygen, silicon, magnesium, iron, sodium, potassium, calcium and aluminum, as well as traces of titanium, lithium and manganese. The very first pulses released hydrogen that formed part of a thin surficial layer. Possibly, hydrogen had been deposited by hydrazine decomposition in the skycrane exhausts. Carbon from the atmosphere was also detected. On the same day, Curiosity imaged the scars as well as Aeolis Mons using the narrow-angle science camera for the first time. Hills at the base of the mountain clearly showed dramatic layering on their flanks. The next day, the arm was unstowed and extended by exercising all of its actuators, as well as the toolkit on its end. The laser next shot three sites in the Goulburn scar. This was confirmed to consist of cemented fragments of basaltic composition. Several other rocks were also fired at. In situ analysis spectrometers were also exercised. More terrestrial air remained in the ovens than predicted, however, which led the pumps to stop. As a result, the instrument ended up analyzing mostly the air of Florida and calibration gas. A follow up test, on Martian air this time, was performed over Sol 22 and 23.

For all of these tests, the rover remained stationary. Then, it was finally time to test the locomotion system. The steering system of the front and rear wheels was exercised on Sol 15, and the following day the rover was commanded to drive 4.5 meters forward, turn 120 degrees to the right and drive 2.5 meters back. It stopped 16 minutes later about 6 meters from the landing spot. Additional narrow-angle images of the base of Aeolis Mons were taken between Sol 19 and 21 in order to better focus the camera. The terrain in front of the dune field seemed to be covered with a sand that was redder than the dark dunes, which hinted at different sand composition. On the hills, a line where the terrain changed its texture seemed to mark the boundary

A mosaic of pictures of the 'underbelly' of Curiosity taken by the arm-mounted camera during Sol 34. The four front hazard-avoidance cameras are clearly visible. (NASA/JPL-Caltech/MSSS)

Golbourn was one of the four areas of the Curiosity landing site where the blast of the Skycrane thrusters removed the surficial dust and exposed a sedimentary pebbly congolomerate. (NASA/JPL-Caltech/MSSS)

between clays and terrains where clays had not been detected from orbit. Remarkably, moreover, terrain above the boundary had layers dipping at a different and more inclined angle, close to the slope of the hills. This could be produced by volcanic activity, or by processes involving either water or winds. Also on Sol 19, the rover examined the rock named Beechey, 3.5 meters away, by firing fifty laser shots at each of five different spots.

Before/after images taken by the camera on the laser spectrometer of the laser target Beechey, examined on Sol 19. (NASA/JPL-Caltech/LANL/CNES/IRAP/LPGN/CNRS)

A number of communications tests were carried out during this time. On Sol 13 the backup relay through the European Mars Express orbiter was exercised, and by Sol 20 over 7 gigabits of data had been returned, most of it through Mars Odyssey and Mars Reconnaissance Orbiter. An adaptive communications strategy was used when communicating with the latter to optimize the data rate to the geometry of its passage over Gale crater, the minimum distance between the two spacecraft, etc. The 'fast relay' was used for a number of public relations stunts as well. The first human voice playback from another planet in the form of a message from NASA administrator Charles Bolden uploaded as an audio file was used to verify some of the communications system capabilities, and several days later the same was done with a song by pop singer will.i.am.

On sol 21 a second drive positioned the rover over the Goulburn scar in order to try to detect hydrated minerals using the Russian neutron activation instrument. The laser was then used to perform further analyses of the scars. Wrapping up the activities at the landing site, Curiosity made a 15-meter drive on the afternoon of Sol 22, where it halted to take more images of Aeolis Mons. When combined with the images taken from the initial position, 10 meters away, it would be possible to generate 3-dimensional views of the base and slopes of the mountain. A 21-meter drive on Sol 24 tested some of the hazard-avoidance algorithms, and wrapped up the month of August.[176]

On Sol 29, having driven a total of 109 meters, Curiosity stopped for a few days while the arm-mounted instruments were assessed and exercised. In particular, on Sol 32 the microscope camera took 'self-pictures' of the mast, and on the next day, after the dust cover was finally opened, of the underside of the rover. The arm then tested the accuracy of its guidance algorithms by positioning itself over the inlets of the deck-mounted instruments. The rover then restarted driving and employed the neutron spectrometer along the way to sound the subsurface. On Sol 37 it observed its first partial eclipse of Phobos, followed a few days later by a transit of Deimos. Two days later, Curiosity passed near a rocky outcrop named Hottah, looking "like a broken sidewalk". This was an exposure of bedrock made of smaller fragments of rock loosely cemented together. The gravel embedded had rounded shapes and was made of relatively large pebbles up to several centimeters across that seemed to have been transported there by a long-lasting flow of water, perhaps part of the Peace Vallis fan, and then become bound into a matrix of sand-like material. A similar bedrock had been observed at Goulburn and at Link, a rock spied around Sol 27. Only remote sensing was carried out at all of these sites but the laser was able to detect hydrated minerals. Based upon the morphology and composition, the scientific consensus was that Link and Hottah, as well as Goulburn, were once the bottom of shallow streams flowing at walking speed.[177],[178] A drive on Sol 43 brought the rover to its first robot arm science target. Scientists picked out a pyramidal rock, 50 cm tall and some 40 cm across, that was named Jake Matijevic (or abbreviated "Jake_M") after a late JPL engineer who worked on Mars rovers. The dark gray rock was examined using the alpha-ray spectrometer, the microscope, and the laser spectrometer over the next days. Fourteen different spots were sampled by the laser and two by the alpha-ray spectrometer, revealing a surprisingly heterogeneous rock. It was classified as an alkali basalt igneous rock with little iron and magnesium and high levels of sodium and potassium relative to other Mars rocks analyzed by Spirit and Opportunity, and could have been formed by a water-rich magma crystallizing at relatively high pressure. Its composition was close to that of uncommon terrestrial rocks found in rift zones and ocean islands such as the Canary Islands.[179] Shortly after noon on Sol 45, the narrow-angle camera snapped the crescent Phobos in the sky – the first images of a Martian moon taken from the surface of the planet to show any detail, specifically the jagged terminator, as well as the night side illuminated by 'Mars-shine'. Finally, after spending five days analyzing Jake Matijevic, Curiosity resumed its drive and ended the day 42 meters closer to Glenelg.[180]

The drives at first were short in order to validate the hazard identification and automatic navigation algorithms, but they became longer as the controllers gained confidence. On Sol 54, the alpha-ray spectrometer was used on the target Bathurst Inlet. Two days later, having driven a total of 484 meters, and reached a point only meters from Glenelg, Curiosity encountered an ancient drift of fine sand, 2.5 × 5 meters wide, where a crust had had time to form. This was named Rocknest. The rover remained in place for several weeks in order to undertake its first scoop test. After a number of samples had been run through the sample acquisition and handling system, the instrument was to be vibrated with dust inside to "sandblast" it and remove any remaining terrestrial contamination before a sample could be delivered to the chemical analysis instruments. The first sample was collected by the 'spoon' on Sol 61. However, the next day activities had to be stopped because images showed a bright object several millimeters long on the ground that would have to be investigated. In the end, it was determined to be a fragment of the rover, possibly a bit of insulation or some wire wrapping. A second scoop for cleaning was collected

The Hottah outcrop, "looking like a broken sidewalk", provided one of the first pieces of evidence of an ancient streambed in Gale crater. (NASA/JPL-Caltech/MSSS)

on Sol 66 and a third, to the right of the previous two, on Sol 69. This sample was then delivered to the rover's observation tray, a flat metallic surface, for scrutiny by the microscope and alpha-ray spectrometer. The sand proved to consist of a finer, light component and a "grainier" darker one with a mineralogy similar to volcanic soil on Earth. The grains were mostly made of plagioclase feldspar, pyroxene and olivine, but there was also volcanic glass present. A number of shiny particles were also spotted in the sand that could be small chips of quartz. The fourth scoop was collected on Sol 74, delivered to the sample analysis instrument and, after sieving to particles less than 0.15 mm across, to the chemistry laboratory.

Other measurements were going on in the meantime. The second atmospheric analysis was carried out between Sol 78 and 81. Whereas the first study had found a

The marks left on the ground after Curiosity took samples of the Rocknest dune using a scoop. The circular observation tray is visible at bottom. (NASA/JPL-Caltech)

strong methane signal, which was discounted for being caused by the remaining terrestrial atmosphere in the analysis cell, the second did not yield any definitive detection of methane down to 3 parts per billion. Almost 96 per cent of the local atmosphere was carbon dioxide, with the rest being argon, nitrogen and traces of oxygen and carbon monoxide. Overall, these measurements were consistent with those made 35 years earlier by the Viking landers, except for nitrogen, which was present in significantly lower percentages. The isotope ratios of argon and carbon indicated a higher ratio of heavier isotopes than on Earth, implying that lighter isotopes had probably been lost to atmospheric erosion. Over its mission, Curiosity would monitor how the composition of the atmosphere changed with the seasons, and in particular as carbon dioxide was being injected into the atmosphere by the thawing of the polar caps in spring and summer. The methane measurements were particularly deceiving, as for some time it had seemed to be the best proxy of some form of biological activity. However, recent telescopic observations by the same team that reported concentrations ten times greater in the early 2000s had yielded similar results to Curiosity. Possibly there had been bursts of methane in the early 2000s and the gas had since dispersed.[181,182,183] The meteorology package was also busy collecting data. Although Gale had seemed to be relatively free of dust devils from orbital observations, more than twenty of them were detected during the first 3 months. This was still a much lower rate than at the Spirit and Phoenix landing sites. East-to-west winds seemed to predominate, although scientists had expected to see more north-to-south winds corresponding to upslope winds on Aeolis Mons. Dominant winds, on the other hand, appeared roughly aligned with the crater floor, with the obstacle of Aeolis Mons to the south and the rim of Gale to the north. A 10 per cent increase in atmospheric pressure during the first 3 months was caused by the thawing south polar cap releasing carbon dioxide into the atmosphere. The radiation fluxes at the surface appeared to correlate with the diurnal pressure cycles, and were actually lower than during the cruise to Mars, possibly posing no threat to future human expeditions. Finally, on Sol 84 and 85 the arm-mounted camera took the first full self-portrait of Curiosity. On 10 November, Mars Reconnaissance Orbiter saw a regional dust storm develop. Atmospheric changes were recorded by the rover's meteorology package, which detected an air pressure decrease as well as an increase of night-time temperatures. More visibly, the storm altered the transparency of the air, so much so that the eastern rim of Gale disappeared from sight within a few days.

With all of the instruments exercised and the most stressful part of the mission over, on Sol 90 the mission team switched back from controlling the rover in Mars time to Earth time. On Sol 91, the alpha-ray spectrometer examined a target dubbed Et-Then. A fifth and final scoop sample was collected from Rocknest on Sol 93. Meanwhile, analyses were being carried out at increasing temperatures, up to 825°C, on the fourth sample. Analyses of the first samples found significant quantities of water vapor, carbon dioxide and sulfur dioxide. Some carbon dioxide was released at high temperature, probably by the decomposition of some carbonate mineral. The water produced on the other hand was sufficient to perform the first measurement of the deuterium-to-hydrogen ratio on Mars. This revealed that water in Rocknest was about five times richer in deuterium than the terrestrial oceans, which might imply that most of the lighter Martian water had been lost from the atmosphere over the ages. On the other hand, isotopic ratios of hydrogen in the atmosphere and soil were compatible, indicating that hydrated minerals in the latter were probably formed by interaction with the former. Comparison with the ancient atmosphere as recorded by gas pockets in the controversial "life on Mars" meteorite ALH84001, which formed more than 3.5 billion years ago, indicated that most of the atmospheric loss had occurred by then. There was no definitive trace of organic compounds, but it was well known that sand was not particularly suitable for seeking organics, as it had been thoroughly exposed to the action of the Martian environment. In their place, the instrument found chlorinated hydrocarbons, carbon- and chlorine-bearing molecules created by the reaction of chlorine (probably from Martian perchlorate salts like those detected by Phoenix) with carbon in the oven. If this carbon was not terrestrial contaminants remaining in the instrument, it could be derived from carbon-rich meteorites, which should be relatively common on the planet, or be of Martian origin. Even if it was Martian, it could very well have originated from non-biological sources like carbonate rocks. The X-ray diffraction analyses revealed the chemical and physical composition of the dust, but somewhat surprisingly failed to detect any trace of clays. In terms of mineralogy, basaltic dust at Rocknest resembled similar soils elsewhere on Mars, implying a global distribution and mixing of dust. Part of the third scoop was examined by alpha-rays on Sol 102 and found to be similar to average Martian soils. By Sol 100, Curiosity had performed eleven alpha-ray, about 425 laser, and 171 neutron analyses, and had taken a total of around 11,000 science images. The laser in particular had fired over 3,600 times, identifying different types of soil and dust. The most important result was the ubiquitous presence of hydrogen representing either hydrated minerals or adsorbed atmospheric water. This hydrated material probably accounted for a large part of the hydrogen that was detected by the gamma-ray spectrometer on Mars Odyssey and by Mars Express.[184,185,186,187,188,189,190] After spending 40 days at Rocknest, on Sol 100 Curiosity took a short drive in order to reposition itself for further analyses. Then two days later it drove eastward 25 meters to a brighter terrain in Glenelg and stopped at an overlook point dubbed Point Lake. In the process, it drove across Rocknest and over its scoop marks in order to collect neutrons from the terrain that it had spent so long studying. On Sol 102 it rehearsed a "touch-and-go" fast analysis using the alpha-ray spectrometer, deploying the robotic arm, integrating a spectrum of the terrain, stowing the arm, and resuming its drive. Once at Glenelg, it took panoramas which would be used to evaluate possible future drives and to identify targets. These images showed some interesting areas, including Shaler, a well-preserved cross-stratified deposit spanning more than 60 meters and at least 80 cm thick. Shaler resembled some of the outcrops examined by Opportunity, preserving in a fossilized form the direction of shallow water flows that carried the grains. A priority was to locate a rock for the first drilling test. Glenelg presented a variety of geological targets: sandstones which had bright quartz-like intrusions, slabs of bedrock, and amazing features resembling fossilized mud bubbles.

On Sol 125, Curiosity entered a low area dubbed Yellowknife Bay, consisting of a succession of layers, each more than one meter thick, of which the lower ones were dubbed Sheepbed and Gillespie Lake. The area looked remarkably different from the landing site. In fact, while the rocks at the Bradbury site were made of a conglomerate of smallish rounded pebbles, approaching Yellowknife Bay they started to exhibit thin layers of mudstone and sandstone. This was reflected in the temperatures measured by the meteorology instrument, which were higher than along the way because of the higher thermal inertia of the rocky terrain. The rover made laser analyses of veins in Crest on Sol 125 and Rapitan on Sol 135, detecting the emission lines of calcium and sulfur indicating calcium sulfate, with the presence of hydrogen implying some form of gypsum. During the Earth holidays, Curiosity remained parked at Yellowknife Bay and imaged its surroundings.

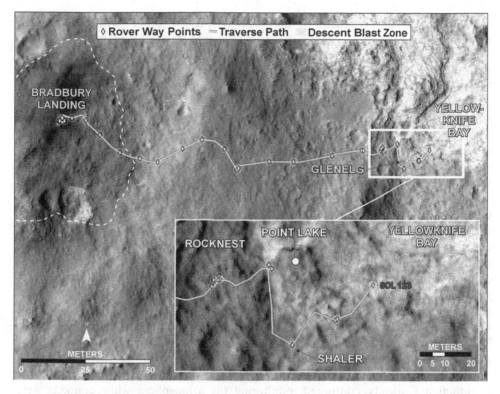

The traverses of Curiosity up to Sol 123 overlaid on an MRO image of the floor of Gale crater. (NASA/JPL/University of Arizona)

To begin its activities in 2013, on Sol 150 Curiosity brushed its first target, a flat rock area of Sheepbed dubbed Ekwir-1, which was afterwards examined by the arm-mounted spectrometer. Microscope images showed very fine grains, and proved it to be a mudstone sedimentary rock. Ten days later, Curiosity drove over the small rock Tintina, only a few centimeters across. The rock broke apart, revealing a bright white interior. On Sol 165 the rover took night-time images of the Sayunei rock using the arm-mounted camera augmented by white-light and ultraviolet light emitting diodes, looking for fluorescent minerals. That same night, Curiosity obtained sky images. An area suitable for the first drill was identified at John Klein, consisting of flat, veined and fractured rocks. At some point, water had filtered through cracks in the rocks, leaving behind veins of hydrated minerals that were rich in calcium and sulfur but poor in magnesium and silicon. In some places, the rock had been eroded to leave the veins sticking up vertically – offering the chance for Curiosity to drive over them and crush some of the minerals for analysis. The most evident of these veins was a dike-like feature that cut through Yellowknife Bay and was dubbed Snake. In other areas of Sheepbed, spherules had become embedded in the mudstone and now appeared as grains several millimeters in size, rounded by abrasion, and looking similar to white crystals. On Sol 171 the rover carried out pre-load tests, placing the drill bit on four different locations and pressing it down. Three days later, a test of the hammering mode of the drill was not completed successfully and was repeated on Sol 176. A drill test to ascertain the characteristics of the powdered rock was performed on Sol 180, drilling a shallow 1-cm hole. Controllers were careful not to leave the drill bit in the rock overnight, as the thermal contraction of the arm could damage it. Finally, on Sol 185 a proper hole 1.6 cm wide and 6.4 cm deep was drilled, straddling a thin vein in the rock. The drill cuttings were of a surprisingly non-Martian dull gray color, suggesting that the interior of the rock had not oxidized like the rest of the surface of the planet. Powdered rock flowed along flutes on the drill to two storage chambers. Then, on Sol 193 a tablespoon of powdered rock was finally transferred to the scoop for sieving and delivery to scientific instruments. Once again, the first dust collected would be used to clean the hardware of potential contamination. A few days later, on Sol 200, however, Curiosity had a flash memory issue on one of its two computers, sending the vehicle into safe mode. While engineers investigated the problem, they switched to the redundant unit to resume routine operations. Just as the recovery was progressing, Mars was hit by a solar coronal mass ejection. Rather than risk a second safe mode while the first one had not been fully solved, Curiosity was commanded to 'sleep' for a few days. Days later, Curiosity had another problem while attempting to send data back to Earth. While this problem was not uncommon, it still delayed full recovery for several more days. The rover was finally coaxed out of safe mode on 19 March, as the teams were preparing for their first solar conjunction in April (during which there would be a precautionary hiatus in communications for two and a half weeks). During the conjunction Curiosity continued monitoring the atmosphere while contacts were interrupted. While continuing analyses of the drill tailings from John Klein, the sample analysis instrument also made the most precise measurement yet of argon isotopic ratios in the atmosphere, finding about four times as much argon-36 than argon-38, a ratio much lower than the one at the formation of the solar system. This implied an atmospheric loss process that favored erosion of the lighter isotope. The result was much more precise than measurements by the Vikings.

After the conjunction, Curiosity was to perform analyses on a second drill sample in order to confirm its results. The second target was Cumberland, less than 3 meters from John Klein. The planning and operation went much faster, resulting in a drill on Sol 279. Cumberland appeared richer in erosion-resistant concretions and granules, and it took the drill 6 minutes to reach a depth of 65 mm. It obtained some 14 cubic centimeters of material, of which less than half made it through the sieve. This time, the composition of the tailings was checked by the laser. Analyses of the John Klein and Cumberland samples revealed them to be made of igneous material but including high percentages of clay-like minerals, which released water vapor when the material was heated to a high temperature. The presence of sulfur-rich and iron-rich clays indicated that they were formed in water that was moderately salty and had a neutral or slightly acidic pH, quite different from the acidic water at Meridiani Planum that was studied by Opportunity. This indicated that the area where Curiosity stood was once a lake bed with a watery environment that lasted for at least several thousand years and was theoretically suitable for microbial life. The

Curiosity's first drill holes in John Klein. The shallow test hole is the one at right. (NASA/JPL-Caltech/MSSS)

once-watery environment was confirmed by the morphology of the rocks in Sheepbed. Perchlorate compounds were again detected, but terrestrial contamination was deemed unlikely because the system had been thoroughly cleaned and the samples purged with helium. Moreover, about the same abundances were detected at John Klein and Cumberland, which was judged unlikely if the sample-handling system was contaminated. Carbon dioxide, as well as more complex organics, were again detected but this time more carbon was released by the drill tailings than in the dust samples and it was released at a lower temperature, as expected for organic molecules rather than carbonate minerals. Once again, organics on Mars would not necessarily imply a biological source and the flux of meteorites to the surface of the planet would be enough to account for all the carbon detected.

Scientists developed a technique to determine the age of the rocks by comparing potassium isotopes measured using the alpha-ray spectrometer with argon isotopes released during sample analysis; argon being a product of the radioactive decay of potassium. Cumberland was determined to be about 4.2 billion years old, matching quite well the age of the area around Gale calculated on the basis of crater counting. Moreover, by measuring noble gases and isotopic ratios caused by the exposure to cosmic rays, scientists were able to determine how long the rocks had been exposed on the surface. It was found that Cumberland had been on the surface for almost 80 million years, having probably been exposed by wind erosion. This result did not bode well for the search of organics, because the bombardment by cosmic rays over such a long time would destroy all organic molecules present in the rock. Scientists made finding and analyzing what appeared to be recently exposed bedrock in which organics would be preserved one of the priorities for the remainder of the mission. Several candidates were identified on the road to Aeolis Mons.[191,192,193,194,195,196]

On Sol 295, Curiosity resumed its drive, making a neutron spectrometer scan of the site in order to correlate the presence of hydrated minerals in the drilling samples with the presence of hydrogen. It was then to head towards Point Lake and Shaler. Point Lake was a dark outcrop that had only been seen from several tens of meters. This displayed centimeter-wide hollows that it was thought might have been gas bubbles known as vesicles in the lava from which it formed. Next, the rover would drive back several meters to investigate the cross-stratified outcrop in Shaler, which was located directly behind Point Lake, to ascertain the watery environment in which it formed. After that, it would resume driving to Aeolis Mons, making only a few science stops along the way. Caution had to be exercised during the drive to Shaler as the tilt of the rover exceeded a limit, and when in place, time was allowed before moving the arm to ensure that Curiosity was firmly in place and would not slip.

While these two sites were being investigated, other observations were made. In particular, in the first 300 Sols, up to June 2013, the radiation monitor had measured an average dose from cosmic rays about half that integrated during the interplanetary cruise. This result was not surprising, as the bulk of Mars would intercept half of the cosmic ray flux. On Sol 242 (11 April) it had recorded a cascade of solar particles that was also observed from Earth, where it was a relatively weak event, and also by one of the STEREO solar probes. Moreover, on Sol 313 the sixth reading of methane in the atmosphere was taken. Once again no methane was detected down to a part per billion level.[197,198]

Wrapping up its activities at Shaler on Sol 326, Curiosity started the long drive to Aeolis Mons, for several days running only a few meters from its route of 9 months earlier. The foot of the mound was almost 8 km away, which engineers expected to be able to cover in 10 to 12 months. Meanwhile, during a drive on Sol 335, almost exactly halfway through the 669-sol primary mission, the vehicle passed the "1 km" traveled mark. Engineers and controllers planned to take advantage of the flat terrain to begin using autonomous navigation as well as to make longer drives. On Sol 340, the rover drove for the first time more than 100 meters in one day and on Sol 385 it was to drive for 100 meters with an autonomous navigation portion at the end of the day. It eventually set a new record of 144 meters in a little more than 4 hours, turning south to a region of interest. It had driven some 400 meters over the last 10 sols and had almost reached the rim of a large depression where driving was to be stopped for several days to investigate some geological features. Meanwhile, it had continued performing "touch and go" science, as well as making astronomical observations. On 1 August, it had taken a sequence of images of Phobos (showing some of its largest craters) occulting the distant Deimos. Also during August and September, it observed a number of Phobos transits of the Sun, eclipses and other

Fossilized cross-bedding in the Shaler outcrop preserving evidence of water flows. (NASA/JPL-Caltech/MSSS)

night-time observations. In particular, it took images of the Andromeda Galaxy in order to ascertain whether the camera would be able to detect comet Siding Spring during its close flyby of Mars in 2014 and other bright comets approaching the inner solar system. Unfortunately, the images failed to detect even the bright galaxy. On Sol 392 the rover halted at the first geological waypoint en route to Aeolis Mons and investigated the Darwin outcrop and rocky veins. It had completed one-fifth of the road to the lower slopes of Aeolis Mons. Darwin, which looked lighter-toned in Mars Reconnaissance Orbiter images, was found to be a cemented conglomerate of pebbly sandstones and younger veins. After some "contact science", the rover was back in motion by Sol 402, heading to the next waypoint, 1.1 km away, which it reached on Sol 439. Dubbed Cooperstown, this was an outcrop where Curiosity spent a few days performing analyses using the laser and alpha-ray spectrometers.

As it transmitted data to Mars Reconnaissance Orbiter on Sol 447 (7 November), Curiosity suffered a computer reboot that put it into safe mode. The rover was then in the middle of a software update and normal operations were restored within 3 days. Just a few days later, observations had to be halted again while the engineering team investigated a voltage change between the chassis of the rover and the power bus. This was found to be a harmless 'soft' short in the RTG and operations were able to resume after a couple of days. One of the main tasks after this break was to

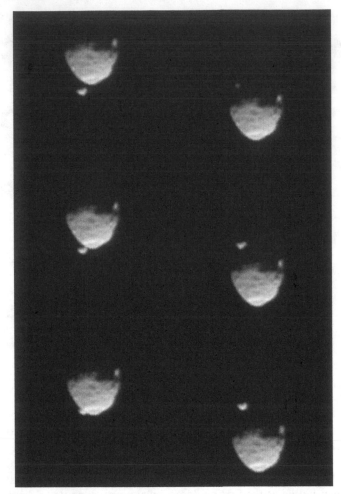

Phobos occulting Deimos, as seen by Curiosity on 1 August 2013. These telephoto images are the best of both moons ever taken from the surface of Mars. (NASA/JPL-Caltech/MSSS/Texas A&M Univ.)

deliver more of the drill sample obtained at Cumberland to the analysis instrument. Curiosity restarted driving on Sol 465. By then, the aluminum wheels had some large tears that, while dramatic, were neither unexpected nor likely to be of any serious concern. Controllers however started considering routes to Aeolis Mons that would minimize this kind of wear. On Sol 487, the remaining dust from the Cumberland drill sample was dumped on the ground, emptying the sample-handling system and simplifying future uses of the robotic arm. By the end of 2013, Curiosity was approaching the third waypoint on the route to Aeolis Mons. After that, the fourth waypoint would be in an area where wind erosion appeared to be still ongoing, raising the prospect of finding rocks that had been exposed only recently and might still contain any ancient organic molecules intact.

The lower reaches of Aeolis Mons and the layered Murray Buttes, which will be explored by Curiosity starting in mid-2014. This picture was taken shortly after landing. (NASA/JPL-Caltech/MSSS)

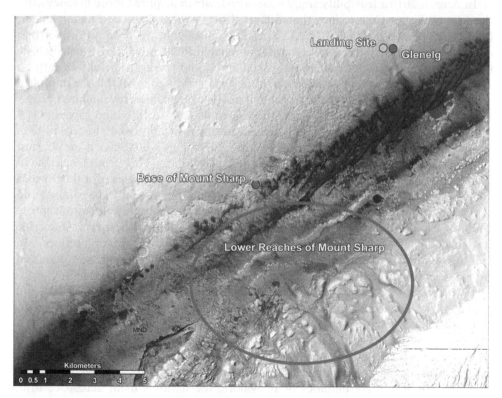

Curiosity's landing site, Glenelg, and the lower reaches of Aeolis Mons in the context of the floor of Gale. (NASA/JPL/University of Arizona)

Approaching Aeolis Mons in mid-2014, Curiosity will have to find a gap in the dune field to gain access to the base of the mound, and to the peculiar layered mesas, dubbed the Murray Buttes after the late JPL director Bruce Murray, where it would probably spend most of the rest of the mission.

SNIFFING THE AIR ON MARS

After its successful Chandrayaan orbiter, which at around $83 million was probably the most cost-effective lunar mission ever, the Indian Space Research Organization (ISRO) began to study using that nation's most powerful rocket, the Geostationary Satellite Launch Vehicle (GSLV), to send an orbiter to Mars to survey the planet's atmosphere and weather system, and its interaction with the solar wind. But as ISRO chairman Madhavan Nair explained, what was needed was a good scientific rationale since Indian scientists would not wish to invest in such a sophisticated and expensive mission merely to repeat investigations being undertaken by other countries.

In August 2010 a feasibility study was started, which acquired more urgency after the loss of Fobos-Grunt since that delayed ISRO's Chandrayaan 2 lunar lander, built in cooperation with Russia. As a result, studies converged on an orbiter that could be launched by the smaller but more reliable Polar Satellite Launch Vehicle (PSLV), a rocket with four alternating solid and liquid stages, hopefully as early as 2013. The Indian press bestowed on the Mars Orbiter Mission (MOM) the unofficial nickname of Mangalyaan (Mars probe) in the style of the Chandrayaan lunar probe. In 2012 the budget for studies was first increased 40-fold to 10 billion rupees ($24.9 million), and then on 3 August the Indian government approved the $73 million Mars Orbiter Mission for launch in November 2013. This would give it the shortest development time for a Mars mission since the 1960s. The project sparked a controversy in the Asian country. Critics focused on the priorities of a third-world country in space, and on the evident haste with which the mission was proposed and approved in the midst of political scandals. Some said India was seeking to leapfrog China in an effectively non-existent "Mars race" by Asian superpowers. One critic dismissed it as "a highly suboptimal mission with limited scientific objectives". In fact, it was conceived from the start as a technology demonstration rather than as a proper scientific expedition.

The spacecraft was based on a proven satellite bus used since the early 2000s, and also used for Chandrayaan, upgraded for the thermal requirements of the cruise to Mars and in orbit around that planet. Chandrayaan itself had been expected to last for 3 years but failed owing to a thermal control problem after only 9 months, by which time 95 per cent of its objectives had been accomplished. The key structure, a central thrust cylinder with composite and metallic honeycomb sandwich panels, was made by Hindustan Aeronautics of Bangalore, the nation's main aerospace firm, within 2 months of the project receiving the go-ahead. The vehicle had a 500-kg dry mass and two tanks inside the central cylinder for 850 kg of nitrogen tetroxide and hydrazine. The total launch mass was 1,350 kg. The propulsion system had a 440-N bipropellant main engine, modified to ensure its restart after the long cruise to Mars, and eight 22-N thrusters for attitude control and course corrections. In addition to these thrusters, four reaction wheels would be used for precise attitude control. Power was provided by a single solar array of three 1.4 × 1.8-meter panels capable of generating up to 840 W at Mars. Communications would be provided by a 2.2-meter high-gain antenna, mounted on the bus opposite to the solar panel, that could ensure data rates of up to 40 kbits per second. A conical medium-gain antenna and low-gain antennas would be used near Earth, for emergencies, and at times when the parabolic antenna could not be used. The 18-meter and 32-meter antennas that India built in Byalalu, a village near Bangalore, in order to support lunar missions would to be uprated for MOM and other future interplanetary flights.

The scientific objectives and payload were discussed at meetings in late 2011 but since it was a technology demonstration mission the payload was restricted to a mere 14.5 kg of essentially experimental instruments. Most of it was provided by the ISRO Space Application Center and was entirely Indian. Chandrayaan, in contrast, had flown a number of international instruments, even including one financed by NASA via the Discovery program. The Mars mission carried a Lyman-alpha photometer to measure the deuterium-to-hydrogen ratio of the atmosphere and estimate the erosion rate, plus a methane sensor with part-per-billion accuracy. The instrument would be working mostly at apoapsis, when the viewing geometry would remain more or less constant, allowing longer scans and higher signal-to-noise ratios. A thermal infrared imaging spectrometer was to map the composition and mineralogy of the surface. A quadrupole mass spectrometer was to sample and directly analyze the composition of the upper atmosphere. The payload was completed by a 2,048 × 2,048-pixel CCD color camera with a resolution of 25 meters at periapsis.

The PSLV launcher was not powerful enough to place MOM directly into a solar orbit bound for Mars, and the relatively low thrust of the spacecraft's engine would make utilizing that to escape Earth impractical in a single burn. The 1.5-km/s escape maneuver was therefore split "à la Fobos-Grunt". The 440-N engine would raise the apogee from 23,000 to over 200,000 km, and finally to escape from Earth. At Mars, a 1.1-km/s insertion burn would place the spacecraft in an elliptical 365 × 80,000-km orbit inclined at 150 degrees to the equator with a period of about 77 hours designed to avoid long solar eclipses. This orbit would also permit observations of Phobos and Deimos, although the latter was not an official target. The primary mission would last 6 to 10 months at Mars, and could be affected by the impending close pass of comet Siding Spring by the planet. On the other hand, the comet could deliver methane and other complex molecules to the Martian atmosphere, which the mission was equipped to study.[199,200,201,202,203]

After launching the Mars Science Laboratory in 2009, the second Mars Scout was to have used the 2011 window. NASA issued an announcement for opportunity in August 2006 and received 26 proposals. Meanwhile, recognizing that it would be almost impossible to fly an entirely new mission successfully within the existing budget cap, the program sought to have this increased above $400 million. One of the proposals, offered by the University of Arizona, was the Tracing Habitability, Organics and Resources (THOR) mission. Inspired by the Deep Impact mission to a

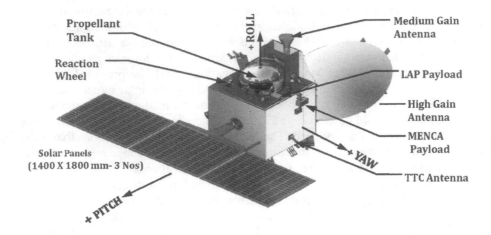

The Indian Mars Orbiter Mission (MOM) spacecraft. (ISRO)

The MAVEN spacecraft during ground tests, with one of its 'gullwing' solar panels deployed.

comet, it would deliver an impactor ('Thor's hammer') into the permafrost at an intermediate latitude to excavate a 10-meter-deep crater while the parent vehicle observed using instruments capable of detecting the signatures of water, methane and other organics. Ames Research Center proposed MATADOR (Mars Advanced Technology Airplane for Deployment, Operations and Recovery). Unlike previous proposals for airplanes, instead of crashing at the end of its flight this one was to make a survivable vertical landing in order to continue providing data. The Mars Autonomous Rovers for Geological Exploration (MARGE) proposal called for a pair of small rovers to investigate known water-modified sites using an instrument suite that would include the life-seeking spectrometer developed for Beagle 2.[204] Yet another proposal was to land on the plain close to one of the craters in which there are gullies, and send a rover to make in-situ observations.

Two proposals were selected for further study, one of which would actually be implemented. These were Mars Atmosphere and Volatile Evolution (MAVEN) and The Great Escape (TGE). Interestingly, both were orbiters to investigate the structure and dynamics of the upper atmosphere and the processes by which the atmosphere was being lost to space.

The original intention was for NASA to select a mission in late 2007 for flight in 2011. Unfortunately, a conflict of interest was discovered: one of the scientists involved in the selection process was also a participant in one of the proposals. As a result the selection board of NASA representatives, space scientists and industry representatives was disbanded, and by the time it was replaced 4 months later the time reserve for a 2011 launch had been eroded and the mission would have to slip to 2013. On the one hand the delay would add $40 million to the program, but on the other hand not flying a Mars Scout in 2011 would free-up funding to cover the escalating cost of the Mars Science Laboratory. Studies for the 2013 window had previously envisaged a Mars Science Orbiter of the Mars Reconnaissance Orbiter-class arriving as the latter was reaching its projected lifespan. This would combine scientific instruments with a communications relay. In the restructuring, however, Mars Science Orbiter was slipped to 2016.

In September 2008 NASA finally selected MAVEN as the second Mars Scout on the basis that it offered the greatest scientific potential at the lowest risk. It was a joint project by the Goddard Space Flight Center, the University of Colorado, and Lockheed Martin, with the company building the spacecraft by exploiting systems and structures used by Mars Odyssey and Mars Reconnaissance Orbiter. The overall budget was $671 million, including the launcher, reserves, and a standard JPL relay package. The 2,454-kg MAVEN had a 2.3 × 2.3 × 2-meter 125-kg primary structure made of carbon composite and aluminum honeycomb, with a 1.3-meter central thrust tube constituting the main structural element and support for the fuel tank containing about 1,640 kg of hydrazine. The tank itself was larger than on Mars Reconnaissance Orbiter in order to accommodate the expected number of orbital changes. Two solar panel wings, 11.43 meters across, were to provide up to 1,700 W of power in Mars orbit. The panels had a distinctive gull-wing shape, angled at 20 degrees, to provide some aerodynamical stability during deep penetrations of the atmosphere. Communications would use a 2-meter high-gain antenna mounted in a fixed position. Since the fixed antenna would require reorienting the vehicle to communicate with Earth, thereby interrupting science activities, communication sessions would be held only twice per week – much rarer than previous Mars orbiters and landers. It having no data-hungry cameras or imaging instruments, this would be more than sufficient to return the data stored in a 32-gigabit internal memory. Two additional low-gain antennas on the forward and aft decks were also available. The propulsion system included six 200-N thrusters for orbit insertion, as well as six 22-N thrusters for smaller burns. Attitude control would be performed mainly by four reaction wheels. Also on board was a DVD containing over a thousand haikus received from the public.

MAVEN carried a 65-kg scientific payload consisting of eight instruments. This included an energetic particle analyzer to determine the energy and direction of solar hydrogen and helium ions, and ion and electron analyzers to measure the parameters of solar wind particles interacting with the Martian upper atmosphere. The electron analyzer was mounted at the end of a 1.65-meter boom. Another instrument was to measure the composition and velocity of high energy ions in the upper atmosphere, a Langmuir probe had sensors on two 7-meter booms, and a magnetometer had two sensors mounted on triangular carbon extensions at the tips of the solar panels. A neutral gas and ion mass spectrometer was to directly sample the upper atmosphere of Mars, its composition, structure and isotopic ratios, and an ultraviolet imaging spectrometer was to map the composition of the upper atmosphere. Three of the instruments were on an articulated scan platform at the end of a 2.3-meter boom. In particular, during periapsis passes, the platform would orient the mass spectrometer inlet in the ram direction and the ultraviolet spectrometer to face the planet. In a sense, MAVEN was to extend the investigation of the history of water on the planet from the surface to its atmosphere. It was designed to measure how fast water, carbon dioxide and nitrogen are escaping from the atmosphere and how this is correlated with solar activity, and so gain insight into how Mars came to lose its early dense and wet atmosphere. The case of the "Halloween" solar flare of 2003 is particularly significant from this point of view, since on that occasion Mars Global Surveyor saw the density of the upper atmosphere increase by an order of magnitude, undoubtedly increasing the rate of erosion for some time thereafter. Unlike Mars Express and Fobos 2, which measured the erosion of atmospheric ions, MAVEN was equipped to study the escape of neutral 'hot' particles such as high energy oxygen atoms. It would also be possible to undertake joint atmospheric studies with Mars Express.

MAVEN was to be launched by an Atlas V 401 during a window running from 18 November to 7 December 2013 and to arrive at Mars on 22 September 2014, at the start of the declining phase of the 11-year solar cycle. A 38-minute maneuver would put it into an initial elliptical orbit inclined at 75 degrees to the equator with a period of 35 hours and a periapsis near 380 km. Over the next five weeks the 150 × 590-km science orbit with a period of 4.5 hours would be established, the booms deployed and the instruments calibrated. Since a circular orbit was unnecessary for the planned measurements, no aerobraking would be carried out. The primary mission was to last half a Martian year (one terrestrial year), during which the spacecraft would perform five "deep dip" campaigns, each lasting 20 orbits, in which the periapsis would be lowered to 125 km in order to directly sample the upper atmosphere and ionosphere. The first dip was scheduled for January 2015. Moreover, owing to the precession of the elliptical orbit, the latitude of periapsis would cycle between 75°N and 75°S, thus allowing the spacecraft to investigate the interactions between the solar wind and the atmosphere for a wide range of latitudes and all local times. (On such a low periapsis, it is possible that the spacecraft may be detectable in images taken by rovers on the ground.) At the end of the primary mission, with the possibility of extending it to a full Martian year, it would serve as a relay. In fact, as the Mars Exploration program was supposed to develop, NASA hoped to have at least two orbital relays available at any time. MAVEN would supplement, and possibly replace, the aging Mars Odyssey and Mars Reconnaissance Orbiter.[205,206,207,208,209,210]

The Indian Mars Orbiter Mission was to be launched first. Its window opened on 25 October 2013, although October, November and December were also the months that the Indian launch facility in Sriharikota was most likely to suffer monsoon rains and hurricanes. The spacecraft and instruments were shipped to Sriharikota in mid-August, where integration of the launcher was already taking place, and were moved to the launch site by truck on 2 October, an Indian holiday that was selected because the traffic on the road would be less. In order to provide critical tracking of the probe during the firing of the fourth and final stages of the PSLV and the separation of the spacecraft, two ships were to be positioned in the South Pacific. Unfortunately, bad weather prevented one of them from reaching Fiji in time, so much so in fact that the launch had to be postponed to 5 November, with Earth escape on 30 November and arrival at Mars on 24 September 2014. The mission was successfully launched on the first attempt on 5 November by the 25th PSLV. The spacecraft was separated about 45 minutes later, having been placed into the planned 250 × 23,500-km phasing orbit with a period of slightly less than 7 hours. It deployed its solar panels and high-gain antenna. Daily burns were conducted at perigee from the 6th to the 8th of November. For the fourth burn, on 10 November, engineers decided to test a redundant method using both the primary and backup fuel valves and lines, but this test failed and the probe received only one-fourth of the planned velocity increment, raising the apogee only by 7,000 km instead of over 30,000 km. A supplementary burn was carried out successfully the following day, increasing the apogee to almost 119,000 km. One last maneuver on the 15th almost reached to 200,000 km, halfway between the Earth and the Moon. Meanwhile, systems were powered on and checked, and communications were tested – the latter involving the NASA Deep Space Network, which was to be made available for the Mars orbit insertion maneuver. Some of the instruments were also powered on for testing and calibration, including the camera. On 19 November, 70,000 km out, the spacecraft took the first color image of Earth, centered on India. The spacecraft remained in the eccentric looping orbit for a fortnight. Finally, on 30 November, the engine restarted as MOM was flying over South Africa and delivered a speed increase of 648 m/s. When the burn was terminated 22 minutes later, with the spacecraft flying somewhere above Bangalore, it was speeding away from Earth and toward Mars in a 0.98 × 1.45-AU orbit. It successfully corrected its trajectory for the first time on 11 December, fine tuning its course to Mars. A total of four maneuvers were planned, the last scheduled only 10 days from orbit insertion.

Assuming the Mars Orbiter Mission is successful, a second Indian probe could fly in 2018. In addition to Mars missions, ISRO hopes to send deep-space missions either to enter orbit around an asteroid or to perform a cometary flyby sometime after 2015. After Mars there could be missions to Venus and Mercury in the 2020s.

The launch of MAVEN was put at risk by the US federal government "shutdown" of October 2013, which could have resulted in delaying it to 2016, and that would have required more fuel to accomplish the mission, jeopardizing the orbital relay phase. As a consequence, after 2 days of halted work, a waiver was issued and processing resumed at Cape Canaveral. The mission was successfully launched on 18 November, on the very first day of the window. After spending about 30 minutes in low orbit, the Centaur stage boosted the spacecraft into a 0.97 × 1.47-AU orbit. Less than 55 minutes after launch, MAVEN deployed its solar panels, and over the next few weeks most of the instruments were powered on and checked. The ultraviolet spectrometer was powered on and tested in early December, snapping

Launch of the Indian Mars Orbiter Mission on a Polar Satellite Launch Vehicle. (ISRO)

The first image of the Earth taken by the camera on the Indian MOM spacecraft. (ISRO)

images of interplanetary hydrogen at a range of different settings, before being used to attempt an observation of comet ISON – or what was left of it after being torn apart during its close perihelion passage. Meanwhile, the spacecraft performed the first of four trajectory corrections on 3 December.

Unfortunately, NASA, faced with severe budgetary constraints, announced that it would halt the Mars Scout series after only two missions in order to concentrate its efforts on expensive surface missions. Some of the Mars Scout proposals may yet fly within the Discovery program, but only if they survive fierce competition from other planetary missions.

A SHOTGUN MARRIAGE?

At the end of 2000 a European 'strategy for space' named Aurora was announced. It called for long-term human exploration of the solar system, beginning with the Moon, Mars and near-Earth asteroids. The concept was fully endorsed both by the European Union Council of Research and by ESA's council, and was kicked off at ministerial level in November 2001. It was to comprise a core program aimed at developing technologies, defining approaches and architectures, and raising public awareness; and the missions themselves, which would mix rapid-development 'Arrow' technical demonstrators with more sophisticated flagships. It was hoped that this approach would enable human missions to such destinations to start in the 2030s or 2040s.

A 3-year preparatory phase was subscribed to by most ESA member states, but the actual program would form a series of consecutive 5-year periods beginning in 2005. It did not form part of ESA's scientific program, participation in which was compulsory for all member states; it was instead to be funded through the agency's human spaceflight directorate, participation in which was optional. Germany did not join Aurora at the start, despite widespread industrial and scientific interest; for example the stereo camera of Mars Express was a German project. A call for ideas received 300 responses and the results were presented at a special workshop. The first missions were approved for further study in October 2002. Two Arrows were selected: an Entry Vehicle Demonstrator to test the technologies needed for a high-speed deep-space re-entry of the Earth's atmosphere, and a small demonstration of 'aerocapture' that would exploit earlier French studies for the PREMIER orbiter. Other proposals included testing techniques for extracting propellant and oxygen from the lunar regolith. There were hundreds of suggestions for the flagships. The first to be selected was the Mars Exobiology mission, or ExoMars, consisting of an orbiter, a descent module and a large rover. It was to fly around 2009 carrying a surface payload devoted to 'biology', plus the Italian drill left over from the 2003 Mars Surveyor sample-return. In a sense ExoMars would be more about 'vertical' rather than 'horizontal' exploration of the surface, with its drill reaching a depth of 2 meters. The mission would seek evidence of past or present life, identify hazards to humans, and improve knowledge of the planet and its environment. Moreover, it would exploit years of European studies of rovers and 'planetary mobility', while also demonstrating entry, descent and landing. To follow up the two Arrow missions and ExoMars, ESA hoped to be able to launch a sample-return flagship as early as 2011. Preliminary contracts for the Arrow missions and ExoMars were placed in September 2003, followed a month later by study contracts on the sample-return mission.[211,212,213]

Also in 2003, the British National Space Center invited national researchers to submit Arrow-class proposals for micro-missions involving ion propulsion. Three concepts emerged. One was a dual-orbiter mission which would study Phobos and Deimos to determine their origin, their relationship (if any) with known asteroid types, and their interactions with Mars and its environment. Ion propulsion would be used to fly between Earth orbit and Mars orbit, then to match the orbits of the two moonlets. In addition to making remote-sensing observations of its assigned target, each orbiter would release a small soft-lander similar to the Philae lander of the Rosetta cometary mission. Another proposal was for a constellation of small orbiters which would carry no scientific instruments, merely ultrastable oscillators and radio systems to provide a vast number of mutual radio-occultations to 'sound' the planet's atmosphere over a wide range of locations and times of day. The radio signals of such a constellation could also provide a rudimentary 'Martian GPS' as a

This is how ESA imagined in the early 2000s the Aurora sample-return lander. (ESA)

navigational aid for surface missions. The third proposal would involve a carrier spacecraft delivering penetrators similar to those developed for the American Deep Space 2 mission to equatorial and polar targets in order to collect data on subsurface water and ice.[214,215]

The plan was to launch ExoMars on a Russian Zenit 3 or Soyuz rocket and use a departure profile similar to that of the Fobos and Mars 96 missions; namely first achieving a highly eccentric Earth orbit and performing the escape burn at perigee. It was intended to fly a mission profile previously used only by the Vikings, with the lander being released after Mars orbit insertion. In this case, however, the fact that the descent module had no propulsion of its own would require the orbiter to perform a de-orbit maneuver, release the lander and then promptly 're-orbit' itself. The Mars Express-based orbiter was not to carry a scientific payload, its role being to serve as a relay for its rover, but it was to release a small subsatellite in order to perform a number of autonomous rendezvous simulations.

Building on years of Russian and European studies, the ExoMars lander would use a huge inflatable heat shield and decelerator. After the first stage provided the heat shield for atmospheric entry, the second stage would inflate to a diameter of 25 meters. By slowing the rate of descent to about 17 m/s, this decelerator would

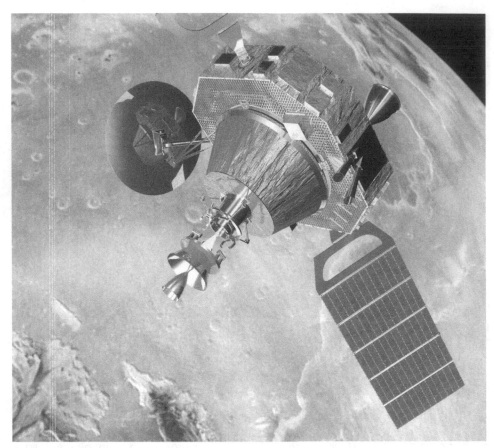

The ESA Aurora Mars sample-return orbiter. It would rendezvous with the sample canister and return it to Earth. (ESA)

effectively replace the parachute. It would also ensure that the rover would land in an upright state, thereby eliminating the need for a self-righting mechanism. But first, tests would be carried out to demonstrate that the wheels could drive over the spent and deflated decelerator without becoming entangled in that fabric. As initially envisaged, the 220-kg rover was to be solar-powered and to have an average speed on rough terrain comparable to its NASA counterparts. It outwardly resembled its US predecessors, having a boxy thermal enclosure, solar panels projecting out from its top deck, six steerable wheels on a chassis similar to the 'rocker-bogie' developed by JPL, a mast-mounted panoramic camera, and a 'patch antenna' on one of the solar panels to communicate with the orbiter.[216] In 2003 ESA invited proposals for the payload, which was to be named Pasteur after Louis Pasteur, the French chemist and microbiologist. A peer review recommended 22 of the 50 submissions for further study, and in September 2005 a science meeting at ESA drew up a preliminary suite of instruments. Also studied (after a recommendation by participant states) was the possibility of using some of the spare mass available to combine ExoMars with the canceled French NetLander project.[217] The result was a geophysical payload named Humboldt in honor of the Prussian geographer, naturalist and explorer Alexander von Humboldt. It would include a dust counter, an ultraviolet spectrometer, an instrument to measure ionizing radiation at the surface and a meteorological package. Incorporated into the base of the lander, it would be left behind when the rover drove off.

After setting the total cost of the ExoMars mission at 400 million euros, ESA opened exploratory talks with Russia and NASA about joining the project. In fact, the assistance of one or other of these nations would be required in designing the radioisotope heaters to keep the electronics warm during the Martian night, as ESA had no prior experience. At the ministerial summit in 2001, ESA sought 40 million euros over a period of 5 years to kick-start Aurora but it received only 14 million euros over 3 years, pending a more detailed definition of the project.[218] Sweden joined in late 2004, when the ESA council also boosted the budget to 41.5 million euros. By late 2005, a dozen countries had committed, including Germany, and the budget had risen to 48 million euros.[219] However, by then the launch of ExoMars had slipped from 2009 to 2011 as a result (amongst other things) of the loss of the Columbia Space Shuttle, which delayed several European programs involving the International Space Station. Furthermore, budget issues prompted the cancellation of the Arrow technology demonstrators, with the implication that future missions would have to rely on existing solutions.[220] With ExoMars postponed, a number of options were considered which would prove the entry, descent and landing system. One such proposal was the Mars Demonstration Lander.[221,222] The less ambitious ExoMars-Lite would have incorporated aspects of the original ExoMars as well as Beagle 2 and Netlander. Following a meeting of the ministers of member states in 2005, ExoMars was allocated 600 million euros. Of the four major partners, Italy would contribute 36.4 per cent, the UK 15.5 per cent, France 14.5 per cent and Germany 13.2 per cent. But ESA had yet to decide between a landed-only mission that would relay through the US Mars Reconnaissance Orbiter, and flying its own purpose-built orbiter for science and communications.[223]

In 2007 ExoMars was given the go-ahead with a configuration that involved an orbiter, the 8.5-kg Humboldt payload on the base and the 16.5-kg Pasteur payload on the rover. The base would be powered by a small RTG supplied by Russia. But by now the projected cost of the mission exceeded 1 billion euros. This slipped the mission to 2013 and meant using a more powerful launcher than the Soyuz-Fregat, namely a Soyuz-2b launched from the European Kourou range. Unfortunately, the 2013 window was ill-suited to a lander because its arrival would occur in the dust storm season. One possible solution was to fly a 'delayed trajectory' which would arrive a little later.

The spacecraft was to be built by a consortium whose prime contractor was the Italian branch of Thales Alenia Space. In the final design, the lander would use a six-segment airbag system that would impact the surface at no more than 25 m/s. The airbags were designed to 'pop' when compressed and rapidly vent to preclude bouncing. By virtue of covering only one side of the lander instead of enclosing it, these airbags would be lighter than the unvented bouncing cocoons used by Mars

Pathfinder and the Mars Exploration Rover missions. On the other hand, ExoMars would be the first to test them. At about 150 kg, the rover was rather lighter than initially envisaged. It was to have six steerable wheels mounted on extendable legs and would negotiate loose sand by 'walking' across it. The vehicle was to drive up to 125 meters per sol, and visit at least seven sites during a baseline mission lasting 180 sols. The payload was based on the recommendations of an ESA exobiology study initiated in 1997. Of course, although ExoMars was not part of the agency's scientific program and was not strictly dictated by scientific objectives, European scientists exploited their first opportunity to conduct a rover mission by specifying an extensive payload to study the surface and subsurface of the planet, in particular seeking evidence of past or present life. By 2007 no fewer than twelve instruments were assigned to the rover and eleven on the lander base.

As on US rovers, a high-resolution panoramic camera was to provide images of the locale. Stereoscopic cameras would be dedicated to navigation and science. An infrared spectrometer would identify water-bearing minerals for sampling, as well as carbonates, clays, sulfates, silicates and organic molecules. An arm similar to the PAW of Beagle 2 would carry several instruments, including a microscope to examine the texture of rocks in close up, a Mossbauer spectrometer to study iron-bearing minerals, and the sampling heads of the Raman and laser spectrometers to analyze the elemental composition, mineralogy and structure of rocks. Fiber-optic cables would link the heads of these instruments to the analyzers in the body of the rover. An infrared instrument mounted on the drill rod would examine the texture of the borehole. A dedicated system would prepare the sample obtained by the drill by crushing and grinding it before feeding the fragments to the instruments. This suite included the Urey payload, which was to seek life using an organics detector and an oxidant analyzer. The organics detector would extract amino acids, amines, sugars and other organic molecules and pass them to a spectrometer that would use fluorescence to identify them. It had been designed to be able to detect organics at a concentration of one part per trillion. It would also determine the chirality of the organic molecules. (On Earth, left-handed organics are more common than right-handed ones.) The oxidant analyzer would study the chemical reactivity of oxides and free radicals in the atmosphere and soil, in particular as a function of depth. A gas chromatograph and mass spectrometer would analyze surface and atmospheric samples. A "life marker chip" would use medical diagnostics methods "to reliably detect present life". An X-ray diffractometer would determine the mineralogy and crystalline structure of the solid material. Also recommended for inclusion were an instrument to measure the size range of dust grains, an ultraviolet spectrometer to assess the radiation levels at the surface, and an ionizing radiation detector.[224,225]

At the same time, the mass of the Humboldt payload had tripled by the addition of instruments such as a seismometer and offloading some of the payload from the rover, including, for example, a ground-penetrating radar to study the stratification of the near-subsurface.

In late 2007 the Aurora program issued a call for proposals for further missions and selected a Mars orbiter and surface network, and a lunar precision soft-lander carrying rovers. The Mars mission, named MarsNext, would launch on a Soyuz in

2016 or 2018 and deliver at least three probes to the surface for NetLander-style research – although these landers would be significantly larger and would provide telemetry during entry, would use retrorockets for braking and would be protected by two toroidal airbags. The accompanying orbiter would address scientific objectives as well as performing autonomous rendezvous experiments.[226,227]

As the development of ExoMars progressed, extensive tests were carried out in 2007 and 2008 of key technologies, and heat shield tests and airdrops of the vented airbags were due in 2009.[228] But as the 2008 gathering of the ministers of member states loomed, the prospects were gloomy. Amongst other things, this meeting was to address the rising cost of the mission. However, in a policy change designed to shift its effort to a national space program and participation in projects such as the Japanese Hayabusa 2, Italy decided not to increase its contribution to ExoMars. A possibility for saving ExoMars was to delete the rover, but to do this would reduce the scientific merit of the mission and severely undermine the existing support. To gain breathing space, ESA considered delaying the launch to 2016 and drawing in partners to share the costs – for example, NASA providing the orbiter in return for a role in the surface mission. In the end, ESA stumped up 850 million euros, some 150 million short of the target, with the remainder being provided by international partners. Although this enabled the mission to continue, it was clear that tradeoffs would have to be made.[229,230,231,232]

While ESA struggled with ExoMars, NASA was facing difficulties: the budget for its Mars Exploration program was frozen for the foreseeable future, the cost of the Mars Science Laboratory was escalating, and the launch of MAVEN had been postponed to 2013. The study for a Mars Science Orbiter had been started in 2005 with a view to launching in 2013. It was to address a number of topics which were not expected to be resolved by the Mars Reconnaissance Orbiter and Mars Science Laboratory missions, in particular atmospheric erosion and trace gas studies, but erosion had been made the focus of the second Mars Scout. Trace gas observations would further investigate the recent discovery of methane in the atmosphere. Not only was the fact that concentrations of this gas existed intriguing, they were seen to change with the seasons in a manner which could not be reproduced by climate models. As molecules of methane could be expected to survive in the atmosphere for several centuries, it should have had time to disperse uniformly over the planet. The very fact that there was methane present meant that it was being replenished, but it was not evident whether this source involved inorganic or organic chemistry. The observations implied a methane lifetime of only several hundred days, which indicated that an efficient mechanism was destroying the molecules. This "would suggest an extraordinarily harsh environment for the survival of organics".[233] Such observations would therefore have a potential bearing on the issue of life on Mars. The new orbiter could also further investigate active gullies and search for shallow subsurface liquid water using improved versions of instruments operated by Mars Reconnaissance Orbiter. Consideration was given to adding a very-high-resolution camera, but this would vastly increase the data processing requirements, mass and cost. A high-inclination orbit would facilitate scans of the atmosphere over a range of latitudes of interest for methane detection. A 3-year 'science emphasis' mission would be flown at an altitude of 300 km, then the orbiter would climb to 400 km and spend 7 years serving as a relay for surface missions.[234]

NASA's options for 2016 and 2018 included another Mars Scout mission, mid-sized rovers similar to the Mars Exploration Rovers, a long-lived surface network, and an Astrobiology Field Laboratory. The Mid-Range Rover, also referred to as the Mars Prospector Rover, would make a relatively precise landing to collect and store samples, thereby demonstrating two technologies needed for a sample-return mission.[235,236] The Astrobiology Field Laboratory, on the other hand, would carry instruments specifically to search for evidence of past or present life on the planet. Reusing hardware created for the Mars Science Laboratory, it would be the more expensive option.[237]

With ESA and NASA pursuing separate strategies for exploring Mars, studies were undertaken in early 2009 to identify whether and how they might collaborate. The result was the Mars Exploration Joint Initiative (MEJI), which addressed the 2016, 2018 and 2020 launch windows. One possibility was to deliver the ExoMars rover in 2016 using a US-supplied carrier based on the Mars Science Orbiter. But because this would severely limit the orbiter's payload the recommendation was to adopt ESA's "enhanced baseline mission" and launch a fully instrumented orbiter in the first window and then follow up with a pair of rovers in 2018, one being the European ExoMars and the other supplied the US. A lander network with the Humboldt payload could be deployed in 2020. Then in 2022 the program could tackle the first elements of the sample-return architecture. Even if it were launched 9 years later than originally envisaged, in this new guise ExoMars would form part of a broader context. Although it would be able to share a skycrane with its US counterpart, this would mean Europe would not need to create its own technology for entry, descent and landing; an outcome that was likely to prompt the European companies to wonder whether the mission, now more heavily focused on science than technology, would benefit them.[238] In mid-December 2009 the ESA council confirmed the 'split' mission for launch in 2018 and a budget of 1 billion euros.

In the 2016 window the joint initiative would dispatch a US orbiter carrying a European lander on a mission named the ExoMars and Trace Gas Mission Orbiter. The orbiter would be based on the design of the Mars Science Orbiter and measure trace gases (mainly methane, but also sulfur dioxide, hydrogen sulfide, hydrogen chloride, ozone, carbon monoxide, nitrogen oxides, etc) in the Martian atmosphere to a sensitivity of 1 part per trillion, monitoring their spatial and temporal variability and correlation with atmospheric parameters such as dustiness. It would have a launch mass of over 3,000 kg, including 125 kg of science payload. After entering orbit it would use aerobraking to circularize at an altitude of about 400 km inclined at 74 degrees to the equator. It would also carry a relay payload to support landed missions through to at least 2022. For this, ESA was to supply a large high-gain antenna inherited from the Rosetta mission with a throughput of 8 Gbits per day. A request for proposals for instruments attracted 19 submissions, of which four US and one European were retained: a trace gas occultation spectrometer plus a high-resolution nadir and occultation spectrometer to detect trace gases and locate their

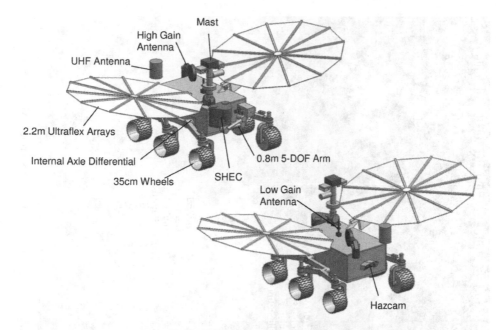

The US MAX-C rover was to be derived from Curiosity, and would be NASA's contribution to the European ExoMars mission.

sources; an infrared radiometer to detect dust and water vapor and correlate their distribution with the presence of trace gases; a high-resolution color stereoscopic camera to provide context for localized sources of methane; and a wide-angle global weather surveillance camera. Responding to the concerns of European industries, the ESA part of the mission would demonstrate some of the proprietary technologies in the first European attempt to achieve a soft landing on Mars since the ill-fated Beagle 2 in 2003.[239,240]

The mission would be launched in January 2016 on an Atlas V rocket supplied by NASA and nominally reach Mars on 19 October of that year. The overall cost of the mission was estimated at $750 million, including launch.

In 2018 ESA was to provide the ExoMars rover with its drill and Pasteur biology suite, and the US was to provide the Atlas V launcher, the cruise stage, entry system, skycrane and second rover. The mission would reach Mars in January 2019.[241,242,243] The US rover was the solar-powered Mars Astrobiology Explorer-Cacher (MAX-C), derived from the Mid-Range Rover concept. During a traverse of about 10 km it would address geology and past or present biology, and collect a cache of samples which might one day be recovered for return to Earth. To do so, it would have a drill to core into rocks to a depth of 5 cm.[244] Of course, carrying two rovers with different capabilities on the same pallet raised questions in landing site selection – was there a site that would be interesting both for subsurface drilling by ExoMars and for collecting a sample cache using MAX-C? Imaging of candidate landing sites, including clay deposits and gullies, began in 2009.

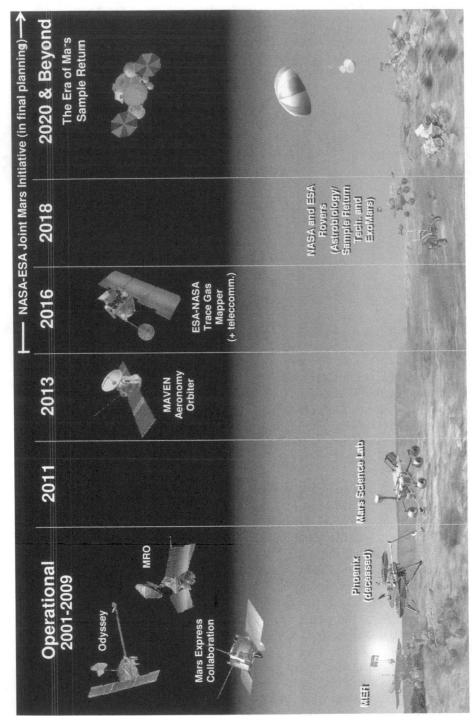

The timeline of future Mars exploration before NASA had to withdraw from the joint ExoMars project with ESA.

The joint plan was again put into doubt in early 2011 with the publication of the second US planetary exploration decadal survey (about which more in the final chapter). The report endorsed the 2016 mission and made MAX-C the top priority flagship mission of the decade, in preparation for the sample-return mission. But independent assessments estimated its cost at $3.5 billion and "a disproportionate share" of the US planetary exploration budget. Costs were mostly associated with adapting the skycrane to land two separate rovers. The survey recommended flying the mission, but only if it could be flown for a cost not exceeding $2.5 billion. It was half-jokingly proposed that the rover should be dubbed 'Austerity'. On the other hand, no alternative Mars exploration architecture was proposed if MAX-C was not flown.

As a result, in an effort to bring NASA's budget for the mission well below the cost cap to around $1.2 billion, plus launch, work on the rovers was stopped on both sides of the Atlantic and a Joint Engineering Working Group was created to design a larger common vehicle. Ideally, the rover would still be built in Europe and fly a mix of US and European instruments, with NASA providing the launcher and the skycrane and entry system, with as few changes as possible in comparison to the Mars Science Laboratory version. This arrangement would capitalize on the years of European research on ExoMars mobility that would otherwise be wasted. The single rover would cost less than two separate ones, but how much less was debatable. It also remained to be seen how work on a common lander could be carried out within the constraints of the ITAR arms control regime. ITAR, it will be remembered, had so limited communications between American and European engineers as to prevent discovery prior to launch of the mismatch between the transmitter on the Huygens probe and the receiver on the Cassini spacecraft.

Recent re-assessments of the methane data, moreover, have cast doubt on their interpretation, suggesting they may be compatible with far lower concentrations. In particular, the quantity and variability of methane in the atmosphere over short timescales reported from Mars Express and ground-based data was incompatible with the known chemistry and dynamics of the atmosphere, since it would require both strong sources and strong sinks of methane. And destroying the estimated amount of methane would rapidly consume the small amount of oxygen present in the planet's atmosphere. An alternative mechanism has been recently proposed that could provide a sizeable portion of the methane in the atmosphere without the need for biological or volcanic processes. Experiments have shown, in fact, that by irradiating samples of organics-rich meteorites, which should be fairly common on the surface of Mars, with ultraviolet mimicking solar radiation and under 'Martian' conditions of temperature and pressure, such meteorites can release substantial amounts of methane. Isotopic ratios of the gas thus produced would be difficult to distinguish from those of metabolic processes, making identification of the origin of methane difficult at best. These results were believed by some US scientists to partly undermine the scientific case for a Trace Gas Orbiter.[245],[246],[247]

Meanwhile, the European side was again struggling with its costs, which were estimated to be about 40 per cent over budget. This included 135 million euros for the orbiter, 145 million for the 2016 lander demonstrator, 240 million for the 2018 rover, with the rest covering program management, launch and initial operations. Also, uncertainties made missing the 2016 launch window a distinct possibility, with downstream repercussions on the 2018 mission for which the Trace Gas Orbiter was to provide telecommunications. To make things more complicated, for ESA the 2016 and 2018 missions were a single budget item, and so the order to cease work on the latter automatically halted the former also. One solution would be to renegotiate an ExoMars contract with the European industries covering only the 2016 orbiter and lander. ESA's human spaceflight directorate restarted work in late May. However, the French government refused to endorse any new spending until the 2018 mission had been better defined, with a more realistic budget. Full-scale development of the rover was stopped when NASA announced that it could not yet fully commit to the 2018 mission. In fact, the budget request for NASA in 2011 did not include funds for Mars exploration beyond 2016. Hence the project was deadlocked, with NASA unable to say that the mission would go ahead and ESA refusing to issue full industrial contracts without a firm American commitment. In the next round of cuts, NASA said that it could no longer provide the Atlas launcher, forcing either a cancellation, a redesign of the spacecraft and mission, or a switch to an alternative launcher. ESA started to consider cost-saving measures such as asking the Russians to donate a Proton launcher in exchange for their scientific participation, or delaying, scaling back or cancelling altogether the 2016 lander demonstrator that was strongly backed by the Italian industry.[248,249]

Finally, in early 2012 NASA backed out of the project. Its 2013 budget, which was announced in February, severely cut planetary science in order to pay for the overruns of the James Webb Space Telescope. In particular, all Mars missions after MAVEN were canceled. The NASA decision, fortunately for ESA, appeared to nicely dovetail with Russia's redesign of its planetary exploration program in the wake of the failure of Fobos-Grunt. Existing projects were stopped or delayed beyond 2015 in order to improve the reliability of the systems and hardware, and resources were concentrated on a joint ExoMars mission with ESA. Russia would provide two Proton heavy launchers as well as scientific instruments to make up for the NASA contribution. At first, Russia also considered flying as many as four Mars 96-like or MetNet surface stations to accompany the 2016 orbiter, but then it was suggested that Russia provide the RTG for the ESA entry, descent and landing demonstrator, turning it into a long-lived, 1-year lander proper, as well as a large part of the 2018 landing system. This plan also collapsed due to restrictions on Russian export of technology and to the technical issues of modifying the short-lived lander to be powered by and RTG, as it would require a heavily modified thermal control system to get rid of the excess heat that would be produced by the generator during the cruise to Mars. The design therefore reverted to the battery-powered short-lived option. In response to a proposal to the Italian Space Agency, the lander was named Schiaparelli after the Italian astronomer who devoted so much of his scientific effort to the Red Planet in the late nineteenth century. This would celebrate the centenary of his death in 1910.[250]

As planned at the time of writing (December 2013), the 2016 mission will consist of the Trace Gas Orbiter with four instruments. A nadir and occultation

A model of the 2016 joint ESA-Russian Trace Gas Orbiter.

spectrometer will be composed of a solar occultation infrared spectrometer, a nadir, solar and limb infrared spectrometer, and an ultraviolet and visible spectrometer, developed by a Belgian-led European consortium and based on a Venus Express instrument. The instruments will scan the limb at sunrise and sunset to detect trace gases molecules in the atmosphere at parts-per-billion level against the bright background of the Sun. They will also be able to scan the surface directly under the spacecraft, although with far less sensitivity. The lack of methane at the Curiosity landing site may mean that the gas is in fact absent on the planet, or at Gale crater, or during some season, or that it may only be found higher in the atmosphere, in which case the occultation spectrometer should be able to detect it. An atmospheric chemistry suite is to be developed by IKI and consist of a mid-infrared and a near-infrared spectrometer, a thermal infrared spectrometer, and a solar camera. A Russian epithermal neutron detector, based on an instrument on the NASA Lunar Reconnaissance Orbiter, will map subsurface deposits of water ice at high resolution and investigate their seasonal cycles. A Swiss and Italian 4.5-meter-resolution color stereo camera will primarily map the source regions of trace gases but could also be used to characterize future landing sites. Also on board the 3-axis-stabilized vehicle will be a redundant NASA-supplied standard UHF relay package as well as the short-lived lander.

After a nominal launch on 7 January 2016 on a Proton-M, with a backup launch window in March, the Briz-M upper stage is to make four burns over a period of almost 4.5 hours to depart Earth orbit. The 4,332-kg launch mass Trace Gas Orbiter will be inserted into a type-2 transfer orbit. The 600-kg entry, descent and landing

A model of the 2016 short-duration Schiaparelli European lander.

demonstrator Schiaparelli is to be released 3 days prior to arrival at Mars, in October 2016, being pushed away by springs and spun up to 2.75 rpm. The orbiter will then perform a maneuver to avoid impacting the planet, and prepare for orbit insertion. The entry and descent demonstrator will have an atmospheric entry module 2.4 meters in diameter. The front heat shield, the technically most critical portion of it, will have a 120-degree forward angle and use an ablative material based on studies for the Netlanders and for Beagle 2. Embedded in the heat shield will be a number of thermocouples, heat flux sensors, and other engineering sensors to determine the performance of the entry system. Sandwiched between the front and back shells will be the circular 300-kg lander, 1.7 meters in diameter. It will be awakened by timers about an hour prior to entry. After being slowed by the heat shield, a single 12-meter supersonic parachute derived from that of Huygens will open. Following the release of the heat shield, the lander platform proper will be released by the backshell when at an altitude of around 1,400 meters and traveling at a vertical speed of 90 m/s. A powered descent phase will then start, using three clusters of three 400-N hydrazine pulsed thrusters. The thrusters will be mounted around the circumference of the platform, each drawing from a separate hydrazine tank, with a total of 32 kg of fuel. A radar altimeter will manage the descent, in the process validating European guidance algorithms. At a height of about 1.5 meters, the thrusters will be switched off and the platform allowed to fall freely onto a crushable honeycomb structure, with the radar altimeter assembly nested at its center. About 6.5 minutes after entry, it will land at about 4.2 m/s, enduring a maximum deceleration of 40 g. The crushable structure should place the base plate of the lander 25 to 60 cm above the ground. Because ExoMars will arrive during the dust storm season, its systems must be designed to survive landing in a severe storm. It will be designed to survive just 8 days on the surface, making use of the NASA orbiters to return data. No direct-to-Earth link is planned. Given its short operating life, no solar panels or radioisotope heaters will be provided, so it will run off batteries. A 3-kg payload will provide a limited opportunity for doing science. This will consist of an entry and descent package using accelerometers to reconstruct the trajectory and the main physical parameters of the atmosphere; a dust characterization and environmental station incorporating anemometers, humidity sensors, barometers and thermometers; and an atmospheric transparency experiment. The surface package will be rounded out by an instrument to measure the electrical field at the surface to detect electrostatic charging of the dust, discharges and electromagnetic noise. Also on board will be a descent camera based on the visual monitoring cameras that ESA has installed on many spacecraft, including Mars Express, that will take at least fifteen images of the descent after the heat shield has been jettisoned. No surface camera is planned. Due to the infrequent orbiter passes and the difficulty in sending commands, the payload must collect its data autonomously. Sites were sought based on landing safety and scientific merit. These had to be at latitudes between 25 and 30°N, at elevations at least 1.3 km below the planetary datum, with minimal large and small-scale slopes and minimal surface roughness in a 110 × 25-km ellipse. A 50-km ellipse centered at 6.1°W, 1.9°S on Meridiani Planum, just west of the Endeavour crater explored by Opportunity, is the prime candidate, but there are three backup sites.[251,252,253]

A model of the European ExoMars rover that is to be launched in 2018. Note the drill box in front.

After releasing Schiaparelli, the spacecraft will enter a preliminary elliptical orbit with a period of 4 days. Eight days later, on its second periapsis, the orbiter will perform a maneuver to lower the period to 1 day, establish the desired 74-degree inclination to obtain a maximum of Sun occultations for atmospheric studies, as well as a good seasonal and latitude coverage, and begin a 6 to 9 month aerobraking phase to achieve a circular orbit at an altitude between 350 and 420 km. The Trace Gas Orbiter will begin its primary mission in June 2017.[254]

The 2018 mission should be launched in May, with a backup in August 2020, for arrival at Mars in either mid-January 2019 or April 2021 respectively. It will consist of a small carrier module supplied by ESA to provide power, trajectory control and communications during the cruise, a Russian descent vehicle and surface module with European input, and a 300-kg ESA rover. Exemplifying the stop-and-go history of the mission, the European rover that had passed a preliminary design review in 2010 before being shelved in favor of a joint ESA–NASA design was resurrected.

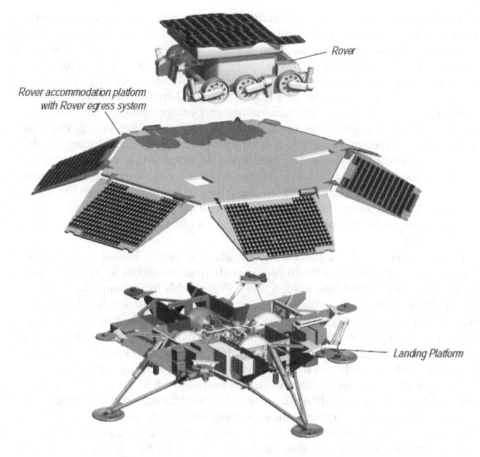

A possible configuration of the 2018 ExoMars lander. (ESA-Roskosmos)

The 1,250-kg descent module is to settle on retractable legs that will later be used to level the platform and allow the rover to drive onto the surface by using the deployable solar panels as ramps. Relay during entry and landing is to be provided by the 2016 Trace Gas Orbiter. The ESA rover will feature a thermally insulated 'bathtub' body structure, electrical and Russian-provided radioisotope heaters for thermal control, and solar panels with a total area of about 2.5 square meters for generating power. The driving system will consist of six wheels on three independent bogies. The payload for the rover was selected in July 2012, and will consist of a number of European and American instruments and two Russian instruments. A deployable mast will carry navigation cameras, two panoramic cameras, and one high-resolution camera as well as a Russian infrared spectrometer. A ground-penetrating radar will obtain stratigraphy of the upper few meters of the surface at centimeter-resolution and locate deposits of interest for subsurface sampling. This will be carried out by a multi-rod drill mounted at the end of an arm with two degrees of freedom. It will be capable of obtaining a 2.8-cubic-centimeter sample from a depth of 2 meters. The drill itself will be equipped with a borehole infrared spectrometer embedded in the drill tool, and a close-up sample imager. Russian scientists will also provide a neutron spectrometer similar to that mounted on Curiosity. Finally, on board the rover will be an analytical laboratory with a sample preparation and distribution system, an organic molecule analyzer including a gas chromatograph to detect past or present biological matter, and mass and laser spectrometers to investigate the mineralogy and composition at the grain size of samples, a Raman spectrometer and a visible and infrared spectrometer. The mission of the rover is expected to last 218 sols (approximately 7 months) during which it is to traverse about 4 km. The landing platform is designed to last for one Martian year and its payload could include a neutron spectrometer, a seismometer, a meteorology package and an instrument (based on the 2016 landing demonstrator) to measure electrical characteristics, a lidar to collect atmospheric profiles up to an altitude of 10 km, an atmospheric composition instrument and a radiation dosimeter and energetic ion flux meter.[255]

In December 2013 ESA and the federal space agency in Russia issued a call for landing site proposals. In order to provide sufficient insolation for the solar panels, the sites must be located between 5°S and 25°N, be able to accommodate the 104 × 19-km ellipse, be safe, and suitable for searches for past or present life. Furthermore, they must be ancient, with sedimentary outcrops recording a watery past in evidence across the entire landing ellipse to maximize the chances of the rover being able to access them, and be covered with very little dust. Finally, they must be relatively low lying to enable the landing system to work properly in slowing the lander.

As a result of NASA's withdrawal, however, the ESA share of the project is expected to grow to 1.2 billion euros, thereby raising once again the specter of its ultimate cancellation. Various solutions were suggested, including switching the 2020s ESA JUICE (Jupiter Icy Moons Explorer) mission from an Ariane 5 to the Russian Proton, allowing the Russian Space Agency to have a larger role on that mission, and diverting funds to ExoMars. New ESA member states Poland and Romania could become involved in the project at least financially, and finally ExoMars could be approved as a scientific "mission of opportunity" to transfer it from the human spaceflight and exploration department. ESA and Russia finally signed a formal partnership agreement in mid-March 2013.[256,257]

In addition to its involvement in ExoMars, Russia is starting to contemplate a Fobos-Grunt 2 mission for 2018 or 2020, possibly with help from ESA. It remains to be seen whether Russia will have the means to carry out a reflight of the failed sample-return mission.

The switch to a skycrane landing system and subsequently to a Russian-provided system meant that the Humboldt payload had to be deleted from ExoMars, but ESA envisaged including it on a later mission. In fact, in the 2020s the lander design demonstrated in 2016 might be used to deliver a network of four small stations that would incorporate NetLander and Humboldt instruments.[258] It is remarkable that establishing a Martian network has been ranked as a top priority planetary mission since the 1980s and reaffirmed in the 2003 'decadal survey' but it has yet to be attempted, probably because of the relatively low public interest in geophysics relative to rovers and the search for life. The irony is that this would be simple to implement, requiring only small Mars Pathfinder-class landers.

The GEMS (Geophysical Monitoring Station) lander, renamed InSight (Interior Exploration using Seismic Investigations, Geodesy and Heat Transport) was one of three finalists for the Discovery 12 mission, consisting of a geophysical network pathfinder. Just two weeks after Curiosity's successful landing, NASA announced the selection of the $425 million InSight over two more ambitious and more risky candidates. Confidence in the mission budget was apparently one key factor in the selection. The JPL mission was the first finalist for a Discovery opportunity since Mars Pathfinder, which probably compensated the scientific community for the cancellation of the Mars Scout program. It is to be launched during a 21-day window opening on 4 March 2016, to land after less than 7 months of cruise on 28 September in a flat, not too rocky and not too sandy equatorial patch of Elysium Planitia. As many as twenty-two 130 × 27-km landing ellipses in Elysium were studied, with high-resolution and context images provided by Mars Reconnaissance Orbiter. These were narrowed down to just four in September 2013. The first 67 Sols will be devoted to deploying and setting instruments on the surface. InSight is then to function for one Martian year to investigate the interior of the planet in terms of its accretion, evolution and differentiation of the core and crust in order to determine the size, composition and state of the core, the thickness and structure of the mantle and crust, and its thermal state. It will do this by measuring the current level of geological activity, the magnitude, rate and geographic distribution of seismic activity and of meteorite impacts.

InSight is based on a New Frontiers proposed that called for three Phoenix-class landers on a single carrier. To accommodate the Discovery cost cap it will be a single, solar-powered Phoenix-type lander, built by Lockheed Martin in Denver, with a highly focused payload of three scientific instruments plus a number of engineering ones. The main payload will be a sensitive seismometer based on the one developed for the canceled Netlanders and able to measure displacements of the order of one billionth of a centimeter. The instrument will be provided by the French CNES with input from institutes in France, the UK, Switzerland, Germany, and also JPL. A German heat flow and physical properties package will be a self-digging, 35-cm-long "mole" at the end of a 4-meter tether which is instrumented every 35 cm for thermal, permittivity and accelerometric measurements. This will be based on the Beagle 2 mole and on studies for the Humboldt payload. A JPL-provided X-band radio science experiment will monitor Mars' rotation and other physical and dynamical parameters, which in turn will place constraints on the planet's internal structure and the size of its core. This will be only the third determination of the direction of the spin axis, 20 years after Mars Pathfinder and 40 years after Viking, and will allow a determination of the precession rate and moment of inertia, in addition to the rather more subtle nutation movements on the basis of 2 years' of tracking. During the first 60 sols, the seismometer and mole will be placed on the surface in selected places using a camera-equipped robotic arm in order to avoid the problems of the Viking seismometer, which mostly recorded vibrations intrinsic to the lander. Engineering instruments based on the Spanish meteorology package flown on Curiosity will measure pressures, winds and temperatures that must to be taken into account in seismometer measurements in order to achieve the necessary accuracy. They may provide additional scientific data. Moreover, the seismometer itself it will be carried inside a conical aluminum structure and insulated by aerogel, covered with a thermal blanket and shielding against wind after being placed on the surface. Although a network of at least three or four landers would be needed to precisely locate quakes and verify, for example, whether they are uniformly distributed or concentrated in the volcanic regions such as the Tharsis bulge, the measurement of just a single station will provide extremely valuable data, albeit that scientists will have to strongly rely upon models of the interior of the planet to compensate. Other than the color camera on the arm's elbow – derived from the Mars Exploration Rovers and Mars Science Laboratory 'navigation cameras' – which will also be used during the first 2 months of the mission to provide a low-priority full panorama of the site as well as to monitor clouds, dust devils and other transient phenomena, the lander will carry a surplus modified Mars Exploration Rover 'hazard camera' under the deck in order to image the area in which the instruments are deployed. Finally, a German three-filter infrared radiometer and a vector magnetometer will be on board. Communications from the lander to Earth will use two medium-gain antennas, and there will be a UHF antenna for dumping data through the orbital relays.[259] Although there are still two unallocated Delta II

A rendering of the US InSight geophysics lander. The robot arm is shown deploying the seismometer, whose thermal protection is still on the lander deck along with the thermal flux 'mole'. (NASA/JPL-Caltech)

rockets in storage, InSight will ride on an Atlas V 401, the least powerful version of this rocket. Remarkably, the launch will occur from Vandenberg Air Force Base in California to achieve a high-inclination parking orbit. It will be remembered that this was initially planned for Mars Odyssey.

Other than InSight, funded from a different budget than the Mars Exploration program, NASA is initiating yet another "integrated" Mars program, dubbed Mars Next Decade. Although this is expected to contribute to ambitions to send humans to Mars, there are no approved or funded plans and in any case such a mission will not be possible until the 2030s at the earliest. The first Mars Next Decade mission could be flown as early as 2018 or 2020, possibly by drawing upon the budgets not only of the scientific mission directorate but also human exploration. A workshop was held in June 2012 to assess architectures for both short-term missions from 2018 to 2024 and plans running into the 2030s. No fewer than 390 abstracts were submitted from ten different nations. Proposals included several variants of the Mars Exploration Rovers capable of precision landing and sample caching for later retrieval by a sample-return mission, a Phoenix-based lander that could test a small ascent rocket, a number of aerial vehicles, as well as an orbiter that could dip into the upper atmosphere and release RTG-powered modules. An instrument that may fly to the surface of Mars on these missions is one of the few which has until now resisted being miniaturized sufficiently to fit within the mass budget of a planetary probe. A geochronometer is an instrument to measure the ages of rocks, a task that so far has been achieved only on Earth, although it has been used to date samples obtained from the Moon by the Apollo and Luna missions. In its present form, a flyable geochronometer would fire a laser to vaporize a sample and a pair of tuned lasers to excite rubidium-87 and its decay product strontium-87. The ratio between the two measured by a mass spectrometer would indicate the age of the rock. Age dating is currently done on Mars using crater counts and cratering ratios, but there are very large error brackets.

From the scientific point of view, it remains to be seen how these missions fit within the priorities of the second planetary science decadal survey, which called for step-by-step progress towards a sample-return mission. In fact, the survey (of which more in the next chapter) called upon NASA to fund a Europa mission if a Mars sample-return could not be supported, and to approve lower priority Mars missions competitively through the Discovery or New Frontiers programs.[260,261] The Mars Program Planning Group delivered its final report in September 2012. It acknowledged that a sample-return mission would not be feasible for years, given the present budget, but it endeavored to stick to the priorities put forward by the decadal survey, namely building toward a return of samples. About $800 million was expected to be available in 2018, which will be the first launch window for the new architecture. This meant that an orbiter would probably be feasible, but not a rover. In fact, 2018, being a 'great opposition' window like that of 2003, would be ideal for a landed mission. On the other hand, waiting until 2020 would probably allow a rover to be launched but it would mean that no orbital relay would be in place. The orbiter options ranged from a simple $200 million single-purpose relay satellite (that could even make it to Mars as a secondary payload) to an evolution of Mars

Reconnaissance Orbiter and MAVEN, as well as electric propulsion sample-return orbiters. Four possible rover architectures of increasing complexity and cost were identified. The simplest would reuse the Mars Exploration Rover structure with updated electronics and instruments. The second would be a scaled-up Mars Exploration Rover with more space for sample caching, but this would require the development of an upgraded airbag and landing system. While the first two options would revert to an airbag-cushioned landing, they would use the guided precision landing techniques pioneered by Curiosity. The third option would be a single Curiosity-class, solar powered rover. The most expensive option was a Curiosity-based rover carrying a rocket to lob samples into orbit. The "wish list" even included a mission, if the budget were available, to land three MER-class rovers to collect samples from three different sites, after which a return mission would land near the most interesting of the three and get the cached samples. Finally, a large, single-shot sample-return mission could be launched during the mid-2020s using the new Space Launch System (SLS) heavy-lift rocket that NASA is developing for human missions beyond Earth orbit. As required by NASA, the report tried to force-fit human missions into the architecture, to the extent of suggesting that astronauts could recover the sample-return module and inspect the samples in space to verify that they would not pose a health risk.[262,263]

Based on this report, NASA retained a Curiosity-class $1.5 billion rover for 2020, so long as there was the budget. The mission would be launched on an Atlas V-class rocket in late July or August 2020 for arrival between January and March 2021. A science team was established in January 2013 to better define the mission and its objectives, and harmonize them with the scientific goals of the decadal survey. The team delivered its report in mid-2013. During a primary mission lasting one Martian year, the still-unnamed rover would seek chemical, mineralogical, and morphological indications of past habitability whilst also demonstrating sample-return technologies. Moreover, it might perform science and tests for future human missions, evaluating the health hazard posed by Martian dust and possibly demonstrating the production of rocket fuel from raw materials available on the surface of the planet. The 950-kg rover would reuse the structure of Curiosity, as well as the skycrane, and although solar-powered versions were evaluated, the baseline mission would use a Curiosity-class RTG for power. The scientific payload was cost-capped at $100 million, and a public tender for instruments was published in September 2013. The rover could use ultraviolet spectroscopy for remotely sensing carbon-bearing rocks, and cache more than thirty samples for a total of several hundred grams using some type of coring tool. The rover should be able to carry a terrain-relative descent camera to compare images with a stored orbital map of the surface and reduce the error in the estimated position at landing to several tens of meters. With prior knowledge of the presence of obstacles, the camera could be used for steering the rover away from hazardous areas during the descent. A lengthy selection process of a landing site is to be expected, the rover having to touch down in an environment suitable for recording and preserving traces of ancient biology in order to accomplish its scientific goals.[264]

NASA has also recently been donated two spy satellite optical assemblies by the

SEEKING SIGNS OF PAST LIFE

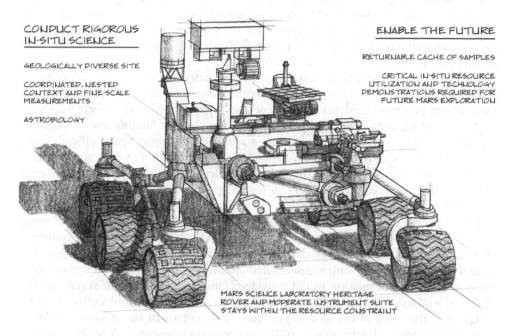

CONDUCT RIGOROUS
IN-SITU SCIENCE

GEOLOGICALLY DIVERSE SITE

COORDINATED, NESTED
CONTEXT AND FINE-SCALE
MEASUREMENTS

ASTROBIOLOGY

ENABLE THE FUTURE

RETURNABLE CACHE OF SAMPLES

CRITICAL IN-SITU RESOURCE
UTILIZATION AND TECHNOLOGY
DEMONSTRATIONS REQUIRED FOR
FUTURE MARS EXPLORATION

MARS SCIENCE LABORATORY HERITAGE
ROVER AND MODERATE INSTRUMENT SUITE
STAYS WITHIN THE RESOURCE CONSTRAINT

An artist's view of the 2020 NASA Mars rover designed to seek signs of ancient life and to collect and cache samples for future return to Earth. (NASA/JPL-Caltech)

National Reconnaissance Office. One of the proposals for their use was to put one of them in Mars orbit to obtain centimeter-resolution images of the surface. It must be noted, however, that a Mars orbiter with such a high resolution has not been ranked as a priority by the scientific community, and there are probably planetary missions more deserving of the $900 million that such a mission would likely cost.

ASIA TO MARS

After the successful launch of the Indian Mars Orbiter Mission, the launch windows of the late 2010s could see two more Asian countries debut (or in the case of Japan, return) in the exploration of Mars.

JAXA is planning to make use of a heavy H-IIA rocket to launch MELOS (Mars Exploration with Lander and Orbiters Synergy) for its fourth planetary science mission. It will recover some of the lost Nozomi science and complement US and European missions with studies of the internal structure of the planet and the erosion of the atmosphere. It will have one satellite in an eccentric orbit to conduct global observations and another in a circular orbit to study the ionosphere, and will send down as many as three landers equipped with seismometers, in addition, possibly, to rovers and balloons or airplanes.[265] JAXA is also studying a dust sample-return for 2018 using technology derived from Hayabusa. Like the US SCIM Scout proposed this would be captured in an eccentric Mars orbit, using aerocapture for the first time, sample the atmosphere and collect dust particles as far down as to an altitude of about 35 km, and then use a bipropellant engine to return to Earth.[266]

According to a 'white paper' on Chinese space activities, the nation will conduct preliminary studies on independent Mars exploration in the first half of the 2010s. A three-phase program modeled on that of the lunar Chang'e is reportedly under study, consisting first of orbital remote sensing, then by soft landing and roving, and finally by robotic sample-return possibly about 2030. Chinese space officials have indicated their desire to launch a wholly national Mars mission as early as 2013, and models of an orbiter and an airbag-cushioned lander similar to Mars Pathfinder were shown in Beijing in 2011. However, no formal go-ahead was given at government level and of course a 2013 mission has not materialized.

Meanwhile, a number of options have been studied and extensive details of some of them have been released. The first option to be described in detail would comprise an orbiter and a semi-hard lander demonstrator, the latter looking remarkably similar to the Schiaparelli entry, descent and landing demonstrator. From a scientific point of view, such a mission would provide chemical analyses of the surface of the planet as well as exploring the local environment, although these objectives may be reviewed in the wake of the loss of Yinghuo. A large part of the mission would be dedicated to technical development of future deep-space missions. After launch on a Long March 3B rocket, the 2,000-kg, 3-axis-stabilized spacecraft would enter a near-polar orbit of Mars with a periapsis at 300 km to make observations for one Martian year. Chinese sources acknowledge that they were investigating the possibility of circularizing the orbit utilizing aerobraking. The lander would be released whilst in an early eccentric Mars orbit, as did the Viking landers. No other spacecraft has used this profile since, preferring to deploy landers in the final phase of the interplanetary cruise. Three flat landing site candidates have been identified in Amazonis, Chryse and Utopia. The 50-kg demonstrator would have a mission lasting no more than 5 days. The orbiter payload would be open to foreign participation, and might well include a ground-penetrating radar, high-resolution cameras, imaging and infrared spectrometers, a gamma-ray spectrometer, and plasma and fields instruments. It has been indicated as a potential carrier for Finland's MetNet stations, and Sweden may provide an ion and neutral atom analyzer.

A simpler alternative would be a Yinghuo-based small orbiter equipped with its own propulsion module rather than relying on a Russian carrier. It would release two penetrators, one targeted at the Martian north pole and the other at the equator. After being released from orbit, each penetrator would be slowed initially by atmospheric drag and then by a parachute. At an altitude of 200 meters, the parachute would be discarded and the penetrator would bury itself in the ground at 80 to 100 m/s. The 50-kg, battery-powered lander would carry a descent camera, a surface camera, and instruments to determine the characteristics of the terrain at the site, including the possible presence of water or organics, and would function for at least 10 days. It is worth noting that China has been working on planetary penetrators since at least the 1990s, probably as a by-product of military studies.[267]

More recently, details of a mission proposed by the Qian Xuesen Laboratory of Space Technology were published. This mission would be launched around 2024 by the Long March 5 and would involve an orbiter and a large entry module carrying three penetrators, a rover, and a balloon. Landing sites recording the presence of either clays or hydrated deposits would be targeted, with Nilosyrtis as the prime site and Gale crater as a secondary one. Chinese studies of missions to Mars seem to be particularly interested in complicated multi-spacecraft concepts. Another such study envisages a main orbiter and a small hitchhiker spacecraft to enter a resonant orbit ensuring hundreds of flybys of Deimos which, owing to its distance from the planet, has been neglected by the orbiters flown by other nations since the Vikings. It was reasoned that such a mission would be of unique scientific interest. Moreover, the subsatellite could also carry out mutual occultation campaigns with the mothership, as was intended for Fobos-Grunt and Yinghuo. Yet another multi-spacecraft concept would involve a constellation of minisatellites flying in formation in Mars orbit with a mothership providing communications with Earth.

A 2018 Mars orbiter and lander demonstrator reportedly passed its design review in 2013, but the actual status of the mission is unclear. As recently as December of that year, in fact, while celebrating the successful landing of Chang'e 3 on the Moon, Chinese space officials pointed out that Mars has still not been made an official goal for future space exploration.[268,269,270,271,272,273]

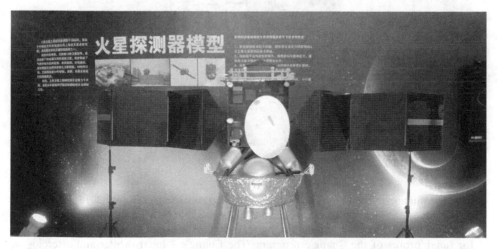

This model of a Mars orbiter was recently revealed by China. It consists of a small, 60-kg short-duration lander similar to the European Schiaparelli (on top), a 590-kg orbiter and a 1,700-kg jettisonable propulsion module. (Courtesy weibo.com)

The first mission of the Long March 5 rocket (at left) is expected around 2015. It will be the Chinese equivalent of the Ariane 5, and will be used for heavy missions to the Moon and elsewhere in the solar system. The Long March 3B and A (at right) have been used for lunar probes of the Chang'e program. The Chang'e 2 lunar-orbiter-and-asteroid-flyby-probe was launched by a Long March 3C, which was a 3B with only two boosters.

AFTER 2020

Barring further budget reductions, during the 2020s space agencies should finally be ready to attempt to retrieve a sample from Mars. A Mars sample-return would have as its principal merits no limit in the instrumentation used for analysis, ideally including instruments not yet invented or perfected at the start of the mission, in addition to the possibility of verifying and duplicating results using different techniques or different samples. Think for example of the difficulty of replicating Viking biology experiments. At the turn of the millennium NASA had hoped to launch its first such mission in 2011, but the freeze on its Mars Exploration program funding in 2005 ruled this out. It will require the development of a number of technologies including pinpoint landing, sophisticated sample handling systems, a Mars ascent vehicle and autonomous rendezvous and docking. These were to have been tested on a series of precursor missions but were placed in abeyance in order to cover the cost overruns of the Mars Science Laboratory. In Europe, the Aurora program also envisaged a similar timeline. A 12-month system study was begun in mid-2006, at which time the necessary funding had appeared likely. The European architecture would have required two Ariane 5 launches, one with the lander, rover and ascent vehicle, and the second with the orbiter and the vehicle that would return to Earth with a 500-gram payload of rocks, soil and atmospheric samples. In parallel with the individual space agencies, an International Mars Architecture for the Return of Samples (iMARS) working group involving representatives of the space science communities of many countries delivered a report in 2008 calling for a program to start in the mid-2010s to develop the technologies to launch a mission around 2022 at an estimated cost of between $4.5 and $8 billion.

All of these studies and proposals were then merged into the Mars Exploration Joint Initiative in which NASA and ESA would work together to launch a sample-return mission in the 2020s. NASA was to provide the entry, descent and landing system – after all, it is the only agency to have successfully soft-landed a working payload on the surface of Mars. ESA wished to contribute the rover, and provide a deep-subsurface sampling system based on the drill devised for ExoMars, which was itself based on the drill designed for the 2003 sample-return. Moreover, ESA has experience in autonomous rendezvous and docking (admittedly using a system designed to work with a Russian counterpart) on the Automated Transfer Vehicle, which first docked to the International Space Station in 2008, and France is interested in developing it into a system to dock with inactive, "non-cooperative" Earth satellites as a further step toward a sample-return mission.[274]

Several technologies have already entered development. NASA and the US Air Force have recently experimented with a small rocket fueled by a mix of powdered aluminum and water ice that could be used to reduce the mass and complexity of a Mars ascent vehicle by exploiting water obtained in-situ. Moreover, in April 2010 NASA issued a contract to the Massachusetts Institute of Technology and Aurora Flight Sciences to develop an optically-guided Mars orbit rendezvous system that could be tested on the International Space Station. A sample-capture basket design has been tested by NASA aboard a low-gravity aircraft. A prototype sample-return

This prototype Mars sample-return canister was developed by French industries under contract to ESA.

capsule 230 mm across that can accommodate eleven separate sample vessels and can withstand a landing shock of 400 g has been built for ESA by a French firm. It is slow progress, but the pieces are beginning to come together.

In an effort to reduce costs, space agencies look set to split the sample-return into three simpler, cheaper missions. In one of the most recent iterations the first mission, the 2018 MAX-C rover, was to collect up to 38 samples in two separate caches, a primary and a backup. The second mission, an orbiter, would arrive 2 years before the Earth-return vehicle and fetch rover, and would provide coverage as well as communications relay for them. The orbiter would be equipped with an optical system capable of detecting the return canister at a range as great as 10,000 km, as well as autonomous rendezvous and docking systems. Fuel would constitute two-thirds of the mass of the orbiter. Aerobraking would be used to reach a circular orbit, so most of the fuel would be for the trans-Earth burn. The final mission would be a lander to deploy a rover to retrieve the cache of samples. A 2-stage all-solid rocket would deliver the canister to an approximately 500-km orbit, where it would be recovered by the orbiter. As a precaution, the return vehicle would be inserted on a trajectory that would miss Earth and a targeting maneuver would be made shortly prior to arrival. Landing would nominally occur in Utah, after which the samples would be kept in a receiving facility until deemed safe for distribution to scientists worldwide.[275],[276],[277],[278] In this scenario, each mission would be small enough to be launched on a relatively cheap rocket. The penalty would be the overall duration of the mission, spread over three windows with the eagerly awaited samples not becoming available for analysis until 2030. The first mission of the scenario looks set to be implemented by the 2020 NASA rover.

A few Mars missions other than MAX-C and the sample-return were assessed by the second decadal survey, released in 2011. These included a geophysical network consisting of at least two InSight-like long-lived landers as well as a number of missions to explore the Martian poles. Discovery- or New Frontiers-class orbiters could carry instruments like context cameras, thermal imagers, spectrometers, extremely accurate altimeters, and radiometers in order to characterize the polar caps and their seasonal cycles. As an alternative, short-lived precision spacecraft could be landed either on the polar layered terrain itself or at its edges to monitor the surface processes and analyze samples. The most expensive solution would be to send a MER-class rover, perhaps nuclear-powered and thus long-lived, able to survive in the polar environment.[279],[280]

At least one team is known to be working on these mission concepts. Space Exploration Technologies (SpaceX) and NASA's Ames are reportedly working on an unmanned Mars lander based on SpaceX's manned Dragon capsule, originally developed to carry cargo and astronauts to the International Space Station. This 'Red Dragon' spacecraft would undertake the Icebreaker mission to drill into the polar terrain, and would be proposed for a Discovery selection later in the decade. Whether modifying a manned capsule into a Mars lander will make better sense than developing a dedicated spacecraft remains to be seen.[281] Meanwhile the Dutch non-profit company Mars One plans to launch in 2018 a Phoenix-based lander to demonstrate some of the technologies required for future manned missions, in particular in-situ production of propellant from available Martian material. It would provide its own relay satellite. However, there is little chance that the company will raise sufficient money to fly anything.[282]

ESA, on the other hand, is studying a European Robotic Exploration Program (EREP) to follow ExoMars and maximally exploit future launch windows. This envisages a three-lander network based on MarsNext for 2022, at the same time as a sample-return mission to a Martian moon. In 2024 it would fly a precision (10 km) lander with a 100-kg rover. This Mars Precision Lander would be a European version of the Mars Science Laboratory, testing guided entry techniques and using a "dropship" similar in concept to the US skycrane to lower the rover down to the surface. The fourth step would be the sample-return orbiter, although its timing would depend on whether a caching rover had already collected the samples to be returned to Earth.[283]

Many alternative and sometimes unorthodox ways of exploring the planet have been considered, some of which may one day reach fruition. These include a drill to penetrate to a depth of 100 meters and demonstrations of the in-situ production of propellant in the context of a human mission. To make rovers more 'intelligent', engineers are experimenting with algorithms to identify and react to novel terrain features by comparing the amount of 'information' present in different images.[284]

These have been tested on terrestrial landscapes which bear a certain similarity to those of other planets, in one case identifying a patch of lichen when this was first encountered and then ignoring lichen when it was no longer a novelty.

Solar hot-air balloons are commonly used as toys on Earth and consist of black plastic envelopes that fly when heated by the Sun. Martian versions, more complex and capable of controlling their altitude, could be particularly useful in the polar regions, which are in continuous sunlight for long periods. A derivative could be a variable-emissivity balloon with a retractable roll-on, roll-off cover of a different emissivity which could 'control' a landing. Balloons could also be used instead of parachutes to land small payloads. Such a system would enable a greater fraction of the mass of an entry vehicle to be landed. A related technology with which JPL has experimented are large rovers with inflatable wheels for negotiating very large obstacles.[285] Inflatable technologies could also be used for heat shields. Russia and ESA have been experimenting with this idea since the 1990s, and the US recently successfully launched a mushroom-shaped Inflatable Re-entry Vehicle Experiment on a suborbital arc. This was initially designed to enable high-elevation landings on Mars.

In the 1990s ESA experimented with a Martian autogyro, a helicopter using an unpowered rotor. Such a device could access locations such as craters, canyons and caves that were inaccessible by other means.[286] Other rugged terrains such as gullies, polar icy-pit depressions, mesas and valleys could be accessed using small penetrator-like instrumented rockets which would be fired from a stationary lander or even from a rover which would not itself be able to reach the site. These would have a much lower impact speed and a much greater accuracy than 'conventional' penetrators dropped from space. Scaling from military analogs, these ought to be able to be delivered within 120 meters of a given point at a range of 5 km.[287] Also studied by the Advanced Concept Team at ESA in collaboration with the Helsinki University of Technology was a tumbleweed-inspired revision of the 1970s 'Mars Ball', using external turbine blades and possibly also mechanisms for steering and overcoming obstacles.[288,289]

One mission that the author would like to see implemented one day, but to the best of his knowledge has not yet been proposed, is a version of the San Marco project to measure the density and other parameters of the outer atmosphere. The San Marco satellites were launched by Italy and the US in the 1960s and 1970s. Each satellite was a spherical shell containing a proof mass. When the outer shell was retarded by atmospheric drag, the inner mass continued essentially unaffected and the relative displacement was measured by flexible arms. The minuscule drag acting on the spacecraft enabled the density of the atmosphere at orbital altitude to be directly measured in a simple, precise and elegant manner.[290] A fascinating, if untested alternative was proposed in the early 1990s in the wake of studies for the 'tethered satellite' that flow twice on the Space Shuttle with mixed results. This consisted of an orbiter connected to a small atmospheric probe by a 100-km cable of high-resistance graphite. Unfortunately, despite having been tested many times over the years, mostly using small satellites, space tethers have not matured to the point of making such a high atmospheric mission possible.[291]

What Mars exploration currently lacks, is coordination and intelligent reuse of technologies. US landers in the 2000s and early 2010s have used, or will use, three different landing techniques, each painstakingly developed from scratch each time. Europe, Russia and China each have their own ideas. Nor is there a true common data relay system. All this is exacerbated by the fact that, almost 20 years after initiatives such as Discovery and Mars Surveyor were initiated in the 'faster, cheaper, better' style in order to bring the cost of planetary missions under control, costs are once again escalating as they did in the 1980s.[292]

HUMANS TO MARS?

Sometime in the 2030s or 2040s, after several successful sample-return missions, it is possible that we will resolve to send people to Mars. The possibility of such a mission was explored even before the space age. Wernher von Braun published his *Das Marsprojekt* as early as 1952. In the 1960s NASA studied piloted flybys and landings and the Korolev design bureau in the Soviet Union was designing heavy Martian spacecraft. These efforts peaked at the time of the Apollo lunar landings, when Mars appeared to be the next logical step. Unfortunately the momentum was lost and in the ensuing decades all projects were stillborn, including the ambitious initiatives of presidents Bush senior and junior.[293,294]

A human flight to Mars poses a number of problems. From a technical point of view it will require the launch from Earth, or the assembly in orbit, of structures massing hundreds of tonnes. Moreover, the issue of how to protect the crew from space radiation has yet to be resolved. There is also the issue of their enduring zero or low gravity for long periods of time. And individuals would react differently to confinement, loneliness and psychological stress. Simulations on Earth have built up to a joint Russia-European experiment lasting 520 days. And, of course, such a mission would cost at least several hundred billion dollars.

It must be said on the other hand that many scientists feel there is no reason to mount a human expedition in the immediate future, because most of the scientific questions can be answered by robotic mission. In fact, the more intelligent robots become, the less need there will be for a human presence. Furthermore, as British space consultant Bob Parkinson has observed, there is a paradox in that if robotic missions fail to find traces of life, Mars will be regarded as far less interesting and a human mission may no longer be deemed worthwhile, whereas if life is discovered it might be decided that humans should not go there![295]

REFERENCES

1 McEwen-2007a
2 Graf-2005
3 Covault-2005a
4 Graf-2005
5 Covault-2005a
6 Covault-2005b
7 Dornheim-2006a
8 Dornheim-2006b
9 Tolson-2008
10 Graf-2007
11 Bishop-2008
12 Covault-2006
13 Ryan-2012
14 Shinbrot-2004
15 McEwen-2011
16 Jaeger-2007
17 Kerr-2007
18 Okubo-2007
19 Lewis-2008
20 Bridges-2012
21 Kok-2012
22 Herkenhoff-2007
23 Hansen-2011
24 For the Vikings see Part 1, page 233
25 For Mars Pathfinder see Part 2, pages 460–461
26 Parker-2007
27 McEwen-2010
28 McEwen-2007b
29 Ehlmann-2008
30 Mustard-2008
31 Carter-2010
32 Seu-2007
33 Holt-2008a
34 Phillips-2008
35 Kerr-2008a
36 Holt-2010
37 Smith-2010
38 Morgan-2013
39 Phillips-2011
40 Thomas-2011b
41 Kerr-2010a
42 Zuber-2007
43 Andrews-Hanna-2008
44 Marinova-2008

45 Nimmo-2008
46 Kiefer-2008
47 Plaut-2010
48 Zurek-2012
49 Adler-2012
50 Zurek-2013
51 Kerr-2008b
52 Ehlmann-2011
53 Leshin-2002
54 Jurewicz-2002
55 ASU-2002
56 Braun-2006
57 Randolph-2003
58 Covault-2007a
59 Garcia-2007
60 Grover-2007
61 Edwards-2006
62 NASA-2008b
63 Covault-2007b
64 Covault-2007c
65 Spencer-2009
66 Garcia-2007
67 Kerr-2008c
68 Covault-2007d
69 Kerr-2008c
70 Grover-2008
71 Covault-2008a
72 Covault-2008b
73 Covault-2008c
74 Covault-2008d
75 For Viking sample delivery problems see Part 1, page 234
76 Niles-2010
77 Kerr-2010b
78 Covault-2008e
79 Covault-2008f
80 Kremer-2008
81 Covault-2008g
82 Covault-2008h
83 Covault-2008i
84 Kerr-2010c
85 Covault-2008j
86 Smith-2009b
87 Boynton-2009
88 Hecht-2009

89 Whiteway-2009
90 Hand-2008
91 Covault-2008k
92 Kerr-2009a
93 Kremer-2009a
94 Galeev-1996
95 Day-2010
96 Korpenko-2000
97 For Mars-Aster see Part 2, pages 123–124
98 Eneev-1998
99 Kuzmin-2003
100 Korpenko-2000
101 Eismont-1997
102 RDIME-2008
103 Zelenyi-2007
104 IKI-2008
105 Coué-2007b
106 Hu-2010
107 People's Daily-2009
108 Zhao-2008
109 Wu-2010
110 Chen-2010
111 Barabash-2007b
112 Chen-2010
113 Kopik-2004
114 Kopik-2003
115 Zak-2008
116 Stone-2009
117 Martynov-2010
118 Morin-2011
119 Kolyuka-2012
120 Harri-2003
121 Polischuk-2006
122 Harri-2007a
123 Harri-2007b
124 Harri-2008
125 Holt-2008b
126 IKI-2008
127 Korablev-2009
128 Marraffa-2000
129 For the UMVL spacecraft and its variations see Part 2, pages 146–147 and 427–437

130 Korablev-2009
131 For the Vesta mission see Part 2, pages 120–123
132 Polischuk-2005a
133 Polischuk-2005b
134 Polischuk-2006
135 Konstantinov-2004
136 Martynov-2009
137 IKI-2008
138 Lyngvi-2004
139 Renton-2006
140 Smith-2000a
141 Smith-2000b
142 For PREMIER see Part 3, pages 367–368
143 Asker-2000
144 Smith-2000a
145 Taverna-2000
146 Morring-2003b
147 Covault-2005c
148 Lawler-2006
149 Covault-2004b
150 Reichhardt-2005
151 Morring-2005c
152 Steltzner-2008
153 Powell-2009
154 Way-2006
155 Buch-2009
156 Morring-2011c
157 Balint-2007
158 Kerr-2006
159 Kerr-2008d
160 Morring-2013b
161 Kremer-2009b
162 Lawler-2008
163 Sietzen-2009
164 Mumma-2009
165 Edgett-2011
166 Hand-2011a
167 Kerr-2011b
168 Zeitlin-2013
169 Kerr-2013a
170 Martin-Mur-2012
171 Abilleira-2012
172 Morring-2012a
173 Norris-2012a
174 Morring-2012b
175 Lakdawalla-2012

176 Norris-2012b
177 Williams-2013
178 Jerolmack-2013
179 Stolper-2013
180 Kelly Beatty-2013
181 Kerr-2012
182 Hand-2012b
183 Kerr-2013b
184 Grotzinger-2013a
185 Meslin-2013
186 Bish-2013
187 Leshin-2013
188 Blake-2013
189 Webster-2013a
190 Grotzinger-2013b
191 Grotzinger-2014
192 Ming-2014
193 McLennan-2014
194 Farley-2014
195 Vaniman-2014
196 Kerr-2013c
197 Hassler-2014
198 Webster-2013b
199 Kumar-2006
200 Rao-2009
201 Morring-2010
202 Lele-2013
203 Goswami-2013
204 Pillinger-2007
205 MAVEN-2008
206 Jakosky-2010
207 Jakosky-2012
208 Witze-2013
209 Morring-2013c
210 MAVEN-2013
211 Schultze-2002
212 Gardini-2003
213 Messina-2003
214 Walker-2006
215 Ball-2009
216 ESA-2002
217 Rouméas-2003
218 Morring-2003c
219 Messina-2006
220 Taverna-2005
221 Peacock-2005
222 Wall-2005
223 Taverna-2006
224 Vago-2006

225 Marlow-2009
226 Taverna-2006
227 Lognonné-2010
228 Morris-2009
229 Clery-2008a
230 Taverna-2008a
231 Clery-2008b
232 Taverna-2008b
233 Lefèvre-2009
234 Calvin-2007
235 Hayati-2009
236 Christensen-2009
237 NASA-2006
238 Lawler-2009
239 Zurek-2009
240 Coradini-2010
241 Lardier-2009
242 Taverna-2009a
243 Taverna-2009b
244 Pratt-2009
245 Zahnle-2011
246 Keppler-2012
247 Kerr-2012
248 de Selding-2011
249 Svitak-2011
250 Messidoro-2014
251 ESA-2010a
252 Van den Broeckc-2011
253 Capuano-2012
254 Vago-2013
255 Cassi-2012
256 Morring-2012c
257 Svitak-2012
258 Lognonné-2010
259 Banerdt-2013
260 Morring-2012d
261 Hand-2012c
262 MPPG-2012
263 Morring-2012e
264 Mustard-2013
265 Sasaki-2009
266 Fujita-2014
267 Tan-1999
268 Huang-2011
269 Yuan-2011
270 Siili-2011
271 Ying-2013
272 Ming-2013
273 Hou-2013

274 Taverna-2008c

275 Syvertson-2010

276 Mattingly-2010a

277 Mattingly-2010b

278 Li-2010

279 JPL-2010a

280 JPL-2010b

281 Hand-2011b

282 Morring-2013d

283 de Groot-2012

284 Kean-2010a

285 Jones-1999

286 Hill-2000

287 Garrick-Bethell-2005

288 Menon-2006

289 For 'Mars Ball' see Part

     1, pages 257–258

290 Broglio-1966

291 Lorenzini-1990

292 Nature-2008

293 Portree-2001

294 Siddiqi-2000

295 Ashworth-2009

12

The future

NASA'S DISCOVERIES

In the wake of its budgetary difficulties, the disappointing failure of CONTOUR and partial failure of Genesis, NASA slowed the selection of Discovery projects from one every 2 years to just three in 11 years. As a result, by the end of the first decade of the millennium Discovery and New Frontiers missions were being selected more or less with the same frequency. After Dawn in 2001, missions were selected in 2007 and in 2012. However, during this interval the agency funded several cheaper "missions of opportunity" under the program by participating in India's Chandrayaan lunar orbiter, in ESA's BepiColombo Mercury orbiter, and the extensions of the Stardust and Deep Impact missions. At the same time, the average cost of a Discovery flight exceeded $450 million.

On the shortlist for the 2007 selection were the Vesper Venus orbiter, which was an updated 1998 finalist, the Origins Spectral Interpretation, Resource Identification and Security (OSIRIS) asteroid mission, and the GRAIL survey of the Moon's gravity field. The OSIRIS proposal was to rendezvous with the 600-meter near-Earth asteroid (101955) Bennu and return a 150-gram sample to Earth in 2017. On this occasion the winner was GRAIL.[1]

For the next competition, NASA wished the Discovery 12 mission to flight test a power source that could supersede RTGs for future outer solar system missions. The technology selected was the Advanced Stirling Radioisotope Generator (ASRG), the first mechanical power conversion system designed for use in space. Developed by Lockheed Martin Astronautics initially for the Prometheus project and managed by the US Department of Energy, it would employ two mechanical Stirling generators to convert some of the heat from a pair of standard plutonium oxide pellets to electrical power. The unit, massing some 20 kg, would provide about 143 W of power at the beginning of the mission with a conversion efficiency of around 28 per cent; four times that of an RTG. Compared to a conventional RTG, the Stirling generator would be lighter, more efficient, have a higher power-to-weight ratio, and be cheaper, using only one-fourth the plutonium of the Mars Science Laboratory's

Multi-Mission RTG, or about 1 kg in total. Its decline being due mostly to the natural decay of the isotopic source, a Stirling generator would suffer less power degradation than an RTG, but it would pose problems of reliability involving moving machinery. For this reason, two identical converters were to be mounted on each generator. However, the failure of one of the converters would create a mechanical imbalance in the moving parts in the form of vibrations. As an alternative, redundant generators could be carried. Another shortcoming was the sensitivity of the mechanical parts to accelerations which, whilst acceptable in the launch environment, could make them unsuitable for hard landers. Conservativeness by engineers and scientists had prevented their being adopted for use in space. To ease their introduction, NASA made available one ASRG to the next round of Discovery missions at no extra cost. Meanwhile, more than 14,000 hours had been accrued in full-scale tests.

Reportedly around one-third of the 28 proposals received by June 2010 envisaged missions to asteroids and comets. No fewer than seven called for missions to Venus, reflecting the renewal of interest in out 'sister planet'. Of these, four were radar mappers. One, the Radar at Venus (RAVEN), sought to improve on the results of the Magellan mission by combining scans of 25-meter-resolution with meter-resolution scans of selected areas, including the landing sites of previous missions. A concurrent proposal by a US-Israeli team would use a lightweight radar developed for an Israeli spy satellite to provide meter-resolution imaging. Candidates also included a long-duration Venus balloon, two lunar polar-ice missions, a Trojan asteroid explorer, a Martian polar lander with a sampling drill, missions to Io and to Titan, a comet coma sample-return, etc. It has not been revealed how many of the proposals would have used the ASRG power source.[2,3] Half of the Discovery proposals were deemed non-viable, and others either would have required expensive technology developments or would not have provided first-rate science.

One of the ASRG-powered missions not selected was the Io Volcano Observer that was to reach Jupiter in 2021 and target its volcanic moon, as well as explore the inner magnetosphere of the planet. It would have made one Io flyby during insertion into an inclined orbit designed to avoid most of the radiation belts, and then six more over the next 10 months at distances between 1,000 and 100 km. Closer flybys would be made during an extended mission that would also have seen the vehicle pass through active volcano plumes prior to impacting. The payload would have been a camera to image key surface features at a resolution of at least 10 meters per pixel, a thermal mapper, and remote and in-situ instruments for the Jovian environment in the vicinity of Io, yielding a hundred times more data on the satellite than the Galileo mission on each flyby. Several mission proposals targeted the Saturnian system. JET (Journey to Enceladus and Titan) was a Saturn orbiter that was to make multiple flybys of those two moons. One of the most publicized proposals was the Aerial Vehicle for In-situ and Airborne Titan Reconnaissance (AVIATR) by the Applied Physics Laboratory. It envisaged an ASRG-powered pusher-propeller aircraft with a 3.5-meter-span delta-wing that would fly for an entire year in slow circles over the day-side at altitudes of several kilometers. The combination of a very dense atmosphere (four times as dense as on Earth) and reduced gravity (one-seventh that of Earth) makes Titan an excellent place for aircraft, requiring far lower power levels and speeds than on Earth. In fact, human-powered flight "à la Icarus" would even be theoretically feasible. Being light-hours from human intervention, AVIATR was required to be capable of autonomously planning its mission and dealing with unforeseen events. Reliability was also a major issue, since its mechanisms had to be capable of working correctly on a flight lasting a year. AVIATR would reach Titan cocooned inside an entry capsule some 4 meters across. Unlike a Mars airplane, however, for which the speed at deployment would be high, an airplane on Titan could be deployed at a speed as low as 10 m/s due to the manner in which the dense atmosphere slowed the entry capsule. The mission was to be launched in 2017 and arrive in 2024 to conduct a flight lasting 23 local days (each of 16 terrestrial days). During this time the aircraft would study a variety of surface and atmospheric targets ranging from Hotei Arcus, Xanadu and equatorial dunes to lakes at both poles using imagers and spectrometers, an atmospheric and aerosol package, a radar altimeter for navigation and a student-built acoustic raindrop detector. A planar, steerable antenna in the nose of the aircraft would provide direct communication to Earth. Even with the ASRG power source, AVIATR would be obliged to glide for the duration of a communication session to provide sufficient power to the transmitters. The total cost of the mission (including the ASRG) was estimated by its proponents at $715 million, making it a good candidate for a New Frontiers mission. It was their hope that AVIATR would be the first step of a Titan Exploration program similar to the Mars Surveyor and Mars Exploration programs, replacing a large and expensive mission that would deliver its results two or three decades after being conceived with a series of smaller, cheaper and more frequent missions.[4,5]

Other notable missions included a Primitive Material Explorer (PriME) to analyze volatiles in situ near a comet in order to investigate the role of primordial bodies in delivering water and other volatiles to Earth. Whipple was a mission to monitor tens of thousands of stars many times per second, to detect occultations by small objects and thereby provide a census of the outer regions of the solar system including those that are unlikely to be explored any time soon, if indeed ever, such as the Oort Cloud. JPL's NEOCam (Near-Earth Objects Camera) was twice proposed for the Discovery program. During a 4-year mission at the inner (L1) Lagrangian point of the Sun-Earth system, it would use an infrared telescope to discover and characterize two-thirds of the population of near-Earth objects larger than 140 meters. The spacecraft would be based on Kepler, with extensive heritage from other space astronomy missions.[6] The mission might also interest the human spaceflight community, since it could help in the search for a suitable target for a piloted asteroid mission. The estimated cost was $425 million.

When NASA announced the final three candidates in May 2011, they included two ASRG-powered missions and one solar-powered Mars lander.

TiME (Titan Mare Explorer) was a proposal by APL and Proxemy Research Inc., to land a floating probe in one of the polar lakes on Titan to address questions raised by the Cassini mission, namely their chemistry, their role in the methane cycle, which is the counterpart of the Earth's water cycle, their origin and their seasonal processes. The mission would launch on an Atlas V 411 in January 2016, make one deep-space maneuver, one Earth and one Jupiter flyby, and reach Titan in July 2023. Since all the scientific instruments would be cocooned inside the descent capsule there would be no cruise science. The flying saucer-shaped probe would splash down in Ligeia Mare, with Kraken as a backup, and would arrive before winter; i.e. before the Sun sets on that pole. Both the Earth and the Sun would need to be above the horizon during the primary mission lasting 6 local days. Possessing no propulsion it would drift with the wind, measuring the temperature, humidity and winds, and observing the cycling of methane between the surface and atmosphere. TiME would be equipped with a mass spectrometer to analyze the lake's liquid content, with the possibility of detecting any peculiar chemical signature of life forms that may have developed in that organic-rich environment. Cameras would produce panoramas during the descent, and thereafter a limited number of images of the sea and sky. The payload would be completed by an instrument package for physical properties and meteorology, including a sonar to measure the depth and the volume of the sea and determine the quantity of organics present. Lacking a relay satellite orbiting either Saturn or Titan, a gimbaled antenna would track the Earth. Nevertheless, the probe would be able to generate a relatively small amount of data. The mission would pioneer low-cost outer planets missions, as well as validating the ASRG in both a deep-space environment and upon a planetary surface.[7]

The Comet Hopper (nicknamed Chopper) was a University of Maryland, Goddard Space Flight Center and Lockheed Martin proposal intended to accompany a comet nucleus around its orbit, landing at different places in order to study the changes as it warmed approaching the Sun. The mission would leave Earth on an Atlas V launcher in November 2015, perform one Earth flyby, and arrive at its target in August 2022 near Jupiter's orbit. The targeted comet would be Wirtanen, the original objective of the Rosetta mission and consequently fairly well characterized. The spacecraft would accompany the nucleus for over 2 years through to perihelion in mid-2024, landing at as many as three sites, several hundreds of meters apart, in order to observe the onset of activity.

The ASRG-powered spacecraft would conduct remote operations at several tens of kilometers from the nucleus for mapping and to select a suitable landing site prior to setting down at less than a meter per second. If safety parameters and thresholds were exceeded, it would be able to autonomously back off to a safe distance. A backup to comet Finlay would see launching in late 2016 and arrival in December 2024. The Comet Hopper mission would build on Deep Impact (including many members of its scientific board) and would provide a more in-depth exploration than was possible on an all-too-brief flyby. It was also thought of as a 'scouting' mission for a later nucleus sample-return. The spacecraft would be equipped with a near-infrared spectrometer, a mass spectrometer, alpha- and X-ray spectrometers, a multispectral camera, a thermal probe and penetrometer, and fixed panoramic cameras positioned behind the circular landing platform and dust shield. Its observations would determine the distribution of elements, minerals, volatiles, organics and ices on the surface, establish relationships between surface features and jets, correlate erosion with changes in the outgassing of volatiles, and study the evolution of the nucleus and the processes at work.[8]

The other Discovery proposal to make it to the final of this competition was JPL's InSight Mars lander, described in the previous chapter.

Each of the three candidates received a $3 million budget to conduct more studies and a preliminary design. The mission would be selected in mid-2012 for launch no later than the end of 2017 with a cost cap of $425 million, exclusive of the launch vehicle.

The Discovery program also decided to finance development of key technologies for three of the non-selected missions, to elevate them to a higher level of readiness. These concerned the NEOCam and Whipple space telescopes as well as a novel mass spectrometer to enable PriME to analyze cometary ices.[9]

In the end, the agency selected the less risky InSight as the Discovery mission that would be launched in 2016.

Further underscoring the budgetary and programmatic problems that face NASA's planetary exploration, the next Discovery call for proposals is to be released in 2014 or early 2015 (barring further budget cuts) for selection in 2015 or 2016 and launch around 2020. It must be said that, contrary to the spirit under which the program was initiated in the 1990s, the rare launch opportunities are boosting complexity and cost and obliging scientists and engineers to endeavor to make the missions less risky.[10]

Preparing an Advanced Stirling Radioisotope Generator (ASRG) prototype for tests.

The Comet Hopper (Chopper) was a finalist for the latest round of Discovery missions. The ASRG-powered probe was to land at several locations on comet Wirtanen in order to study its activity as a function of heliocentric distance.

NASA'S NEW FRONTIERS

In 2007, after the selection of New Horizons and Juno as the first and second New Frontiers missions, NASA asked the National Research Council for an assessment of the progress of the program and to suggest further priority medium-class missions in additions to those remaining from the decadal survey. The five candidates identified were a 'primitive' asteroid sample-return, a Ganymede and Io observer, a Martian or lunar lander network, and a Trojan or Centaur reconnaissance mission.[11]

At about the same time, there was a call for proposals for the third mission of the

The Titan Mare Explorer (TiME), another ASRG-powered Discovery proposal that was not eventually selected.

series. In late 2009 three were each given $3.3 million to conduct a feasibility study. These were the MoonRise South Pole-Aitken basin sample-return, the OSIRIS near-Earth asteroid sample-return, now "upgraded" from the Discovery program candidate as OSIRIS-REx (Origins, Spectral Interpretation, Resource Identification, Security, Regolith Explorer), and the Venus SAGE (Surface and Atmosphere Geochemical Explorer). Only one would be selected for implementation.

The SAGE mission had been under study at JPL since the early 2000s and was a joint proposal with the University of Colorado. It would be launched by an Atlas V or Delta IV in December 2016, reach Venus in May 2017, and study the nature of the surface and the climate and history of the atmosphere. A flyby bus would release the probe. The Pioneer Venus-class capsule would release a pressurized spherical lander capable of operating for 1 hour during the descent and then at least 3 hours on the surface. The spherical module would have a crushable cylindrical 'skirt' shock absorber, and five outriggers both to cushion the impact and to right the probe. The first part of the atmospheric descent would be made using a parachute. Like the old Soviet Veneras, this would be released at high altitude and the lander would fall freely to the surface, being slowed and stabilized by a drag plate installed just above the spherical pressure compartment. Imaging the surface would begin at an altitude of about 15 km, as the capsule broke through the lower haze layers. Touchdown would occur at 10 m/s. The target was on the edge of the volcano Mielikki Mons, which is over 300 km wide and 1,500 meters tall and is believed on the basis of Venus Express observations to be the site of a recent eruption. The atmospheric descent profile would not be the only thing in common between SAGE (the first US atmospheric probe for that planet since the 1970s) and the Soviet Venera landers, because lithium nitrate salt would serve as a thermal buffer to the titanium pressure vessel carrying the electronics and avionics by absorbing large quantities of heat as it changed from solid to liquid phase. During the descent the probe would measure pressures, temperatures, and wind speeds. Once on the surface, a high-temperature robotic arm and drill would scrape into the rocks to a depth of 10 centimeters to expose "unweathered" material. Rocks would be imaged by a microscope and fired upon by two tunable lasers, with the liberated gases being fed to gamma-ray and neutral mass spectrometers. Four cameras would take pictures during the descent as well as on the ground. As a direct link with Earth would require too much power, the data would be relayed via the carrier spacecraft, derived from Mars Reconnaissance Orbiter and the twin lunar GRAIL orbiters. It would store and retransmit the data to Earth. It would also be equipped with a Russian ultraviolet and near-infrared camera to take contextual images of the entry site whilst performing its flyby.[12,13,14,15]

In May 2011, however, it was OSIRIS-REx that was selected. This joint project of the University of Arizona and NASA's Goddard Space Flight Center was based on Hera, a Discovery sample-return proposal from the late 1990s.[16] It is to be launched in September 2016 on an Atlas V, fly by Earth in September 2018, arrive at its target asteroid in October or November 2019, collect a sample, depart in March 2021 and then return to Earth in September 2023. The target is the near-Earth object (101955) Bennu, also known by its preliminary designation 1999 RQ36, a very dark, primitive carbonaceous body with a diameter of about 550 meters that has been cataloged as a potentially hazardous object owing to its relatively high chance (about 1-in-1,800) of hitting Earth in the second half of the 22nd century. It was imaged using radar shortly after discovery by the LINEAR survey in September 1999 and, surprisingly, found to be more or less spheroidal and featureless, completing one rotation in just 4.3 hours. More observations were made by the Herschel infrared space telescope and by one of the 8.2-meter elements of the European Very Large Telescope (VLT) and revealed the extensive presence of a fine regolith on the surface as well as a relative low average density. Variations in the surface roughness might imply that regions will be found on the surface that expose very fresh material. The thermal inertia resembled that of the asteroid Itokawa studied by the Hayabusa mission, hinting that it too may be a rubble pile.[17] The asteroid had not yet been named when the mission was selected but, as in the case of Deep Space 1's target, asteroid (9969) Braille, a competition was held by NASA and the Planetary Society and the winning proposal, Bennu, is a reference to a sacred Egyptian heron and one of the symbols of the god Osiris.

Unlike the Discovery OSIRIS proposal, which focused on sample-return only, the $800 million New Frontiers OSIRIS-REx mission will accompany its target for 15 months and be equipped to characterize it in detail. In particular, it should provide a better determination of the orbit, thereby directly measuring the Yarkovsky effect and other small perturbations. This data will be of value to the possible future deflection of dangerous objects, as well as for planning possible human missions to small near-Earth objects.

The 1,530-kg spacecraft will be based on the 2-meter-wide basic structural bus of Mars Reconnaissance Orbiter, and will employ solar arrays which have an active area of 8.5 square meters. Four 200-N thrusters will provide trajectory control during the entire mission. It will carry three cameras: a 20-centimeter telescope will be used to acquire images of the target at long range as well as 1-meter-resolution views at close range; a four-color mapping camera will search for satellites and signs of activity, and take high-resolution imagery of the sampled site; and a downward-looking sampling camera will document every phase of the acquisition at millimeter-scale resolution. A student-provided X-ray imaging spectrometer will provide not just spectra but also composition maps of the surface in sixteen different bands in order to identify, among other things, the most interesting sites to sample. A visible and infrared spectrometer will produce spectral maps with 20-meter resolution, as well as high-resolution maps of candidate sampling sites. A thermal emission spectrometer and radiometer similar to that carried by Mars Odyssey will identify the surface composition and measure the thermal budget of the asteroid. A laser altimeter provided by the Canadian Space Agency will map the entire surface of the asteroid from distances closer than 7.5 km. Learning a lesson from the Hayabusa mission, OSIRIS-REx will not collect samples immediately upon arrival, but will characterize the asteroid in detail, map it in order to select the best landing sites, and validate a sound sampling strategy by rehearsals. After several sites have been selected, it will descend and carry out its sampling runs. Instead of the target markers and stroboscopic lights used by Hayabusa, OSIRIS-REx will use the laser altimeter to control its horizontal speed by assembling a synthetic image of the terrain. It won't actually land on the asteroid to collect samples but will approach at a rate of about 10 centimeters per second, taking a picture every second, and extend a robotic arm. At the tip of the arm will be a collection disk similar to a car's air filter. The disk will fire an annular jet of ultra-pure nitrogen onto the surface in order to stir up regolith, dust, dirt and gravel. Gases will pass through the filter and escape to space but the regolith material will be retained. The spacecraft will rebound off the surface using springs on the robotic arm and move away by firing its thrusters. It will have sufficient maneuvering propellant for three sampling attempts. Tests in microgravity indicate that the system should be able to capture at least 60 grams on each run, and the mission will hopefully return a total of 2 kg to Earth. After finishing all the sampling runs, the collection disk will be stored inside a sample-return capsule derived, like the robotic arm, from the Stardust mission.

At the end of the mission, the sample capsule will land on Earth in the same Army range in Utah used by Genesis and Stardust. The spacecraft will remain in solar orbit to conduct an extended mission if NASA can fund one. The Johnson Space Center in Houston, where most of the lunar rocks returned by the Apollo missions are stored, as well as samples recovered by Stardust and Genesis, and Martian meteorites etc., will store, document, and distribute the samples to researchers. Primitive material from carbonaceous asteroids is not well represented in meteorites because it is altered by heat during atmospheric entry, even in those rare cases where it survives to reach the ground. Samples will be allocated to the science team starting six months after return, once they have been cataloged. However, the largest portion (perhaps three-quarters) will be archived for future analyses.[18] On the topic of asteroid sample-return missions, it is worth noting that a different, inexpensive kind of sample-return was accomplished in Sudan in October 2008 when pieces of a small object only meters across, discovered hours before by telescopes, were recovered from the desert. This was the first time an object in space was established to be on a collision course with Earth so that its point of impact could be calculated and a team sent to examine the site.[19]

To provide additional scope for OSIRIS-REx, a proposal has been put forward by JPL to fly ISIS (Impactor for Surface and Interior Science) as a secondary payload of NASA's InSight Mars lander. Employing a heavy launcher like the Atlas V for that mission would provide about 1,000 kg of spare payload mass. ISIS could be mounted on the standard adapter that supports InSight on the launcher without significantly influencing the design of the primary payload. This would add a propulsion module, an avionics module, six small solar panels and four ballast masses to the adapter and make it a fully independent, 420-kg dry-mass spacecraft. After flying by Mars for the second time in January 2019, ISIS would reach Bennu in March 2021 (after OSIRIS-REx has completed its primary mission) and strike it at about 14.9 km/s. The energy of the impact would be comparable to 14 tonnes of TNT. OSIRIS-REx should be able to observe the formation of a crater several tens of meters across from a safe distance and monitor the seismic effects of such a collision. Furthermore, in as little as two or three weeks it would be able to precisely measure the velocity change imparted to the asteroid. ISIS was estimated to cost

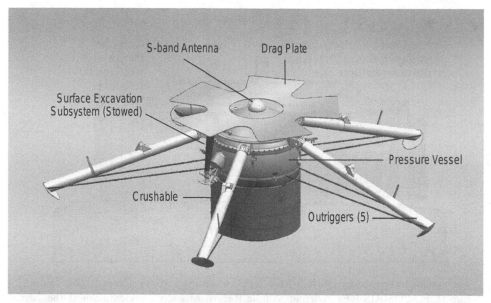

The SAGE New Frontiers finalist would have been the first US Venus lander, if selected.

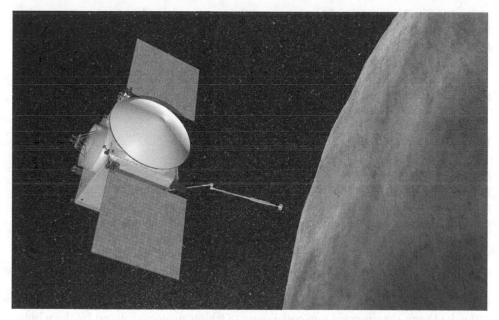

A depiction of the OSIRIS-REx spacecraft orbiting the small, near-Earth asteroid Bennu, with its sampler arm deployed.

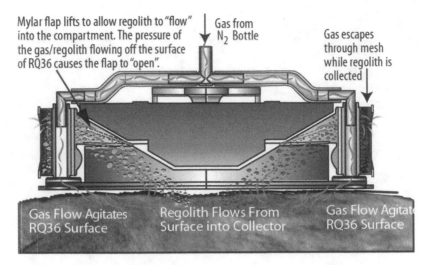

The system to be used by OSIRIS-REx to collect samples from asteroid Bennu. (From Lim, L.F., "OSIRIS-REx Asteroid Sample Return Mission", presentation at the 7th Small Bodies Assessment Group (SBAG) meeting, Pasadena, July 2012)

about $100 million and would have needed to be approved in late 2013 to be ready to launch with InSight. As the present budgetary difficulties of NASA prevented this, ISIS may eventually fly as a secondary payload on a telecommunications satellite launch in 2017 or 2018.[20,21]

After OSIRIS-REx, the announcement of opportunity for the fourth mission of the New Frontiers program is expected to be issued by NASA in 2016.

EUROPA OR TITAN?

Around the mid-2000s NASA and ESA started to study, independently of each other, future missions to the solar system giant planets, Jupiter and Saturn and their moons, to follow up the surveys of those planetary systems by the Galileo and Cassini missions and the brief but highly successful Huygens mission to Titan.

Ever since the Galileo spacecraft's observations of the Jovian satellite Europa and the likely discovery of a liquid water ocean beneath its surface, scientists yearned to devote a mission entirely to the icy moon, which had been described by some as "a cryological wonderland". In fact, based on its geology and chemistry Europa may be the most likely place to find life elsewhere in the solar system. The 1990s and early 2000s had seen the canceled low-cost Europa Orbiter project replaced by the wholly unrealistic Jupiter Icy Moon Orbiter (JIMO).[22] The primary objective of a mission to Europa would be to use a ground-penetrating radar to verify the presence of the ocean and measure its extent. A recent study has speculated that liquid water may be present only a few kilometers beneath the peculiar chaotic terrains such as Thera and

Thrace Macula, depressions many hundreds of meters deep. Pockets of lower-volume liquid water would cause the surface above to break, collapse, and create these structures. In turn, these pockets of water could be formed over hot spots such as volcanic vents or hydrothermal sources on the ocean floor. A terrestrial analog is Grimsvotn, a volcano in Iceland. It is covered by an ice cap, the base of which melts during an eruption and causes the surface to collapse. As the water pocket on Europa refroze and expanded, it would cause the chaotic terrain to bulge up and create a different type of chaotic terrain, like the intriguing region of Conamara. It is also possible that active hot spots would intermittently cause surface features to shift and undergo changes over very short timescales, perhaps so brief that it would be possible to detect differences since the Galileo mapping.[23],[24]

A radar orbiting Europa would also supply insight into the mechanism by which craters are obliterated, possibly discovering buried impact structures. At present, only 24 craters larger than 10 km are visible on the surface. A probe in orbit would provide other valuable scientific data. Doppler tracking could resolve ocean floor features and structures like ridges that would indicate an active interior. Moreover, mapping of the gravitational field would determine the amplitude of the tidal bulge on Europa, which in turn would reveal whether the thin icy crust is entirely disconnected from the rocky core via a global ocean. A decoupled surface would create a tidal bulge tens of times larger than one connected to the inner structure. Spectrometers would seek to determine the composition of non-ice materials on the surface, specifically any salts that might once have been dissolved in the ocean. This could imply that the easiest way to sample the ocean would be to sample an appropriate portion of the surface. The Galileo infrared spectrometer unfortunately did not have the spectral resolution necessary to detect the signature of salts, but recent ground-based observations of the trailing hemisphere of Europa revealed the presence of magnesium sulfate that might represent magnesium chlorate from the ocean which, on reaching the surface, reacted with a sulfur 'paint' originating from Io's volcanoes. Sodium and potassium chlorates are also believed to be present on the surface. Sulfuric acid is present in quantity, having been detected by Galileo. This is the result of the reaction between water ice and sulfur ions. Hydrogen peroxide was detected in abundance. It would be made by the irradiation of surface water ice. Interestingly, when hydrogen peroxide is dissolved in water it can liberate free oxygen, a critical molecule for the rise of complex forms of life.[25] The prospect of an ocean on Europa was boosted by Hubble Space Telescope observations of an Enceladus-like water vapor plume near the south pole. Signs of activity such as this were sought by Galileo but not found, indicating that they may be intermittent. Such variability has also been reported at Enceladus, with the plumes being more active at apoapsis than at periapsis. Unfortunately, the origin of Europa's plume appears to be in an area that was poorly imaged by previous missions and so there is no insight into its source. However, the very fact that Europa emits plumes implies that a probe that passes through one should be able to sample the composition of the ocean.[26]

Another Jovian moon to attract the interest of scientists was Ganymede, the largest satellite in the solar system and the only one to have an intrinsic magnetic field. From Galileo's gravity data, Ganymede appeared to be a strongly differentiated body with an 800-km-thick icy shell, a rocky mantle, and possibly also an iron core that would generate the intrinsic magnetic field. The ice layer was possibly a 100-km-thick shell of crystalline ice overlying an ocean of liquid water, followed by additional layers of crystalline and amorphous ice. Moreover, Galileo found Ganymede to be interacting in peculiar ways with the Jovian magnetosphere, including, where the magnetic field of the moon is open at the poles, possibly modifying the chemistry of the surface ice.

On the other hand, there was an urge to continue exploring Titan and the Saturnian system. After Cassini, our knowledge of Titan was comparable to that of Mars after Mariner 9 in 1972 in that we had a preliminary medium-resolution geological survey. After all, Cassini spent only a few days in total in the vicinity of the moon. Scientific questions unanswered by Cassini and Huygens include details of the prebiological (or perhaps protobiological) chemistry, the distribution and evolution of surface organics, and the atmospheric dynamics and meteorology. To address these questions it will be necessary to send landers, perhaps to roam the surface or to drift on the hydrocarbon lakes, and in particular either a balloon or a 'heavier than the air' vehicle to achieve a wide-area remote sensing capability.

In America, one of the first phases of mission planning involved most of the major players of planetary exploration, including JPL, APL, and the Goddard and Langley NASA centers, and assessed whether a scientifically worthy mission involving Titan and Enceladus could be flown within the cost cap of a New Frontiers-class mission. One such mission might be a Stardust-based Enceladus sample-return, flying by that moon and collecting icy particles sprayed by its geysers. However, the mission would be overly complex and, Enceladus being deep inside Saturn's gravity field, the flyby speed would be too fast for the samples to preserve much of their original chemistry. In the end, studies found that no significant results could be expected from a low- or moderate-cost mission.

The absence of viable low-cost architectures for either Europa or Enceladus set the stage for the 2007 NASA flagship study in which four groups were given 6 months to formulate possible missions to Titan, Enceladus, Europa and Ganymede.

The Titan studies focused on balloons and other aerial vehicles. Studies of "Titan-intensive" post-Cassini missions were carried out at a reduced pace during the 1980s and 1990s, while scientists and engineers were busy developing the latter. During the 1980s a reference mission was fashioned in the style of Pioneer Venus and envisaged an orbiter and as many as four entry probes which would release "buoyant stations" designed to fly at different altitudes. One large probe would deploy a powered blimp that would make long, controlled excursions and deliver surface packages.[27]

Exploration of Titan from the air was the focus of a workshop in 2001. Concepts discussed on that occasion included balloons, airships, helicopters, vertical take-off aircraft, and an "aerover". Building on studies for the "Mars Ball", the "aerover" was a rover with large inflatable tires which could also provide balloon-like buoyancy. A concept that Soviet engineers devised for the exploration of Venus was also revived. This would consist of two tethered balloons, one filled with a lightweight gas such as hydrogen or helium and the other with a condensable gas. As the balloon rose, the gas would condense in the cooler atmosphere and reduce the buoyancy. And then when it lost altitude, the gas would evaporate and restore the buoyancy. This would enable it to drift with the wind in a given altitude range. An alternative form of aerial vehicle could be a twin-hulled buoyant glider that employed the asymmetric displacement of masses to perform maneuvers in a similar manner to a well-proven robotic oceanic vehicle.[28] Proposals had been put forward since the 1970s to generate buoyancy by burning atmospheric methane with onboard oxygen. However, this strategy would be inefficient because the high molecular mass of oxygen would make the mass that had to be launched from Earth unacceptable. Furthermore, combustion products including carbon dioxide and water would immediately freeze in the cold Titanian atmosphere, clogging the exhausts. An airship could be delivered to Titan by a JIMO-like Saturn Prometheus spacecraft, which could in turn release a small amphibious tracked rover. In those years, preliminary work was also undertaken on Titan aerocapture missions. Some consideration was given to proper fixed-wing or rotary-wing airplanes. Flying at a few meters per second, an airplane would take about 6 months to travel from one pole to the other, providing access to any point on the surface thanks to the presence of strong zonal winds in longitude. And it could possess aerodynamics such that it could land at a gentle and survivable speed simply by stalling. As an alternative, a lander could make a several-hours-long flight at atmospheric entry suspended under a parafoil wing.[29] Unfortunately, flying a bona fide 'heavier than the air' aircraft on Titan did not seem a viable option because of the low power-to-weight ratio of RTGs, the only power source available at that time for a long-duration mission so far from the Sun. Better ratios would be provided by ASRGs (as envisaged for AVIATR) but this technology remains untested in flight.

As a result, airships were part of the Titan flagship study which was led by APL, with substantial input from JPL, NASA Langley, and private companies. Separate studies were carried out at NASA Goddard for an Enceladus mission. Three possible mission architectures were studied: an orbiter and lander, an Enceladus orbiter alone, and a Saturn orbiter with Enceladus lander. Studies were soon stopped when it was established that getting into orbit around an object so small and so deep into Saturn's gravity field, or landing on it, would be extremely difficult using present-day technology.

The APL study converged on a Titan Explorer orbiter, lander, and balloon mission carried in separate entry or aerocapture vehicles on a single cruise stage that would be launched by a single Atlas V-class rocket. It was argued that this modular approach would allow elimination of one component during development with minimal impact on the others.

The orbiter would perform a mission lasting 4 years, as compared to 1 year for the lander and balloon. It would arrive at Titan inside a heat shield and deeply penetrate the atmosphere on arrival to achieve orbit by aerocapture, eliminating the need for a retrorocket. Aerocapture would be easier and more efficient at Titan than at any other solar system body, but it would not be possible to perform any cruise or approach observations since the instruments would be inside the aerodynamic shell until after orbit insertion. Furthermore, no new data could be collected in close to Enceladus. In order to fit inside the heat shield, the orbiter would have a squat body to which were attached an articulated high-gain antenna and five Stirling generators. The adoption of aerocapture would allow the 1,800-kg spacecraft to carry a 170-kg payload all the way to Titan. As a result of the aerocapture maneuver, the spacecraft would first enter a 1,100 × 1,700-km orbit around Titan at an inclination of 85 degrees. A short time later, this would be circularized at 1,700 km. The orbit would be perturbed by Saturn throughout the mission, thereby allowing a number of different geometries for remote sensing and magnetospheric observations. Near the end of the mission, the periapsis would be lowered over the north pole in order to sample the atmosphere using a mass spectrometer. The observations would include remote sensing by cameras, ultraviolet, microwave and infrared spectrometers, and in-situ particles and fields measurements. Unfortunately, as in the case of Cassini, very high resolution imaging at Titan would be impossible due to the hazy atmosphere and the need to remain above an altitude of 1,000 km to minimize friction with the atmosphere. This constraint would also make measurements of the finer components of the gravity field impossible.

The lander and balloon would be released on the approach to Titan and separately enter the atmosphere. Including heat shields, they were allotted masses of 900 kg and 600 kg respectively. The former would be a Mars Pathfinder-like airbag-cushioned lander, although airbags were not strictly needed since the thick Titanian atmosphere would end with an impact at less than 4 m/s. Solar panels would be impractical due to the distance of Titan from the Sun and also the screening of the thick atmosphere. A short-lived, battery-powered lander was ruled out early on in the study as being of insufficient scientific value. It would therefore be powered by a Stirling generator. It was envisaged the lander would carry meteorology instruments and magnetometers, deploy seismometers on the surface, and release an aircraft or a Beagle 2-like 'mole'. A 1-kg vertical launch battery-powered aircraft carrying meteorology instruments and cameras could fly for several hours and provide high-resolution imaging of the area with a radius of several kilometers from the point of landing.[30] The provisional target was Belet, an equatorial region rich in sand dunes that were attractive for a study of the organic chemistry of the surface. Meanwhile the balloon would study atmospheric chemistry and dynamics, make meteorological observations, image the surface below at high resolution and make radar subsurface soundings. Its payload would also have a chemistry package considerably more advanced than Huygens' gas chromatograph and mass spectrometer, and a magnetometer both to locate magnetic anomalies and to confirm the presence of a liquid water ocean under the surface. It would consist of a Montgolfière hot-air balloon flying at 10 km altitude, drifting with the prevailing wind, dropping atmospheric profiling probes en route. Its latitude would probably vary as it was influenced by less predictable near-surface winds. Furthermore, because it would suffer none of the stresses of terrestrial balloons (frequent heating cycles, ultraviolet degradation of the inflated fabric, an aggressively oxidizing atmosphere and so on) it could easily drift for years. On a nominal mission of 1 year it would circumnavigate Titan once or twice. A hot-air balloon was preferred to a gas balloon because it would be particularly well suited to Titan owing to the very low atmospheric temperature, it could use air heated by the radioisotope generator for buoyancy, and it would be less susceptible to small leaks and tears in the envelope. The Stirling generator would be mounted at the neck of the balloon to deliver hot air into the envelope. It remained to be demonstrated that the envelope was capable of inflating in midair following entry. Data from the lander and balloon could be returned either directly to Earth by a small high-gain antenna at a very slow data rate or through the orbital relay at much greater speed.[31],[32],[33]

In parallel with American studies, a Titan and Enceladus Mission (TANDEM) was proposed in 2007 by a large European team led by scientists of the Paris Observatory in response to the ESA "Cosmic Vision" 2015–2025 call for ideas. It was intended in 2011 to select a cornerstone mission at 650 million euros and a medium mission at 300 million euros for launch in the late 2010s or early 2020s. No fewer than fifty proposals were received. The eight that were shortlisted included TANDEM and the Laplace Europa and Jupiter mission.

TANDEM called for two medium-sized spacecraft, one to enter orbit of Saturn, fly by Enceladus several times and then be captured in orbit around Titan, and the other to deliver a hot-air balloon and several landers to Titan. But it was judged to have a lower level of technical maturity than the US Titan Explorer, for which engineering studies had been underway for years.

Building on the results of Titan studies by NASA and ESA, in 2008 the American agency directed JPL to study a joint Titan and Saturn System Mission (TSSM) targeting not only Titan but also Enceladus, which the TANDEM study had shown would be "explorable" without interfering with the Titan exploration objectives.

The Titan and Saturn System Mission would consist of a Saturn orbiter and a Titan balloon and lake lander.

The Titan and Saturn System Mission orbiter.

Conventional technologies rather than aerocapture would be used to enter orbit around Saturn. The long-duration balloon was retained, and the dune lander was replaced by a probe that would splash down in one of the hydrocarbon lakes near the north pole.

The 6,200-kg spacecraft would launch in September 2020 on an Atlas V. It would have a solar propulsion stage which had solar panels to provide no less than 15 kW of power and a trio of NASA Evolutionary Xenon Thrusters (NEXT) developed by the Glenn center. They would provide a thrust of 0.236 N and have a total supply of 450 kg of fuel. Five years into the mission, and after three Earth flybys and one at Venus, the propulsion stage would be jettisoned. The mission was to reach Saturn in October 2029. The orbiter was to have a dry mass in excess of 1,600 kg, including the 165-kg payload of ten instruments. It would be powered by five Stirling generators capable of providing at least 540 W at the end of the mission. Communications would be by a gimbaled 4-meter high-gain antenna allowing real-time acquisition and downlink of scientific data. Insertion into orbit around Saturn would be achieved using a chemical engine similar to that of Cassini. The planned 2-year orbital tour would provide seven encounters with Enceladus, directly sampling the southern plumes, and no fewer than sixteen Titan flybys. By that time, the vehicle would have slowed sufficiently for it to use chemical propulsion to achieve Titan orbit. It would initially fly an elliptical orbit with a periapsis as low as 600 km in order to directly sample the complex molecules produced by photochemical reactions at altitudes much lower than any Cassini flyby. Aerobraking over a period of 2 months would lower the apoapsis and circularize the orbit at 1,500 km with an inclination of 85 degrees. The orbital reconnaissance phase of the mission would last an additional 20 months, with specific campaigns dedicated to mapping, to atmospheric dynamics and composition, and to ionospheric studies. At the end of the mission the whole of Titan would have been mapped by radar at 100-meter resolution, in addition to 1-meter-resolution coverage of the areas overflown by the balloon.

ESA would provide the balloon and possibly also the floating lake probe. The 600-kg Montgolfière would be released by the orbiter while approaching Titan on the first revolution. It would enter the atmosphere at a latitude of about 20°N, inside a heat shield which could be instrumented in order to undertake an atmospheric and surface mission similar to that of Huygens. During the descent, air heated by a US-provided Multi-Mission RTG similar to that of the Curiosity rover would fill the envelope to achieve buoyancy. The 10.5-meter-diameter balloon would then ascend to its cruising altitude of 10 km to start a mission of at least 6 months during which it would fly over 10,000 km, circumnavigating Titan at least once. A valve on top of the canopy would allow it to hold a more or less fixed altitude. The 144-kg balloon nacelle would carry 25 kg of instruments, including cameras able to return meter-resolution images, spectrometers, magnetometers, and a ground-penetrating radar sounder to locate deposits of methane beneath the surface. It could also carry several surface probes which it would release above selected sites. Precise tracking would enable reconstruction of the wind fields. The data from the balloon would mostly be transmitted to Earth via the orbiter. The 190-kg lake probe would be released prior to the second encounter with Titan and be targeted at Kraken Mare. As did Huygens, it was to report on the atmosphere while descending. The primary scientific objective was to determine the composition of the liquid in the lake but other objectives would be addressed by a five-instrument suite, including surface imaging. Batteries would ensure a total duration of at least 9 hours. Given the requirement to protect the surface of Titan, and in particular its lakes, every effort would be made to prevent contamination by terrestrial biology. In view of its scope and complexity, the Titan and Saturn System Mission was baselined at around $3.7 billion.[34]

The mission to Europa had a similar development with independent studies being undertaken in Europe and in the US. In the mid-2000s ESA started to contemplate a mission to Jupiter as one of its Technology Reference Study concepts. The original scenario involved two "minisatellites" with dry masses of no more than 200 kg which would set off on a single Soyuz-Fregat and use Venus and Earth flybys to reach Jupiter. The first spacecraft was the Jovian Europa Orbiter that would perform a tour of the Jovian moons prior to entering orbit around Europa. The second spacecraft was the Jupiter Relay Satellite that would enter a high orbit in order to avoid most of the radiation belts. It would study particles and fields, make remote observations of the moons, and serve as a relay for its partner. Since the mission in orbit around Europa would last no more than 2 months before either perturbations caused the spacecraft to crash onto the moon or the intense radiation damaged its electronics, its copious data would have to be stored by the relay orbiter and transmitted to Earth over more than a year. Both spacecraft would have Rosetta-heritage solar cells augmented with mirror 'concentrators' for greater output. The Europa orbiter would carry a highly integrated sensor suite, the most important item being a subsurface radar. ESA also investigated the possibility of mounting a miniaturized hard lander with a seismometer to detect any cracking of the icy crust and acoustic waves at the boundary between the ice and the liquid water beneath, but the need to stabilize the lander and provide it with at least a basic ability to brake its fall would have made it far too heavy. A second phase of the study addressed additional scientific objectives of a multiple-spacecraft particles and fields mission and multiple Jovian probes to penetrate the planet's atmosphere to the depth where the pressure was 100,000 hPa, or 100 atmospheres, as compared to the 23,000 hPa zone that was reached by the Galileo entry probe.[35,36]

In parallel with these concept studies, a proposal that rejected the small spacecraft architecture was submitted when ESA issued a call for missions in 2007. This was dubbed Laplace after the French scientist Pierre-Simon de Laplace whose studies of celestial mechanics also covered the origin of the solar system. It ideally envisaged a coordinated three-spacecraft mission involving a dedicated Europa orbiter, a Jupiter planetary orbiter optimized for remote sensing, and an orbiter to study the planet's magnetosphere. If fuel reserves and mission design permitted it, the planetary orbiter would later enter orbit around Ganymede. One of the themes of the "Cosmic Vision" program was the origin of life in the solar system and the conditions conducive for its emergence, including the possibility of habitable moons around gas giants. For this, Laplace would single out Europa and Ganymede as examples of icy moons harboring subsurface oceans and, for Ganymede, a sizeable magnetic field as well. The mission was designed from the start as a cooperation between ESA and NASA.[37]

In the US, JPL had continued its studies of a Europa orbiter carrying a radar after the mission was canceled in 2002. Moreover, a Galileo-like Jupiter system mission had been studied in parallel. Talks were held between US and European scientists as a preliminary to performing definition studies and proposing missions to the space agencies. As occurred in the case of Titan, in 2008 the various studies were regrouped in a Europa and Jupiter System Mission (EJSM) that envisaged two spacecraft, each with dry masses of about 1,700 kg: namely a US Jupiter Europa Orbiter (JEO) and a European Jupiter Ganymede Orbiter (JGO). The orbiters would carry complementary payloads massing in excess of 100 kg each. The notional budget called for $2 billion to be spent by NASA and $1 billion by ESA, with the possibility of contributions by Russia and Japan – the former might provide a Europa lander and the latter a Jupiter magnetospheric orbiter.

The US Jupiter Europa Orbiter was to characterize the thin icy shell, obtain further evidence of a subsurface ocean and measure its extent, its relationship to the moon's deeper interior, its response to the planet's magnetosphere and gravitational tides, and characterize the non-ice materials on the surface and the processes by which ridges and fissures form on the surface. The vehicle would be built by JPL and powered by Multi-Mission RTGs like that of the Curiosity rover, providing 540 W at the end of mission. Communications would use a 4-meter high-gain antenna in order to ensure a rapid download of the enormous amount of data generated by the radar. The notional payload included three cameras with different fields of view and spatial resolutions to image selected sites at a resolution better than 1 meter per pixel and stereoscopically at 10-meter resolution, as well as color coverage of at least 80 per cent of the surface at 100 meters. A visible and infrared spectrometer, an ultraviolet spectrometer, and a thermal mapper would form the spectrometric portion of the remote sensing payload and investigate in particular the composition of the non-ice material. A laser altimeter would potentially revolutionize knowledge of Europan topography in much the same way that a similar instrument on Mars Global Surveyor did for Mars, in particular measuring the tidal bulge caused by gravitational tides. As with other proposals for orbiters, the ice-penetrating radar was the most important instrument. In this case it would profile the electrical characteristics of the subsurface with a vertical resolution of 10 meters or better in the topmost several kilometers and 100 meters down to a depth of 30 km. The payload would be completed with magnetometers, mass spectrometers, and instruments for particles and fields investigations. Finally, a radio science experiment would make use of precise tracking data to further characterize the deep interior of the moon.

As ESA had never developed its own equivalent of an RTG, the European Jupiter Ganymede Orbiter would be solar powered, with large panels providing between 600 and 700 W at Jupiter. It would investigate the internal structure of the largest Jovian moon, specifically its deep ocean and magnetic field. To address these objectives, the payload would resemble that of the Europa orbiter, with two cameras, a visible and infrared imaging spectrometer, an ultraviolet spectrometer, a microwave radiometer, a laser altimeter, an ice-penetrating radar, a magnetometer, a mass spectrometer, and a radio and plasma-wave sensor. A 3.2-meter high-gain antenna would return at least 1 gigabit of data on a daily basis. And of course the tracking data would facilitate the radio science experiment.

The addition of a Europa lander was a possibility. Two of the problems that such a lander would face were the absence of an atmosphere to slow its fall and the potential for the engine exhaust to contaminate the surface, possibly invalidating biological results. To obviate these issues, a small Deep Impact-like projectile was proposed so that the cloud of debris that this created could be scanned for organics by the spacecraft and by ground-based and near-Earth telescopes. However, unlike Deep Impact the debris cloud would be short-lived owing to the much greater gravity of Europa, as against a small cometary nucleus, requiring the analyses to be conducted in a matter of seconds or minutes instead of hours.[38]

In addition, a Japanese miniature magnetospheric orbiter could fly either attached to the European orbiter or reach Jupiter as part of JAXA's planned Jupiter and Trojan asteroid solar sail demonstration mission. What is more, the Lavochkin Association and IKI in Moscow were willing to participate by using either a Europa penetrator or the Fobos-Grunt-derived Laplas-P (Laplace-P for "pasadka", landing) that would consist of an electric propulsion module, an orbit insertion stage, an orbiter, and a 1,210-kg soft lander for Europa that would be similar to the E-8 lunar landers of the 1970s.

The US probe would be launched first, in 2020, using an Atlas V or Delta IV-class rocket and use Venus and Earth flybys to reach Jupiter in 6 years. After making flybys of all four Galilean satellites (three times with Io) it would enter orbit around Europa in 2028. The reconnaissance of Europa in a high inclination orbit only a few hundred kilometers above the surface would last only a couple of months.

The European orbiter would be launched one month later using the ECA heavy-lift version of the Ariane 5. On arrival in 2026 it would spend 2 years touring the Jovian system involving no fewer than nine flybys by Callisto, prior to entering orbit around Ganymede in late 2028. After observing the moon from high-, mid-, and low-altitude orbits, the latter as low as 200 km, it would end its primary mission in July 2029.

For both missions, there were options for performing asteroid flybys on the way to Jupiter, and some time would be dedicated to distant observations of one or more of the outer irregular satellites of Jupiter. The Jovian ring and the nearby small satellites would be targeted as well.[39]

This was the first time that a competition occurred between two flagship missions, one to Saturn and Titan and the other to Jupiter, Europa and Ganymede. In its report the 2002 decadal survey had prioritized missions to the outer solar system, recommending a Jupiter-interior mission (flown as Juno) and the Europa orbiter. But since then the data from Cassini had raised Titan and Enceladus to the same priority of a Europa orbiter.

As previously, it was NASA that made the selection, with ESA destined to follow. Owing to the fact that Europa orbiter missions had been under development since the late 1990s and the requisite technology was more mature, NASA opted for the Jovian mission.[40] Whether the agency could afford to start a new flagship project (especially such an expensive one) within its shrinking budget remained to be seen. The announcement of opportunity to select the instruments was expected at the end of 2010. The European contribution would have to be confirmed in 2011 by ESA selecting a large scientific mission against other valid candidates.

With New Horizons en route to Pluto and the Kuiper Belt and Voyager sampling the termination shock of the heliosphere in progress, there are only a few regions of the solar system remaining to be explored. These include the environment within 0.3 AU of the Sun, the corona, and the region where the solar wind is accelerated. Thus far, these regions have only been "remotely sensed" by Earth satellites and missions either orbiting the Sun or at the Lagrangian points of the Earth-Sun system.

A number of missions to explore this zone have been proposed in the past, starting at least from 1958, and including NASA's and ESA's "Sun-diving" missions of the 1980s and 1990s described in previous parts of this series.[41] A Sundiver mission has even been proposed by Australian scientists as the foundation of that nation's decadal plan for space sciences.[42] More recently, a solar-polar mission was proposed to both NASA and ESA that would fly within 0.2 AU of the Sun and continue the work of the Ulysses mission. This project was dubbed Telemachus after Ulysses' faithful son in Greek mythology. After a Venus flyby, additional flybys of

The US orbiter part of the joint Europa and Jupiter System Mission with ESA.

Earth and Jupiter, and a number of propulsive maneuvers, Telemachus would enter a 0.2 × 2.5-AU orbit that was inclined perpendicular to the Sun's equator in order to pass over both solar poles at distances less than 0.4 AU every 1.5 years. The solar-powered spacecraft would be equipped to image the Sun's photosphere (its 'visible surface') and inner corona in order to study in detail the flows of matter and magnetic fields at the poles, active regions, stellar winds, flares, mass ejections, etc.[43]

A low-perihelion Solar Probe was again identified as one of the highest priorities of the 2003 solar physics decadal survey, prompting NASA to undertake a feasibility study during 2004 and 2005. The mission was determined to be technically feasible, although a number of modifications were recommended. For more than 20 years the Solar Probe was to be powered by RTGs and was to be launched first toward Jupiter so that the gravity of the giant planet would rob the probe of most of its energy and cause it to 'fall' towards the Sun in an orbit with its perihelion at several solar radii. Moreover, scientists had always indicated a preference for a polar orbit in order to fly over the poles at a very close range to investigate the corona over as large a range of latitudes as possible. But this mission architecture was deemed too expensive and a study was requested by NASA to determine whether a simpler mission design could achieve the core science goals at a reasonable cost.

Exactly 50 years after it was first proposed, NASA finally gave the go-ahead to the Solar Probe Plus (or "Solar Probe+") mission in 2008 under the auspices of "Living With a Star", a program of the agency's heliophysics division. The mission would be developed by APL and managed by the Goddard center. The single Jupiter flyby was replaced by multiple encounters with Venus, the minimum perihelion distance was increased to almost 10 solar radii, the polar orbit was substituted by one that would remain within a handful of degrees of the ecliptic plane, and the RTGs were replaced by solar panels. This orbit would rule out some observations but it would yield more frequent data because the probe would return to perihelion every 3 months instead of every 5 years in the case of the Jupiter-flyby architecture. Moreover, some of the lost observations could be recovered by the contemporaneous European high-inclination Solar Orbiter.

The main scientific objective of the Solar Probe Plus mission is to understand the role of the solar magnetic field in heating the corona to millions of degrees whilst the photosphere is just 5,800 K, and to understand how the solar wind is accelerated out of it. To achieve this, the mission will make in-situ observations of the structure and dynamics of the solar magnetic field, measure the density, velocity, and temperature of electrons, protons and alpha-particles, sample both the slow and fast solar portions of the wind as well as coronal mass ejections, make visible-light observations to map the morphology and density of structures in the inner solar wind, and determine the intensity and spectrum of energetic electrons, protons, and heavy ions in the regions where these particles are being accelerated. The importance of being able to observe the fine structures in the corona has been recently proved by the flight of an extreme-ultraviolet telescope mounted on a sounding rocket. This resolved details as small as 150 km on the photosphere, seeing bundles of magnetic field lines that were wrapped around each other. The reconnection and dissipation of these 'braids' is deemed the most likely mechanism for heating the corona.[44,45]

Solar Probe Plus is to be launched during a window that extends from 31 July to 18 August 2018 using an Atlas V 551 of the same configuration as for New Horizons but utilizing a more powerful solid-fueled third stage developed for the mission. The flyby of Venus 2 months into the mission will produce a perihelion at 0.16 AU (i.e. about 35 solar radii) a month later. Six further Venus flybys will gradually reduce this range. The mission design envisages a total of 24 perihelia within 35 solar radii, no fewer than 19 within 20 solar radii, and finally, beginning in December 2024, three within 10 solar radii (7 million km). It will collect data mainly while it is within 0.25 AU of the Sun, and will spend a total of at least 25 hours within 10 Rs and 950 hours within 20 Rs. The 665-kg vehicle will draw upon hardware developed for other APL missions, including the high-temperature solar panels of the MESSENGER Mercury orbiter. The roughly conical probe will be 2.9 meters high and span 2.3 meters across its forward thermal shield. As the most critical component, the shield must withstand an insolation 500 times that at 1 AU and temperatures reaching almost to 2,000°C on its sunward side. The 12-centimeter-thick shield consists of a core of carbon foam, a sandwich of carbon sheets, and an aluminum coating on the sunward side. The shield is connected using a 'thermal choke' to almost eliminate the transfer of heat to the body. With only a few exceptions, most of the vehicle will hide in the shadow cone of the shield at perihelion. In addition to a couple of instruments, these exceptions will include two small solar arrays mounted at the tip of the primary array to collect solar light at perihelion and supply 377 W of power. Hence the solar panels will consist of a primary and a secondary section. A mechanism will vary the angle of the primary during an orbit, ranging from nearly perpendicular to the Sun at aphelion to almost parallel at perihelion, at which time only the secondary will be illuminated. The latter will use solar cells optimized to operate at very high temperatures with an active, liquid-circulating cooling system, as well as large radiators mounted between the struts that separate the heat shield from the spacecraft bus to radiate excess heat. It was planned to have two separate strings of deployable solar panels, one for use at large distances from the Sun and the other for perihelion passes, with associated mechanisms, but in the end it was decided to adopt a simpler architecture. Attitude control of the 3-axis-stabilized spacecraft will be particularly critical, because to avoid overheating at perihelion it must be pointed within 2 degrees of tolerance. The attitude will be controlled by four reaction wheels augmented by a dozen hydrazine thrusters, the latter also being used for trajectory control. Solar limb and other sensors will warn of an angle violation. In the event of a major malfunction, the safe mode will automatically adopt an attitude that prevents the spacecraft from being exposed to the Sun.

The payload for the spacecraft was selected in early September 2010 and includes a detector for solar wind electrons, protons and helium ions, and a mass spectrometer and particle and ion monitor that will inventory the chemical elements in the corona. An electromagnetic fields, radio, and shock wave package will use three niobium antennas that extend several meters beyond the thermal shield. This instrument will double up as a dust detector by recording the clouds of plasma created by dust specks striking and vaporizing on the spacecraft as it zooms by the Sun at a speed of almost 200 km/s. The final instrument will be a wide-field telescope

The Solar Probe Plus leaving Earth bound for Venus and several perihelia at a small heliocentric distance.

that looks to the side of the heat shield and on the forward-facing, ram side of the spacecraft to provide scans of the Sun's atmosphere and images of the corona and its structures as the spacecraft passes through them. A 60-centimeter high-gain antenna and other antennas mounted at fixed positions in the shadow of the heat shield will transmit up to 128 gigabits of data during each perihelion.

The Solar Probe Plus mission is currently being developed for an estimated cost of about $1.4 billion including launch and operations. The preliminary design review is scheduled for the second quarter of 2014. Hopefully it will then receive the final go-ahead.[46,47,48]

The 2020s may also see another NASA-sponsored solar mission, the Solar Polar Imager. This is designed to fly a circular 4-month orbit 0.5 AU from the Sun at an inclination of 60 degrees to the ecliptic to conduct observations of the polar regions. It could be an ideal opportunity to use solar sails for orbit control.

LARGER AND LARGER

In January 2004, US President Bush launched the "Vision for Space Exploration" (VSE) meant to provide an alternative to the Space Shuttle as the nation's means of human access to the International Space Station and then, in the late 2010s, to return to the Moon fully half a century after the Apollo missions. This would then provide a stepping stone for an eventual mission to Mars. VSE gave rise to the Constellation program, for which NASA and its industrial partners started to develop medium and heavy rockets named Ares I and Ares V respectively. The former would insert the Apollo-like Orion capsule into Earth orbit, either for missions to the space station or to rendezvous with a lunar lander and propulsive stage launched by the Ares V, which would reuse technology and components originally developed for the Space Shuttle and have a lifting performance similar to the discontinued Saturn V of the Apollo era or the Soviet N-1 and Energiya rockets. Its first stage would be structurally based on the external tank of the Shuttle and its thrust would be augmented with Shuttle-style solid boosters. Ares I would use a lengthened Shuttle booster as its first stage, with an uprated Apollo-era engine to power its cryogenic upper stage. A prototype Ares I with an inert upper stage was launched in 2009, but developing the Ares V would have obliged the US to seriously commit to returning humans to the Moon and allocate the immense budget that this would require. After months of debate, the Obama Administration halted work on the Constellation program and supported commercial alternatives for transporting humans and cargo to the space station and other possible missions in low orbit.

But Congress, urged on by members from the states that would suffer the most job cuts in the wake of the cancellation of Constellation and the retirement of the Shuttle, ordered NASA to continue development of the Orion manned spacecraft as the Multi-Purpose Crew Vehicle (MPCV) and initiate parallel development of a Space Launch System (also dubbed the "Senate Launch System") as a replacement for the Ares V. The eventual task of this heavy rocket will be to place 130 tonnes of payload into low orbit to facilitate human missions into deep space. Considering the harsh budgetary realities facing NASA, it is debatable whether this rocket, whose requirements were effectively "designed" by politicians and presently lacks any approved missions, will ever fly. A lightweight version with a 70-tonne capacity should fly unmanned as early as 2017 if NASA receives the $18 billion required to complete both its development and that of the Orion spacecraft. Developing the 130-tonne version and assigning it a viable human objective in deep space would require many more years of comparable investment.

In 2007, while the Constellation program was still active, NASA, concerned with the indifference of scientists, asked the US National Research Council to carry out an assessment study of the scientific missions and opportunities that its heavy launchers might facilitate. A total of fourteen concepts were evaluated on topics which included astronomy, astrophysics, solar physics, and planetary sciences. The committee sought to identify both whether a concept promised significant scientific advancements and would benefit from the Constellation infrastructure. This would mainly consist of the Ares V heavy launcher. The Ares I was dismissed because even with a Centaur upper stage it would not offer a performance very different from that of existing launchers. Its only advantage would be its 'man rating', since the necessary redundancies would make it more reliable than equivalent unmanned rockets.[49] The Ares V was attractive because it had a huge payload fairing capable of accommodating payloads that were more than 8 meters wide and up to 860 cubic meters in volume. Several missions of interest to this book were identified by the study.

An interstellar probe would surely benefit from the use of the Ares V. One with an additional Centaur upper stage could boost a conventional-propulsion probe to such a speed that it could achieve 200 AU in just 23 years. Ares V could launch 10 tonnes to Uranus or Neptune, and much more to Jupiter or Mars, facilitating Neptune or Titan missions using conventional engines rather than aerocapture for orbit insertion. This was of significance because NASA still has no firm program to test the feasibility of the aerocapture technique.

Palmer Quest was devised at JPL in 2004 as a "vision mission" for nuclear power and received grants to increase its technological maturity.[50] It called for a Mars polar lander to deliver a nuclear-powered cryobot (derived from studies for the exploration of Europa) to melt its way to the base of the permanent ice cap in order to determine whether microbial life ever existed on the planet. Along the way, it would analyze the stratigraphy of the ice and seek organic molecules embedded in it at different epochs. The baseline mission could be launched by a Delta IV in the 2020s and consist of a lander with a small atmospheric station, a nuclear reactor and cryobot, and an RTG-powered rover with inflatable wheels. However, the mission would require too many technological advances and the Ares V would not offer any significant benefit.

Solar Polar Imager would travel in a 0.48-AU orbit inclined at 78 degrees to the ecliptic and observe the Sun using both remote sensing and in-situ particles and fields instruments based on heritage from the Solar and Heliospheric Observatory (SOHO), STEREO, Ulysses, and many others. The baseline mission envisaged launching on a Delta IV and using a solar sail to maneuver into the operating orbit, taking 7 years in the process. It could be greatly simplified by using the Ares V because it could use proven solar electric or even chemical propulsion instead of the solar sail. Alternative scenarios that would be enabled by use of the Ares V might include inserting two or more smaller spacecraft into equally spaced polar orbits for continuous monitoring of the Sun.

An Ares V-launched Solar Probe 2 mission would use Solar Probe Plus hardware to approach within 4 solar radii, essentially as envisaged for the 1980s Starprobe and for the 1990s Fire & Ice mission.[51,52] In comparison to Solar Probe Plus, this mission would require less time to attain its final orbit, with a perihelion closer to the Sun and a period of only 100 days.

At the same time, JPL carried out a parallel study on how a Mars sample-return mission might benefit from the Ares V. It proposed a 40-tonne spacecraft that would use aerocapture to deliver two landers and an orbiter. The landers would be equipped with rovers, deep drills, in-situ propellant production demonstrators, and no fewer than three ascent vehicles with sample canisters. The 7-tonne orbiter would retrieve and deliver the three samples to Earth, sending each one down separately. This would demonstrate a heat shield for entering the Earth's atmosphere at high (interplanetary) speed. Many of the technologies that this mission would test and demonstrate would be applicable to human missions to the Red Planet.[53]

The heavy lifting capabilities of the Space Launch System, on the other hand, have been touted for sample-returns from Mars, or even Europa or Enceladus.

The study by the National Research Council stated the reasonable concern that the Ares V-enabled missions would all be extremely costly flagship-class, or perhaps even larger, and that each could put a substantial strain on NASA's scientific budget. Although no cost estimate for the launcher itself was available, it could be expected to be very high and would greatly inflate the overall cost of any mission that used it. In a sense, there was no place for either the Ares V or the Space Launch System in a balanced space science program. Moreover, the study cautioned that the overwhelmingly expensive missions enabled by the Ares V could replicate the history of two planetary heavyweights: the 1960s Voyager Mars lander (that was to be launched by Saturn V) and the deep-space JIMO, both of which exceeded the $10 billion price tag in current dollars and were axed during development.[54,55]

After the cancellation of Constellation, new programs were drafted which would develop technologies and acquire knowledge that could be applied to sending human missions beyond Earth orbit. These included robotic planetary ventures consisting of an Exploration Precursor program of missions costing $500 to $800 million each and more narrowly focused, higher risk Scout missions costing $100 to $200 million. The focus would be on technology, but some worthwhile science would be performed.

Possible precursors included asteroid rendezvous missions by multiple spacecraft, lunar landings, and a variety of Mars missions. The early Mars proposals included a Phoenix-class lander to demonstrate the production of propellant by the processing of soil, various rovers, orbital resource mappers, aerocapture demonstrators, atmospheric and dust sample-returns, and Phobos and Deimos rendezvous.[56] An orbiter was proposed with a camera which had a resolution of just 7 centimeters per pixel, even better than that of Mars Reconnaissance Orbiter. Other possibilities under study included a high-power solar electric propulsion demonstrator to fly by a 'dead' geostationary satellite and then visit a near-Earth asteroid. Asteroids were the focus of several precursors in preparation for human missions. They were to scout likely targets years in advance of a human mission, reporting data such as surface texture and ruggedness, the radiation environment, etc. One such mission was proposed by APL at a cost of $500 million. This NextGenNEAR was an updated version of the Near-Earth Asteroid Rendezvous Discovery mission utilizing low-cost instruments and sensors. The simplest proposal, submitted by the Ames center, called for a flyby. At the other end of the spectrum the Boeing Corporation proposed a Near Earth Object Exploration System consisting of a modified, ion-propelled communications satellite carrying a small lander. The latter would have cameras, a laser obstacle detection system, strain gauges, penetrometers, and other instruments and would be powered by Stirling generators.

But the Exploration Precursor program was canceled after just one year, having failed to gain the interest of the human spaceflight community.[57] Also unfunded were a number of technology demonstration missions, including one that would see NASA cooperating with the Department of Defense to demonstrate a number of Earth orbit maneuvers prior to leaving for Mars to conduct a rendezvous with either Phobos or Deimos. Fueled with 1,000 kg of xenon, the spacecraft would be able to fly to Mars and return to Earth within 1,100 days.[58]

Meanwhile, the manned asteroid mission had evolved in an unexpected manner. A study released in 2012 by the Keck Institute for Space Science and performed at

The canceled US Ares I and Ares V launch vehicles, part of the Constellation program whose objective was to return astronauts to the Moon.

JPL described a method for retrieving a small object of about 1,000 tonnes using existing technology and maneuvering it into a stable high lunar orbit where it could be studied fairly easily and reached by a human mission. The study concluded that a high-power electric propulsion spacecraft massing some 18 tonnes, most of which would consist of xenon fuel, could be launched by an existing rocket and retrieve an asteroid within a decade at a cost not exceeding $3 billion. The spacecraft would have cameras and spectrometers for navigation and science, and for characterizing the target, as well as an inflatable cylindrical-conical catching bag some 15 meters across with a crushable foam pad to absorb the shock when the asteroid, itself only a few meters in diameter, was collected. A similar technique has apparently been studied by the US military as a means of recovering failed and "uncooperative" satellites.[59]

Based on this study, the NASA budget request for 2014 included $100 million to start a mission to recover a carbonaceous asteroid several meters in size and deliver it either into lunar orbit or to the L2 Lagrangian point of the Earth-Moon system in the 2020s, where it would be accessible to humans launched by the SLS. As the launch mass would be about 15 tonnes, this Asteroid Return Mission (ARM) would require either a long-duration propulsive maneuver to spiral out of Earth orbit or a heavy-lift rocket which currently does not exist. Meanwhile, a survey of the catalogs

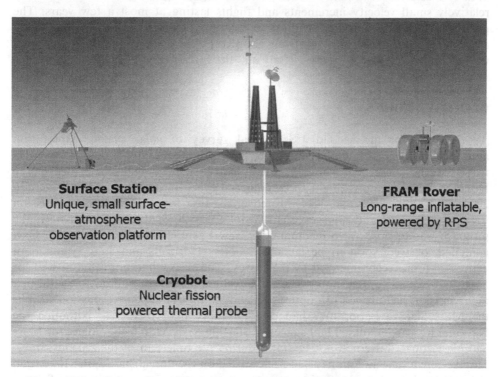

The ambitious Palmer Quest Mars mission would be enabled by a heavy launch vehicle like either the Ares V or the SLS that has replaced it.

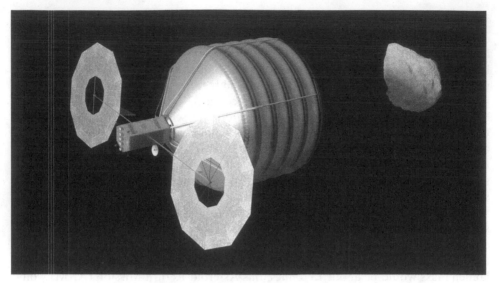

The Asteroid Return Mission about to capture its target asteroid.

of known asteroids identified a dozen potential targets which could be retrieved with relatively small velocity increments and flights lasting at most a few years. The project (which was initiated without consulting NASA's scientific small body assessment group) has been tentatively adopted by the agency, and talks are under way to open it up to ESA, ASI of Italy and the Indian ISRO.[60,61]

GLOOMY VISIONS

As the date for the second US decadal survey of solar system exploration approached, scientists had due cause to be pleased by the results of the first, but also reasons to be concerned. The 2003 survey had concluded that small, inexpensive Discovery-class missions ought to be launched every 18 months, and they were in 2004, 2005, 2007, 2009 and 2011. Moreover the first Mars Scout low-cost mission was flown in 2007 and another was scheduled for 2013; then, unfortunately, the program was to close. Two medium-class New Frontiers missions had either been launched, or were soon to be launched – New Horizons and Juno – and both addressed topics recommended by the survey. The survey had not fared as well for large and expensive 'flagships'. The initial estimate of $650 million for a Europa radar mission had proved too optimistic, and an already expensive mission such as the Mars Science Laboratory had cost far more than predicted. Although the National Research Council had suggested starting to develop technologies for Mars Sample Return, this could not be flown during the timespan of the second survey. Finally, much time and money had been wasted on the unrealistic Jupiter Icy Moon Orbiter. In view of the dramatic cost increases of projects such as the Mars Science Laboratory and the James Webb Space Telescope, scientists and engineers required a better method of estimating costs in order to give their priorities a more realistic outlook and to prevent overruns from eating funds that were intended for other projects which were no less worthy.[62]

At the same time, the US planetary exploration program was facing shortfalls not only from programs whose budgets were out of control but also from the escalating cost of launchers. A set of contracts awarded to commercial companies for launch services in 2010 proved significantly more expensive than hitherto. Moreover, the relatively inexpensive Delta II rocket was being retired with no real replacement. Equivalent launchers were being developed and pressed into service, including the privately funded Falcon 9 and Antares, but NASA had no plans to use them until they had flown a significant number of successful missions.[63]

Work on the second decadal survey was started in 2009 by NASA and the National Science Foundation. Whereas the first survey had merely provided a list of missions that were worth implementing, the second emphasized recommending missions that were consistent with the predicted budget and funding levels. To create a realistic list of prioritized missions, the panels and steering groups included engineers and experts in program management and cost estimation. Furthermore, independent cost analyses were performed in parallel. This put much greater emphasis on the actual feasibility and technological maturity level of each single proposal.[64] Moreover, NASA directed the National Science Foundation to treat Mars just like any other target for the survey, stating that they would not budget Mars exploration independently, as had been so in the previous decade.

The survey rapidly identified 24 sufficiently mature mission candidates from JPL, APL, and the Goddard center for evaluation and prioritization based on the input of the planetary science community. They included a Mercury lander and no fewer than three Venus missions, a sign that interest in our 'sister planet' was reawakening after 20 years of neglect – at least in the US. Two missions were dedicated to the Moon: a geophysical network and a polar volatiles explorer. The Mars missions were aimed at following up on the discoveries of the 2000s. They included the Trace Gas Orbiter, a polar mission, a lander network, and of course the sample-return mission. In addition to the flagships for Europa and Titan, missions to the giant planets included probes to Io and Ganymede, an atmospheric probe for Saturn, a Titan lake lander, and a mission dedicated to Enceladus. Missions to Uranus and Neptune and their satellite systems were proposed by APL and JPL. And interest in the minor bodies of the solar system took the form of a main belt asteroid lander, a Chiron orbiter, a Trojan asteroid tour, and a comet surface sample-return.

The survey report was published in March 2011 as *Visions and Voyages for Planetary Science in the Next Decade*, and it was drastic. It prioritized a number of missions but recognized that most of the flagships would have to be "descoped" and their costs severely reduced before they could be considered realistic.

From a programmatic point of view the Discovery program was supported, as in 2002, but it was deemed to lie outside a strategic plan, so the survey made no specific recommendation other than that it should continue at the current (or increased) budget level and cadence. Discovery missions would ideally be selected and flown every two years, rather more frequently than they are selected nowadays. On the other hand, the report made no recommendation on continuing the discontinued Mars Scout series of dedicated small missions beyond MAVEN. The New Frontiers program should also continue, with a fourth and a fifth mission being selected in the coming decade. The report suggested raising the New Frontiers cost cap from $1 billion including launch to $1 billion *excluding* it in order to protect against "volatile" launch vehicle costs. In any case, even if the planetary science budget were to be reduced, the survey recommended keeping the Discovery and New Frontiers programs alive and countering by cutting flagships. Concepts for the fourth New Frontiers mission included a cometary surface sample-return, a lunar south pole sample-return, a Saturn atmospheric probe, a Trojan asteroid tour and rendezvous, and a Venus in-situ explorer with no particular priority being expressed. For the fifth mission, an Io observer and a lunar geophysical network were to be added to any concept not selected on the previous round. In the meantime, the Mars Trace Gas Orbiter joint mission with ESA was endorsed and was recommended for launch in 2016.

The survey prioritized three flagships. The most important was a caching rover on Mars as the first step toward returning samples from the Red Planet. However, it was recommended that the joint MAX-C mission with ESA be descoped to cost no more than $2.5 billion. At the time of the survey its cost was estimated at $1 billion more than that, making it "a disproportionate share" of the budget. In fact, this project was later unilaterally canceled by NASA. The Europa orbiter was ranked second, but it would have to be reduced in both scope and cost, as the estimated $4.7 billion budget was considered unsustainable. However, even if its cost could be reduced, the survey recommended its adoption only if the planetary exploration budget was increased as otherwise it would take too great a share of the budget and preclude too many other opportunities. In other words, for the mission to proceed it would have to be reduced in cost *and* the planetary science budget would have to be increased. The survey gave the third highest flagship priority to a 'new entrant' in the form of a Uranus orbiter with an atmospheric probe. Most mission studies were oriented toward Neptune, the other 'ice giant', but Uranus was preferred for reasons of launch window, flight time, and cost. The Saturn, Titan, and Enceladus flagships did not make the final shortlist because, with the Cassini mission scheduled to run to 2017, it was felt that more time would be needed to properly interpret its discoveries before initiating another project aimed at the Saturn system. Nevertheless, a fourth flagship was recommended in the unlikely event of an increase in the planetary science budget. This would be either an Enceladus orbiter or a Venus climate mission. The report recommended significantly increasing the funding for research, development, and analysis, in particular to cover such technologies for future flagships as the Mars ascent vehicle and aerocapture for a Neptune orbiter and probe mission.

The survey also addressed suggestions for (as yet unfunded) human missions to asteroids or other targets beyond low Earth orbit. The National Science Foundation committee believed that unmanned science missions to the asteroids should retain their scientific focus and not act merely as human precursors. Hence, if any of the data from such missions would assist human exploration, the relevant manned program should pay for its analysis.

As for ground facilities, the report endorsed the Large Synoptic Survey Telescope, a planned wide-field telescope in Chile that will enter service in the early 2020s. The telescope, it was thought, would contribute to planetary science by finding hundreds of small near-Earth asteroids, Kuiper Belt objects and comets.

Two potential 'showstoppers' for future planetary exploration were the rising cost of US launchers and the availability of plutonium-238 to power missions to the outer solar system. The US was facing a shortage of space-grade plutonium oxide to use in RTGs and Russia wished to price its stockpiles as high as possible. Therefore, unless the production of plutonium-238 was restarted, future missions would require to use alternative sources for power. The Department of Energy was the only entity in the US allowed to produce and store nuclear material. Congress had repeatedly refused requests to resume production, which was expected to cost $75 to $90 million over a period of 5 years, reportedly because the beneficiary would be NASA but the burden of production would entirely fall on the Department of Energy.[65,66] Alternative power sources were studied in parallel. In particular, spurred by the recommendations of the decadal survey's Giant Planets Panel, NASA's Glenn Research Center led a study on the feasibility of a small fission reactor to facilitate planetary missions. This was a miniaturized concept having a solid-block uranium-molybdenum core that would use liquid metal for cooling, with thermoelectric converters generating electricity. As this system would generate only 1 kW of power out of 13 kW of heat, its efficiency was relatively low. For space-rating, the design was meant to remain 'subcritical' under a variety of scenarios including launch accidents, and would be activated only in space. If the program were to be pursued (itself an unlikely event) it would require 10 years of development.[67]

It was unfortunate that the decadal survey was released before the NASA budget requests for 2012 and 2013 were unveiled. The proposed budget reduced funding for planetary science from $1.5 billion to $1.2 billion yearly, and even less up to 2017. It provided no money at all for Mars missions after 2016, nor for the Jupiter-Europa mission, it canceled the US participation in ExoMars, and included a number of drastic steps such as reviewing programs that were in the extended-mission phase for termination. Even Cassini became a target for cancellation. In later months NASA rescheduled a Mars sample caching rover to 2020 and received funds specifically intended to start work on a simplified Europa mission. As a result of the current financial squeeze, the agency will be able to launch only five or six planetary missions in the 2013 to 2022 timeframe addressed by the decadal survey, as against a dozen in the previous decade. In particular, five Discovery missions were launched in 2003–2012, as against only one or two. And all of NASA's missions are to be directed toward the Moon, Mars or near asteroids; the outer planets and their satellites will be totally neglected.

On a more positive note, after a 25-year hiatus NASA-funded test production of plutonium-238 was restarted in early 2013. Whereas plutonium-238 was previously obtained as a by-product of the fabrication of fissile material for nuclear weapons, this time it will be created by the irradiation of a target of neptunium-237. At 1.5 to 2 kg per year, it will take 5 years to produce sufficient plutonium for a single mission, to be mixed with older, partly decayed material to create the proper energy density for space applications. The cost of developing the ASRG has significantly increased and still no mission has been approved to use it. Therefore, in view of NASA's shrinking budget and the resumption of production of plutonium-238, the agency's planetary science division announced in November 2013 that it was to discontinue the procurement of ASRGs and would instead continue using the traditional, albeit less efficient RTGs. Nevertheless, development of the Stirling generators was to continue at the Glenn center. The production rate of plutonium, as well as the existing supply, should allow the 2020 Mars rover to proceed, as well as an RTG-powered mission to either Europa or Uranus.

After the publication of the decadal survey and the abandonment of the Jupiter-Europa mission, JPL started a 1.5-year evaluation of low cost missions to Europa that would still provide high scientific returns. One option explored early on involved two separate and independent spacecraft: one having cameras and a laser altimeter, which would have the radiation shielding required to operate in orbit around Europa; the other would enter a resonant Jupiter orbit and make dozens of flybys of the moon to construct radar maps of its subsurface in much the same manner as Cassini did for Titan. The data would be replayed to Earth during idle times between flybys. These vehicles would reuse hardware developed for the New Horizons and Juno missions. This would enable each spacecraft to be launched in the 2020s on either an Atlas V or a Delta IV and fit within a $1.5 billion cost cap, thereby cutting the total cost of the Europa mission by one-third.[68,69,70]

However, the savings expected from such a mission were still not enough to grant its approval. As a result, studies soon focused on two one-spacecraft architectures: an orbiter proper and a Jupiter orbiter that would make frequent flybys, with both being fully compliant with the scientific objectives of the decadal survey. At the request of NASA headquarters, a study was initiated with the task of developing a scientifically compelling lander mission costing $1.5 billion. It produced a design for a legged soft-lander of 500 kg carrying a payload of about 50 kg that would survive on the surface of Europa for at least three Jovian orbits, performing seismometry of the icy crust.[71] However, the lander was deemed overly ambitious in the short term because some of its technologies were not yet available and the mission would in any case be far too expensive.[72,73,74,75]

As a result of these comparative studies, JPL adopted the Europa Clipper multiple-flyby architecture as its new baseline mission for Europa. This would achieve most of the scientific objectives of a dedicated orbiter at an estimated cost of about $2 billion. Moreover, the 2013 NASA budget, as finally approved, included $75 million to start formulating a Europa mission. As currently envisaged, the mission would perform 45 low altitude flybys of Europa (typically at 100 km) during three and half years with simple, repetitive operations and observation sequences. Each flyby would facilitate gravity studies, topographic imaging at low and high resolutions, infrared scans, mass spectrometer samplings of the atmosphere, and radar swaths while at altitudes below 1,000 km. At the end of the mission the radar data would form a dense and globally distributed network of intersecting swaths that probed the subsurface of most of the moon. In addition to proving the presence of a subsurface ocean, the data could also be exploited to identify areas where the ice was thin enough to be targeted by a future lander. A topographic camera, on the other hand, would yield images of the surface with a spatial resolution as fine as Cassini's best images of Enceladus, and provide a near-global map at 100 meters per pixel. A high-resolution camera might characterize potential landing sites. Also part of the baseline scientific payload would be a neutral mass spectrometer, an infrared spectrometer to identify the signature of non-ice material on the surface, and a magnetometer and Langmuir probe to investigate the presence of a liquid water ocean by detecting its response to the Jovian magnetic field. A dedicated gravity science antenna would complete the payload. Gravity science would measure the shape of Europa over a wide range of orbital positions and monitor the response of the icy shell to the varying tidal effects. Moreover, it would also enable scientists to determine the structure of the rocky core of the moon and in particular whether it has a partially molten mantle. The presence of volcanic activity on the floor of the ocean would increase the likelihood of Europa having given rise to life.

The Europa Clipper was initially envisaged as a squat, modular, 3-axis-stabilized spacecraft with a Juno-style titanium avionics vault buried deep inside and protected by the propulsion module in order to allow the use of radiation-hardened electronics employed by Earth-orbiting satellites. Power was to be provided by ASRGs or RTGs but the use of solar panels inherited from Juno was also studied. The use of

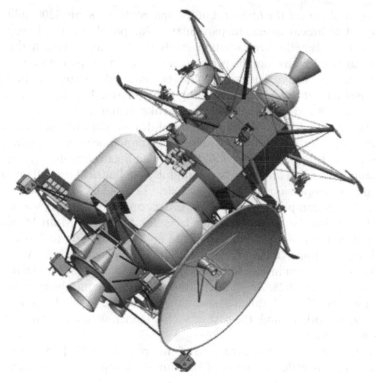

A proposed Europa lander, part of the redesigned US effort to explore this Jovian moon.

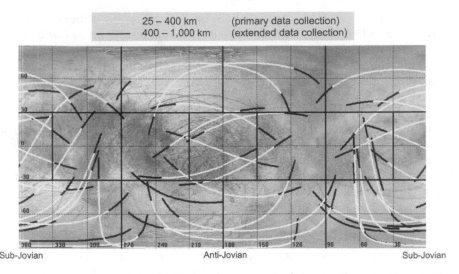

| | 25 – 400 km | (primary data collection) |
| | 400 – 1,000 km | (extended data collection) |

Sub-Jovian   Anti-Jovian   Sub-Jovian

The radar coverage of Europa enabled by the Europa Clipper mission that would function in orbit around Jupiter.

solar panels would lower the total cost of the spacecraft by some $200 million, but at the penalty of technical issues. In particular, solar panels on the Europa Clipper would have to endure higher radiation levels than Juno, although the exposure would be minimized by having the periapsis no closer to the planet than the orbit of Europa. Moreover, heavy batteries would be required to sustain the spacecraft during eclipses and to provide power during flyby phases. The design has since been modified to a stacked configuration that looks remarkably similar to Cassini. Although this resulted in a simpler mechanical design, the fact that the electronics were no longer protected inside the structure demanded heavier radiation shielding. Like Cassini, the Europa Clipper would have a high-gain antenna on top, then an avionics, communication and instrument module, a cylindrical propellant tank in the middle, and a propulsion unit at the rear. The ASRG-powered option was abandoned for its complexity, just before procurement of ASRGs was ended. The power will therefore be provided either by solar panels or Multi-Mission RTGs containing more than 14 kg of plutonium-238 and providing up to 440 W of power.

If the Europa Clipper mission is approved and financed by NASA, the baseline launch window will open in November 2021 and after Venus and Earth flybys it will reach Jupiter in April 2028. As an alternative, the probe could be launched directly to Jupiter using the Space Launch System and the voyage would last only two and a half years. On the other hand, the cost of such a launch would probably exceed that of the entire mission![76]

Of course, given the current shrinkage of the space agency's budget for planetary exploration, there is little likelihood of the Europa Clipper flying anytime soon.

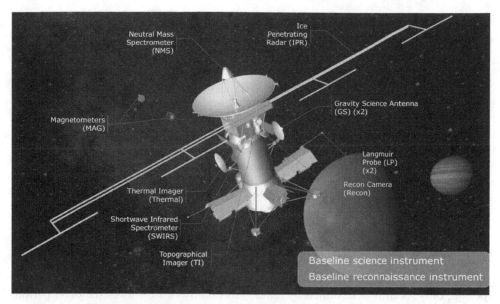

A possible configuration of the RTG-powered Europa Clipper.

ESA'S 'CORNERSTONES'

In comparison to NASA, the status of planetary exploration at the European Space Agency can be said to be in good shape: four planetary and deep-space missions were launched between 2003 and 2013, including one experimental Moon mission, and at least five are expected in the next decade with JUICE and BepiColombo expanding its reach to more challenging destinations.

ESA began to study Mercury orbiters using solar electric propulsion as early as the mid-1980s, and a detailed design of one such mission was finally kicked off in 1997. This settled at the beginning on a two-vehicle architecture including a 3-axis-stabilized orbiter to observe the planet and a smaller spinning magnetospheric orbiter for in-situ studies of particles and fields. A lander was also envisaged.[77] More conventional ideas were also considered: a Mercury orbiter was a candidate medium mission early in the decade, and there was also a Mercury Express 'flexi' mission. Better known as LUGH (after the Celtic deity equivalent to Mercury) for Low-cost Unified Geophysics at Hermes, this would consist of one flyby mothership and two polar microprobes. However, the mission was deemed of little interest, particularly because NASA had just approved the more comprehensive MESSENGER mission.

In 1996, acknowledging the significance of Mercury in studying the origin of the solar system, ESA's Horizon 2000+ scientific program included an orbiter as one of its 'cornerstone' missions; equivalent to NASA's flagships. Four costly, world-class missions were picked to fly as soon as practicable. These included the 'planet-finder' Darwin space telescope, the Gaia mission to carry out a precise survey of more than a billion stars of our galaxy, the LISA (Laser Interferometric Space Antenna) mission to detect gravitational waves passing through the solar system, and of course a Mercury orbiter. It was decided that Gaia and the Mercury orbiter required the least technology development.

The Japanese space agency ISAS was also studying a Mercury orbiter at that time, and a working group was set up in 1997 to perform a feasibility study. This mission would use NASDA's H-IIA launcher and a combination of a solar electric propulsion stage and intermediate flybys to enter an elliptical polar orbit of the planet tailored for particles and fields observations. The baseline scenario called for launch in the fall of 2005 and arrival in early 2008. The spacecraft would carry a number of instruments dedicated to studying the surface and interior structure of the planet, its atmosphere, magnetosphere, and local environment.[78]

The Science Program Committee of ESA selected the Mercury orbiter in late 2000 and definition studies were put out to tender to European industries. The mission was named BepiColombo in honor of Giuseppe Colombo (Bepi being short for Giuseppe in northeastern Italy) who contributed so much to planetary exploration (he was "the father" of the Giotto Halley mission) and especially to Mercury science. For example, in the 1960s he was one of the first to point out that Mercury's rotation period, only recently determined by radar, was exactly two-thirds of the orbit period, providing an explanation for this spin-orbit resonance. And then it was he who suggested to JPL that by precisely steering Mariner 10, the mission would be able to encounter the planet repeatedly at intervals of 6 months.[79,80,81]

In 2000, ESA and ISAS reached an agreement that the Japanese would provide the magnetospheric orbiter for the mission.

BepiColombo is to investigate the geological evolution of Mercury; seek an explanation for its high density; provide details of the internal structure, looking in particular for a liquid metallic core and for the origin of the magnetic field; map the composition of the surface, including the polar deposits; map the magnetic field and investigate the nature of its interactions with the solar wind and interplanetary environment; study the exosphere to determine its structure and sources; study particles in the vicinity of the planet; and conduct experiments into fundamental physics and general relativity.

The rationale for flying to Mercury in the wake of MESSENGER was debated and deemed absolutely compelling because a two-orbiter team was capable of much more specialized and in-depth investigation. In particular, BepiColombo will have a much more extensive payload suite, with instruments for studies not directly addressed by MESSENGER, and overall it could be expected to provide up to 80 times as much data. Moreover, the mission would not be geared toward making new discoveries but to collecting data that will bring knowledge of Mercury to the same level as the other rocky 'terrestrial' planets. In a way, MESSENGER was for Mercury what Mariner 9 was for Mars, and BepiColombo will be the counterpart of the Viking orbiters. And of course it will fill high-resolution imagery and altimetry gaps left by MESSENGER over the southern hemisphere owing to the American probe's distant apoapsis at high southern latitudes.

The initial proposal included a small lander that was to set down poleward of 85 degrees and near the terminator, where the thermal loads would be more benign, in order to conduct physical, chemical, mineralogical, and remote sensing observations. This 'surface element' would have comprised either a hard-landing station and a penetrator, or a soft-lander that would be braked by retrorockets and airbags and deploy either a minirover or a Beagle 2-style 'mole'. ESA had been studying microrovers for some time. A 'nanokhod' prototype envisaged a tethered, tracked robot of only 2.5 kg that would be able to orientate its payload cab to undertake a variety of analyses. A larger version might carry a deep drilling and sampling system.[82] The overall payload would include instruments to measure the physical and thermal properties of the surface, instruments to determine the composition of rocks, and magnetometers and cameras. ESA reportedly gave some consideration to including a capability to retrieve samples from a penetrator in order to return them to Earth.[83] Although the lander was a very popular element of the mission, it was discarded early on as being too expensive. The problem with a Mercury lander was that, like the Moon, the planet has no atmosphere to brake a lander's fall, with the added complication of one of the most challenging thermal environment in the solar system.

After the mission was initially assessed at ESA a great number of trajectories and mission designs were investigated before a profile which combined solar electric and chemical propulsion as well as planetary gravity assists was selected. Trade-offs were also carried out on whether all of the elements could be launched as a single stack on an Ariane 5 or split across two identical propulsion stages and launched on Soyuz-Fregat rockets. In the single launcher mode the mass at the time of launch would be in the range 2,500 to 2,800 kg, whereas for a dual launch each stack would be almost 1,500 kg. Both solutions were found to be quite complex and expensive. However, a new version of the Soyuz with an uprated Fregat stage was under development to be launched from Kourou to exploit the near-equatorial location to increase the payload. In 2004 a single Soyuz-Fregat architecture was thus adopted and industrial definition started. To compensate for the missing thrust of the Ariane 5, a circuitous profile was necessary. Flybys of the Moon and Earth were included. The fact that a smaller cruise stage had to be used meant that the available ion engine thrust was halved.

On this baseline mission, BepiColombo was to have been launched from Kourou in August 2013. The launcher was to put the stack into a transfer orbit not dissimilar to that of a satellite heading for geostationary orbit. From there, the spacecraft would use its chemical engine to raise the apogee to the orbit of the Moon. After a number of phasing Earth orbits it would fly by the Moon for a slingshot into a solar orbit that was similar to Earth's, but more eccentric. Then 6 months of electric thrusting would begin in order to ensure a return to Earth for a flyby that would reduce its heliocentric velocity and enable it to penetrate more deeply into the gravitational well of our star. Two Venus flybys, performed one orbit apart, would not only reduce the perihelion to Mercury's orbit but also increase the inclination to 7 degrees in order to match that of the innermost planet. Two years and 6.5 orbits around the Sun would follow, during which BepiColombo would simply thrust and cruise between Venus and Mercury. The result would be two Mercury flybys 44 days (and half a Mercurian orbit) apart, with the first occurring at perihelion and the other at aphelion.

After the second flyby, BepiColombo would be in a solar orbit that would return it to Mercury in March 2019. Since orbit insertion would not require an engine burn, the cruise stage would be jettisoned 2 months prior to arrival. The 'gravity capture' technique pioneered by the SMART 1 (Small Missions for Advanced Research in Technology) lunar orbiter would be used to achieve orbit. In this, the small relative velocity between the spacecraft and the planet, in combination with perturbations by solar gravity, would ease the vehicle into a distant orbit after a slow pass through one of the colinear Lagrangian points of the Sun-Mercury system. Flight controllers on Earth would then have a number of opportunities to establish a stable orbit before the spacecraft would be perturbed back into a solar orbit.[84] BepiColombo would slip into a highly eccentric 400 × 180,000-km orbit that would be sufficiently stable during an entire Mercurian year of 88 Earth days. Chemical thrusters would lower the apoapsis first to the magnetospheric orbiter's altitude of about 12,000 km. There, the Japanese orbiter and its sunshade would be set free. The European spacecraft would descend to its own polar orbit ranging between 400 and 1,500 km with a period of 2.3 hours and an equatorial periapsis. After instrument calibration, observations would begin and last nominally for 1 terrestrial year, with a possible 1-year extension. Due to the large number of flybys and thrusting arcs, detailed analyses and simulations were carried out to verify that the mission would be able to continue if any of these events were missed. In February 2006 the Science Program Committee confirmed the mission. Its cost was estimated at 1 billion euros, of which ESA would provide about 665 million euros. The remainder would be provided by JAXA for the magnetospheric orbiter, by Russia and NASA for several instruments, and by single European states sponsoring instruments and experiments.

Built by NEC, the JAXA-supplied Mercury Magnetospheric Orbiter (MMO) is a 250-kg octagonal drum 180 centimeters across and 90 centimeters tall, and it will be spin-stabilized at 15 revolutions per minute to scan its environment. The structure of the 'drum' consists of two decks connected by a central thrust tube and four internal bulkheads. On the outside, the upper portion of the spacecraft is covered by a mix of 50 per cent solar cells to provide up to 450 W of power and 50 per cent mirrors. The lower portion is entirely covered by mirrors. There is a despun parabolic antenna 0.8 meters in diameter on top of the cylinder to provide communications with Earth at an average rate of 16 kilobits per second and a medium-gain antenna on the lower deck. Sun sensors and a star scanner will facilitate attitude determination, and the attitude will be controlled by six 0.2-N gas thrusters that can draw on a total supply of 4 kg of compressed nitrogen. Its payload of about 25 kg includes seven sensors to undertake plasma and particle investigations: two electron analyzers, a mass spectrometer, an ion analyzer, high energy particle detectors for electrons and ions, and an energetic neutral particle analyzer. The mass spectrometer provided by the Southwest Research Institute in the US and funded through the Discovery program is intended to study the exosphere of Mercury. Four sensors and three receivers covering different frequency ranges will study plasmas, radio waves and electromagnetic fields. Sensors include a wire probe antenna, an electric field sensor, and a pair of search-coil magnetometers mounted at the tip and halfway along a 5-meter-long boom. The package will also use perpendicular wire antennas having a tip-to-tip span of 32 meters. A spectral imager will provide full-disk images of the planet and its exosphere in the emission line of sodium. Finally, a dust monitor will provide data about the environment of the inner solar system.

During the interplanetary cruise, the magnetospheric orbiter will be protected by a sunshade and connecting interface to the vehicle. The lightweight sunshade is conical in shape with a diagonal cut out, and is made of carbon fiber that has a heat rejection finish. The interface consists of a lightweight cruciform structure that has a four-point attachment.

EADS Astrium of Germany (now Airbus Space) was selected in 2007 as the prime industrial supplier for the Mercury Planetary Orbiter (MPO) developed by ESA, with the Italian branch of Thales Alenia Space as 'co-prime'. The 3-axis-stabilized spacecraft will map the surface of the planet in detail and study its structure, composition, and gravity field. With a dry mass of 1,075 kg, it resembles a 3.9 × 2.2 × 1.7-meter box with aluminum honeycomb surface panels. Five of the six sides will be illuminated by the Sun at one time or another. The sixth houses a 2 × 3.6-meter radiator that is to remain shadowed from the Sun all of the time. A double-H interior structure will support the dynamic loads of launch. Inside are two tanks for 790 kg of hydrazine and nitrogen tetroxide. Power will be provided by a single three-panel array that has 70 per cent of one side covered by high-efficiency solar cells and the rest mirrored. Moreover, the panel will usually be kept at a tilted angle, since to directly face the cells to the Sun would cause them to overheat, lose efficiency and perhaps fail. Nevertheless, the array will be able to produce 1,515 W at perihelion. The Ka-band communications system is capable of transmitting 1,550 gigabits of data every year. This is about ten times the return of the magnetospheric orbiter, which will use X-band telemetry. A gimbaled titanium high-gain antenna will be placed directly in sight of the Sun at the end of a boom. In order to prevent the dish from deforming in the heat, a special coating was developed that ought to limit its maximum temperature to 300°C. A steerable medium-gain antenna and two low-gain antennas fixed at the edges of the radiator will guarantee contact at any time during the mission, including the cruise phase. Attitude will be determined using gyroscopes, three stellar trackers and rough Sun sensors, and controlled by four reaction wheels or 1-N hydrazine monopropellant thrusters. Maneuvers in orbit around Mercury will be made by four 22-N bipropellant thrusters.

As in the case of MESSENGER, the plane of the orbits of the two BepiColombo vehicles should remain fixed in space as Mercury travels around the Sun. Counter-intuitively, the mission was designed so that apoapsis would lie between the Sun and Mercury when at perihelion. This is because the solar flux will change relatively little over an orbit but the infrared heat radiated back to space by the planet, whose surface can reach 400°C, will be dramatically reduced at apoapsis. In fact, the heat radiated by the surface was a main driver for the design of the thermal shielding. To further complicate things, the deck on which most of the instruments are mounted will face the planet more or less continuously. This was one of the most complex parts of the spacecraft to be designed, owing to the possibility of heat 'leaking' from it to the rest of the structure.[85] Selection of the high-temperature materials started in 2001, over a decade in advance. The low-altitude BepiColombo orbiter will be subjected to a heat flux from the Sun of 14 kW/m², ten times greater than at Earth, and to no less than 6 kW/m² from Mercury, which is "worse than a hot plate on a cooker". Thermal control will be performed by a mix of passive and active means. The vehicle will be covered by a total of 66 kg of insulation which comprises ten layers of conventional thermal blankets covered by thirty layers of high-temperature ceramic cloth. The outer layers must withstand temperatures in excess of 360°C, and to minimize heat transmission they will be offset from the inner layers by spacers. Liquid flowing inside 93 sealed pipes embedded in the structural panels will transfer heat from the Sun-facing side to the shadowed radiator. Titanium louvers will prevent the radiator from directly seeing the hot surface of the planet when facing it. The louvers will reach 400°C, whilst the radiator will operate at a relatively cool 60°C. In order to test BepiColombo, ESA had to modify its thermal-vacuum test chamber to simulate the conditions in orbit around Mercury.

The planetary orbiter payload will total about 50 kg and have eleven instruments. Two cameras will provide stereoscopic images of the entire planet at a resolution of 30 meters and of selected areas at 10 meters. They might also be used to search for Vulcanoids closer to the Sun than Mercury. A visible and infrared spectrometer will map the entire surface, yielding remote sensing composition support to the X-ray and gamma-ray spectrometers. The latter two instruments will determine the composition of the surface and map ice deposits in permanently shadowed craters at the poles. The gamma-ray spectrometer will be supplied by the Russian Space Agency as a derivative of that developed for Fobos-Grunt.[86] A second infrared mapping spectrometer with radiometric capabilities will measure the temperature of the surface under a range of illumination conditions, as well as providing data concerning its texture and thermal conduction characteristics. A laser altimeter will generate a global topographic map to a vertical accuracy of 1 meter. An ultraviolet spectrometer will map the exosphere, both monitoring the known constituents and searching for new ones, including noble gases. A neutral and ionized particles analyzer will investigate the exosphere and the Mercurian energetic particle environment. A magnetometer on a 3.2-meter-long boom will map the structure and dynamics of the inner magnetosphere. A precise oscillator will allow radio science to investigate the rotational state of the planet and its gravity field and mass concentrations. Furthermore, by precisely tracking the motion of the spacecraft as it travels with Mercury around the Sun and by timing the propagation of electromagnetic waves during the frequent solar conjunctions, it will be possible to obtain a good determination of some relativistic parameters, measure the oblateness of the planet and, for the first time, that of the Sun.

The Mercury transfer module will be mounted at the bottom of the stack at launch and will provide propulsion, power, and attitude control for the interplanetary cruise. It is a thermally controlled enclosure that has three radiators and a central pyramidal carbon fiber structure to support the planetary orbiter. Its huge five-panel solar wings will span over 30 meters when fully deployed and have a surface of 48 square meters capable of providing up to 14 kW of power at Mercury using the same high-temperature cells as on the main orbiter. Mounted at the bottom of the module will be a cluster of four ion thrusters supplied by Qinetiq in the UK, each yielding a maximum thrust of 145 mN. European electric propulsion and its integration with a deep-space probe has already been tested on the SMART 1 lunar orbiter. At Earth's distance from the Sun only one thruster will be able to operate at any given time, delivering a thrust between 100 and 130 mN. Inside the orbit of Venus the power provided by the solar arrays will enable two thrusters to run simultaneously and deliver a combined thrust of 290 mN, which is a force roughly equivalent to the weight of a 30-gram object on Earth. Over the duration of the mission, electric thrusters will provide at least 4.5 km/s of total velocity change. Maneuvers during the interplanetary cruise will be assisted by chemical thrusters.

Development proceeded smoothly until 2008, when the solar panels were found to be incapable of coping with the expected high temperatures and ultraviolet flux. They had to be redesigned for a substantially greater area. This both drove up their cost and increased their mass and that of the supporting structure. As a result of these and other modifications the launch mass of the stack increased by 1 tonne to 4 tonnes, to the point that the mission was no longer viable using any version of the Soyuz-Fregat launcher. The project reverted to the Ariane 5 ECA. The extra performance provided the spacecraft with a generous mass margin but increased the cost of the mission for ESA from 665 million euros to 970 million euros. Although it was widely expected that BepiColombo would be canceled, it was again endorsed by the agency's Science Program Committee. Following the redesign, launch was initially expected in July 2014 and then in August 2015 but development and testing problems with the thrusters, high temperature solar panels and antenna meant it had

The minuscule Nanokhod rover initially planned to be part of a BepiColombo Mercury lander.

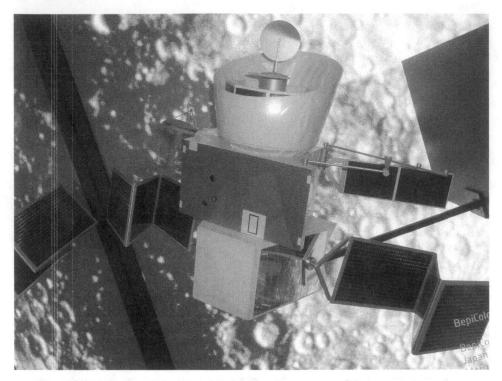

A model of the BepiColombo stack. From the top: the Japanese Magnetospheric Orbiter stowed inside its thermal protection shield, the European Planetary Orbiter, and the solar electric propulsion module.

to be slipped to a backup opportunity. On 9 July 2016 an Ariane 5 will directly insert BepiColombo into a solar orbit, eliminating the lunar flyby. On the new interplanetary cruise it will fly by Earth in July 2018, by Venus in September 2019 and May 2020, and by Mercury five times in July 2020, April 2021, July and December 2022 and February 2023. Arrival in orbit is expected on or about 1 January 2024. The primary mission will last fifteen months until April 2025.[87,88,89,90] As of the time of writing (December 2013), the mission is proceeding to the final phases of development. The Japanese orbiter was subjected to its first thermal tests at a heat flux equivalent to 0.3 AU in late 2010, and the European orbiter in 2011.

The early 2010s selection for the next ESA large mission saw two candidates out of three put forward to fly in deep space. One was a proper planetary mission, being the European contribution to a joint Jupiter project with NASA, and the other was the joint LISA mission to seek out gravitational waves in deep space. The third candidate was International X-Ray Observatory (IXO), a joint astronomy mission with JAXA. Although the Jupiter-Europa mission was ranked second in the US planetary science decadal survey, the astrophysics decadal survey rated LISA and IXO second and third respectively.

The objective of LISA is to investigate Einstein's concept of gravitation, which

The components of the BepiColombo spacecraft. From the left: the propulsion module, the Japanese Magnetospheric Orbiter, that spacecraft's thermal shield, and the European Planetary Orbiter. (ESA)

requires the existence of waves to propagate the effect of this force at the speed of light. In his theory, gravity is a deformation of the space-time fabric and under some circumstances a moving massive body can produce ripples in the fabric that travel as gravitational waves. Such waves have been detected indirectly, earning the physicists a well-deserved Nobel prize, but they have yet to be measured directly owing to their minuscule effect. It will be remembered that attempts were made to detect them using widely spaced probes like Galileo, Ulysses and the ill-fated Mars Observer, amongst others. LISA was first proposed in 1993 by a team of US and European scientists as ESA's third medium-sized project. A single Delta-class launcher would deploy three 460-kg spacecraft which, once in position, trailing 20 degrees behind the Earth in its solar orbit, would create the vertices of an equilateral triangle which spanned several million kilometers. Each spacecraft would contain two optical assemblies, consisting of identical 30-centimeter Cassegrain telescopes, one to send a laser beam to the next spacecraft and the other to receive a beam from the previous one, tying them together. Each vehicle would contain a 5-centimeter cubical 'proof mass' of platinum and gold which was shielded from any external disturbances capable of imparting upon it an acceleration. Lasers would track the position of each mass and the relative positions of the vehicles, with low-thrust ion engines maintaining a stable triangular formation and zeroing the spacecraft hull on the unperturbed position of the proof masses. This arrangement would make the array an antenna sensitive to gravitational waves in the millihertz to decihertz frequency range. As a gravitational wave traveled through the solar system it would displace the sides of the triangle by less than the diameter of an atom, but the system would be able to measure it. From the data, it would be possible to determine the amplitude, direction, and frequency of the perturbing wave. Similar experiments are presently being performed on Earth, but are obliged to use much less sensitive baselines of at most several kilometers.[91]

It was acknowledged early on that the scale of the mission, and in particular the technology that would have to be developed, would cause it to exceed the cost cap of a medium mission. Therefore, soon after having been proposed, it was reformulated as a cornerstone project for the Horizon 2000+ scientific program, along with the BepiColombo Mercury orbiter. By 1997 LISA had become a collaboration between ESA and JPL/NASA, with ESA providing the spacecraft set and NASA donating launch and mission operations. The payload would be shared between the two agencies.[92] In view of the many technological unknowns, JPL argued that a technical demonstration mission should be flown to validate various systems, including proof-mass tracking and 'drag free' centering of the vehicle around it. This proposal became SMART 2 or LISA Pathfinder, the second in ESA's series of rapid-development Small Missions for Advanced Research in Technology. Initially budgeted at 185 million euros and with a launch around 2005, LISA Pathfinder has inflated to over 400 million euros and will not be launched before 2014. Of course, such a long development cycle is not what would be expected for a SMART mission! In the meantime, astrophysicists have been using supercomputers to simulate the gravitational wave signatures of events such as the fusion of two supermassive black holes in order to be ready to interpret the LISA data when it eventually becomes available.[93]

Although the candidates for the next ESA large missions were announced in 2010, the fact that two of the three would involve NASA participation meant that the final selection of the first mission to pursue would be delayed to early 2012 because of the US agency's financial issues. A reduction in the budget available for its astrophysics program obliged NASA to withdraw support for LISA in April 2011 and disband its science team. LISA was then redesigned as the Europe-only New Gravitational wave Observatory (NGO) involving one mothership and two daughters to be inserted into solar orbit by a pair of Soyuz-Fregat launchers. Only the mothership would transmit data to Earth, the two daughters being connected to it via laser links.

With its participation in LISA ruled out and IXO requiring expensive technology development, NASA also canceled its part of the joint Europa and Jupiter System Mission even though it was deemed a low risk, needing no critical new technology. It was thus no surprise that in May 2012 the primarily European Jupiter Icy Moons Explorer (JUICE) was unanimously adopted as the next ESA large mission.[94] Early estimates put it at 830 million euros plus 240 million euros for instruments, with the latter figure possibly including a $100 million contribution from NASA.

Broadly based on ESA's Jupiter Ganymede Orbiter proposal, the JUICE mission will characterize in detail the extent of that moon's ocean and icy shell, as well as its relationship to the deep rocky interior. It will map the composition, distribution, and evolution of the surface material, investigating the geology of the surface for signs of past and/or current activity. Furthermore, it will characterize the particles and fields environment near the moon, and how this interacts with the planet's magnetosphere and the solar wind.

As currently planned, JUICE will launch in June 2022 (with a backup opportunity in November 2023) and, after a number of Earth and Venus flybys, enter orbit around Jupiter in January 2030 and then around Ganymede in September 2032. The mission will nominally terminate in June 2033. Inbound to Jupiter, the spacecraft will fly by Ganymede for the first time shortly prior to orbit insertion, to receive a gravity assist. Additional Ganymede flybys will then reduce the apoapsis of the capture orbit. After NASA abandoned its Jupiter Europa Orbiter, the mission for JUICE was modified to include two flybys of Europa as close as radiation protection will permit, in order to recover some of that lost science. These will be made 36 days apart and will provide the first measurements of the thickness of Europa's ice shell. JUICE will also make a study of Callisto, seeking evidence for why Ganymede's interior is so differentiated whilst Callisto's is not. But the primary role of the Callisto flybys will be to establish the high-inclination orbits around Jupiter in which JUICE will spend 260 days mainly studying the particles and fields environment. After a total of 12 flybys of Ganymede and 13 of Callisto the vehicle will enter a polar orbit around the former. It will first spend 30 days in an eccentric orbit with the apoapsis at 10,000 km, then 90 days in a circular orbit at 5,000 km, then 30 days in a second elliptical orbit and 102 days in a 500-km circular orbit before finally spending a month in a 200-km circular orbit.

A 100-kg payload of eleven instruments based on instruments that have flown on many European and international missions was selected in February 2013. Wide- and narrow-angle cameras will take images having spatial resolutions of up to 2.4 meters on Ganymede and about 10 km on Jupiter using a multitude of filters. A visible and infrared imaging spectrometer will analyze the chemistry of the planet's atmosphere and the ices and minerals on the surfaces of the icy moons. An American ultraviolet imaging spectrometer will exploit the Rosetta, New Horizons, Lunar Reconnaissance Orbiter and Juno heritage. A submillimeter radiometer will measure the temperature, structure, and composition of the Jovian atmosphere, as well as the exospheres and surfaces of the moons. A magnetometer will study the magnetospheres of Jupiter and Ganymede and their interactions, as well as the magnetic fields which are induced in the salty oceans beneath the icy surfaces of the large moons. Radio and plasma-wave sensors will record electric and magnetic fields. A Swedish particle, plasma and ion package and mass spectrometer (with input from APL) will characterize the particle environment of Jupiter, as well as neutral gases, plasmas and neutral atoms. A laser altimeter will map Ganymede with a vertical resolution of 10 centimeters while in the 200-km orbit. A 10-kg ice-penetrating radar with a 16-meter-diameter antenna will be provided by the Italian Space Agency with input from JPL and will penetrate the icy shells of Ganymede, Callisto and Europa to a depth of about 10 km with a vertical resolution of 30 meters. Finally, two radio science experiments will investigate the gravitational fields of Jupiter and its moons, as well as provide radio-occultation data for the atmosphere of the planet and the tenuous envelopes of its moons.

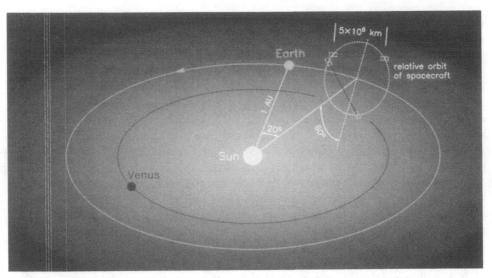

The formation orbit initially envisaged for the LISA gravitational wave mission. (ESA)

The European JUICE Jupiter and Ganymede orbiter. (ESA)

The JUICE spacecraft will have a mass of some 1,900 kg dry, and with 2,900 kg of fuel it will be almost 5 tonnes at launch. Nevertheless, it will have relatively light shielding to protect its electronics from the Jovian radiation, far lighter in fact than a Europa orbiter proper. The spacecraft will be 3-axis-stabilized and will use large low-intensity solar arrays based on those of Rosetta, with an active area of about

A rendering of the Ganymede lander proposed by Russia to accompany JUICE. (Lavochkin Association)

65 square meters to generate up to 700 W of power at Jupiter. A body-fixed 3-meter antenna will relay in excess of 1.4 gigabits of data during a daily standard 8-hour tracking pass.

Following the selection of JUICE, two competing industrial studies were started at Airbus and Thales Alenia Space. The selection of the prime contractor and initiation of industrial work is expected in late 2014. The mission is baselined for launch by an Ariane 5 ECA, but following talks with Russia on the joint ExoMars mission it might be reassigned to a less expensive Proton in exchange for flying a Russian Ganymede lander. On the other hand, a lander would probably require a dedicated orbital relay, which might be impracticable for this already complex mission.[95]

Looking farther ahead, the third ESA large mission (after the X-ray telescope) is likely to be an updated version of the New Gravitational wave Observatory proposal called eLISA that will fly in the mid-2030s.[96]

ESA'S MEDIUM MISSIONS

To follow up Mars Express, in 1999 ESA issued a call for proposals for two 'flexi' missions to fly in the late 2000s, each capped at 200 million euros. No fewer than 49 proposals were received. Of six selected for further assessment, two were deep-space

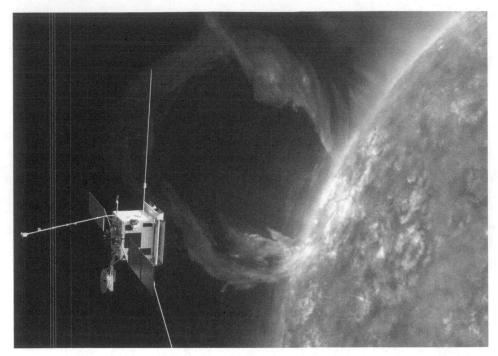

The European small-perihelion and high-inclination Solar Orbiter. (ESA)

missions: Master (Mars + Asteroid) and Solar Orbiter (SOLO). Master would reuse the Mars Express spacecraft bus to carry an array of four Beagle 2-class or Netlander-class landers to Mars and would then proceed to make a low-speed flyby of Vesta. SOLO was a low perihelion out-of-ecliptic mission in the European tradition of solar probes and satellites.[97]

The Solar Orbiter mission was first discussed at a European solar physics meeting in March 1998. After a pre-assessment study in 2000, it and European participation in NASA's Next Generation Space Telescope (later the James Webb Space Telescope) were the missions recommended for adoption, with launch of the solar probe at some point between 2008 and 2013. The probe was expected to build on ESA's experience with solar electric propulsion which, along with several flybys of Earth and Venus, would slowly increase its inclination to almost 30 degrees relative to the ecliptic and draw its perihelion within 50 solar radii. It would continue Ulysses' observations of solar particles and fields, sampling the different 'tastes' of the solar wind at a much closer distance to our star. An interesting feature of the trajectory was its speed at perihelion, which would match the rotation rate of the Sun itself. In fact, for several days around perihelion the spacecraft would essentially 'hover', like a geostationary satellite of Earth. Using an extensive array of optical and remote sensing instruments, the mission would exploit this fact. High resolution and full-disk telescopes, working in many different regions of the electromagnetic spectrum, would be able to make out details on the photosphere as small as several tens of kilometers to address some of the mysteries concerning how solar magnetic fields interact with the hot atmosphere. Meanwhile, a coronagraph would image the innermost portions of the corona. With the spacecraft matching the angular speed of the Sun, the coronagraph would be able to follow the evolution of single coronal features, unaffected by the distortions of the feature rising over the limb, transiting the disk and setting over the far limb, thereby expanding on the work of the ESA/NASA SOHO mission which had been observing the corona almost continuously since the mid-1990s.

As with BepiColombo, particular care would have to be taken when designing the Solar Orbiter to ensure that its structure and instruments would survive the heat loads at perihelion, which would be almost 25 times greater than at 1 AU. For this reason the vehicle was to hide behind a large sunshield which provided instrument apertures, and the optical instruments would incorporate mirrors, aperture blocking devices, and suitable materials to minimize deformation at high temperatures. Magnetometers and plasma-wave sensors requiring booms and antennas would be mounted on the anti-Sun side to obviate problems of thermal warping such as experienced by Ulysses, and the solar panels would rotate to minimize their exposure to heat at perihelion.

Because it would rely on technology developed for the BepiColombo mission, the Solar Orbiter followed the vicissitudes of the former, slowly accumulating delays on its optimistic schedule. Meanwhile, a thorough assessment evaluated alternative propulsion solutions, including chemical or hybrid propulsion. The results reaffirmed the solar electric option. The mission was to be launched by a Soyuz-Fregat from the new pad in Kourou. Interest in the mission was confirmed, and it was included in the Horizon 2000+ scientific program in June 2004, with a launch date between 2013 and 2015.[98,99,100,101,102,103]

Meanwhile, as the cost of the mission increased, a joint scientific and technology definition team was established from ESA and NASA personnel to consider how the former's Solar Orbiter might be combined with the latter's Solar Sentinels in order to reduce the financial burden. NASA's proposed Solar Sentinels would consist of four identical spacecraft resembling the 1970s Helios probes, which would be launched by a single Atlas V and use Venus flybys to enter two different Mercury-crossing orbits. Initially expected to launch in 2015, this mission was to provide a 3-dimensional reconstruction of the plasma and magnetic environment of the inner heliosphere at a cost of about $875 million.[104]

After these studies, the Solar Orbiter became a joint European-US mission and a call for instruments was issued. A total of 14 proposals were received. Meanwhile, an 18-month industry study of the spacecraft was kicked off. In addition to the launcher, NASA would provide two instruments and parts for two others. However, in 2008 the US agency decided to develop Solar Probe Plus to fly closer to the Sun whilst staying near the ecliptic, flying the missions simultaneously in order to make complementary observations. The initial instrument selection therefore had to reaffirm the validity of the European small perihelion mission. Ten investigations would find places on Solar Orbiter for in-situ and remote sensing of the Sun and inner heliosphere. An extreme-ultraviolet camera to provide images of the outermost layers of the Sun's atmosphere and corona would give the first out-of-ecliptic views of our star. There would also be a coronagraph to image the corona in the visible and ultraviolet; a visible imager and magnetographer to provide high resolution measurements of the magnetic field and gas velocities of the photosphere in order to investigate the convective motions of the Sun; an energetic particle detector; a magnetometer provided by UK scientists; and a radio and plasma-wave sensor to measure the magnetic and electrical fields. NASA would supply an imager for high resolution views of the Sun and track the evolution of coronal mass ejections; an extreme-ultraviolet spectrometer; and a suprathermal ion spectrograph. The solar wind would be analyzed by a dedicated instrument and an X-ray imager would complement the remote sensing instruments. The payload would therefore span the electromagnetic spectrum from visible wavelengths to X-rays.

However, in November 2008, with industrial contracts awarded and a preliminary payload selection in place, the Science Program Committee decided that the Solar Orbiter was a candidate for the larger and more expensive medium-class missions of the "Cosmic Vision" program for 2015–2025, with a launch opportunity in 2017. In February 2010 the committee recommended it as one of three medium-class mission candidates that should proceed to the definition phase; the other two being Plato (to observe planetary transits) and Euclid (to study dark matter and dark energy). As a further complication, in March 2011 budgetary factors obliged NASA to reduce its payload contribution to one instrument and part of another. The unfunded instrument would be replaced by a European-led one. Despite these difficulties, Euclid and the Solar Orbiter were selected as the next ESA medium-class missions in late 2011. The contract to build the Solar Orbiter went to Astrium UK. By then, however, the overall cost of the mission had increased to almost 500 million euros.

The Solar Orbiter will be a 3-axis-stabilized spacecraft that will point at the Sun, with a BepiColombo-heritage heat shield for protection at perihelion, two-sided solar arrays, and a high-temperature high-gain antenna. There were baseline launch windows in January and March 2017, with backups in July 2017 and in August and October 2018. The mission will be launched by an Atlas V 401 from Cape Canaveral, with the Ariane 5 as a more expensive backup. At the end of 2013 difficulties with the high-temperature solar panels and some of the instruments dictated the switch to the July 2017 launch window. After launch, the Solar Orbiter will fly by Earth twice in July 2018 and June 2020 and Venus twice in May 2020 and January 2021, spending some of its time at heliocentric distances of almost 1.5 AU. Since the spacecraft is not required to operate that far from the Sun, it will adopt a 'light hibernation' for several months. At the end of this first series of flybys, it will be in a 180-day orbit with the perihelion as low as 0.28 AU (60 solar radii or about 42 million km from the photosphere). This was to have been at about 0.22 AU, but was increased to a range which would be marginally safer without impacting on the scientific objectives. Fast flybys of Venus at almost 20 km/s of relative speed will be exploited to deflect the orbit out of the ecliptic. This will start with the third and fourth Venus flybys in June 2023 and April 2025, which will raise the inclination to 21 degrees. The nominal mission will conclude at the fourth Venus flyby. A fifth and a sixth Venus flyby in July 2026 and October 2027 may form part of an extended mission and increase the inclination to almost 27 degrees relative to the ecliptic, equivalent to as much as 34 degrees relative to the Sun's equator. While both the Solar Probe Plus and Solar

Orbiter are solar observation missions, their orbits and operations will be similar to planetary encounters with bursts of activity at perihelion and then long quiet cruise periods during which data will be downloaded, instruments calibrated and trajectory corrections performed.[105,106,107,108,109]

A proper planetary mission has been competing for a position as a medium-class mission since at least the mid-2000s. The Marco Polo asteroid sample-return mission was initially to be a joint ESA-JAXA venture, the name itself being a reference to the Venetian merchant and explorer who traveled to China under Kublai Khan in the late thirteenth century and provided the first information in Europe about the existence of Japan and the nature of its culture.

Marco Polo would sample a primitive asteroid in order to study the conditions and evolution of the nebula from which the solar system was born and to seek evidence of the organics that could have provided Earth with the 'building blocks' of life. The target was to be the dormant comet Wilson–Harrington, with launch in April 2018, rendezvous in 2022, and then return to Earth in 2026. Alternatively, it could launch in 2017 or 2018 to the unnamed asteroid (162173) 1999 JU3 and return to Earth in 2024 after spending 17 months orbiting its target. A number of different architectures were evaluated, including chemical and ion-propulsion. A Japanese-led Marco Polo would envisage a Hayabusa-based spacecraft carrying a European static lander derived from Philae as well as several rovers and hoppers. Alternatively, an ESA-led Marco Polo would exploit sampling technologies developed in the context of the Deimos sample-return study, and employ a small combined orbiter and lander to sample as many as three different sites and lift off in the weak gravity of the asteroid. As an even more complex alternative, a lander could be used for subsurface sampling and then lob the sample canister towards the mothership, hovering nearby. Although more risky, this would collect unweathered material from beneath the surface and demonstrate some of the technologies needed for a Mars sample-return. In any case, the mission would return at least 30 grams (and ideally up to 100 grams) of asteroidal material to Earth.

Marco Polo was a candidate for the "Cosmic Vision" program of 2015–2025, but at over 600 million euros it was far above the 475 million cap for a medium-class mission.[110,111,112,113,114] While JAXA turned its attention to the Hayabusa 2 mission, European scientists developed Marco Polo-R in an effort to reduce the envisaged cost to 470 million euros without sacrificing too many of the scientific objectives. Marco Polo-R would target (175706) 1996 FG3, a rapidly rotating, primitive, carbonaceous binary object with a diameter of 1.4 km and a small satellite some 400 meters across. It is believed that rapidly spinning asteroids such as this would readily form satellites by shedding some of their material if they were loosely bound 'rubble piles'. 1996 FG3 was discovered in March 1996 by the Australian Siding Spring Observatory and was soon recognized to be a suitable target for a space mission, requiring only a small velocity change to be reached from Earth orbit.[115] Several windows were identified between 2020 and 2024 for launch by a Soyuz-Fregat and Earth return dates as late as 2029. On reaching the target, the ion-propelled Marco Polo-R would enter orbit in a plane perpendicular to the direction of the Sun where it would remain for 6 months, flying about 10 km above the asteroid. In this orbit, the probe would be continuously illuminated by the Sun for its solar panels to generate power. It would identify and characterize up to five suitable landing sites before a landing was attempted at one of them. Meanwhile, cameras and a laser altimeter would determine the shapes of the two components, investigating their orbital motions, rotations, and gravity fields in sufficient detail to infer their internal structure.

Marco Polo-R was proposed to ESA as a medium-class scientific mission in 2012, with JAXA contributing one instrument and NASA possibly another. A total of 47 proposals were received by ESA on this occasion, including probes to Venus, Uranus, and Trojan asteroids. Four missions were selected to undergo an initial assessment prior to one being picked for implementation in February 2014. Marco Polo-R was again one of the finalist, with the other candidates being a mission to characterize the atmospheres of exoplanets in a search for evidence of life, an X-ray observatory to investigate rapidly changing phenomena, and an experiment in fundamental physics to test the principle of equivalence. Two competitive industrial studies were started, and smaller companies were asked to propose a sampling system. The technologies under study include rotary brush wheels and grab buckets, both of which are to be tested in microgravity starting in 2014. While awaiting the final selection, the Marco Polo-R team changed its target to asteroid (341843) 2008 EV5, a 400-meter-wide spheroidal C-class object which radar imaging had revealed to sport a ridge running parallel to the equator and a 100-meter cavity, probably a crater. There will be launch opportunities to 2008 EV5 in 2022 and 2023 for a 4.5-year mission duration. There is a backup in 2024, but with a duration of 6.5 years. Compared to 1996 FG3, the new target offers a shorter mission and requires a smaller velocity change. Moreover, the entire mission would be carried out nearer the Earth and in a more circular solar orbit that would simplify thermal control, and, finally, the sample capsule would return to Earth at a slower speed. Unfortunately, however, it was not selected.

Together with OSIRIS-REx and the two Hayabusas, Marco Polo-R would have provided a better understanding of the diversity of near-Earth objects.[116,117,118,119]

RUSSIAN PLANS

NASA and ESA are not the only agencies working on low-perihelion solar missions. The Interhelioprobe (or Intergeliozond) that was studied in Russia is similar to ESA's Solar Orbiter. It was conceived in 1995 by the Russian Academy of Sciences, by IKI, and by the German Max Planck Institute. In 1998 the Russian Space Agency funded a preliminary mission study. The mission would investigate the mechanisms of coronal heating, the structure and dynamics of the solar atmosphere and the solar wind, and the origin and propagation of certain solar phenomena. As first envisaged, the 430-kg spacecraft would be provided by Lavochkin and launched on a Soyuz. It would be protected from the solar heat by a conical carbon shield and draw power from a mix of foldable and jettisonable solar panels. It would carry around 70 kg of instruments, including several solar telescopes and coronagraphs, instruments to study the solar wind, magnetometers, and other instruments dedicated to sampling the interplanetary medium. Like the European Solar Orbiter, Interhelioprobe would

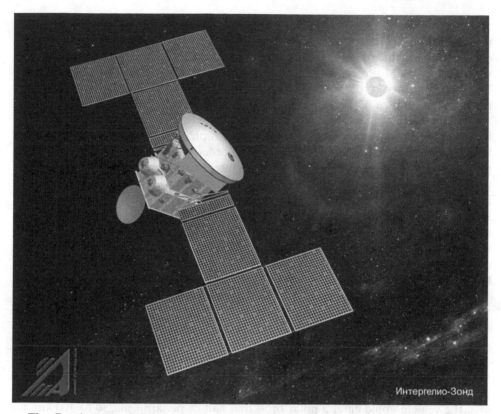

Интергелио-Зонд

The Russian Interhelioprobe (or Intergeliozond) electric propulsion solar orbiter. (Lavochkin Association)

make use of Venus flybys to reach its final orbit, with the help of ion thrusters. An initial series of flybys would reduce the orbital period and the perihelion distance to 30 or 40 solar radii (subsequently revised to 60 to 70 solar radii or some 40 to 50 million km). A second series of flybys would then increase the orbital inclination to almost 30 degrees out of the ecliptic. After many delays, and discussions about the possibility of launching two identical spacecraft, Interhelioprobe appears to be funded and proceeding to full-scale development for launch on a Proton in 2019.[120],[121]

Venus exploration, at which the Soviet Union excelled, still features prominently in Russian plans despite the fact that not a single spacecraft has been flown since the mid-1980s. Unlike Europe and Japan, which have both flown Venus orbiters recently, landers will be the focus of Russia's return to that planet. Given the technical skill of Russian scientists and engineers, two of the most important planetary projects under development are Venera-D (Dolgozhivyshii, long-duration) and Venera-Glob.

The objectives of the Venera-D mission would feature analyzing the atmosphere to measure its trace gases and isotopic ratios, as well as the composition, structure, and chemistry of the clouds, super-rotation, and the heat balance and runaway greenhouse effect. It would also investigate the geological structure and chemical composition of the ancient terrains, measure the isotopic ratios of the rocks, and seek indications of present volcanic and seismic activity. The priority targets for a Venus lander would be tesserae like Fortuna, Clotho and Tellus, and tessera transition terrains, followed by shield and lobate plains, and plains with wrinkle ridges. In fact, one of the objectives of the next wave (if any) of Venusian landers will be to seek rocks that are older than the "catastrophic resurfacing" that appears to have occurred around 500 million years ago in order to find out what Venus was like prior to that. For example, it would be interesting to find granite-like rocks predating the resurfacing, because granite forms in the presence of water. This could imply that, as some scientists suspect, the planet was once more Earth-like.[122,123] Knowing the isotopic ratios of Venusian rocks would also assist scientists trying to refine our understanding of the formation of our Moon. The current theory states it was formed when the proto-Earth was hit by a Mars-sized object with a different composition. This assumption is based on the isotopic ratios of Martian material, but the model may be overly simplistic, and some objections would be relieved if Venus were found to have similar isotopic ratios to Earth, enabling the impactor to be assumed to have been more Earth-like.[124]

For Venera-D, an atmospheric entry module would be launched along with a small orbiter. The technical details and specifications of the spacecraft have evolved over the years since the first proposal in the 2000s. It would use a Lavochkin Fobos-Grunt Unified Transport Module weighing about 600 kg for the orbiter and carry a payload of 40 to 60 kg. To study the exosphere, ionosphere, and atmospheric mass loss, the vehicle would enter a polar orbit that had its periapsis at 250 to 300 km, its apoapsis in excess of 60,000 km, and a period of about 24 hours. The nominal duration of the orbital mission would be 2 years (subsequently increased to 3 years). The place of the Fobos-Grunt Earth-return rocket and capsule would be reassigned to a spherical entry module (other sources depict a trunco-conical capsule). Inside the capsule would be a 150-kg lander carrying 15 to 20 kg of payload, some of it possibly inside a thermally controlled enclosure for a longer life. The primary objectives of the lander would be to take infrared images of the surface during the descent over the night-side from an altitude of 40 to 45 km, analyze the composition and physical characteristics of the atmosphere from 60 km, measure the concentration of water below 20 km, measure the isotopic ratios of oxygen and xenon, and carry out compositional measurements of the surface.

Drawing upon studies for the DZhVS (Dolgozhivushaya Veneryanskaya Stanziya; long-duration Venusian probe) project of the 1980s, the Venera-D lander was initially designed to operate on the surface for 30 days to undertake detailed seismic studies, possibly using water ice as a heat sink to facilitate the use of essentially conventional electronics operating at temperatures of the order of 200°C. And like the DZhVS, the payload was to have short-life and long-life instruments in separate compartments.[125] But these objectives were deemed overly ambitious and the survival requirement for the lander was reduced to 24 hours (still a significant engineering achievement over the several hours of the old Venera landers) and then

A drawing of the Venera-D Russian Venus orbiter and lander mission.

to just an hour to an hour and a half in addition to an hour spent making the descent. The revised lander would have an array of panoramic and close-up cameras with resolutions better than a millimeter, and instruments to analyze the atmosphere and surface. Two mission profiles were studied: one where the lander would be released during the approach phase to enter the atmosphere at interplanetary speed, and another where it would be released by the orbiter near apoapsis on one of the early orbits in order to enter the atmosphere at the next periapsis at a speed of less than 7 km/s.

Like the Vega missions, Venera-D would carry at least two aerostats, each having a 20-kg payload, either on the same entry capsule as the lander or on separate capsules with inflatable heat shields. One balloon was to fly at an altitude of some 60 km and the other one 10 km lower to study meteorology and the chemistry of the clouds and atmosphere. Both balloons would survive for at least 8 days, sufficient for the super-rotating atmosphere to complete two circumnavigations of the planet. Moreover, they could carry small 'drop zonds' to be discarded as ballast and to report atmospheric profiles as they fell. The high apoapsis was designed to provide long periods in which to track the aerostats, and after the end of the mission of the latter the apoapsis would probably be lowered to 10,000 km.

After first being proposed to the Russian Academy of Sciences in 2003, Venera-D was included in the 10-year Federal Space Budget for 2006 to 2015. A conference at IKI in Moscow in 2009 was to have marked the end of mission studies and the start of actual development. If the estimated $55 million (or 300 million euros according to a different, more realistic source) was made available, the mission could have been launched in early December 2016 to reach Venus in mid-May 2017. In any case, the project would rely on a large measure of international cooperation, possibly flying instruments from the UK, Italy, France and Hungary. Reflecting the many changes to the mission and spacecraft design, the launch mass was reported as being between 4,580 kg and no less than 8,100 kg, and the launcher could be a Proton-M or either the new Angara multirole rocket or the Soyuz-2. The mission eventually entered preliminary design in 2011 with the balloons deleted in favor of a small subsatellite that would carry plasma and radio science instruments for ionospheric studies for which orbital periods of 12, 24 and 48 hours were being considered.

The second Russian project is Venera-Glob. This consists of an orbiter carrying a high-resolution radar to characterize prospective landing sites, and up to six landers and aerial probes. A number of studies of long-duration landers were carried out, one of which would possess as many as eight wheels with hermetic motors, transmissions, and brakes. A panoply of flying probes have also been proposed, varying from long-duration 30-day balloons to variable altitude balloons, as well as genuine aircraft. A small kite-like paraglider 'vertolyet' flying probe was initially designed for the Vesta mission in the 1980s. It was to remain aloft for up to a month at an altitude of 50 km carrying a 20-kg payload. The aircraft could carry an Italian subsurface radar like that of Mars Express or Mars Reconnaissance Orbiter.

Following the loss of Fobos-Grunt in 2011 during the Earth-escape maneuvers by a cause that is disputed, Venera-D was delayed to the early 2020s at the earliest due to its complexity and the need to finalize its mission design and objectives, and Venera-Glob was postponed indefinitely. Unfortunately, the launch windows in 2021 or 2024 are incompatible with a ballistic landing in one of the scientifically desirable tessera regions.[126,127,128,129,130,131,132,133,134]

Lavochkin has also made a feasibility study of a Mercury lander dubbed Merkur-P (Pasadka, landing) which would investigate the chemistry, geology and seismicity of the surface. With a launch mass in excess of 8,000 kg, Merkur-P would consist of an orbiter and a small lander similar to the 'eggs' of Luna 9 and the old Mars missions. Russian engineers recognize that having the lander operate on the illuminated surface of Mercury for a long time would be extremely difficult owing to the heat, so only a brief mission would be possible. To reach Mercury, the spacecraft would employ the Dvina electric propulsion transfer module that is being developed by Lavochkin (for possible use by the Mars-Grunt sample-return). Although the envisaged launch date of 2016 was clearly unrealistic, the Merkur-P mission was adopted as one of the top priorities for Russian planetary exploration, and the Russian Space Agency considers it a worthy addition to US, European and Japanese orbiters. After the loss of Fobos-Grunt, it too was delayed to no earlier than 2026.

SON OF HAYABUSA

Having learned lessons from its first asteroid sample-return mission, Japan's JAXA set out to fly a successor, Hayabusa 2.[135] The new spacecraft, which draws upon the design of its predecessor to save time and money, is to investigate a more primitive body of the taxonomic C-class that may possibly be rich in organics. It will focus on both technology and science while perfecting the method for sample collection that is likely to return no more than 1 gram of primitive material. The preferred target (on a shortlist of ten) is the near-Earth asteroid (162173) 1999JU3, which was extensively observed in preparation for the mission. It is a dark object some 900 meters across, or about twice the size of Itokawa, rotating in somewhat less than 8 hours. The original baseline schedule was to launch in September 2010, reach 1999JU3 in July 2013, and return the samples to Earth in January 2016. In 2007 the Italian Space Agency began talks with JAXA to consider its participation in the mission by providing an ESA-developed VEGA (Vettore Europeo di Generazione Avanzata) advanced launcher, but in the end it was decided to employ a national H-2A202-4S rocket. The underfunded program was promised increased resources from the science and technology ministry in the wake of the successful return of Hayabusa and ESA's decision not to proceed with the joint Marco Polo sample-return. On the revised schedule, Hayabusa 2 will go in December 2014, fly by Earth in December 2015, and arrive at 1999JU3 in June or July 2018. Unlike Hayabusa, which had to conduct a hurried reconnaissance prior to attempting to collect its sample, Hayabusa 2 will arrive near perihelion and perform proximity operations for about 18 months, or more than one asteroidal year, to enable all operations to be adequately tested and rehearsed. Sampling will not be carried out until after a global characterization phase is completed in February 2019, followed by a cratering experiment nominally set for August 2019. It will then leave the asteroid in December 2019 and return to Earth in December 2020, entering the atmosphere at around 11.6 km/s. There are backup launch windows in June and December 2015 but for those the ion engines would need to thrust 96 per cent of the time (rather than a nominal 80 per cent) in order to reach the target.

Hayabusa 2 will have a total launch mass of 600 kg, some 90 kg heavier than its sibling – this mostly representing redundancy and performance increases. The boxy spacecraft bus will also be slightly larger at 1.0 × 1.6 × 1.25 meters. Solar panels will generate 1.4 kW of power when near the 1.4-AU aphelion. Like Akatsuki and unlike Hayabusa, the probe will be equipped with two planar high-gain antennas: one for X-band communication, and the other for the Ka-band capable of returning data at up to 32 kilobits per second. Prior to returning to Earth, all scientific data will be stored on a 1-gigabyte data recorder. For propulsion, the spacecraft will be equipped with four uprated 10-mN-class ion engines fueled by a total of 50 kg of xenon, capable of an overall velocity change of 2 km/s. A conventional chemical hydrazine and nitrogen tetroxide propulsion system and twelve 20-N thrusters will provide attitude control as well as more responsive trajectory control, with an improved plumbing configuration to preclude the leaks and problems which plagued Hayabusa. Improved autonomous guidance systems should prevent a repeat of the rough landing on the asteroid that is believed to have been partly responsible for the fuel leak and subsequent problems. Furthermore, these systems should enable flight controllers to leave the spacecraft to fly unattended for days at a time while it is thrusting.

The instruments on the main spacecraft will be a mix of Akatsuki and Hayabusa heritage. There will be three CCD cameras, including a fisheye to show the landing site from horizon to horizon, a laser altimeter, a near-infrared spectrometer operating at different wavelengths to Hayabusa and tailored to identify hydrated minerals, and a thermal infrared imager. The sampling system will be a heavily modified version of the original that is theoretically capable of returning a larger sample. Four projectiles will be carried instead of three, as well as five target markers to enable the system to obtain three separate samples instead of two. The sampling horn itself will be able to collect millimeter-sized particles on its rim even if a projectile is not fired. When the vehicle lifts off, dislodged particles will make their way up the funnel into the sample chamber. The system should collect about 100 milligrams of soil on each sampling run. In addition to the chambers for the samples proper, the sample container will be able to retain noble gases released by asteroidal dust for further analysis.

As in the case of Hayabusa, most of the scientific operations of the mission will be undertaken near solar conjunction, when the asteroid, probe, Sun and Earth are more or less lined up, since that will simplify the designs of both the solar panels and the communications system. At the actual time of solar conjunction in December 2018 the spacecraft will retreat to a safe distance while communications are suspended for one full month. Operations will be divided into a distant global mapping phase and a low altitude phase prior to releasing a number of landers and undertaking sampling. Like Hayabusa, the sampling runs will involve flying a descent profile over the day-side of the chosen point on the asteroid and will be fully autonomous from 100 meters down – when the target marker, made of highly reflective fabric and filled with soft material to prevent it rebounding, will be released. The target markers will be identified by the camera and used to measure and null the horizontal velocity relative to the surface of the rotating asteroid, in order to hover. Again like Hayabusa, the main spacecraft will carry deployable payloads. The most interesting of these will undoubtedly be a small impactor with an explosive charge that should create a crater several meters wide and at least 50 centimeters deep. The spacecraft's final touch down will attempt to sample this new crater in the hope of finding organic molecules and hydrated minerals that should have remained unchanged since the formation of the solar system. In any case, the experiment should provide information on the strength and internal structure of a near-Earth asteroid. The impactor will be a 30-centimeter-diameter cylinder massing 18 kg, of which 9.5 kg is explosive. After releasing the impactor at a height of 500 meters the spacecraft will maneuver to the other side of the asteroid for safety. Some 45 minutes into the operation, the shaped charge will detonate, deforming a 2.5-kg disk of copper into a bullet and accelerating it to 2 km/s in less than 1 millisecond. It will strike the asteroid with a kinetic energy much lower than Deep Impact's at comet Tempel 1, but will do the job. The impact will be monitored by a deployable camera similar to that used by the IKAROS solar-sail demonstration mission.

Also deployed will be three MINERVA (Micro/Nano Experimental Robot Vehicle for Asteroid) Japanese robots based on Hayabusa's single hopper, massing about 1 kg each, and a Mobile Asteroid Surface Scout (MASCOT) hopper supplied by Germany. MASCOT was conceived for the joint Marco Polo mission, and the DLR (Deutsche Zentrum für Luft- und Raumfahrt; the German center for flight and spaceflight) was invited by JAXA to contribute it to Hayabusa 2. It will be a 30 × 30 × 20-centimeter shoebox-sized battery-powered robot with a mass of 9.4 kg, and will be released by springs at a height of 100 meters during either a dedicated maneuver or a sampling run. It will nominally operate fully autonomously, with the possibility of intervention from Earth in case of anomalies. It is expected to explore up to three different sites during two asteroidal 'days' (about 16 hours) before its battery expires. An eccentric spinning mass will be used to initiate hops and to right MASCOT after each landing, with each hop peaking at a height in excess of 200 meters and lasting about an hour. Four scientific instruments will be mounted on the hopper, with the data returned to the spacecraft for relay to Earth. An infrared hyperspectral microscope will investigate the mineralogy and composition of the surface, a radiometer will measure its thermal properties, and a wide-angle camera equipped with LED lights to illuminate the scene during night-time will take panoramas. The payload will be completed by a flux-gate magnetometer.

Despite financial difficulties, especially following the catastrophic earthquake and tsunami of March 2011, the 31.4 billion yen ($367 million) required for the mission was assigned by the Japanese government in January 2012. The basic structure of the spacecraft was completed in late 2012 and systems tests began early in 2013.

Given the excellent performance of the H-IIA rocket, JAXA invited proposals for several small 'piggyback' payloads to be released in solar orbit. At first, it seemed an aerocapture demonstrator would be sent to enter orbit around the Red Planet, but it now appears that no fewer than three minisatellites will fly. Weighing in at 59 kg, PROCYON (PRoximate Object Close flYby with Optical Navigation) developed by the University of Tokyo in cooperation with JAXA will demonstrate high-efficiency X-band transmission and precise radio navigation. It will also demonstrate imaging technology for very close flybys of small objects. On returning to the vicinity of Earth one year into the mission, and using a miniature electric thruster, PROCYON will be redirected to pass within about 30 km of a near-Earth asteroid in early 2016 with the option of two more possible flybys later on. The 30-kg Artsat 2 was developed by the University of Tokyo in conjunction with the Tama Art University. It is a spiral-shaped artsy structure realized using 3-dimensional printing techniques that will demonstrate telemetry transmission using a unidirectional antenna and communications over the social networks. Interestingly, it is to generate audio samples from data obtained by its sensors. Being battery powered, Artsat 2 is expected to transmit for no more than a week, during which time it will recede 3 million km from Earth. Finally, if the mass budget permits it, the H-IIA will also carry the 15-kg Shin'en 2 microsatellite for the University of Kagoshima to test long-range communications and demonstrate the use of thermoplastic carbon composite in a spacecraft structure.

JAXA has also made preliminary plans for an 'operational' sample-return,

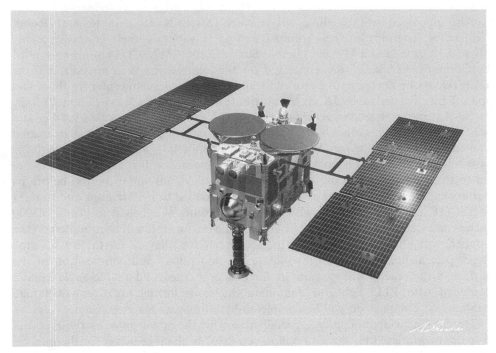

A rendering of the Hayabusa 2 Japanese near-Earth asteroid sample-return mission. (JAXA)

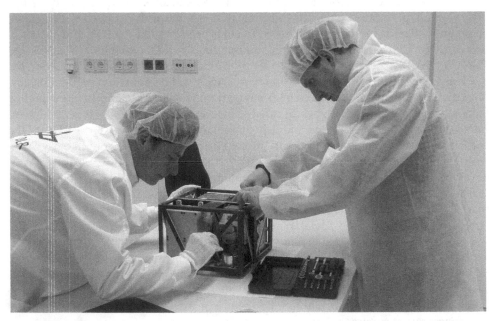

The minuscule German MASCOT hopper that is to fly with Hayabusa 2. (DLR)

A test specimen of the Hayabusa 2 asteroid impactor. (JAXA)

dubbed 'Hayabusa Mk-2' that will represent a complete departure from the original concept by being several times larger. It will be sent to sample the extinct comet-cum-asteroid Wilson–Harrington. However, this will not fly until after Hayabusa 2, which means, if successful, sometime in the 2020s.[136,137,138,139,140,141,142,143,144]

Other potential Japanese missions, discussed elsewhere in this volume, include the MELOS Mars probe, the Jupiter solar sail, and the SELENE 2 lunar lander.

THE INNER PLANETS

After BepiColombo, there are no currently funded missions to the innermost planet. A lander would probably be the top priority for Mercury, since it could address some scientific questions better than an orbiter; in particular, the composition of the rocks, internal structure, magnetic field and geological history.

A landing on Mercury would be the objective of the Russian Merkur-P, mentioned above, and it was also the focus of the only Mercury mission study for NASA's 2011 decadal survey. Proposed by APL, this was squarely in the flagship cost range, being estimated at in excess of $1.5 billion regardless of whether it flew an entirely ballistic trajectory like MESSENGER or made use of electric propulsion like BepiColombo. The study recognized that the mission would be extremely complex, needing not only to reach Mercury, in itself a difficult target from an energy point of view, but to do so at as low a relative speed as possible, and to land on a massive planet that lacked an atmosphere for braking. The probe was to decelerate using a large solid-fuel rocket in the style of the Surveyor lunar landers of the 1960s. A landing site at a high latitude and several days prior to sunset would probably have to be selected in order to reduce the thermal stresses to which the lander would be subjected. In fact, the mission was expected to last only a few weeks, with operations occurring mostly during the night and ending the following Mercurian morning when the Sun was high enough to cause hardware failures. Ironically, even though the planet is so near the Sun, the lander would probably have to be powered by an isotopic source.[145]

As for the second planet of the solar system, after Venus Express and Akatsuki scientists reckon (perhaps a little too optimistically) that the orbital reconnaissance of that planet will essentially be complete. Consequently, most studies for medium-term Venus exploration again envisage a variety of atmospheric and landing missions such as the Russian Venera-D. But these stand no chance of flying before the mid-2020s. Of course, the preference given by space agencies to Mars over Venus derives in part from the fact that while human missions to the Red Planet are very likely to occur by the middle of this century, it would be remarkable if humans ever set foot on the hot surface of our 'sister planet'. And whilst missions to Mars take longer to complete, it is much easier to obtain data from there using orbiters and landers than it is from the surface of Venus. Nevertheless, the discovery of many hot planets orbiting in close to other stars has renewed interest in studying Venus.

As already briefly mentioned, in 2012 the Indian space agency ISRO revealed that in parallel with a Mars probe it had been carrying out preliminary studies of a Venus orbiter. This could be launched in May 2015 using either a PSLV or the larger GSLV Mark III, and arrive at Venus five months later. The mission would study the atmosphere and surface using various scientific instruments, possibly including a mapping radar. In addition, preliminary studies of Venus orbiters have been carried out in China, which has reportedly expressed its willingness to contribute to an international exploration program. In 2005 Japanese engineers and scientists started to study a post-Akatsuki water-vapor balloon to cruise at the relatively low altitude of 35 km, just beneath the main cloud deck. The balloon would be deployed by a low-drag, fast-descent capsule based on Hayabusa hardware that was delivered by a drum-shaped carrier that would employ aerocapture to achieve orbit. Balloon deployment would initiate at an altitude of about 50 km and be complete on attaining its cruising altitude. The main challenges would be to endure a temperature in excess of 200°C for a minimum operating life of 14 days and to use high-temperature solar cells for power. The 1-kg payload would include a Very Long Baseline transponder for interferometric tracking by terrestrial radio-telescopes, as was done for the Vega balloons in 1985.[146,147]

ESA has investigated a Venus Entry Probe (VEP) for its own planetary Technology Reference Studies. While not part of the formal science program, such studies sought to identify technologies that would be required in order to develop low-cost missions focused on science. The VEP study (which was one of the most advanced of its kind) envisaged a three-spacecraft mission consisting of a high altitude science orbiter and relay satellite, a polar orbiter, and a small long-duration balloon. The balloon was to fly for at least a fortnight at an altitude held constant at around 55 km by periodically releasing 'ballast' in the form of 15 to 20 microprobes, each the mass of a cellphone, that would fall to the surface. They would have about 10 grams of payload that might include miniaturized thermometers, barometers, and a light-flux-meter. In addition to receiving the transmitted data, the balloon would 'poll' each probe to reconstruct its trajectory. To increase the survival time of a probe, the electronics would be mounted on an insulated section into which heat would penetrate only slowly during the rapid descent. The technologies to be developed for this mission included long-lived power generation systems for the gondola of the balloon, lightweight structures, advanced propulsion systems, etc.[148,149,150]

Another possibility considered was the Lavoisier mission proposed by the French Service d'Aéronomie and bearing the name of a French scientist associated with the study of gases. Like the Eos proposal of the 1960s by the same Service d'Aéronomie, Lavoisier would consist of a flotilla of balloons to study the dynamics and chemistry of the lower Venusian atmosphere. After one day, each balloon would deflate and the payload would descend for a brief surface mission. It was proposed to ESA in 2000 as a 'flexi' mission and again to CNES in 2002.[151]

Building on the VEP and Lavoisier studies, European and Russian scientists came up with a proposal for ESA's "Cosmic Vision" program for 2015–2025. In its original form this very ambitious European Venus Explorer (EVE) was to involve an orbiter, the VEP balloon and microprobes, JAXA's low-altitude balloon, four small descent probes, and perhaps even an atmospheric sampler that would return to Earth. A more reasonable scenario would be a Venus Express-derived orbiter, a balloon for a 1-week flight, and a short-duration descent module and lander whose design drew on Russian expertise. Although EVE was not selected for Cosmic Vision, it was well received in terms of its scientific objectives. Several Venus missions have recently been proposed to ESA as future large missions, including balloon platforms, radar-equipped orbiters, descent probes, short-duration landers, and aircraft.[152,153,154,155]

NASA has also considered balloons for Venus exploration. In the early 2000s JPL and Arizona State University proposed VEVA (Venus Exploration of Volcanoes and Atmosphere) as a Discovery mission that would involve a Pioneer Venus-style entry

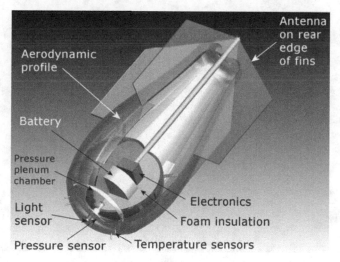

One of the European Venus Entry Probe (VEP) instrumented microprobes which would be released all over the planet from a balloon. (ESA)

A rendering of the European Venus Explorer (EVE) balloon. (ESA)

capsule carrying an atmospheric probe, a 1-week-duration balloon, and a number of 'drop sondes' to take aerial imagery of geologically interesting sites. The objectives were to better determine the composition and isotopic ratios of the atmosphere, to investigate any remanent magnetic field in the crust, and to obtain 'airborne' optical imagery for correlation with the Magellan radar data.[156] The first decadal survey also recommended the adoption of a Venus In-Situ Explorer to address the question of the varied evolutions of the 'terrestrial' planets and their atmospheres and volatiles, plus a number of other questions specific to Venus including the origin of its runaway greenhouse, the possible presence of an ocean early in its history and the mechanism that apparently resurfaced the planet in geologically recent times. With these issues in mind, NASA Glenn studied a mission involving both high-altitude aircraft and long-duration rovers similar in size and capability to the Mars Exploration Rovers. Like the Soviets and their DZhVS concept in the 1970s, the US engineers were faced with the alternatives of either using high-temperature resistant hardware for their rovers or creating a 'thermal enclosure' for electronics and all temperature-sensitive hardware. They devised an interesting 'third way' in which only temperature resistant hardware would be on the rover, because all of the computational and control electronics would be in a solar powered aircraft. Glenn tested a high-performance electronic chip which was designed to be used in the exhaust of jet engines and it proved capable of working for 1,700 hours at a temperature of 500°C. This could facilitate long-duration Venus missions, along with high-temperature radio systems and wheel motors.

Venus has been deemed the best location in the solar system to fly a solar powered aircraft because: the intensity of sunlight at high altitude is similar to (if not actually greater than) that on Earth, and the clouds are so reflective that extra panels could be installed on the underside of the vehicle; the pressure at high altitude is Earth-like, yielding a well-understood flight environment and wind speeds; and the slow spin of the planet would enable long flight times in continuous daylight at precisely the altitude at which the largest fraction of solar energy is absorbed, thereby allowing direct measurements of greenhouse processes. The options include a small aircraft with a foldable wing configuration that could fit inside the heat shield of one of the 'small probes' of the Pioneer Venus mission. Simpler aircraft have been studied for Discovery missions.

The mission in the Glenn study would comprise four radioisotope powered rovers, each of which had its own airplane carrying the computer and other electronics, four additional scientific aircraft, a relay satellite, and a number of seismometers and drop probes released by the rovers and airplanes. Owing to the large mass to be launched from Earth, the mission would use a nuclear electric propulsion stage.[157,158,159,160] A more recent and rather more plausible study called for a wind-driven landsailer. This 3-wheeled cart would have high-temperature electronics onboard and a 5-meter sail that was covered with high-temperature solar cells on both sides for propulsion and power. The cart would need only two motors: one to steer the airfoil and another for the front wheel. The notional mission would last almost a month (9 a.m. to 3 p.m., local solar time), driving for 15 minutes per day and covering a distance of about 30 meters each time. Of course, the landing site would have to be flat, debris free, and windy. The Venera 10 site appeared suitable in these terms. Interestingly, landsailers could be adapted to the Martian and Titanian environments.[161]

Simpler Venus In-Situ Explorer studies at JPL and NASA Ames have focused on smaller, more conventional landers that would provide access to the highland tesserae that are believed to be the oldest exposed terrains on the planet, predating the recent resurfacing. But they are also some of the roughest sites and would probably present steep slopes and other obstacles to a successful landing. Hence Ames engineers have proposed a probe that has an auto-rotor and an obstacle avoidance system to make a helicopter-like soft landing.[162] The result of these studies was the SAGE finalist for the New Frontiers program, which is likely to be proposed again in the second half of the 2010s.

No fewer than three mission concepts were studied for the second decadal survey, hopefully indicating a resurgence of interest in our 'sister planet' after 20 years of neglect, at least in the US. Of these, the simplest was the SAGE-like VITaL (Venus Intrepid Tessera Lander) proposed by the Goddard Space Flight Center. Another was the Venus Mobile Explorer which would analyze the chemistry of the surface at two sites located several tens of kilometers apart and determine the origin, evolution, and interactions of the atmosphere with the surface. Moreover, the mission would search for evidence of an ancient water-rich surface. The probe would consist of a gondola for instruments and a metallic helium-filled bellows for buoyancy. After finishing its rock analysis during 20 minutes at the first landing site, the bellows would be inflated and, after discarding the helium tanks as ballast, the lander would fly about 10 km to another location. Imaging from an altitude of several kilometers would be carried out during the ascent, flight and descent. The whole surface mission would last at most 6 hours, including 3 to 4 hours in flight. As in the case of the Soviet Veneras and other proposals, the spacecraft bus would perform a flyby of the planet and relay the data from the surface probe to Earth. The Venus Mobile Explorer, however, would require extensive technology development which would put it beyond the cost cap of a New Frontiers mission. A more conventional mission to deliver multiple landers to make such measurements might offer an alternative with a lower risk which could sample a larger number of locations for the same cost.

Finally, the Venus Climate Mission would use an orbiter, a long-lived balloon, two drop sondes and a miniprobe to undertake a 3-dimensional study of the atmosphere. The orbiter would carry essentially only an atmospheric monitoring camera and relay hardware. The battery-powered 8.1-meter balloon would float within the moderately warm band at a height of around 55 km, well above the cloud base. During its 21-day science campaign the super-rotating wind would carry it around the planet five times. The balloon would be a scaled-up form of a prototype developed at JPL. During the entry and descent inside a Pioneer Venus-like capsule, the balloon would release the miniprobe, a titanium vessel only several tens of centimeters in diameter that would transmit data during its 45-minute fall to the surface. During its flight, the balloon would release the drop sondes, each about half the size of the miniprobe. The mission was judged to be fairly mature and worthy of a low-budget flagship. In fact, it was one of the flagship missions recommended if there were to be an increase in NASA's budget for planetary exploration.[163,164,165]

Balloons ought to provide important experience for developing a Venus sample-return mission, as envisaged by both NASA and ESA during the late 1990s.

The ESA approach consisted of a two-launch architecture using Ariane 5 rockets to send an ion propelled orbiter and a large lander to Venus during successive launch windows. The orbiter would arrive first, enter a high apoapsis capture orbit and then use aerobraking to lower and circularize its orbit. Then the lander would be launched. Despite probably exploiting Russian expertise, the spacecraft would be quite different from the Venera landers. In fact, it would do without a pressurized vessel and instead employ heat exchangers to admit 'air' at the ambient pressure and at an acceptable temperature. As with the Veneras, heat sink materials would prevent the electronics from overheating during the necessarily brief surface mission. Atmospheric samples would be obtained at various altitudes during the descent to assess the presence of water vapor, the variability of chemical composition with height, and the role, if any, of present day volcanism. The surface aspect of the mission would last only an hour. During this time, the lander would perform experiments and collect about 200 grams of samples using two different devices. A drill would reach a depth of 50 centimeters to obtain a rock core. In addition, a 'vacuum cleaner' would be lowered to the ground. Helped by the inrush of the Venusian high pressure atmosphere into a vacuum tank, this apparatus would gain an atmospheric sample, dust and pebbles. Once the in-situ operations were complete, the sample canister would depart. Of course, to attempt to reach orbit directly from the surface would be prohibitive. The thick atmosphere and almost terrestrial gravity would require a rocket even more massive than is required for leaving Earth. The mission would use a 'rockoon' launcher, consisting of a small rocket suspended from a high altitude balloon. After inflation, using the same helium as was used to cool air inside the lander, the balloon of temperature-and-acid-resistant fabric would rapidly ascend to the launching altitude of 54 km, where the environment is much more benign. A launch canister carrying the solid-propellant ascent rocket (and providing

The bellows-like US Venus Mobile Explorer that would visit several sites during a period of a few hours.

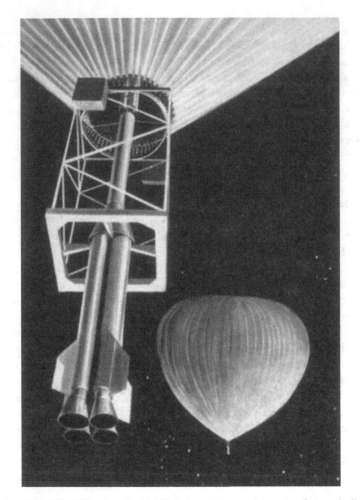

The US Air Force Farside was a multistage sounding rockoon (rocket + balloon) that reached altitudes in excess of 6,000 km in 1957. A similar concept would be used to place a sample canister in Venus orbit for ESA's sample-return mission.

thermal protection for it) would be correctly oriented to place the samples in the vicinity of the orbiter, and the rocket would be fired. Rockoons have been used for decades as sounding rockets and were envisaged for Canada's proposed da Vinci suborbital tourist spacecraft. But no orbital launch has ever been carried out from a balloon platform and it can be expected to require overcoming non-trivial guidance problems. Upon reaching orbit, the sample container would be tracked and recovered by the orbiter, acting autonomously, for return to Earth. A compromise would be for the samples to reach high altitude in a balloon for analysis over hours or days instead of just minutes.

The ESA study, completed in 1998, listed technological 'mission enablers' which would require a specific development effort. The cost of the mission was estimated at almost twice that of a cornerstone mission such as Rosetta but could be significantly reduced by employing a single, more powerful Ariane 5-derived launcher.[166]

The architecture of the NASA sample-return would be extremely similar, with the possibility of the ascent vehicle flying at high altitude on a balloon, blimp or airplane and having samples delivered to it from another balloon.[167]

A much simpler sample-return mission would involve a Stardust-class spacecraft flying through the upper atmosphere of Venus to obtain a gas sample. Although this material would be greatly affected by aerodynamic heating as a result of the probe's hypersonic velocity, it ought still to provide a precise measurement of isotopic and possibly also elemental ratios.[168]

GODDESSES OF DESTRUCTION

Having been the subject of several exploratory missions during the 1990s and early 2000s, near-Earth asteroids continue to stimulate the interest of scientists and space engineers. As several of these objects are being explored and sampled, space agencies are studying the best means of establishing a census of the dangerous ones, and are in preliminarily discussions about how a catastrophic impact might be prevented. The number of known near-Earth asteroids started to pick up in the late 1970s when two American teams began to photograph the sky to better define the probability of an impact with Earth. In the 1980s the first digital-camera surveys began. These were mostly replaced by fully automated surveys during the 1990s.

In 1998 NASA's Science Director of Solar System Exploration testified before the US Congress that the agency was "committed to achieving the goal of detecting and cataloging 90 per cent of the near-Earth objects larger than 1 km in diameter within 10 years". With 5,500 near-Earth asteroids of all sizes known by 2008, a hundred-fold increase with respect to 1978, this goal can be said to have been achieved. The residual threat is smaller objects which could do regional damage such as tsunamis if they hit the oceans or kilometer-sized craters on land. Consequently, there has been a call to find 90 per cent of the asteroids exceeding 140 meters in size, of which about 20,000 are believed to exist. A large number of 100-meter near-Earth objects, as well as a few remaining kilometer-sized ones, were located by the Wide-field Infrared Survey Explorer (WISE) satellite which was launched by NASA in 2009. The first space telescope entirely dedicated to the search for near-Earth objects, the Near-Earth Object Survey Satellite (NEOSSat), was launched by Canada in 2013 and will target objects exceeding 500 meters. JPL is proposing NEOCam for the Discovery program. Finally, large survey telescopes scheduled to enter service in the 2010s should be able to complete the census of this population.

There has recently been a growing urge to discover and catalog even smaller near-Earth objects. The airburst over Chelyabinsk on 15 February 2013 resulted from a 15-meter object exploding in the atmosphere high above the Russian city with the energy of a 500-kiloton bomb. While the meteor itself was harmless, thousands of people were hurt by the shock wave shattering glass. This airburst was the largest recorded since the 1908 Siberian Tunguska 'event'. It is worth noting, however, that the 2013 object was one full order of magnitude smaller than those expected to be discovered by future surveys and a full census of all the objects of this size, of which there are probably millions, would be impossible.

In parallel with surveys and Earth-based studies of the late 2000s and early 2010s, there has been a call for space agencies to stage a mission to the 320-meter asteroid (99942) Apophis, appropriately named after the Egyptian goddess of destruction. It was for some time the only object known to pose a non-negligible risk of impacting the Earth in the near future. Discovered in June 2004, Apophis was initially considered to have a 2.6 per cent chance of hitting Earth on 13 April 2029, until observations over a longer time span ruled this out. On that date it will pass just 30,000 km away, briefly becoming a naked-eye sight for observers in Europe and Africa.[169],[170],[171] But this flyby could have long term consequences because, owing to the uncertainties of its orbit, the range of possible positions where the asteroid could pass relative to Earth included a series of 'keyholes' that could cause it to return between 2034 and 2037 and pose a 1-in-5,500 chance of impacting in 2036. With a mass "equivalent to about 200 aircraft carriers", an impact with Apophis would cause a regional catastrophe.

The B612 Foundation, a non-profit organization devoted to protecting Earth from impacts, derives its name from the asteroid in Antoine de Saint Exupéry's novel *Petit Prince*. It drew Apophis to the attention of NASA, saying that the risk might warrant the launch of a small spacecraft equipped with a transponder to fly to the asteroid and enable tracking by radio-telescopes to either eliminate or confirm the possibility of an impact with ample forewarning.[172],[173],[174] Also, the Planetary Society in collaboration with several other associations and institutes, including NASA and ESA, announced an Apophis mission design competition to develop a mission which could rendezvous with and 'tag' the asteroid. Of the dozens of proposals, the top three were announced in February 2008.

The winning proposal, the $140 million Foresight by US Space Engineering Inc., would be launched between 2012 and 2014 and rendezvous with Apophis about 10 months later. It would use a US Minotaur IV launcher, a derivative of the Minuteman ballistic missile. For about one month after entering orbit around Apophis, Foresight would characterize it utilizing a multispectral camera and then recede to a formation-flying orbit several kilometers off and precisely measure the distance with lasers for 300 days while the vehicle was tracked from Earth in order to determine the orbit of the asteroid and facilitate analysis of its future trajectory. The runner-up was the $388 million A-Track of the Spanish firm Deimos Space. It was to start off by determining the mineralogy of Apophis employing a navigation camera, a multispectral camera, a visible and near-infrared spectrometer, and an infrared radiometer. The orbit of the asteroid would be reconstructed by precisely tracking the spacecraft and modeling the perturbations affecting both it and the asteroid. The third proposal was the Apophis Explorer, priced at about $495 million and advanced by the British Astrium branch of the European EADS aerospace giant.[175]

A number of other Apophis missions have been proposed by space agencies across the world. In 2009 Deimos Space proposed the Proba IP (InterPlanetary) mission to ESA as the next in its series of Proba technology demonstration missions.

This would test autonomous guidance, and deep-space navigation and control technologies using a miniaturized vehicle to reach Apophis.[176] Russia is working on an Apophis-tagging mission. This was first proposed by Lavochkin during a conference to mark the 100th anniversary of the Tunguska impact. The mission, based on Fobos-Grunt hardware, is currently planned in the early 2020s and appears to be funded by the Russian Space Agency. As an alternative, young engineers at various Russian space industries and institutes have recently proposed a small technology demonstration mission dubbed Anapa, after the Black Sea resort, to use solar electric propulsion to rendezvous with Apophis and another small asteroid. Finally, CNES in France has proposed a mission carrying a thermal and a visible camera to characterize the asteroid, a radio altimeter for mapping, and a radio transponder for precise tracking.[177] Interestingly, there was a Venus flyby trajectory for the auxiliary payloads launched together with the Japanese Akatsuki mission that would have resulted in an Apophis flyby in July 2012 but this opportunity was not pursued.[178]

Meanwhile, orbital tracking of Apophis up to 2009 has reduced the likelihood of an impact in 2036 to 1-in-250,000. Optical and radar observations in 2012 and early 2013 effectively ruled out this impact. However, a collision in 2068 is still remotely possible. Nevertheless, the close flyby in 2029 will be a unique opportunity to study a small asteroid and its internal structure because if the body is a 'rubble pile' there is every chance that Earth's attraction will alter its spin state, triggering internal failures, quakes, and mass adjustments. With this in mind, CNES initiated another Apophis mission study, this time calling for an ion propulsion vehicle to rendezvous with the asteroid 6 months ahead of its Earth flyby to deliver a network of small seismometer-equipped landers to study the tidal effects during the close pass of our planet, as well as transponders to further refine its orbit.[179]

As regards 'planetary protection', or the ability to deflect a dangerous asteroid on a collision course with Earth, the first theoretical study was Project Icarus carried out by students at the Massachusetts Institute of Technology in the 1960s. This imagined that the near-Earth asteroid Icarus was destined to collide with Earth in 70 weeks and considered how the impact might be prevented. The study recommended the use of six Saturn V rockets, as were then being manufactured for the Apollo lunar program. Launched one at a time during the period ranging from 72 to 5 days in advance of the projected date of impact, each rocket would detonate a 100-megaton bomb within 30 meters of the surface of the asteroid. Optical and radar sensors would be used to identify and target the asteroid.[180] But scientists were concerned that a project intended to use nuclear weapons to protect Earth could also open the door to the 'weaponization' of space. This became particularly evident when the first major workshop on protecting against near-Earth asteroids was held in the early 1990s and was attended by nuclear weapon designers as well as scientists, with the former, in the person of the "father" of the hydrogen bomb, Edward Teller, seizing the opportunity to present asteroids as sufficiently threatening to justify the development of much bigger bombs.[181,182,183]

The simplest means of protection (and one of the few that is doable with existing technology) would be to deflect the asteroid by exploding a nuclear weapon several tens of meters above its surface. The objective would not be to destroy the object, as the fragments could still either hit Earth or pass through 'keyholes' and return later. Instead, the blast would vaporize a surface layer, which would propel the asteroid like a rocket in the opposite direction and alter its heliocentric velocity by several centimeters per second. The explosive could be delivered by a Deep Impact-like kinetic impactor. The earlier a deflection is achieved, the better. Ideally, it would be done tens of years in advance. Hence the need to catalog all potentially hazardous objects.

Using less mature technologies, solar sails or mirrors could concentrate solar heat to 'boil off' rock and create jets of gas. The same effect could be achieved by a laser beam of the kind being developed to shoot down ballistic missiles. But the prototype weapon was so large that in trials it filled a Boeing 747! Other technologies would involve changing the reflectance of the asteroid either by painting it or by creating an impact that would splash out brighter ejecta, the intention being to modify its course by the Yarkovsky effect. Of course the method would have to be tailored to the nature of the threatening object, its surface texture, the orientation of its spin axis and rotational speed, its mass and a number of other variables, and the likelihood of success would depend on how much time was available.[184]

The prospect of Apophis hitting the Earth in 2036 prompted a number of studies of how it might be deflected. The simplest concept was an 'asteroid tugboat' relying on a nuclear fission reactor and equipped with relatively high-thrust plasma engines. After drawing up alongside the asteroid, the tug would anchor onto it using harpoons and guy lines. The first task would be to reorient the spin axis to the most convenient direction. Then the tug would impose an orbital velocity change of a magnitude that would depend on how much time was available. In the case of Apophis, if a velocity change of several centimeters per second could have been achieved in the 2010s it would probably have been sufficient. An asteroid deflection was one of the possible missions envisaged for the Prometheus space nuclear power initiative.

An elegant alternative to the anchored tugboat would use only gravity, making it a 'gravity tractor'. A vehicle with a mass of several tonnes would rendezvous with the asteroid and use ion engines to maintain a position several tens of meters above the surface, as Hayabusa did in station-keeping with Itokawa. Over a period of months (if not years) the center of mass of the asteroid-spacecraft system would be progressively displaced and the trajectory deflected. If the asteroid and the deflection requirements were small, then the tractor need be no more massive than a typical communications satellite. The B612 Foundation hired JPL to make a detailed analysis of the concept against a notional 140-meter asteroid. This concluded that the operation was feasible. Of course, reliable thrusting will be the key to stable station-keeping when flying in such close proximity to an irregular, massive rotating object.[185,186,187,188,189]

Having established in 2004 an external near-Earth object mission advisory panel, ESA has investigated the feasibility of mounting inexpensive missions to test kinetic techniques for orbit deflection. The agency has also funded six preliminary studies of near-Earth asteroid missions including three space telescopes to detect them and three rendezvous missions. The rendezvous missions were SIMONE (Smallsat Interception Mission to Objects Near-Earth) with a fleet of five microsatellites exploring objects of different spectral types; ISHTAR (Internal Structure High-resolution Tomography by Asteroid Rendezvous) using radar tomography to study the interiors of two near-Earth objects, one carbonaceous and the other stony; and Don Quijote to characterize a near-Earth object and test deflection techniques.

The highest priority was assigned to Don Quijote, and JAXA contributed to the preliminary industrial studies. The mission would consist of two separate spacecraft. The first, named Sancho, would enter orbit around a hundred-meter-sized asteroid with the usual payload of cameras, radiometers and spectrometers to characterize the surface of the asteroid and determine its mass. The second spacecraft, Hidalgo, would arrive months later and use an optical autonomous navigation system to strike it at a speed of about 10 km/s while Sancho observed the impact, possibly using a network of seismometers. Although kinetic impacts for deflection purposes could be effective against small objects, they would depend on such unknowns as the density, structure, and composition of the target. Don Quijote would provide real insight into how the characteristics of an asteroid would influence the orbital velocity change. A precise radio tracking system would enable the orbit of the asteroid to be determined prior to the test, and be able to measure any perturbation within several months. Hidalgo was estimated to be capable of changing the semimajor axis of such an asteroid's orbit by more than 100 meters by purely kinetic means.

The two spacecraft would be basically identical, although Sancho would include provisions for instruments and penetrators, while the 500-kg Hidalgo impactor would have simplified power, communications and propulsion systems. Two scenarios were studied. Initially, Sancho and Hidalgo were to be launched on a

A rendering of the proposed Discovery NEOCam (Near Earth Object Camera) to complete a census of small asteroids that fly close to our planet.

The European Hidalgo, part of the Don Quijote mission, impacting its asteroidal target under the watchful eye of the Sancho orbiter. (ESA)

single Soyuz-Fregat with an Earth flyby to separate their trajectories and a second flyby of an inner planet to delay Hidalgo's arrival until 6 months after its partner had achieved orbit around the asteroid. In the second scenario, the two spacecraft would be launched separately on low-cost missiles, possibly the Dnepr decommissioned

Russian ballistic missile. Sancho would set off in March 2011 and use plasma thrusters to reach the 380-meter unnumbered object 2002 AT4 in June 2015. With Sancho safely in orbit around the target, Hidalgo would be launched in December of that year to impact at 9 km/s in June 2017.[190,191] Unfortunately, Don Quijote was ruled too expensive and was not pursued further.

The space agencies are still interested in testing kinetic deflection techniques. The current AIDA (Asteroid Impact and Deflection Assessment) study actually consists of two independent missions. One is the Double Asteroid Redirection Test (DART), led by APL with input from many NASA centers, including JPL. It would target (65803) Didymos, a well-studied binary asteroid with a primary about 800 meters in size and a 150-meter satellite. The satellite has enabled scientists to estimate the density of the primary as almost twice that of water, indicating it is a 'rubble pile'. The spacecraft would have a mass of 300 kg, including ballast, and would be a simple design with a fixed high-gain antenna and extensive heritage from other APL missions, including a high-resolution camera based on the New Horizons telescope for both targeting and imaging. Costing no more than $150 million, it would be launched by a lightweight Minotaur V rocket. On reaching Didymos in October 2022, it would smash into the satellite at a speed in excess of 6 km/s. This solution is extremely clever, because the orbit change of the satellite following the impact would be far easier to measure than if a lone near-Earth asteroid were targeted in order to alter its orbit around the Sun. In fact, the impact would change the orbital period of the moon by roughly 10 minutes, enough for the variation to be easily detectable even from Earth. Of course, far better measurements would be obtained by a spacecraft that was in the vicinity, for example the European Asteroid Impact Mission (AIM). Initially envisaged as a two-spacecraft mission like Don Quijote, AIM is now a 150 million euros asteroid characterization mission for a single small, ion propulsion spacecraft launched by the VEGA medium-lift rocket. It would arrive at Didymos at least 2 months ahead of DART in order to study the two bodies and then take measurements of the impact, plume, and cratering process.[192,193,194]

Relatively simple rendezvous missions to asteroids are opportunities for emerging 'space powers' to venture into deep space and provide unique and worthy science. As mentioned, India and China plan comet flybys and asteroid rendezvous in the late 2010s. Brazil's INPE (Instituto Nacional de Pesquisas Espaciais, National Space Research Institute) has started preliminary studies of a low-cost project to be carried out in cooperation with IKI in Russia. The Brazilian researchers intend to send a 100-kg spacecraft to rendezvous with one of the near-Earth asteroids known to possess at least one satellite. It would use solar electric propulsion and its payload might include a multispectral camera, a laser altimeter, a near-infrared spectrometer, and a mass spectrometer. The primary target would be the unnamed asteroid (153591) also listed as 2001 SN263; with (136617) 1994 CC as a backup. Radar imaging has shown each of these asteroids to have at least two satellites. The primary target is a true 'ternary asteroid', 2.8 km across and with satellites 1.2 and 0.5 km in size, while the backup is smaller at 700 meters and the satellites are a mere 50 meters across. The mission design could even include orbits which would jump from one component of the triple system to the other. Brazil has long been developing a small national launcher, but it has failed all launches so far and the pace of development was dramatically set back by an accident that cost the lives of many technicians. Brazil is working on a project that would see larger Ukrainian rockets use its equatorial facility, and it is cooperating with China on Earth-observation satellites. However, no deep-space missions are included in the Brazilian space agency planning for the entire decade of the 2010s.[195,196]

South Korea has been developing a series of indigenous launchers for some years with mixed success, and is reportedly working on a lunar orbiter and lander for the 2020s. A simple deep-space probe to a near-Earth asteroid would be a logical follow on to that, although no plans have been officially announced.[197] Of course, as was to be expected when a South Korean lunar exploration program was made public, North Korea announced that it had started work on lunar probes. It successfully launched its first satellite in December 2012 but apparently the spacecraft itself never transmitted any data. A lunar mission would use a larger version of its Unha (Galaxy) ballistic missile and space launcher, a rocket which could be used for geostationary satellites and manned flights and could easily be adapted for deep-space missions.

In an interesting new development, the late 1990s and 2000s saw the private sector express an interest in demonstrating the feasibility of commercial exploration of near-Earth asteroids, including staking claims for future mining and resources utilization. The most active company in this field was SpaceDev of Colorado, involving students and teachers of the Space Institute of the University of California at San Diego and industry teams from Lockheed Martin. It envisaged developing a small, Near Earth Asteroid Prospector (NEAP) with a scientific payload consisting of cameras, neutron, alpha- and X-ray spectrometers, and other instruments, as well as four 'drop cans' that it would deploy. The company hoped either to have some of the payload funded through NASA's Discovery program or to sell some of its data to the agency. In fact, financing would mainly come from selling space on the probe for instruments and 'novelty' payloads, such as sending personal items to the asteroid. The 700-kg NEAP was initially to be launched by a Russian Rokot converted ballistic missile, but it was redesigned as a $50 million spacecraft which could fly as a secondary payload on an Ariane 5. Based on the experience of NEAP, SpaceDev hoped later to be able to offer a small bus for deep-space missions to the Moon and Mars. A numerical search was made of the 416 near-Earth asteroids then known to meet a number of constraints, including a maximum mission duration of 550 days, but all the feasible targets turned out to be small, poorly observed bodies. The nominal targets were (65717) 1993 BX3 or (249603) 1999 FO3, with flight times of 9 to 15 months. In 1998 the mission was retargeted to (225312) 1996 XB27, and subsequently to (4660) Nereus with launch in April 2001 and rendezvous in the following year. But Nereus was dropped because at that time it was the intended target of the Japanese MUSES-C mission (later renamed Hayabusa). As a result, the target was moved back to the unnamed 350-meter asteroid (65717) 1993 BX3. On approaching its target NEAP would reduce its relative speed over an interval of one month as a preliminary to matching the orbital velocity of the asteroid. In the characterization phase it would observe the asteroid with its instruments and drop packages onto the surface. About a month later the entire spacecraft would land. Unfortunately, the NEAP proposal collapsed in 1999 when it was realized that the available payload space would not be subscribed, not even by NASA as a mission of opportunity for the Discovery program.[198,199,200] In fact, the idea of selling mission data or instrument carriage on probes assumes a market that does not really seem to exist. In particular, no institute or university is actually able to afford to pay millions of dollars for an instrument unless it is helped financially by a national space agency.

Nevertheless, no fewer than two corporations and one foundation are now actively promoting private asteroid missions. A company called Planetary Resources Inc. was established in the US in 2012 by Internet and information-technology billionaires and certain other 'big names', with the goal of discovering and developing the means to mine near-Earth asteroids. As a first step, it wants to launch a swarm of 22-centimeter-telescopes into orbit to discover small nearby asteroids. These would be based on the Arkyd-100 bus, a tiny 11-kg satellite which uses as many off-the-shelf components as possible and can be launched as a secondary payload by many different rockets. The telescopes would perform a preliminary spectral characterization of the objects they find. Next, the company would fly small Arkyd-200 probes to some of the asteroids to assess their mineralogies, and swarms of Arkyd-300 spacecraft to multiple targets before starting mining operations in the 2030s for rare metals like platinum and water for fuel production.[201] In January 2013 a competitor emerged. Starting in 2015 Deep Space Industries intends to launch a series of small, low-cost asteroid prospectors as secondary payloads. The first step involves Firefly spacecraft that are some 25 kg in mass and employ technologies developed for Earth-orbiting 'cubesats' to explore near-Earth asteroids by imaging and spectral analyses. The Fireflies would be followed by Dragonflies, about 30 kg in mass and able to return up to about 65 kg of asteroidal material to near-Earth space

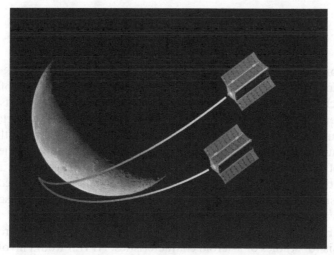

The Interplanetary NanoSpacecraft Pathfinder In Relevant Environment (INSPIRE) with a mass of 5 kg will be the first US nanosat to be inserted into solar orbit.

on trips lasting up to 4 years. They would be followed by true mining vehicles. The company also plans to address the technology required to produce metal parts from raw material in space by using 3-dimensional printers. But deep-space probes the size of Fireflies have not yet been successfully tested and the only miniature spacecraft launched into solar orbit, Shin-en by Japan, failed just after launch.

For this reason, NASA is funding JPL to develop a pair of satellites with masses of less than 5 kg for Interplanetary NanoSpacecraft Pathfinder In Relevant Environment (INSPIRE). These will be inserted into an Earth-like solar orbit sometime after 2014 to undertake simple science experiments using a magnetometer and a camera, and test long-range communications.[202]

Of course, the type of missions envisaged by Planetary Resources and Deep Space Industries raise the key issue of ownership, because the 1967 United Nations Outer Space Treaty states that celestial bodies cannot belong to any nation. Moreover, and more importantly, there are serious doubts over the economics of asteroid mining and in particular its financial viability and the likelihood of it generating a positive cash flow.[203]

A less controversial mission has been announced by the B612 Foundation. It is to launch a privately funded 50-centimeter infrared telescope in a Venus-like solar orbit to spot potentially hazardous objects on a mission similar to JPL's NEOCam. This Sentinel spacecraft may be launched as early as 2016 or 2017 for a 6.5-year mission, provided funding is forthcoming. The objective would be to detect 90 per cent of all near-Earth objects larger than 140 meters, 50 per cent of those larger than 50 meters, and well over half a million objects larger than 25 meters. The spacecraft would be based on Spitzer and Kepler heritage and use a single Venus flyby (which would not need to be exceedingly accurate) to achieve a 0.6 × 0.8-AU orbit from which it would be able to detect asteroids having perihelia within the Earth's orbit.[204,205]

MAIN BELT ASTEROIDS, HILDAS, TROJANS AND CENTAURS

In comparison to the near-Earth asteroids, those in the main belt seem set to continue to receive relatively little attention. With the exception of the Dawn mission, which has spent some time orbiting Vesta and is now on its way to do likewise at Ceres, the exploration of this class of object has been undertaken mostly by vehicles making opportunistic flybys en route to other targets. On the one hand this has meant that the visited asteroids represent a random sample of the population but on the other, with only a handful visited, our knowledge of the main belt is sparse to say the least. Only five types of the 24 asteroidal types identified so far have received a visit. Extending the concept of random flybys, European scientists and engineers have investigated an Asteroid Population Investigation and Exploration Swarm (APIES, 'bee' in Latin) mission that would provide a more significant statistical basis for studies of asteroid characteristics. APIES would involve flying an ion-propelled 'hub spacecraft', known as the Hub and Interplanetary Vehicle (HIVE), and a swarm of as many as nineteen identical Belt Explorers (BEE) on a trajectory through the main belt, with the BEEs distributed in a cloud some 0.1-AU in radius. This architecture would provide flybys of a hundred small objects during 6 years. Each BEE would be equipped with a low-thrust engine to maneuver to set up a flyby, a multispectral camera, and a radio science experiment for asteroid mass determination. The HIVE would provide the necessary propulsion into the main belt and then relay data between the BEEs and Earth.[206]

A 2013 proposal for a future large mission by ESA was INSIDER (Interior of Primordial Asteroids and the Origin of Earth's Water), which would explore several diverse large main belt asteroids, possibly representing primordial objects unchanged since the formation of the solar system. It would rendezvous with each one and enter orbit. On discovering a water-rich candidate it would release a Philae-style lander to investigate its composition and measure the deuterium-to-hydrogen ratio in order to determine whether water-rich asteroids could have provided the water in the Earth's oceans. The common wisdom that comets carried water to Earth is in doubt because the isotopic ratio of hydrogen to deuterium in the spectra of most examples studied to date does not match that of terrestrial water. Extensive deposits of water ice have also been found on the surfaces of main belt asteroids including Ceres itself, (24) Themis and (65) Cybele, and these would be interesting targets for a lander mission to carry out chemical and isotopic analyses.[207] The mission was baselined to orbit one or two water-rich asteroids, a multiple object with one or two moons, and a metallic object. Stony asteroids, being well represented among the objects visited (and also being the majority of the known population) would be deliberately avoided. Water-rich targets were (10) Hygiea and (24) Themis, where the lander would be released. The triple asteroid (87) Sylvia was also a candidate, particularly because its spectrum resembles that of Trojan and trans-Neptunian objects. And finally (16) Psyche, the largest of the known metallic asteroids, would be targeted. In this case, INSIDER would ascertain whether the object represents the core of an ancient differentiated object or was formed in a metal-rich state. Its internal structure would be studied to reconcile the relatively low density of metallic asteroids with the high density of metallic meteorites, the idea being that asteroids are significantly porous. To achieve this, the probe would employ a radar and a magnetometer (two instruments never previously flown on a dedicated asteroid mission) in addition to the usual remote-sensing cameras and spectrometers. It would require either a high-performance electric propulsion system for at least 12 km/s of velocity change, or advanced chemical propulsion augmented by flybys of Earth and Venus. The propulsion requirements could be significantly reduced by deleting (87) Sylvia, eliminating the need to achieve a high orbital inclination.[208] However, since the first new ESA large mission to be selected was the JUICE Jupiter orbiter, it was not unexpected that the second and third missions would investigate topics other than planetary science.

Following the Dawn mission, interest is expected to grow for revisiting the largest asteroids and possibly returning samples to Earth. The latest decadal survey judged a mission to Vesta with penetrator-mounted seismometers as well as explosive charges as triggers worthy of the New Frontiers program but it was not mature enough. For asteroid sampling, ESA has designed a subsurface penetrator and sampler inspired

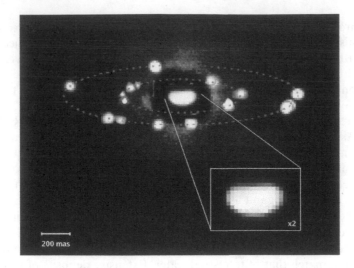

A composite telescopic image of asteroid (87) Sylvia and its twin satellites Remus and Romulus. (ESO)

by the mechanism utilized by wood wasps to drill the holes in which they lay their eggs. The mechanism consists of two teethed valves sliding one against the other to enable the drill to progress. The advantage of this drill compared to a conventional rotating one is that it does not require to be pushed against the surface of the material, since in the weak gravity of a small asteroid the applied force could tip the lander over.[209]

Orbiting at the outer rim of the main belt, and completing three orbits of the Sun in the time it takes Jupiter to complete two orbits, is the currently poorly known and unexplored family of asteroids named after the first and largest member discovered, (153) Hilda. About 1,500 of these Hildas are known, and most are dark objects with surfaces resembling cometary nuclei.

No outer solar system mission has yet visited the L4 and L5 Lagrangian points of the Sun-Jupiter system. Only the Ulysses mission has approached the L3 point at the antipode of Jupiter, and that occurred at aphelion in 1997–1998. Consequently, the Jovian Trojan asteroids are unexplored. Of an estimated population of 600,000 larger than 1 km, only 5,000 are known. Two contradictory theories have been advanced to explain the origin of the 'Trojan clouds'. In the first theory, these asteroids formed in orbits approximating that of Jupiter and were then captured at the Lagrangian points by the mass and gravity of the planet. In this case, they would represent a sample of the material from which Jupiter, Saturn, and their moons formed. In the recent, highly successful 'Nice model' of the formation of the solar system, all of the giant planets formed fairly close to the Sun and then their orbits became chaotic, causing them to move away from the Sun. The migration of Uranus and Neptune in particular would have scattered a broad disk of comet-like objects, flinging some into Jupiter-crossing orbits. The majority would be expelled from the solar system by the intense gravity of the giant planet. A certain percentage would become distant, irregular satellites of the planet. Some would be captured at the Lagrangian points to become the Trojans, and others would find their way to the main belt to form the Hildas. If this were the case, the Trojans and Hildas would be very similar to other 'cometary objects', such as the Centaurs and trans-Neptunians by virtue of having a common origin.

The Trojans are known to be quite dark, reflecting only a few per cent of the light of the Sun, and coated with reddish, possibly carbon-rich and organic-rich materials. Their spectra are similar to those of some short-period comets and they are said by some to be one of the sources of Jupiter-family comets. But the link is debatable and there are hints of two different populations within the Trojans. Ground-based studies have so far discovered only two members to be accompanied by satellites, thereby enabling a determination of their mass and average density. The first, (617) Patroclus, proved to have a mean density close to that of water, suggesting a highly porous object that is probably water ice-rich like a comet. On the other hand (624) Hektor, which is the largest of the Trojans, is more than twice as dense and therefore resembles the main belt asteroids.[210,211,212] Together with the Centaurs, the Trojans and Hildas will be the last major unexplored population of solar system bodies after New Horizons inspects Pluto and other targets in the Kuiper Belt. A Trojan tour and rendezvous mission that could fly by several such asteroids and then travel alongside one for an extended period of exploration is one of the highest priorities of the New Frontiers program. A proposal of the mid-2000s called for a Dawn-style mission to explore Hektor and at least another object. But unlike Dawn, it would use radioisotope electric propulsion in which an RTG powered both the spacecraft and its ion engines.[213] The Odysseus Trojan mission was proposed to NASA in 2008 for New Frontiers, but not pursued.

As already mentioned, however, an Atlas V-launched Trojan tour and rendezvous mission has been put forward as the fourth or fifth New Frontiers mission in the 2011 decadal survey. APL considered two mission concepts. The first was a purely ballistic opportunity, whereby the spacecraft would fly by Jupiter approximately 2 years after launch and be perturbed into an orbit that was external to the main asteroid belt with its aphelion in the Trojan cloud at the L4 leading Lagrangian point, which it would reach a decade after launch. The second option was for an ion-propelled craft which would reach the target directly, without gravity assists, after an 8-year cruise. It would then spend 5 years crossing the Trojan cloud. In both cases, the mission would target a large object for rendezvous, baselined as the 167-km (911) Agamemnon, where it would spend up to a year conducting an orbital mission similar to that of Eros by the NEAR mission. Before reaching the primary target, however, the probe would fly by other smaller Trojans along the way. No specific targets were listed, but the Trojan cloud was considered sufficiently populated, and the time it would take to fly through it sufficiently long, that suitable objects would be able to be found. In the case of the ion-propelled option the available margins might facilitate either rendezvousing with more than one object or landing on one. The baseline mission would characterize the bulk composition, physical properties, and surface morphology of a particular Trojan and survey several others. It would measure the abundances of key elements and the composition of the surface using a gamma-ray and an infrared imaging spectrometer, determine the presence of ice using a neutron spectrometer, and look for evidence of outgassing akin to cometary activity using an ultraviolet spectrometer. In addition, it would carry multispectral cameras possessing narrow and wide fields of view, a laser altimeter, a thermal infrared imager, and a radio science experiment. The spacecraft for both the chemical and the electric propulsion forms would use Sterling generators for power (up to six for the latter configuration). The chemical propulsion would use an advanced bipropellant engine that is currently under development at NASA. This was indicated as the lowest risk option achievable with existing technology and was estimated to cost less than $1 billion.[214]

In parallel with the US studies, a Trojan Odyssey mission has been proposed as a medium-class mission for ESA's "Cosmic Vision" program. It would be dedicated to the flyby of at least five Trojans and one Hilda, without entering orbit around any of them. The probe would be launched by a Soyuz rocket, make several flybys of Venus and Earth, and reach the leading L4 Lagrangian point some 7 years into the mission. It would make relatively slow flybys of one or two objects larger than 90 km that might be primordial, unchanged objects, and several smaller ones believed to be collisional fragments of larger asteroids. After leaving the vicinity of the Lagrangian point and heading back sunward the spacecraft would fly by a Hilda object. Like the US mission, the ESA proposal would require no significant technology development. It would inherit systems, instruments and structures from Mars Express, Rosetta and JUICE. Moreover, it would take advantage of the surveys planned for the 2010s both by ground-based and space telescopes that should significantly increase the number of Trojans and Hildas known. Unfortunately, the mission was not selected by ESA.[215]

Jupiter is not the only planet to possess Trojan asteroids at its Lagrangian points. A handful of Neptunian and Uranian Trojans are known, as well as, rather more interestingly, companions of Mars and indeed Earth.

The largest of the Martian Trojans is (5261) Eureka, a 3-km object that has a small moon. It was discovered in 1990. In many cases, these asteroids have orbits that are stable over the age of the solar system and may represent a sample of the primordial material from which Mars and the other 'terrestrial' planets formed. Eureka, its moon, and the other five objects at the Martian L5 point would also appear to be fragments of a larger, differentiated body. Unfortunately, despite their scientific interest, all of these objects have relatively large inclinations and hence are not suitable targets for a rendezvous mission. It has also been suggested that the Jovian and Martian Trojans would be ideal locations for an observatory whose task was to detect small asteroids on a collision course with Earth.[216,217,218]

Earth has at least one small Trojan, preliminarily designated 2010 TK7, currently following a chaotic orbit around the leading Lagrangian point. This 300-meter object would make a difficult target for exploration because of its high orbital inclination of 21 degrees.[219] There is also the curious case of asteroid (3753) Cruithne. This 5-km object orbits the Sun in an eccentric 364-day orbit. Although it appears to follow a bean-shaped horseshoe trajectory as seen from Earth, it is not, as it is sometimes said to be, the Earth's second moon. Its relatively high inclination makes it a challenging target for exploration. Nevertheless, a small, ion propulsion probe would be able to reach Cruithne in less than a year.[220]

There is a family of minor bodies beyond the orbit of Jupiter. These Centaurs are believed to be dormant cometary nuclei that are in the process of migrating into the inner solar system from the cometary reservoir of the Kuiper Belt. The first to be discovered was (944) Hidalgo in 1920 but they were not recognized as a group until the discovery of (2060) Chiron in 1977. At the time of its discovery, this had asteroidal characteristics and an orbit that took it from aphelion just outside the orbit of Uranus to perihelion inside that of Saturn. Over the following years it was seen to develop a coma as it headed to its first (and so far only) observed perihelion.[221,222] The Centaurs are hybrid objects capable of displaying asteroidal and cometary characteristics, and one of them is the smallest object known to possess a ring system.

A Discovery mission to fly by Chiron was proposed in the early 1990s using the backup Pluto Fast Flyby spacecraft. Employing RTGs for power generation, however, it would not match that cost-capped program. A Chiron Orbiter was a New Frontiers concept evaluated by the Goddard center for the most recent decadal survey. It would study the interior, surface, and atmosphere of this Centaur, monitoring its outgassing and outbursts. Five possible propulsion systems were considered, including different combinations of chemical, solar, and radioisotopic ion propulsion. In any case, even including Earth, Jupiter or Saturn flybys, the cruise phase would last 11 to 13 years in advance of 3 years of observing the target. All of the options either arrived at Chiron with too great a speed, or had small mass margins, or insufficient payload masses, or were short of power. The conclusion was that a Chiron Orbiter was not viable within a New Frontiers mission cost cap. But all was not lost, because other Centaurs like (52872) Okyrhoe or (60558) Echeclus, orbiting closer to the Sun, would address the same scientific objectives.[223]

THE GAS GIANTS

With the Juno mission due to arrive at Jupiter in the mid-2010s, JUICE to launch in the early 2020s and possibly a Europa mission as well, the Jovian system has become one of the top priorities for solar system exploration. All of these missions will target either the planet and its magnetosphere or the three outermost Galilean moons, but Io, the innermost moon, and the most volcanically active object in the solar system, will receive only distant observations. To fill this gap, an Io Observer was suggested by the National Research Council in 2007 as a New Frontiers-class mission and again as part of the second decadal survey. Moreover, an Io Volcano Orbiter was an ASRG-powered Discovery mission proposal.

An Io Observer mission for the New Frontiers program would seek to determine the internal structure of that body, the mechanisms of tidal heating and dissipation, its tectonics, chemistry, composition, atmosphere and ionosphere. The baseline mission would be launched in 2021 and, after several Venus and Earth flybys, arrive at Jupiter more than 6 years later. The probe would then enter a highly inclined elliptical orbit of the planet to spend most of its time outside the radiation belts except when passing through the plane of Io's orbit. This would ensure that a 'mild' radiation protection like that of Juno would suffice. The nominal mission would provide six to ten flybys at altitudes ranging from 500 to 90 km passing over the day-side and night-side of Io, with the final flyby in 2030. During each encounter, cameras would map the volcanic surface at a resolution of better than 1 km and in selected spots at 10 to 100 meters. A thermal mapper would locate and monitor volcanic hot spots when Io slipped into the planet's shadow, and a magnetometer would seek evidence that the moon had either an induced or intrinsic magnetic field. In order to look for rapid changes, at least two flybys would ideally pass over a given location. It might be possible to fly through a volcanic plume, the composition of which a neutral mass spectrometer would sample. And of course a spacecraft in such an orbit could also monitor the planet and make routine observations of the other large moons. Unlike the Discovery-class Io Volcanic Observer, which envisaged using an advanced radioisotopic power source, the New Frontiers mission would be powered by Juno-like solar panels.[224]

A Ganymede Orbiter very similar to the European JUICE mission was the focus of a JPL study for the 2011 decadal survey. No new technology developments would be required.

In the longer term, it is possible China may send missions to the outer solar system sometime during the 2030s.

In addition to the aforementioned missions to Titan and Enceladus, no fewer than four concepts involving the ringed planet and its satellites were assessed by the 2011 decadal survey. One, which was also recommended as a possible fourth or fifth New Frontiers-class mission, was a Saturn atmospheric probe to determine the structure of its interior, the abundances of noble gases, and the isotopic ratios of hydrogen, carbon, nitrogen and oxygen. Employing mostly existing technology, this would be a low-risk option. The only item that was not available off the shelf was the Stirling generator. On the other hand, the mission would be difficult to achieve in the 2020s because the planet would be as far out of the ecliptic plane as it ever gets, requiring substantial gravity assists at Venus and/or Earth to deflect the spacecraft out of the ecliptic. What is more, it would be the mid-2030s before Jupiter would be able to provide a gravity assist like it did for the Voyagers and Cassini. Orbital mechanics would thus dictate a launch no sooner than 2027, with arrival in either 2033 or 2034 after a single flyby of Earth. The mission would consist of a carrier and relay spacecraft that would have no instruments onboard, and one or two atmospheric probes similar to that of the Galileo mission with a mass spectrometer and an atmospheric structure package consisting of temperature, pressure, and density sensors. Each entry probe would be delivered on a collision course to hit the atmosphere at about 27 km/s, a significantly lower velocity than Galileo. It would descend by parachute to a depth corresponding to terrestrial sea-level pressure. The parachute would be jettisoned, and the probe would either use a smaller parachute or fall freely to an overall depth of 250 km. The other spacecraft would collect and relay the probe data whilst performing a fast flyby of the planet on a solar system escape trajectory.[225] A Saturn entry probe and flyby was also one of the concepts considered by ESA for a large mission to be launched in the late 2020s and mid-2030s.

Two decadal survey concepts were dedicated to Enceladus. One would fly through the water-ice plumes of the moon and use an aerogel collector like that of Stardust to return samples to Earth. The other, an Enceladus orbiter, was judged to have a greater scientific value and to be more attractive for an inaugural mission to this destination. It was ranked as the fourth-priority flagship-class mission. As a fairly conventional mission requiring no particular new technology development, it would take 8.5 years to reach Saturn in 2031 employing several flybys of Venus and Earth. It would then spend at least 3.5 years making repeated flybys of Titan, Rhea, Dione, Tethys and Enceladus itself prior to entering orbit around that moon. This phase of the mission would last 6 to 12 months with the spacecraft approaching the surface of the moon at distances of several tens to several hundreds of kilometers. It would map the magnetic and gravity field to a precision that was impossible for Cassini in order to investigate the internal structure and the presence of the subsurface ocean, monitor the temporal and spatial variability of plumes and hot spots and the conditions at their sources, and analyze samples of the plume material to determine the composition of large molecules, their chemistry, the chirality of organic molecules, the isotopic ratios, etc. It would also be possible to study how the plumes interact with the E ring, and the plasma and water tori within the Saturnian system. Finally, it would characterize the surface to assist in selecting future landing sites. The addition of a lander was considered, but it was apparent that even a small one would probably make the mission too complex and too risky, especially as the characteristics of the surface could range from loose snow to solid water ice.[226]

Titan figured prominently in the second decadal survey. JPL was directed to study a lake probe, seeking architectures that would fit either a New Frontiers or a flagship mission. Its mission would be to study the role of lakes in the moon's methane cycle, the composition and chemistry of the target lake, in particular the organic molecules and the isotopic ratios of the dissolved noble gases, to gain insight into the interior of the moon and its evolution over time. It would also be able to study the interior by measuring the response of the moon to Saturnian tides, repeating every 16-day orbit. The best way to address these objectives was deemed to be a mission that combined a buoyant lander and an independent submarine, performing a nominal mission lasting two orbits by the moon around Saturn. The architectures would differ primarily in the carrier vehicle, with the New Frontiers mission delivered to Titan by a 'dumb' carrier stage and the flagship option delivered by a TSSM-like orbiter. And because the basic New Frontiers option would communicate with us directly from Titan's surface, the target lake would require to have both a line of sight to Earth and to the Sun for the duration of the operation, which would pose significant constraints on its location. This option would also require the lander to have a large antenna, gimbaled to keep track of Earth in every sea state. And to target the northern lakes with a launch in the early 2020s an interplanetary cruise lasting no more than 6 years would be necessary, and that would mean a powerful and hence expensive launcher. In fact, it would be winter from 2025 to 2038 at the northern lakes and for a large part of this period the Earth would not be visible from that area of Titan. Southern hemisphere targets like Ontario Lacus would likely be too small for the predicted landing ellipse. Another possibility for a New Frontiers mission would be for the flyby carrier to serve as a relay. This would require it to have a large antenna in order to return the data to Earth. However, such a mission would have a significantly shorter operational time and the science would probably be severely constrained. Common to all of these options would be a payload consisting of a mass spectrometer to analyze the lake chemistry, a wide-angle camera to return several images every day showing waves, or floating material such as methane ice rafts or icebergs, or any detail of the shore if the probe were to drift close to that, and an imaging sonar to characterize the lake morphology and sediment layers. Unfortunately, all of the New Frontiers-class missions were deemed likely to exceed the cost-cap and a submarine would probably not be accommodated by these solutions. On the other hand, a flagship mission could include a lake lander and a passive submersible probe to reach the bottom of the lake. The battery-powered submarine would report data to the lander using a VHF link and then surface at the end of the mission to communicate with the mothership in case the lander had drifted out of sight. It would consist of two modules, one for instruments and the other for electronics. Upon reaching the bottom of the lake it would sample and analyze sediments. At the end of the mission, the instrument module would be let go and the other would resurface. Unlike the New Frontiers concept, a flagship using an orbital relay would be able to function during the winter season in the north polar zone.[227],[228]

A Titan mare lander, Titan balloons, as well as Saturn and Titan orbiters and Enceladus missions were under consideration for a future ESA large mission. The German space agency DLR conceived of an Enceladus Explorer that would land near one of the sulci terrains and send a tethered mole to melt its way through the ice into the liquid water reservoir so that this could be sampled and examined. On the other hand, a paddled boat for a Titan lake is under investigation by the Spanish aerospace firm SENER.

A further mission to Saturn was studied by JPL with input from the Glenn center as a possibility for the third decadal survey, for the 2020s. This Saturn Ring Observer would hover several kilometers above the plane of the ring system, essentially 'flying in formation' with ring particles in order to undertake a high-resolution study of the dynamics and interactions of particles in the rings, as well as the gaps, density waves, propellers, and edge-scalloping, to collect data that cannot be adequately simulated in Earth laboratories. Moreover, studying the dynamics of Saturn's rings would provide insights into the processes operating in the disks around young stars where planetary systems are believed to form. The mission would ideally target several sites along the span of the 'classical' A, B and C rings, in addition to ring edges, with several orbits dedicated to each site. Two properties make such a mission very challenging. One is the need to enter a circular orbit at the distance of the rings, which would require an insertion and circularization maneuver three to four times longer than a conventional propulsion system could provide. This is one reason why nuclear electric propulsion is considered particularly suitable for this kind of mission. The other issue is that a circular orbit with its radius within the ring system would involve flying through the ring plane twice per orbit, and that would probably

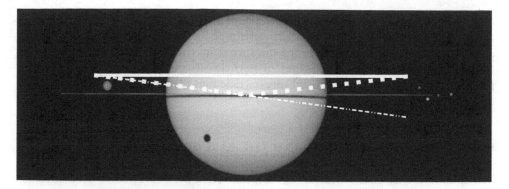

Three possible tight orbits around Saturn seen edge-on. The dashed-dotted line is for a Keplerian orbit that cuts through the ring plane twice per revolution. The continuous line is for a hovering orbit like the one of an ion-propulsion Saturn Ring Observer. The dotted line is for a chemical-propulsion Saturn Ring Observer that changes the plane of its orbit twice per revolution. The distances and the inclinations to the ring plane are exaggerated, and the vehicle would always be within several kilometers of the ring plane. (Background image: Erich Karkoschka of the University of Arizona Lunar & Planetary Laboratory and NASA)

destroy the spacecraft. Engineers might attempt to circumvent this problem by using a 'non-Keplerian' orbit in which the focus was displaced several kilometers from Saturn's center of mass. This would involve making propulsive maneuvers to continuously adjust the orbital plane, which would be readily feasible using electric propulsion. Another, more conventional possibility would be to enter an orbit at a low inclination to the ring system and make precisely timed engine burns just before ring plane crossings to modify the orbital plane. The spacecraft would appear to bounce up and down on the same side of the rings, never passing through. So long as the inclination was sufficiently small, the burns would be no larger than trajectory corrections; just much more frequent (at least twice on each orbit). Once the fuel was exhausted, the spacecraft would be doomed to cross the ring plane and in being smashed to smithereens it would add a new swarm of particles to the rings. Although this orbit design is attractive, achieving it would be a non-trivial task.

Such a mission was first proposed in the 1970s using nuclear electric propulsion, and several times later, most recently for the 2011 planetary science decadal survey. Various propulsion technologies were evaluated for achieving an orbit around Saturn, including electrical or high performance chemical engines, aerocapture in the planet's atmosphere, and a combined gravity assist and aerobraking at Titan. Even a relatively simple probe using an electric propulsion module and Stirling generators and launched by a relatively lightweight Atlas V could achieve most of the goals of the mission, although it would not have sufficient fuel to spiral all the way to the tightest orbits hovering over the inner rings; it could migrate inward only as far as the Cassini division that separates the A and B rings. Chemical propulsion would be significantly less capable. Only nuclear electric propulsion could achieve all of the scientific goals. The baseline mission would carry only two instruments. A narrow-angle camera with a spatial resolution of about 10 centimeters at the rings to image and permit analyses of individual particles for their spin, collisions, and behavior, and a laser altimeter to determine the distance from the ring plane, the thickness of the ring, and the off-plane velocities of particles. If mass margins permitted, the probe could release a small artificial ring particle that would enable it to better study impact dynamics.[229],[230]

URANUS, NEPTUNE, PLUTO AND THE KUIPER BELT

Missions to the ice giants Uranus and Neptune have been formulated ever since the fast flybys of the Voyager 2 mission in 1986 and 1989 respectively. An Outer Planet Science Working Group established by NASA in 1991 deemed Neptune to be more scientifically interesting than Uranus or Pluto owing to its more dynamic atmosphere and to the large moon Triton that undergoes geyser activity. Scientists were eager to know why the meteorology of Neptune was so different to (and more dynamic than) that of Uranus. In addition to the chemistry, dynamics and structure of the atmosphere, there was the issue of the internal heat source. The displaced magnetic field was a feature it shared with Uranus. There were the unexplained structures of the ring system. And of course the history, capture, tenuous atmosphere, activity, and geological evolution of Triton were (and still are) high on the list of scientific objectives for a mission to Neptune. And in the early 1990s Triton became attractive as an easy-to-access Kuiper Belt object. But lacking the benefit of nuclear propulsion, all orbiter missions to the outer solar system must deal with long flight times. Theoretically, a direct flight from Earth to Neptune on a low-energy transfer orbit will take 30 years if the vehicle is to arrive with a sufficiently slow relative speed to enter orbit around the planet. Flights using a Jupiter gravity assist could cut the travel time to between 14 and 19 years but such opportunities only occur on three or four consecutive years in each 12-year orbit by Jupiter around the Sun. Adding other flybys, for example of Earth and Venus, will not greatly shorten the flight time but it does permit the mass of the spacecraft to be increased and will expand the launch opportunities.

In the early 1990s the NASA advisory Solar System Exploration Committee came down in favor of a Mariner Mark II Neptune orbiter and probe. The recommendation of course was rendered inapplicable by the collapse of the Mariner Mark II concept in 1992.[231] The spacecraft would have been similar to Cassini, carrying an atmospheric probe similar to that of Galileo instead of Huygens, with uprated thermal protection and batteries. The payload of the orbiter would be similar but the optical instruments, and in particular the camera, would have been substantially uprated to operate in the much lower light levels so remote from the Sun. The mission was envisaged as being launched by a heavyweight Titan IV-Centaur in July 2002 and using flybys of Venus, Earth and Jupiter to reach Neptune after 19 years, whereupon it would spend 4 years touring the planet and its rings and satellites. An 'aggressive' science plan during the cruise would include observations of Jupiter and its satellites, a 2-year excursion into the magnetospheric tail of Jupiter, flybys of asteroids, and a distant observatory phase of the Centaur prototype Chiron over several years. The tour design for Neptune was unusual. The orbiter would initially be placed in an orbit revolving around the planet in the same direction as its spin. Upon reaching apoapsis, 100 days later, a propulsive maneuver would reverse the direction of travel against the rotation of the planet so as to match the retrograde orbit of Triton and provide slow flybys of it. Orbital tours of the system would be made possible by the massive moon Triton, much like Titan has facilitated Cassini's tour of the Saturnian system. Dozens of flybys of Triton carried out during the mission would far exceed the limited coverage achieved by Voyager 2, which mapped only half of the moon at a good resolution. Non-targeted flybys of the smaller satellites would also be carried out, but not of mysterious Nereid because that orbits too far out. The price tag for the baseline orbiter and probe was estimated at about $1.5 billion, which was enough to erode interest in pursuing it.

Following the cancellation of the Mariner Mark II concept, a lightweight Neptune orbiter using technologies developed for the 'Pluto 350' fast-flyby study was briefly considered. Using a cheaper rocket than the Titan IV, such a Neptune orbiter could be launched in 2002 and reach Neptune in 2018. Although it could not carry a probe or a lander and would be much less capable in terms of the data returned, such a mission could be flown for a fraction of the cost of a Mariner Mark II orbiter. A more sensible proposal, given the limitations of conventional propulsion, would be a Neptune flyby that dropped a Galileo-like probe into the atmosphere of the planet and performed a close flyby of Triton to re-examine it using state-of-the-art instruments.[232]

The first decadal survey of the National Research Council ranked the exploration of Neptune quite highly in its scientific priorities. The mission architecture envisaged in that context was to use aerocapture, possibly in conjunction with a Triton flyby, to enable the spacecraft to achieve orbit using as little fuel as possible. Probes could be dropped into the planet's atmosphere, one at the equator and the other at high latitude in order to measure elemental abundances and isotopic ratios. The orbiter would have the usual instruments to study the planet, its rings and satellites, and characterize the magnetic field, dust environment, etc. Of course, much of the mission time would be devoted to observing Triton, for which there would be a lander. Because a gravity assist by Jupiter would almost certainly be required, after the launch windows of the mid-2010s there would be a lengthy gap until the second half of the 2020s.

In the meantime, scientific interest in Uranus was picking up. In 1986 Voyager 2 revealed it to possess several unique characteristics. Its axial tilt was almost parallel to the plane of its orbit, creating extreme seasons. In addition, its magnetic field was not only tilted at 60 degrees to the spin axis but also displaced approximately 70 per cent of the way out from the center. In addition, its internal heat source proved to be a full order of magnitude weaker than that of Neptune.[233] In later years a number of large telescopes on Earth and in space revealed a wealth of new facts about Uranus, including hints of a more active meteorology than the bland globe seen by Voyager, new small moons and a second ring system.[234,235] And recently Israeli scientists have used Voyager data to convincingly demonstrate that the winds and atmospheric

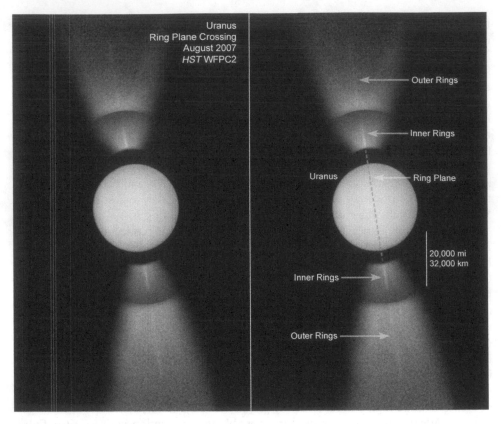

A Hubble Space Telescope view of the edge-on rings of Uranus in August 2007. (NASA, ESA, and M. Showalter of the SETI Institute)

dynamics on both Uranus and Neptune are confined to an outer layer no more than 1,000 km thick.[236],[237]

Scientists were also keen to re-examine the rings of Uranus, whose composition is unknown because Voyager 2 was unable to detect them at infrared wavelengths. Even more interesting were the satellites, because planetary scientists identified similarities between the Uranian and Saturnian systems. Titania and Oberon, the largest satellites, are sufficiently large that they would have retained a liquid water ocean beneath their icy surfaces if ever they had been melted. This would be easy to confirm by the magnetic field that would be induced by the uniquely shaped, asymmetric Uranian field. Ariel could be the Uranian equivalent of Enceladus, because having undergone large tidal flexing and melting it would have a rocky core overlaid by ice. Its relatively smooth surface could indicate that it was resurfaced in geologically recent times, with icy flows burying craters. What is more, the large ovoidal features on Miranda could be ancient, inactive counterparts of the currently active 'tiger stripes' on Enceladus.

Finally, Uranus and Neptune, being the ice giants of the solar system, could be considered as analogs of a certain type of extrasolar planet. Their satellites provide an environment where ices which are too volatile to occur on the satellites of Saturn and Jupiter, such as methane ice, can condense on the surface.

A proposal for a second fast reconnaissance of Uranus was advanced as the New Horizons Pluto mission was being prepared for launch in mid-2002. Scientists and engineers investigated whether there was scope for a scientifically worthwhile 'New Horizons 2' mission. The most interesting profile elaborated would require a launch in 2008 and a Jupiter flyby in 2009. After that, New Horizons 2 would head for an October 2015 flyby of Uranus, just 3 months after its sibling encountered Pluto. The Uranus flyby would take place relatively close to the planet's 2007 equinox, at a time when in the past observers had reported seeing a complex system of equatorial clouds and belts (and indeed they duly reappeared on cue). Furthermore, a flyby at this time would have facilitated imaging the hemispheres of the moons that were unavailable to Voyager 2 at the southern summer solstice. Afterwards, New Horizons 2 would be retargeted to make a September 2020 flyby of (47171) 1999 TC36, the first 'binary' Kuiper Belt object discovered (other than Pluto), consisting of a 600-km primary and a 200-km satellite, both of which were larger than any Kuiper Belt object that New Horizons was expected to encounter. After that, it would be retargeted to investigate a few of the smaller trans-Neptunian objects. The mission would thus provide a second reconnaissance of Uranus as well as backup for the Kuiper Belt, ranked by the first decadal survey as a key scientific objective for planetary exploration. Furthermore, it would mark the return to the 'traditional' NASA dual-spacecraft approach for outer solar system exploration.[238,239] Due to the interest in such a mission, NASA created a panel to determine whether it could be developed and flown within a short timeframe and at a cost considerably less than New Horizons. The panel came to the conclusion that the mission would have scientific merit but was unlikely to produce a 'paradigm shift' in planetary science. Because it would likely encounter financial, planning, and technical issues, it would not be able to be flown for much less than the cost of New Horizons itself. And finally, the earliest that a refurbished or new RTG could be made available was estimated to be 2011 whereas the Jupiter-Uranus launch window closed in 2009.[240]

A recent proposal for continuing the exploration of Neptune was JPL's Argo New Frontiers candidate to fly by Neptune and Triton and go on to a Kuiper Belt object. It was to exploit existing technology, mostly based on New Horizons, be powered by RTGs, and pursue a simple Voyager-like mission profile. After launch in 2019, it was to fly by Jupiter and Saturn in 2020 and 2022 respectively, Neptune in either 2028 or 2029, and a large Kuiper Belt object within 5 years of that. Overall, as in the case of Uranus, there was a scientific consensus that a second flyby of Neptune decades after that of Voyager 2 would yield a justifiable amount of new information. It would fill in before a future Neptune flagship mission, which was not realistically expected before the 2030s, by monitoring the evolution of the rings and ring arcs, and mapping the hemisphere of Triton seen only at low resolution and from a distance by Voyager 2, as well as the high northern latitudes that were in winter darkness at that time. Finally, it would monitor the atmosphere of Triton and the activity of its geysers and plumes. Ground-based observations had shown that the moon had warmed significantly since

The JIMO-like Neptune mission delivering probes at Triton.

1989. A flyby of the southern hemisphere of the planet was particularly desirable in order to complement Voyager's magnetospheric observations during its passage over the north pole. Beyond Neptune, the cone of space accessible to Argo would include dozens of large Kuiper Belt objects, including several binary ones.[241]

A similar mission to fly by Neptune and distant Kuiper Belt objects was proposed to ESA as a medium-class mission by the French ONERA (Office National d'Études et de Recherches Aérospatiales, national aerospace study and research office). This OSS (Outer Solar System) mission would combine the planetary sciences objectives of Argo with fundamental physics experiments in deep space. In particular, it would have an experimental 2-way laser link and extremely sensitive space accelerometers developed at ONERA since the 1970s to repeat Cassini's relativity experiment and seek deviations from general relativity on long distance scales. It would also address the so-called Pioneer Anomaly.[242] But this proposal was not pursued.

Another proposal to ESA from that time which had no more success was the Uranus Pathfinder mission. In 2021 a Soyuz-Fregat would launch a spacecraft derived from Mars Express and Rosetta. Jupiter being unavailable for a gravity assist, the vehicle would use flybys of Venus and Earth and possibly Saturn to arrive at Uranus 15 years later, precisely 50 years after Voyager 2's reconnaissance. Arriving at Uranus close to the northern summer solstice of the planet's 84-year orbit, the mission would provide a different view from Voyager 2, which observed that planet at its southern summer solstice. The vehicle would enter a polar orbit around Uranus with its periapsis at less than 2 planetary radii and its apoapsis at several hundred radii in order to enable it to collect data at periapsis and transmit it to Earth at a low bit rate during the long climb to apoapsis. The mission would use a radioisotopic generator powered by americium-241 rather than the plutonium-238 used by the US. Being a waste product of nuclear reactors, this would be readily available in Europe. However, americium yields only around 25 per cent of the heat per unit mass as plutonium, and the fact that it emits more neutrons and gamma-rays meant it would require heavier shielding.[243]

Flybys of Uranus, Neptune, Triton and Kuiper Belt objects figured prominently in the mission studies of the second decadal survey. For Neptune, the options included four classes of increasingly complex missions: New Frontiers flyby probes like Argo, a minimal orbiter, a small constellation of orbiters, and a flagship-class orbiter to provide a leap in knowledge for that planetary system comparable to Cassini at Saturn.

Studies were undertaken on Uranus and Neptune flagships consisting of an

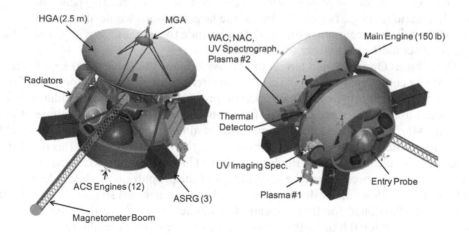

The 2012 decadal survey Uranus Orbiter.

orbiter and a shallow atmospheric probe that would survive to a depth equivalent to at least one Earth atmosphere and ideally up to five atmospheres. The Neptune mission was soon judged too complex. It would be more risky than a Uranus mission because it would probably require the use of aerocapture, there was no suitable launch window in the 2020s, and the flight time would be considerably longer. The Neptune flagship study was thus dropped early on. On the other hand, Uranus could be reached using solar electric propulsion and Earth flybys after a 13-year cruise and the annual launch windows were unconstrained by a requirement for a Jupiter flyby. A Uranus Orbiter flagship was recommended as the third-top priority in this category. At between $1.5 and $1.9 billion, it was also the cheapest of the flagships. The mission would investigate the dynamics and composition of the atmosphere, including measuring the noble gas abundances, study the peculiar magnetic field, determine the internal structure of the planet and the structure of its atmosphere. It would also observe the northern aspects of the satellites and the rings. The baseline mission, studied by APL, would launch on an Atlas V-class rocket in late July 2020 and return to Earth in June 2024 for a gravity assist that would stretch the aphelion of the spacecraft's orbit beyond Saturn. A solar electric propulsion module would then extend this to Uranus before being released. The vehicle would reach its destination in August 2033, close to the northern summer solstice. One month prior to arrival, a small probe would be released on a collision course to enter the atmosphere of the planet. The spacecraft would enter a polar orbit with its periapsis around 1.3 planetary radii so that it would not be required to travel through the main ring system. Because the Uranian system has similar satellite/planet mass ratios to the Jovian system, a Galileo-style tour of the four largest moons Ariel, Umbriel, Titania and Oberon was feasible if the spacecraft flew in an orbit that was more or less in the plane of the satellites but almost impossible from a polar orbit.[244] It would essentially fix the periapsis over the southern hemisphere, opposite Earth and the Sun, making it difficult to perform Doppler tracking at periapsis to study the gravitational field of the planet. After 2 years and 20 orbits, mostly spent observing the planet and its rings, targeted flybys of the largest moons would start, resulting in at least two close encounters with each. To conclude the mission, as with Cassini, the spacecraft could be sent closer to the planet.

The Uranus Orbiter would have a total mass at launch of some 4,200 kg, of which only 900 kg constituted the dry mass of the orbiter proper. The largest portion of the mass would be allocated to the solar electric propulsion stage that would have three redundant ion thrusters and two large fan-shaped solar panels similar to those of the Phoenix Mars lander but scaled to a diameter of almost 7 meters. The module would thrust for about 4 years until 5 AU from the Sun, then be jettisoned. At this point the spacecraft would be placed into hibernation. The only major technical advance which the mission needed was the solar arrays of the propulsion module, which were to be similar to those under development for the Orion manned spacecraft (and have since been abandoned for that vehicle). The orbiter would have three ASRGs for power, a conventional bipropellant engine, and smaller monopropellant thrusters for trajectory and attitude control. Its scientific payload would consist of a wide-angle camera for imaging at moderate resolutions, a visible and infrared mapping spectrometer, and a magnetometer to investigate the planetary magnetic field and its interactions with the satellites to seek evidence of intrinsic satellite magnetic fields and subsurface oceans. Plasma instruments and a narrow-angle camera would constitute an enhanced orbiter payload. A 2.5-meter-diameter high-gain antenna would return the data to Earth. The entry probe would sound the atmosphere of the planet to a depth equivalent to at least 1 atmosphere of pressure, determining the noble gas abundances, isotopic ratios, and atmospheric structure during a descent lasting around an hour. The battery-powered probe would be based on the 'small probe' of the Pioneer Venus mission, with a mass spectrometer, a nephelometer, and sensors to measure the temperatures, pressures and accelerations.[245]

A Neptune orbiter that would make multiple Triton passes was also under study at ESA as a large mission for the 2020s or 2030s. Another proposal would have flown twin spacecraft to Uranus and Neptune in order to conduct comparative studies of the two ice giants.

Very little mission planning has been carried out on post-New Horizons missions to Pluto. A technology assessment by ESA's Advanced Concepts Team investigated a Pluto orbiter carrying an RTG to provide power for both the probe and its ion engine. The 18-year interplanetary cruise would make the mission a good test of the Pioneer Anomaly.[246,247,248]

It also remains to be seen whether the handful of recently discovered dwarf planets in the outer solar system will be visited in the near future. Missions to these objects would undoubtedly require very long flight times but could complement and provide context for New Horizons' observations at Pluto. Of course, missions to such distant worlds would require significant technological progress to enable their instruments to operate effectively at light levels thousands of times fainter than at 1 AU.

The best known of these bodies is (136199) Eris, which was briefly saluted as the 'tenth planet' of the solar system when its discovery was reported in 2005. It was the existence of Eris that precipitated the controversial demotion of Pluto from

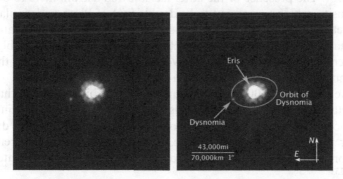

A Hubble Space Telescope view of the Kuiper Belt object Eris, together with its satellite Dysnomia. Comparable in size to Pluto, Eris is the largest known unexplored surface of the solar system, but it is unlikely that a probe will visit it before the 2050s. (NASA, ESA, and M. Brown of the California Institute of Technology)

planetary status, establishing conclusively that Pluto is actually one of the largest members of the Kuiper Belt and, as such, the easiest to observe. Eris is known to travel in an even more extreme and inclined orbit than Pluto, with its perihelion at 38 AU and aphelion at 98 AU in a plane at the very steep angle of 44 degrees to the ecliptic, with a period of 557 years. Surprisingly, at the time of its discovery Eris was close to aphelion and far off the ecliptic. Its diameter has been fixed rather precisely at 2,326 km, making it almost as large as Pluto. In fact, the diameter of the airless Eris is better known than that of Pluto, whose measurement is made difficult by the presence of an atmosphere. The similarities between Pluto and Eris are only skin deep. Thanks to the discovery of its small moon, Dysnomia, Eris is known to be significantly denser, and probably has a large rocky core. And because Eris reflects 96 per cent of the sunlight it receives, its surface must be covered with bright, pure, and probably young ice. It is the second brightest object in the solar system after Enceladus, which derives its high reflectivity from the plume activity that deposits snow on its surface. Perhaps Eris has an active surface too, despite being one of the coldest objects known in the whole solar system. Or perhaps its bright surface is evidence of a transient atmosphere that sublimates at perihelion (the most recent occurring in the late 17th century) and then snows out as it approaches aphelion.[249,250]

The strangest large object discovered in the outer solar system is (90377) Sedna, which completes an orbit around the Sun in 11,400 years. Even at its perihelion of 76 AU it is beyond most of the Kuiper Belt. Its aphelion is at 975 AU. When identified, it was at 90 AU from the Sun inbound for perihelion in 2076. Given its strange orbit, it is not even considered to be a Kuiper Belt object, but from the Oort reservoir of long-period comets. With a diameter 75 per cent that of Pluto, Sedna is one of the reddest objects in the solar system and it is even possible that, on nearing perihelion, it could develop a thin atmosphere as nitrogen ice on its surface starts to sublimate.[251]

Quaoar is unusual in that it is a 900-km object (as large as Ceres) with a satellite named Weywot. The presence of the latter has allowed astronomers to determine its mass and density, which turned out to be double that of Pluto and similar to that of a rocky planet or asteroid rather than an icy body. However, ices of water, methane and ethane have been spectroscopically detected on the surface, probably forming a thin, frozen crust. It is possible that Weywot formed during an impact that stripped Quaoar of most of its ice, reducing it to a rocky core.

Makemake is about 1,500 km across and has an orbit similar to that of Pluto. With a reflectance of about 77 per cent, it is the second brightest object in the Kuiper Belt. No atmosphere has been detected around it but sublimation may develop near the sub-solar point to brighten its surface. Lacking a satellite, measurements of its density are preliminary and an estimate of less than twice that of water has been disputed.[252,253]

The other giant of the Kuiper Belt, Haumea, is unusual in having a rapid rate of axial rotation, completing a full revolution in less than 4 hours. It is also known to be markedly non-spherical, being an ellipsoid with axes of $2,000 \times 1,500 \times 1,000$ km. Its two satellites, Hi'iaka and Namaka, could have formed in a giant impact which also caused Haumea to spin sufficiently rapidly to elongate its shape. Once again, its surface is known to be covered with ice. Like Eris and Quaoar, it appears to have a relatively high density and is probably a rocky core with a thick crust of ice.

Preliminary studies of New Horizons-like reconnaissance flybys of some of these objects have been carried out. Using a similar launcher to New Horizons and assisted by Jupiter flybys, a mission to Quaoar could take as little as 13.5 to 14 years with the launch window in 2016, but it is too late for that and the next window lies between 2027 and 2030. There is a window in 2033 for Sedna, but the voyage would last 25 years. A Jupiter-Eris launch window occurred around 2011, when Uranus was also in a good position for a gravity assist, but this alignment repeats only every 80 years. The next window for Eris is 2032. Interestingly, missions to Sedna and Eris by way of Jupiter would have fast encounters of their targets within 5 months of each other in 2051. While the Sedna encounter would occur at 78 AU from the Sun, Eris would be almost 93 AU away and still near aphelion.[254,255]

A more detailed exploration of the Kuiper Belt than would be possible using fast flybys will require advances in propulsion technology, in particular the introduction of more efficient systems. Radioisotope Stirling generators, and advanced engines such as NASA's Evolutionary Xenon Thruster and long-life Hall electric thrusters, could facilitate low speed rendezvous missions with Centaurs or Kuiper Belt objects with relatively brief flight durations. A flagship-class Kuiper Belt orbiter was recently studied by JPL and the NASA Glenn center to validate radioisotopic-powered electric propulsion. This showed that a Delta IV-Heavy-class rocket, radioisotopic electrical propulsion, and a Jupiter flyby, would enable a Cassini-class spacecraft to reach and enter orbit around a Kuiper Belt object located 32 AU from the Sun in as little as 16 years, which is comparable with ESA's Pluto orbiter proposal.[256]

COMETS FOR ALL TASTES

Scientists would dearly wish to send spacecraft to 'transition objects' such as (3200) Phaethon and (4015) Wilson–Harrington, which show, or have shown, characteristics of both comets and asteroids. Another case of a comet-turned-asteroid is the unnamed kilometer-sized object (196256) 2003 EH1 that appears to be the same as (or at least a fragment of) a comet reported in Chinese, Korean and Japanese annals, then briefly observed in 1491 and now realized to be the source of an annual meteor shower. But the high inclination of its orbit, more than 70 degrees, makes this mysterious object a difficult target for anything except a fast reconnaissance mission.[257] Another class of transition objects is represented by comet 133P/Elst–Pizarro, which shows a narrow tail but orbits entirely within the main asteroid belt. At the time of its discovery in 1996 this was speculated to be a mundane asteroid which had recently undergone an impact exposing volatiles at its surface. However, that explanation was challenged by renewed cometary activity over the years. It is now known that Elst–Pizarro is the prototype of a class of main belt comets. These seemingly ordinary main belt asteroids periodically develop comas and tails. Main belt comets were the focus of a proposal to ESA for a large mission because determining the isotopic ratios of their ice could establish whether they are the source (or one of the sources) of water in the terrestrial oceans.[258]

As regards Jupiter-family comets, an interesting possibility for rendezvousing with one or more would be to exploit the characteristics of some 'jumping' objects. These are comets that can become very weakly gravitationally bound to Jupiter, often while performing a 'hop' between an orbit unbound to Jupiter and one bound to it. A good example is comet Gehrels 3. This was in an 18-year solar orbit prior to 1969, became bound to Jupiter, and after some highly eccentric orbits of the planet, escaped into an 8-year solar orbit in 1973. If it had been possible to launch a probe into a very weakly bound elliptical orbit around Jupiter with its periapsis occurring in August 1970, that vehicle would have been able to make a slow speed approach to the comet using only 15 per cent of the braking burn that would have been necessary to rendezvous with it in solar orbit. To put this into context, the braking burn would have been just 25 per cent of that needed by the European Rosetta comet orbiter. Given the relatively minor propulsive maneuvers, this profile would be of particular interest for a sample-return mission. Although the obvious drawback is the need to predict which comets will be performing appropriate 'hops' in the near future, the catalogs show a number of viable candidates.[259],[260]

Another intriguing outer solar system object is periodic comet 29/Schwassmann–Wachmann 1. This is estimated to have a large nucleus tens of kilometers across, and it has the peculiarity of orbiting the Sun in an almost circular orbit just beyond that of Jupiter. It also experiences frequent outbursts that can increase its brightness tens-, if not hundreds-fold, for which no firm cause has ever been shown. Other comets reside in similar orbits, but none shows a similarly mysterious behavior.[261] Other outbursts, such as that of comet Holmes in October 2007, which brightened 400,000-fold within several hours and was visible to the naked eye as a fuzzy patch, are believed to be the explosive release of water vapor temporarily trapped by a thick 'air-tight' crust.[262],[263]

Given the interest in comets, there is a mounting urge to fly a mission to return a sample to Earth, as the original concept for the Rosetta mission planned to do in the 1990s. This would enable scientists to investigate the chemistry of complex organics and the isotopic ratios of volatiles, including water, that would not be preserved by an aerogel sample collector, as used by Stardust, due to the high speed of the encounter. Hence a New Frontiers comet surface sample-return mission was one of the studies of the second US decadal survey, with the objective of returning about half a liter of material to Earth. This would build upon the Stardust and Deep Impact missions, and serve as a pathfinder for a more complex flagship mission to collect a core sample of a cometary nucleus and preserve it at cryogenic temperatures for return to Earth. Such a spacecraft would use visible and infrared cameras to characterize the nucleus and the sampled locations, and determine whether material was collected in an active area or not. Samples would be collected from a depth of the order of the tens of centimeters by drills or other devices. The results of the Deep Impact mission, and the number of comets that have been observed to break apart, suggests that obtaining material from the surface of a comet should be a relatively straightforward task. (The Philae lander of the Rosetta mission will hopefully confirm this impression.) The New Frontiers spacecraft would return to Earth using conventional or solar-electric propulsion and release a capsule of a type that exploits over a decade of studies for the entry vehicle of a Mars sample-return mission, with the addition of a system to keep the sample at a temperature below $-10°C$ at all times during the return, entry and landing, in order to prevent its alteration.

While most short-period comets are believed to originate in the Kuiper Belt, long-period comets, as well as objects such as Sedna, appear to come from the Oort Cloud whose exploration is another 'Holy Grail' for astronautics. It is recognized that long-period comets must be the most pristine material in the solar system, as they rarely, if ever, undergo solar heating at perihelion, but there are many factors that make them some the most difficult targets to explore:

- Their apparitions are unpredictable. The comet possessing the longest known period is 153P/Ikeya–Zhang. Its period is 341 years, only two perihelia have been observed, and its return had not been predicted.
- They have random inclinations relative to the ecliptic. Encounters designed to occur near their orbital nodes may involve flying a spacecraft far from both the Sun and Earth.
- The Earth-comet-spacecraft geometry at the encounter may be unsatisfactory.
- Several months of observations are needed before the comet can be verified to be a true long-period one, and its orbit determined with sufficient accuracy to compute an intercept mission.
- Because they are usually discovered only a few months prior to perihelion, a mission would have to be launched soon and its flight time would need to be short.
- For the same reasons, planners cannot rely on planetary gravity assists.

Despite all of these constraints, studies since the 1980s have shown that even at the rate at which comets were being discovered at the time at least one suitable flight opportunity every year should be expected.[264] Given the much higher discovery rate nowadays, many more opportunities can be expected. For example, no fewer than 25 new long-period comets were discovered in 2004, as against just five in 1984. Small probes are probably particularly suited to flyby missions to long-period comets. They could be kept in storage waiting for a target to appear, with storage costs estimated at about $100,000 per year. A small launcher like the US Pegasus or a Russian recycled ballistic missile that was actually designed for long-term storage could be used.[265]

One of the best opportunities for such an intercept mission was presented in 1997 with comet C/1995 O1 Hale–Bopp. With a nucleus estimated at 40 km in diameter, it was one of the intrinsically brightest comets ever observed. It was discovered in July 1995, almost 2 years prior to its perihelion of May 1997, and has an orbital period in excess of 4,000 years. One of the nodes of its orbit was at a heliocentric distance of 1.1 AU, which was conveniently just a little more than the distance of Earth from the Sun. Profiles have been computed for a hypothetical mission that could have set off in early 1996 to intercept the comet at that node. The mission would have used a tiny 150-kg spacecraft carrying only a wide-field camera and a magnetometer in order to collect data on one of the most scientifically interesting bodies ever discovered.[266] It

Long-period comet Hale–Bopp, which passed perihelion in 1997, was one of the intrinsically brightest comets known, and would have been a perfect target for a reconnaissance mission to such an object.

will also be remembered that the failed CONTOUR could easily have been retargeted to encounter a long-period comet if a suitable target had appeared after the spacecraft had achieved its main mission. Had CONTOUR been in space in 1997, it could have encountered Hale–Bopp. Another good candidate for an encounter would have been comet Lulin, which was discovered in July 2007 and had perihelion in early 2009. Its orbit showed little indication of ever having interacted with planets, so it might have been making one of its earliest forays into the inner solar system. It would have been an easy target since its orbit is almost coincident with the ecliptic, but its retrograde direction would have yielded a very high relative speed at encounter.

A mission to a long-period comet would probably answer one of the fundamental questions in the study of the solar system, namely whether the different classes of comets – Jupiter-family objects, Centaurs, objects from the Kuiper Belt or the Oort Cloud – have different chemical, isotopic, and physical characteristics. We now have obtained a significant amount of data for a large number of objects by ground-based and space-based observatories, and a handful of encounters in space, but as yet there is no correlation between the dynamics of a comet and its chemistry.[267]

The 2003 decadal survey Comet sample-return concept.

Missions to long-period comets were mentioned in the first decadal survey, where their necessary synergy with ground-based telescopes was remarked. The report noted that either the discovery capabilities of comets at large distances from the Sun would need to be dramatically improved or a ready-to-go spacecraft would have to be stored on the ground or in solar orbit. A Discovery-class mission was mentioned again in the second decadal survey.

BEYOND THE HELIOSPHERE

In addition to investigating the Sun and the bodies that orbit around it, there is an urge for a mission to travel in the wake of the Voyagers and explore the boundaries of the heliosphere and the interstellar medium on the far side of the heliopause, to carry out in-situ sampling of the 'local bubble', a region of space over 400 light-years wide that may have been formed by a close supernova explosion. Measurements of cosmic rays, plasmas, neutral particles, and dust imply that the Sun has been traveling within this feature for the last 3 million years. Scientists also believe that the density of the bubble and of the local interstellar medium has an influence on the cosmic ray fluxes that reach Earth, influencing the atmosphere of the planet and in particular the ozone layer.

The idea for a heliopause mission was endorsed in the 2003 heliospheric science decadal survey, and two years later was included in NASA's "heliophysics roadmap". Since then, the IBEX (Interstellar Boundary Explorer) satellite has revolutionized our knowledge of the heliospheric boundary. It found an unexpected ribbon of emission from the heliopause that seems to indicate the orientation of the interstellar magnetic field. It also revealed that the interactions of the solar magnetosphere with the local interstellar medium are different and weaker than expected. The Sun is traveling at a significantly slower speed through the interstellar medium and in a different direction to what had been believed. In fact, it is slower than the interstellar medium, meaning that no supersonic bow shock forms ahead of the heliosphere, instead there is a broad and weaker bow wave.[268] An in-situ mission would directly study the structure of the heliopause and determine how the interstellar medium influences the dynamics of the heliosphere and, of course, vice versa, how the presence of the Sun affects interstellar space. To transit the heliopause and fly as far as 180 or 200 AU in a reasonable time, a mission would have unique propulsion requirements. The possible options include conventional propulsion, solar sails, nuclear electric propulsion, and RTG-powered ion engines. The NEP option would resemble the 1980s Thousand Astronomical Unit proposal, with a 'tug spacecraft' carrying the reactor and engines plus smaller probes that would be released after the thrusting phase. Solar sails and relatively close flybys of the Sun have been identified as a realistic technology to accelerate an interstellar probe for a reasonable launch mass and flight duration.[269,270,271]

During the mid-2000s, ESA studied an Interstellar Heliopause Probe in the context of its Technology Reference Studies. It would travel a distance of 200 AU along the solar apex over a period of 25 years. Given the mass and cost constraints, engineers identified solar sails as the most promising propulsion technology. The probe would first spiral inwards to a heliocentric distance of 0.25 AU in order to benefit from the intense pressure of sunlight as it began to accelerate away, and would jettison the sail on reaching 5 AU having attained the desired speed.[272] A solar-sail-propelled probe to reach 200 AU in 25 to 30 years has recently been proposed as an ESA large mission.

In the US meantime, APL, with substantial input from JPL, was studying an Innovative Interstellar Explorer using RTGs and ion engines as well as a Jupiter flyby. It would consist of a 'low risk', 500-kg spinning lookalike of Pioneer 10, dominated by a 3-meter-diameter high-gain antenna to return data at several kilobits per second from a heliocentric distance of 200 AU. The spacecraft would be propelled by at least two or three redundant ion thrusters and draw power from as many as six RTGs or Stirling generators. Protruding from the spacecraft would be 25-meter-long antennas for radio and plasma-wave detectors and a shorter boom for the magnetometer. A key issue for the mission would be the asymptotic speed to be reached on exiting the solar system, which in turn would affect the mass allocated to the payload. The baseline called for launch on a Delta IV Heavy rocket fitted with two additional solid fuel upper stages, or the most powerful version of the Atlas V, or the privately developed Falcon Heavy. Trajectories to the bow of the heliosphere with Jupiter flybys exist in 2014 and then approximately every 12 years (i.e. one Jovian year). After this flyby, the ion thrusters would need to fire for 15 years,

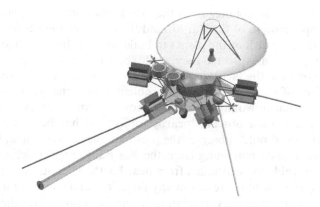

A rendering of the RTG-powered Innovative Interstellar Explorer.

enabling the Innovative Interstellar Explorer to reach 100 AU in 17 years and 200 AU in 29 years. Using the other giant planets for gravity assists would result in longer flight times: 33 years in the case of Saturn and 69 years for Neptune (owing in part to the fact that the latter would not be favorably placed to hurl the spacecraft to the bow of the heliosphere until later in the 22nd century). The probe would have a magnetometer, a plasma-wave receiver, and a plasma instrument to collect and analyze the solar and interstellar plasmas as well as ions 'picked up' by the heliosphere. Other instruments would include energetic particle detectors, dust sensors and analyzers, and a neutral atom imager. It was decided not to carry cameras for imaging during the Jupiter flyby. It is also to be noted that, in order to collect the extremely faint signal from the probe, arrays of modified deep-space antennas would be needed. The mission was estimated to cost about $1.6 billion, including the Delta IV Heavy launcher.[273],[274],[275] Of course, given the present lack of plutonium for the RTGs the mission may not be feasible for many decades.

The heliocentric distance of 550 AU is scientifically interesting as the first focus of 'solar lensing'. The gravitational lensing of electromagnetic waves by a large mass was first noted by Albert Einstein in applying his general theory of relativity, but was virtually forgotten until the late 1970s, when a 'twin quasar' image produced by the gravitational field of an interposed galaxy was discovered by English astronomers. It was then realized that electromagnetic waves grazing the Sun would be focused at 550 AU, with waves at increasing grazing distances being focused farther out. Hence, a suitably positioned spacecraft would have at its disposal a 'telescope' possessing a lens 1.5 million km in diameter (that of the Sun) which would focus electromagnetic waves arriving from beyond the Sun. SETI (Search for ExtraTerrestrial Intelligence) researchers have noted that this focal point would be a sort of 'magical location' for listening at radio frequencies for signals from other civilizations and for conducting interstellar communications. Another use of the focal point would be for kilometer-resolution imaging of extrasolar planets. Solar lensing of an Earth-size planet which was at a distance of 10 parsecs would produce an image no less than 3 km across, so enormously magnified, in fact, that a focal-point mission would require some means of propulsion in order to counter both the orbital motion of the planet around its star and the proper motion of the star around the center of the galaxy. Moreover, it would have to scan the disk of the planet to build up a complete image. As stringent as these requirements may seem, they could be met by existing ion propulsion technology. Apart from the fact that such a telescope would be able to observe only along a given line of sight, the technical problem is that the positions of both the spacecraft and the image would have to be known to an accuracy of a few meters. Consequently, rather than observe 'nearby' objects, it has been suggested that a focal-point mission should observe the cosmic microwave background, small temperature fluctuations remaining from the Big Bang, with a definition a billion times better than is able to be attained from near Earth. The rationale for this is that the microwave background is present in any direction, and so would not require the spacecraft to navigate accurately in relation to the focal point. But the corona close to the Sun would affect the image, reducing its definition until a much larger distance is reached. For the microwave background, the minimum distance is estimated to occur at 763 AU. There is presently no compelling scientific interest in making this kind of observation.

If ever such a mission is attempted, the challenge will be to design a vehicle that is capable of reaching the focal point in an acceptable time. A Voyager-class mission would require more than 150 years just to reach 550 AU. In 1993 a study called Focal was proposed to ESA as a medium-sized mission. It called for an 800-kg spacecraft powered by RTGs that would have an inflatable antenna and solar sail more than 50 meters across, and the objective would be to reach the nearest focal point within 25 to 50 years.[276,277,278,279,280]

PHYSICS MISSIONS

This survey of future solar system missions would not be complete without mention of several proposals which have addressed physics experiments and problems in deep space. The best known example is LISA, described elsewhere in this chapter.

Lots of papers were written in the 2000s on the so-called Pioneer Anomaly, the apparent constant sunward acceleration of about 0.00000008 centimeters per second per second inferred from radio tracking of the two Pioneer missions to the outer solar system. A variety of explanations were offered, ranging from a massive 'tenth planet' to new theories of gravitation, but the anomaly has since been nicely explained as an unaccounted thermal radiation effect on the spacecraft. Nevertheless, missions were devised to investigate this phenomenon in the outer solar system. The vehicle would have needed an extremely accurate radio-tracking system, spin stabilization, precisely calibrated thrusters, a careful choice of the placement of the RTGs supplying power, and a carefully designed thermal control system. A passive laser reflector 'test mass' to fly in formation with the spacecraft would have given an accurate measurement of the separation between the objects and a measurement of accelerations acting on the pair to a sensitivity three orders of magnitude better than was possible for the Pioneer data.[281,282,283]

Similar to LISA, but to improve our understanding of relativistic parameters, was the ASTROD (Astrodynamic Space Test of Relativity using Optical Devices) mission recently proposed by Chinese and European researchers. This would address many of the objectives of the 1970s European SOREL (Solar Orbiting Relativity Experiment) proposal, involving in particular the precise timing of signals between the spacecraft and a terminal on Earth and the tracking of a 'proof mass' that was subjected only to gravitational forces in order to derive some relativistic parameters to a high accuracy. Bending of electromagnetic waves in the vicinity of the Sun could be measured to an accuracy two orders of magnitude better than achieved by Cassini in 2002. Various options were studied, including a spacecraft which could be launched by a Chinese Long March 4B medium-lift rocket. ASTROD would be placed into a short-period solar orbit by a series of Venus flybys, during which the mass and other properties of the planet could be accurately measured. The project was proposed to ESA as part of its "Cosmic Vision" program, along with a similar SORT (Solar Orbit Relativity Test) mission, but neither was pursued. Since then, Chinese scientists have concentrated on a Chinese-led LISA-like mission called ASTROD-GW, optimized for the detection of gravitational waves. It would consist of three spacecraft located at the L3, L4 and L5 Lagrangian points of the Sun-Earth system, forming an equilateral triangle with arms some 260 million km in length; L3 being the point at a heliocentric distance of 1 AU on the opposite side of the Sun from Earth.[284,285]

IN CONCLUSION

Several of the projects and wild ideas presented in this last chapter are approved and funded and a few of the others might be realized in coming decades, but the rest are probably destined to remain in the realm of science fiction for the foreseeable future. Although some of the technological advancements envisaged by these concepts may never be attained, just thinking about them serves to broaden our outlook. Fifty years ago the very idea of sending probes to explore deep space was science fiction. On the one hand we are exploring a 'new frontier' in science, yet on the other were are still in our backyard, which is what the solar system really is. All the resources we could ever wish for are out there. What is needed, especially in times of financial crisis, is an appreciation of this fact, political commitment and public support.

Addendum

3MV AND ZOND SOVIET PROBES

The recent declassification of Soviet government documents has revealed the true nature of the failed deep-space probes of 11 November 1963 and 19 February 1964. The former had been described as a test of the 3MV spacecraft intended for missions to Venus and Mars, in this case including a lunar flyby similar to that which was later performed by Zond 3, and the latter as a demonstration flight to a distance equivalent to a Venus encounter which was launched just prior to the opening of the March and April 1964 window for that planet. Zond 1 would set off during this window. The spacecraft bus for both flights was described as a 3MV-1A; i.e. similar to the 3MV-1 Venus landers like Zond 1 and Venera 3.

It has now been revealed that the 3MV-1A were engineering Zonds designed as pathfinders for the Venus and Mars flights of 1964. They were to be inserted into heliocentric orbits at approximately 1 AU in a plane inclined at least 5 degrees to the plane of the ecliptic, with a period of a year. During their 6 months of operation, they would be north of the ecliptic and would reach their maximum distance of 12 to 16 million km from Earth some 3 months after launch. They would collect data on solar ultraviolet and X-rays, as well as on the interplanetary environment, and by virtue of their positions, remain in almost continuous contact with radio antennas in the Soviet Union. After two course corrections, the first probe was to release its 270-kg re-entry capsule and attempt a landing on Earth at escape speed in order to validate the design of the capsule and its parachutes. In fact, the spacecraft became stranded in low Earth orbit and was named Kosmos 21. The second did not even reach orbit. It is not clear whether it would have returned to Earth or would have performed deep-space tests to Venus' distance. It should be noted, however, that an Earth re-entry would not have occurred before August 1964, which was after the Venus landers reached their target, and so would have provided no engineering data to assist them.

A further engineering Zond was also proposed. This 3MV-4A was to have taken test images of Earth at increasing distances and tested deep-space communications at up to 2 AU (300 million km). The mission was to have flown in April or May 1964, just after the Venus window closed, but it is not clear what happened. One possibility is that it was never launched; another is that it was launched toward Mars as Zond 2 the following November. In the latter case, it is also possible that consideration was given to targeting Zond 2 to impact Mars to deliver a Soviet pennant to its surface.

THE MARINER MARS 69 MISSION

The extended missions of Mariners 6 and 7 were funded up to 31 December 1970. Telemetry from both was last received on 21 December. Both spacecraft were to be placed in a "final state", turning off the transmitters, between 23 and 30 December. The signal from Mariner 7 was lost because the spacecraft began tumbling, evidently when its supply of attitude control gas became depleted.

THE VOYAGER INTERSTELLAR MISSION

Recent years have seen Voyager 1 cross the heliopause, which defines the boundary between the heliosphere and interstellar space.

The first indication of this was that beyond a heliocentric distance of 113 AU the radial velocity of solar particles dropped essentially to zero. This was something that was expected at the heliopause, where particles were expected to be deflected to flow on the surface of the boundary itself. Therefore, in order to verify whether some sort of boundary had been reached, starting in March 2011 the vehicle was commanded to roll several times every 2 months in order to scan the environment. What it found was that the particles had practically zero speed in any direction, and it looked like the heliosphere had not yet been reached.

On 28 July 2012 Voyager 1 recorded a dramatic drop of solar particles and at the same time an increase of galactic rays, another phenomenon predicted to occur at the boundary of the heliosphere. For the first time, low-energy galactic cosmic rays were able to reach the spacecraft's sensors unhindered. Five similar events were recorded until 25 August, at about 122 AU, when the flux of solar particles dropped more than 1,000-fold and then remained constant. Analysis of the data over the ensuing months indicated that the intensity of the magnetic field had also varied in phase with these events, although not its direction. The latter was expected to change at the heliopause and mimic the direction of the interstellar field. Scientists therefore announced that Voyager 1 had not yet exited the heliosphere, and was actually in an unexplored and unpredicted connection region between solar-dominated space and interstellar space, dubbed the "magnetic highway" or "heliosheath depletion region". Not all scientists agreed with this, and many felt that, in spite of the missing change of the magnetic field, all the other evidence indicated that Voyager 1 was in interstellar space.

What was missing to resolve the quandary was data about the plasma, which was expected to increase in density at the boundary. Unfortunately, the experiment which could directly measure this had long since failed. The proof was supplied by a solar eruption on 9 April 2013. The plasma wave instrument recorded waves that implied an electron density tens of times greater than the spacecraft had ever recorded inside the heliosphere, and close to the density expected in interstellar space. Moreover, just as expected, the density appeared to be increasing as Voyager 1 continued away from the Sun. In September 2013 NASA made the announcement that Voyager 1 had left the heliosphere on 25 August 2012.

It seems that the interaction between the magnetic fields of the heliosphere and of interstellar space is more complicated than predicted, accounting for the absence of a change in direction. New models and simulations seem to show that the structure of the heliopause is such that the interstellar magnetic field intersects the solar field at only a shallow angle, and therefore a change of direction would not be expected.

Voyager 2, which has a functioning plasma instrument, is trailing its partner by several billion kilometers and will probably exit the heliosphere sometime in the late 2010s.

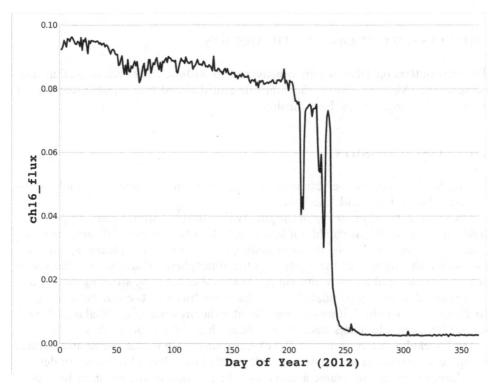

Fluxes of solar protons recorded by Voyager 1 during 2012, showing the dramatic decline as it crossed the heliospheric boundary on 25 August.

THE INTERNATIONAL COMETARY EXPLORER (ICE)

A radio carrier signal was received from ICE on 18 September 2008, after 9 years of silence. At that time proposals were being prepared for submission to NASA to fund another mission extension after the spacecraft's return to the vicinity of Earth on 10 August 2014. The most interesting proposal, estimated to cost about $22 million over a 6-year period, would involve ICE spending 4 years at the L1 Lagrangian point of the Earth-Moon system prior to using a lunar flyby in August 2018 to return to solar orbit and pass through the tail of comet 46P/Wirtanen on 14 December at a distance 0.08 AU from Earth. The comet could be observed and characterized simultaneously by the James Webb Space Telescope (presuming it to be operational by then). Little science would probably be supplied by the 40-year-old vehicle – the exercise would mostly serve as an educational tool for engineering students, who would be involved in the design, planning, and implementation of the trajectory and maneuvers. At the time of writing, a mission extension is neither approved nor financed.

THE ULYSSES OUT-OF-ECLIPTIC MISSION

The transmitters on Ulysses were deactivated on 30 June 2009 because at that time, being over 5 AU from Earth, the data rate had declined to a barely useable level. This was the longest-lived ESA mission.

THE CASSINI MISSION

Cassini is still making observations of Saturn, its magnetosphere, rings and moons, in particular of Titan and Enceladus.

On Titan, detailed modeling of the gravity field and its interior seems to point at a rigid, thick icy shell that would not be compatible with a geologically active surface. This would rule out the possibility of finding traces of present volcanism, and leave unresolved the mystery of the methane in the atmosphere, which ought otherwise to be rapidly destroyed by solar ultraviolet. On the other hand, by studying how Titan's shell responded to varying tidal effects as the moon traveled its eccentric orbit during six close 'gravity flybys', it was possible to infer the presence of a global liquid ocean several hundred kilometers thick located beneath a 100-km thick crust.

On Enceladus, analysis of over 200 observations made by the visible and infrared mapping spectrometer in the interval 2005 to 2012 has allowed scientists to detect a correlation between the plume activity and the position of the moon in its slightly eccentric orbit. The plume appeared to be several times brighter when Enceladus was at apoapsis, when the tiger stripes were undergoing greater tension than at periapsis.

The 3.2-km 'egg moon' Methone, seen by Cassini during its 1,900-km flyby on 20 May 2012.

MARS EXPRESS

ESA's Mars Express had another series of close flybys of Phobos in December 2013. On 29 December it made its closest pass ever at just 45 km. Cameras could not be used at such a small distance, and only tracking data were obtained to further refine our knowledge of the gravity field and internal structure of the moonlet.

THE MARS EXPLORATION ROVER OPPORTUNITY

Studying an outcrop at Matijevic Hill located at Cape York of the crater Endeavour, Opportunity discovered tiny embedded spherules that had nothing in common with the blueberries found earlier, except for their shape. These spherules were only up to 3 mm across and could be lapilli from a volcanic eruption or spherules formed during impacts, or indeed have been formed by other geological processes. It then examined the light-toned Esperance outcrop. This seemed to have been in contact with substantial amounts of water, and its clay-like minerals had compositions unlike any other rock previously examined by the rover.

After the April 2013 conjunction, Opportunity resumed driving south, heading for Solander Point, on the northern tip of the next promontory on the crater's rim, where the orbital signal of clays was stronger. On Sol 3,303 it broke the driving record set by the Apollo 17 astronauts on the Moon, having traveled 35.760 km. This made it the second most traveled planetary vehicle after the Soviet Lunokhod 2, which according to a recent estimate traveled between 42.1 and 42.2 km (almost a Marathon) on the Moon. Opportunity arrived at Solander Point in August with plenty of time available to explore the area prior to parking for the winter. On 25 January 2014 it celebrated a remarkable 10 years of operations.

Glossary

ACE: Advanced Composition Explorer

Aerobraking: A maneuver where a spacecraft's orbit is changed by reducing its energy through repeated passages in a planet's atmosphere.

Aerocapture: A maneuver where a spacecraft is inserted in orbit around a planet by slowing it down through a passage in the planet's atmosphere.

Aerogel: A silicon-based foam in which the liquid component of a gel has been replaced with gas or, for use in space, effectively with vacuum, resulting in a very low density solid.

AIDA: Asteroid Impact and Deflection Assessment

AIM: Asteroid Impact Mission

Aphelion: The point of maximum distance from the Sun of a solar orbit. Its contrary is the perihelion.

APIES: Asteroid Population Investigation and Exploration Swarm

APL: Applied Physics Laboratory

Apoapsis: The point of maximum distance from the central body of any elliptical orbit. This word has been used to avoid complicating the nomenclature, but a term tailored to the central body is often used. The only exceptions used herein owing to their importance were for Earth (apogee) and the Sun (aphelion). The contrary of apoapsis is periapsis.

Apogee: The point of maximum distance from the Earth of a satellite orbit. Its contrary is the perigee.

ARES: Aerial Regional-scale Environmental Survey

ARM: Asteroid Return Mission

ASI: Agenzia Spaziale Italiana (Italian Space Agency)

ASRG: Advanced Stirling Radioisotope Generator

ASTROD: Astrodynamic Space Test of Relativity using Optical Devices

Astronomical Unit: To a first approximation the average distance between the Earth and the Sun is 149,597,870,691 ($\pm$30) meters.

AU: Astronomical Unit

AVIATR: Aerial Vehicle for In-situ and Airborne Titan Reconnaissance

BEE: Belt Explorer

Booster: Auxiliary rockets used to boost the lift-off thrust of a launch vehicle.

Bus: A structural part common to several spacecraft.

CAST: Chinese Academy of Space Technology

CCD: Charge Coupled Device

CE: Chang'e Chinese lunar probes

CHON: Carbon, Hydrogen, Oxygen, Nitrogen rich molecules

Chopper: Comet Hopper

CMOS: Complementary Metal-Oxide Semiconductor

CNES: Centre National d'Etudes Spatiales (the French National Space Studies Center)

Conjunction: The time when a solar system object appears close to the Sun as seen by an observer. A conjunction where the Sun is between the observer and the object is called 'superior conjunction'. A conjunction where the object is between the observer and the Sun is called 'inferior conjunction'. See also opposition.

CNSA: Chinese National Space Administration

CNSR: Comet Nucleus Sample Return

CONTOUR: Comet Nucleus Tour

COROT: COnvection, ROtation and planetary Transits

Cosmic velocities: Three characteristic velocities of spaceflight:

First cosmic velocity: Minimum velocity to put a satellite in a low Earth orbit. This amounts to some 8 km/s.

Second cosmic velocity: The velocity required to exit the terrestrial sphere of attraction for good. Starting from the ground, this amounts to some 11 km/s. It is also called 'escape' speed.

Third cosmic velocity: The velocity required to exit the Solar System for good.

COSPAR: the United Nations' Committee on Space Research

CRAF: Comet Rendezvous/Asteroid Flyby

Cryogenic propellants: These can be stored in their liquid state under atmospheric pressure at very low temperature; e.g. oxygen is a liquid below $-183°C$.

DART: Demonstration of Autonomous Rendezvous Technology

DART: Double Asteroid Redirection Test

Deep Space Network: A global network built by NASA to provide round-the-clock communications with robotic missions in deep space.

DFH: Dong Feng Hong, "East is Red" Chinese satellites

Direct ascent: A trajectory on which a deep-space probe is launched directly from the Earth's surface to another celestial body without entering parking orbit.

DIXI: Deep Impact eXtended Investigation

DLR: Deutsche Zentrum für Luft- und Raumfahrt; the German center for flight and spaceflight

DS: Deep Space

DSN: Deep Space Network

DUNE: Dust Near Earth

DZhVS: Dolgozhivushaya Veneryanskaya Stanziya; long-duration Venusian probe

Ecliptic: The plane of the Earth's orbit around the Sun.

Ejecta: Material from a volcanic eruption or a cratering impact that is deposited all around the source.

EJSM: Europa and Jupiter System Mission
EPOCh: Extrasolar Planet Observation and Characterization
EPOXI: EPOCh + DIXI
EREP: European Robotic Exploration Program
ESA: European Space Agency
Escape speed: See Cosmic velocities
ESRO: European Space Research Organization
EVE: European Venus Explorer
EXCEED: EXtreme ultraviolet spectrosCope for ExosphEric Dynamics
Flyby: A high relative speed and short duration close encounter between a spacecraft and a celestial body.
GEMS: Geophysical Monitoring Station
GPS: Global Positioning System
GRAIL: Gravity Recovery and Interior Laboratory
GSFC: Goddard Space Flight Center
GSLV: Geostationary Satellite Launch Vehicle
HiRISE: High-Resolution Imaging Science Experiment
HIVE: Hub and Interplanetary Vehicle
HST: Hubble Space Telescope
Hypergolic propellants: Two liquid propellants that ignite spontaneously on coming into contact, without requiring an ignition system. Typical hypergolics are hydrazine and nitrogen tetroxide.
IAU: International Astronomical Union
IBEX: Interstellar Boundary Explorer
ICE: International Cometary Explorer
IKAROS: Interplanetary Kite-craft Accelerated by Radiation Of the Sun
IKI: Institut Kosmicheskikh Isledovanii (the Russian Institute for Cosmic Research)
iMARS: International Mars Architecture for the Return of Samples
INPE: Instituto Nacional de Pesquisas Espaciais, the Brazilian National Space Research Institute
INSIDE Jupiter: INterior Structure and Internal Dynamical Evolution of Jupiter
INSIDER: Interior of Primordial Asteroids and the Origin of Earth's Water
InSight: Interior Exploration using Seismic Investigations, Geodesy and Heat Transport
INSPIRE: Interplanetary NanoSpacecraft Pathfinder In Relevant Environment
IRAS: Infrared Astronomy Satellite
ISAS: Institute of Space and Astronautical Sciences
ISHTAR: Internal Structure High-resolution Tomography by Asteroid Rendezvous
ISIS: Impactor for Surface and Interior Science
ISO: Infrared Space Observatory
ISON: International Scientific Optical Network
ISRO: Indian Space Research Organization
ISS: International Space Station
IXO: International X-Ray Observatory
JAXA: Japanese Aerospace Exploration Agency

JEO: Jupiter Europa Orbiter
JET: Journey to Enceladus and Titan
JGO: Jupiter Ganymede Orbiter
JIMO: Jupiter Icy Moon Orbiter
JPL: Jet Propulsion Laboratory (a Caltech laboratory under contract to NASA)
JUICE: Jupiter Icy Moons Explorer
LADEE: Lunar Atmosphere and Dust Environment Explorer
Lagrangian Points: Five equilibrium points for a gravitational system comprising
    two large bodies (e.g. the Sun and a planet) and a third body of negligible mass.
Lander: A spacecraft designed to land on another celestial body.
LaRC: Langley Research Center
Launch window: A time interval during which it is possible to launch a spacecraft to
    ensure that it attains the desired trajectory.
Lidar: laser radar
LINEAR: Lincoln Near Earth Asteroid Research
LISA: Laser Interferometric Space Antenna
LOCO: Long period Comet Observer
LUGH: Low-cost Unified Geophysics at Hermes
MAI: Moscow Aviation Institute
MARGE: Mars Autonomous Rovers for Geological Exploration
MARVEL: Mars Volcanic Emissions and Life
MASCOT: Mobile Asteroid Surface Scout
Master: Mars + Asteroid
MATADOR: Mars Advanced Technology Airplane for Deployment, Operations
    and Recover
MAVEN: Mars Atmosphere and Volatile Evolution
MAX-C: Mars Astrobiology Explorer-Cacher
MEJI: Mars Exploration Joint Initiative
MELOS: Mars Exploration with Landers and Orbiters Synergy
MER: Mars Exploration Rovers
MESSENGER: Mercury Surface, Space Environment, Geochemistry and Ranging
MetNet: Meteorological Network
MINERVA: Micro/Nano Experimental Robot Vehicle for Asteroid
MIT: Massachusetts Institute of Technology
MMO: Mercury Magnetospheric Orbiter
MMRTG: Multi-Mission RTG
MOM: Indian Mars Orbiter Mission
MPCV: Multi-Purpose Crew Vehicle
MPO: Mercury Planctary Orbiter
MRO: Mars Reconnaissance Orbiter
MSL: Mars Science Laboratory, Mars Smart Lander
MUSES: MU [rocket] Space Engineering Satellite
NASA: National Aeronautics and Space Administration
NASDA: National Space Development Agency
NEAP: Near Earth Asteroid Prospector

NEAR: Near-Earth Asteroid Rendezvous

NEOCam: Near-Earth Objects Camera

NEOSSat: Near-Earth Object Survey Satellite

NEP: Nuclear Electric Propulsion

NEXT: NASA Evolutionary Xenon Thrusters

NGO: New Gravitational wave Observatory

NIMO: Neptune Icy Moons Orbiter

Occultation: When one object passes in front of and occults another, at least from the point of view of the observer.

ONERA: Office National d'Etudes et de Recherches Aérospatiales, the French national aerospace study and research office

Orbit: The trajectory on which a celestial body or spacecraft is traveling with respect to its central body. There are three possible cases:

Elliptical orbit: A closed orbit where the body passes from minimum distance to maximum distance from its central body every semiperiod. This is the orbit of natural and artificial satellites around planets and of planets around the Sun.

Parabolic orbit: An open orbit where the body passes through minimum distance from its central body and reaches infinity at zero velocity in infinite time. This is a pure abstraction, but the orbits of many comets around the Sun can be described adequately this way.

Hyperbolic orbit: An open orbit where the body passes through minimum distance from its central body and reaches infinity at non-zero speed. This describes adequately the trajectory of spacecraft with respect to planets during flyby maneuvers.

Opposition: The time when a solar system object appears opposite to the Sun as seen by an observer.

Orbiter: A spacecraft designed to orbit a celestial body.

OSIRIS: Origins Spectral Interpretation, Resource Identification and Security

OSIRIS-REx: Origins, Spectral Interpretation, Resource Identification, Security, Regolith Explorer

OSS: Outer Solar System mission

Parking orbit: A low Earth orbit used by deep-space probes before heading to their targets. This relaxes the constraints on launch windows and eliminates launch vehicle trajectory errors. Its contrary is direct ascent.

Periapsis: The minimum distance point from the central body of any orbit. See also apoapsis.

Perigee: The minimum distance point from the Earth of a satellite. Its contrary is apogee.

Perihelion: The minimum distance point from the Sun of a solar orbit. Its contrary is the aphelion.

POSSE: Pluto and Outer Solar System Explorer

PPCO: Planetary Protection Coordination Office

PREMIER: Programme de Retour d'Echantillons Martiens et Installation d'Expériences en Reseau, Mars sample-return and network experiment establishment program

PriME: Primitive Material Explorer

PROCYON: PRoximate Object Close flYby with Optical Navigation

PSLV: Polar Satellite Launch Vehicle

'Push-broom' camera: A digital camera consisting of a single row of pixels, with the second dimension created by the motion of the camera itself.

RAVEN: Radar at Venus

Rendezvous: A low relative speed encounter between two spacecraft or celestial bodies.

REP: Radioisotope Electric Propulsion

Retrorocket: A rocket whose thrust is directed opposite to the motion of a spacecraft in order to brake it.

Rj: Jupiter radii (approximately 71,200 km)

ROLAND: Rosetta Lander

Rover: A mobile spacecraft to explore the surface of another celestial body.

RTG: Radioisotope Thermal Generator

SAGE: Venus Surface and Atmosphere Geochemical Explorer

SCIM: Sample Collection for Investigation of Mars

SECCHI: Sun–Earth Connection Coronal and Heliospheric Investigation

SEP: Solar Electric Propulsion

SETI: Search for Extra Terrestrial Intelligence

SIMONE: Smallsat Interception Mission to Objects Near-Earth

SLS: Space Launch System

SMART: Small Missions for Advanced Research in Technology

SOFIA: Stratospheric Observatory for Infrared Astronomy

SOHO: Solar and Heliospheric Observatory

Sol: A Martian solar day, lasting 24 Terrestrial hours, 39 minutes, and 35.244 seconds

Solar flare: A solar chromospheric explosion creating a powerful source of high energy particles.

SOLO: Solar Orbiter

SOREL: Solar Orbiting Relativity Experiment

SORT: Solar Orbit Relativity Test

Space probe: A spacecraft designed to investigate other celestial bodies from a short range.

SPC: the ESA Science Programme Committee

Spectrometer: An instrument to measure the energy of radiation as a function of wavelengths in a portion of the electromagnetic spectrum. Depending on the wavelength the instrument is called, e.g. ultraviolet, infrared, gamma-ray spectrometer etc.

Spin stabilization: A spacecraft stabilization system where the attitude is maintained by spinning the spacecraft around one of its main inertia axes.

SPORT: Sky Polarization Observatory, Solar Polar Orbit Radio Telescope

STEREO: Solar Terrestrial Relations Observatory

TANDEM: Titan and Enceladus Mission

TEGA: Thermal and Evolved Gas Analyzer

Telemetry: Transmission by a spacecraft via a radio system of engineering and scientific data.

TGE: The Great Escape

TGO: Trace Gas Orbiter

THEMIS: Time History of Events and Macroscale Interactions during Substorms

3-axis stabilization: A spacecraft stabilization system where the axes of the spacecraft are kept in a fixed attitude with respect to the stars and other references (the Sun, the Earth, a target planet etc.)

THOR: Tracing Habitability, Organics and Resources

TiME: Titan Mare Explorer

TSSM: Titan/Saturn System Mission

UHF: Ultra-High radio Frequency

UMVL: Universalnyi Mars, Venera, Luna; Universal for Mars, Venus and the Moon

UNITEC: University space engineering consortium Technology Experiment Carrier

UTC: Universal Time Coordinated (essentially Greenwich Mean Time)

VCO: Venus Climate Orbiter

VEGA: Vettore Europeo di Generazione Avanzata, advanced generation European launcher

VEP: Venus Entry Probe

VEVA: Venus Exploration of Volcanoes and Atmosphere

VISE: Venus In-Situ Explorer

VITaL: Venus Intrepid Tessera Lander

VLBI: Very Long Baseline Interferometry

VLT: Very Large Telescope

VSE: Vision for Space Exploration

WISE: Wide-field Infrared Survey Explorer

WSB: Weak Stability Boundaries

YORP: Yarkovsky–O'Keefe–Radzievskii–Paddack effect

Appendices

Chronology of Solar System Exploration 2004–2013

Date	Event
4 July 2005	Deep Impact impacts comet Tempel 1
28 February 2006	New Horizons flies by Jupiter
10 March 2006	the Mars Reconnaissance Orbiter enters orbit around Mars
11 April 2006	Venus Express enters orbit around Venus
25 May 2008	Phoenix lands on the Martian arctic
8 June 2010	IKAROS is first solar sail to be successfully deployed
10 July 2010	Rosetta flies by asteroid Lutetia
4 November 2010	Deep Impact flies by comet Hartley 2
18 March 2011	MESSENGER enters orbit around Mercury
16 July 2011	Dawn enters orbit around Vesta
6 August 2012	Curiosity lands in Gale crater, Mars
13 December 2012	Chang'e 2 flies less than 2 km from asteroid Toutatis

APPENDIX 2

Planetary Launches 1960-2013

Launch Date	Name	Main Target	Launcher	Launch Site	Country	Volume
11 March 1960	Pioneer 5	Solar orbiter	Thor Able IV	Cape Canaveral	USA	1
10 October 1960	(1M No.1)	Mars	8K78 Molniya	Tyuratam	USSR	1
14 October 1960	(1M No.2)	Mars	8K78 Molniya	Tyuratam	USSR	1
04 February 1961	(1VA No.1)	Venus	8K78 Molniya	Tyuratam	USSR	1
12 February 1961	(Venera 1)	Venus	8K78 Molniya	Tyuratam	USSR	1
22 July 1962	(Mariner 1)	Venus	Atlas Agena B	Cape Canaveral	USA	1
25 August 1962	(2MV-1 No.1)	Venus	8K78 Molniya	Tyuratam	USSR	1
27 August 1962	Mariner 2	Venus	Atlas Agena B	Cape Canaveral	USA	1
01 September 1962	(2MV-1 No.2)	Venus	8K78 Molniya	Tyuratam	USSR	1
12 September 1962	(2MV-2 No.1)	Venus	8K78 Molniya	Tyuratam	USSR	1
24 October 1962	(2MV-4 No.1)	Mars	8K78 Molniya	Tyuratam	USSR	1
01 November 1962	(Mars 1)	Mars	8K78 Molniya	Tyuratam	USSR	1
04 November 1962	(2MV-3 No.1)	Mars	8K78 Molniya	Tyuratam	USSR	1
11 november 1963	(3MV-1A No.1)	Earth return	8K78 Molniya	Tyuratam	USSR	1
19 February 1964	(3MV-1A No.2)	Solar orbit	8K78 Molniya	Tyuratam	USSR	1
27 March 1964	(3MV-1 No.3)	Venus	8K78 Molniya	Tyuratam	USSR	1
02 April 1964	(Zond 1)	Venus	8K78 Molniya	Tyuratam	USSR	1
05 November 1964	(Mariner 3)	Mars	Atlas Agena D	Cape Canaveral	USA	1
28 November 1964	Mariner 4	Mars	Atlas Agena D	Cape Canaveral	USA	1
30 November 1964	(Zond 2)	Mars	8K78 Molniya	Tyuratam	USSR	1
18 July 1965	Zond 3	Lunar fly-by	8K78 Molniya	Tyuratam	USSR	1
12 November 1965	(Venera 2)	Venus	8K78M Molniya	Tyuratam	USSR	1
16 November 1965	(Venera-3)	Venus	8K78M Molniya	Tyuratam	USSR	1
23 November 1965	(3MV-4 No.6)	Venus	8K78M Molniya	Tyuratam	USSR	1
16 December 1965	Pioneer 6	Solar orbiter	Thor Delta E	Cape Canaveral	USA	1
17 August 1966	Pioneer 7	Solar orbiter	Thor Delta E1	Cape Canaveral	USA	1

Date	Name	Target	Launch vehicle	Launch site	Country	
12 June 1967	Venera 4	Venus	8K78M Molniya	Tyuratam	USSR	1
14 June 1967	Mariner 5	Venus	Atlas Agena D	Cape Canaveral	USA	1
17 June 1967	(4V-1 No.311)	Venus	8K78M Molniya	Tyuratam	USSR	1
13 December 1967	Pioneer 8	Solar orbiter	Thor Delta E1	Cape Canaveral	USA	1
06 November 1968	Pioneer 9	Solar orbiter	Thor Delta E1	Cape Canaveral	USA	1
05 January 1969	Venera 5	Venus	8K78M Molniya	Tyuratam	USSR	1
10 January 1969	Venera 6	Venus	8K78M Molniya	Tyuratam	USSR	1
25 February 1969	Mariner 6	Mars	Atlas-Centaur	Cape Canaveral	USA	1
27 March 1969	(2M No.521)	Mars	Proton-K/D	Tyuratam	USSR	1
27 March 1969	Mariner 7	Mars	Atlas-Centaur	Cape Canaveral	USA	1
02 April 1969	(2M No.522)	Mars	Proton-K/D	Tyuratam	USSR	1
27 August 1969	(Pioneer E)	Solar orbiter	Thor Delta L	Cape Canaveral	USA	1
17 August 1970	Venera 7	Venus	8K78M Molniya	Tyuratam	USSR	1
22 August 1970	(4V-1 No.631)	Venus	8K78M Molniya	Tyuratam	USSR	1
09 May 1971	(Mariner 8)	Mars	Atlas-Centaur	Cape Canaveral	USA	1
10 May 1971	(3MS No.170)	Mars	Proton K/D	Tyuratam	USSR	1
19 May 1971	Mars 2	Mars	Proton K/D	Tyuratam	USSR	1
28 May 1971	Mars 3	Mars	Proton K/D	Tyuratam	USSR	1
30 May 1971	Mariner 9	Mars	Atlas-Centaur	Cape Canaveral	USA	1
03 March 1972	Pioneer 10	Jupiter	Atlas-Centaur	Cape Canaveral	USA	1
27 March 1972	Venera 8	Venus	8K78M Molniya	Tyuratam	USSR	1
31 March 1972	(4V-1 No.671)	Venus	8K78M Molniya	Tyuratam	USSR	1
06 April 1973	Pioneer 11	Jupiter	Atlas-Centaur	Cape Canaveral	USA	1
21 July 1973	Mars 4	Mars	Proton K/D	Tyuratam	USSR	1
25 July 1973	Mars 5	Mars	Proton K/D	Tyuratam	USSR	1
05 August 1973	Mars 6	Mars	Proton K/D	Tyuratam	USSR	1
09 August 1973	Mars 7	Mars	Proton K/D	Tyuratam	USSR	1
03 November 1973	Mariner 10	Mercury	Atlas-Centaur	Cape Canaveral	USA	1
10 December 1974	Helios 1	Solar orbiter	Titan IIIE	Cape Canaveral	USA/FRG	1
08 June 1975	Venera 9	Venus	Proton K/D	Tyuratam	USSR	1
14 June 1975	Venera 10	Venus	Proton K/D	Tyuratam	USSR	1
20 August 1975	Viking 1	Mars	Titan IIIE	Cape Canaveral	USA	1
09 September 1975	Viking 2	Mars	Titan IIIE	Cape Canaveral	USA	1

Launch Date	Name	Main Target	Launcher	Launch Site	Country	Volume
15 January 1976	Helios 2	Solar orbiter	Titan IIIE	Cape Canaveral	USA/FRG	1
20 August 1977	Voyager 2	Jupiter	Titan IIIE	Cape Canaveral	USA	1
05 September 1977	Voyager 1	Jupiter	Titan IIIE	Cape Canaveral	USA	1
20 May 1978	Pioneer Venus Orbiter	Venus	Atlas-Centaur	Cape Canaveral	USA	1
08 August 1978	Pioneer Venus Multiprobe	Venus	Atlas-Centaur	Cape Canaveral	USA	1
12 August 1978	International Cometary Explorer	Comet	Delta 2914	Cape Canaveral	USA	2
09 September 1978	Venera 11	Venus	Proton K/D	Tyuratam	USSR	1
14 September 1978	Venera 12	Venus	Proton K/D	Tyuratam	USSR	1
30 October 1981	Venera 13	Venus	Proton K/D	Tyuratam	USSR	1
04 November 1981	Venera 14	Venus	Proton K/D	Tyuratam	USSR	1
02 June 1983	Venera 15	Venus	Proton K/D	Tyuratam	USSR	2
07 June 1983	Venera 16	Venus	Proton K/D	Tyuratam	USSR	2
15 December 1984	Vega 1	Venus + Comet	Proton K/D	Tyuratam	USSR	2
21 December 1984	Vega 2	Venus + Comet	Proton K/D	Tyuratam	USSR	2
07 January 1985	Sagigake	Comet	Mu-3SII	Kagoshima	Japan	2
02 July 1985	Giotto	Comet	Ariane 1	Kourou	ESA	2
18 August 1985	Suisei	Comet	Mu-3SII	Kagoshima	Japan	2
07 July 1988	(Fobos 1)	Mars	Proton K/D	Tyuratam	USSR	2
12 July 1988	Fobos 2	Mars	Proton K/D	Tyuratam	USSR	2
04 May 1989	Magellan	Venus	OV 104 + IUS	Kennedy Space Center	USA	2
18 October 1989	Galileo	Jupiter	OV 104 + IUS	Kennedy Space Center	USA	2
06 October 1990	Ulysses	Solar orbiter	OV 103 + IUS	Kennedy Space Center	ESA/USA	2
25 September 1992	(Mars Observer)	Mars	Titan 3 Commercial	Cape Canaveral	USA	2
25 January 1994	(Clementine)	Moon + Asteroid	Titan II SLV	Vandenberg AFB	USA	2
17 February 1996	NEAR	Asteroid	Delta 7925-8	Cape Canaveral	USA	2
07 November 1996	Mars Global Surveyor	Mars	Delta 7925A	Cape Canaveral	USA	2
16 November 1996	(Mars 8)	Mars	Proton K/D	Tyuratam	Russia	2

Date	Mission	Target	Launch vehicle	Launch site	Country	
04 December 1996	Mars Pathfinder	Mars	Delta 7925A	Cape Canaveral	USA	2
15 October 1997	Cassini-Huygens	Saturn	Titan 401B	Cape Canaveral	USA/Italy/ESA	3
03 July 1998	(Nozomi)	Mars	M-V	Kagoshima	Japan	3
24 October 1998	DS1	Asteroid	Delta 7326	Cape Canaveral	USA	3
11 December 1998	(Mars Climate Orbiter)	Mars	Delta 7425	Cape Canaveral	USA	3
03 January 1999	(Mars Polar Lander-DS2)	Mars	Delta 7425	Cape Canaveral	USA	3
07 February 1999	Stardust	P/Wild 2	Delta 7426	Cape Canaveral	USA	3
07 April 2001	Mars Odyssey	Mars	Delta 7925	Cape Canaveral	USA	3
30 June 2001	WMAP	Solar orbit	Delta 7425-10	Cape Canaveral	USA	4
08 August 2001	(Genesis)	Sun probe	Delta 7326	Cape Canaveral	USA	3
03 July 2002	(CONTOUR)	Comet	Delta 7425	Cape Canaveral	USA	3
03 May 2003	Hayabusa	Asteroid	M-V	Kagoshima	Japan	3
02 June 2003	Mars Express-(Beagle 2)	Mars	Soyuz-FG	Tyuratam	ESA/UK	3
10 June 2003	Spirit	Mars	Delta 7925	Cape Canaveral	USA	3
08 July 2003	Opportunity	Mars	Delta 7925H	Cape Canaveral	USA	3
25 August 2003	SIRTF	Solar orbit	Delta 7920H	Cape Canaveral	USA	3
02 March 2004	Rosetta-Phylae	Comet	Ariane 5+	Kourou	ESA	4
03 August 2004	MESSENGER	Mercury	Delta 7925H	Cape Canaveral	USA	4
12 January 2005	Deep Impact	P/Tempel 1	Delta 7925	Cape Canaveral	USA	4
12 August 2005	Mars Reconnaissance Orbiter	Mars	Atlas V 401	Cape Canaveral	USA	4
09 November 2005	Venus Express	Venus	Soyuz-FG	Tyuratam	ESA	4
19 January 2006	New Horizons	Pluto	Atlas V 551	Cape Canaveral	USA	4
25 October 2006	STEREO A-STEREO B	Solar orbiters	Delta 7925	Cape Canaveral	USA	4
04 August 2007	Phoenix	Mars	Delta 7925	Cape Canaveral	USA	4
27 September 2007	Dawn	Asteroid	Delta 7925H	Cape Canaveral	USA	4
07 March 2009	Kepler	Solar orbit	Delta 7925-10L	Cape Canaveral	USA	4
14 May 2009	Herschel-Planck	Solar orbit	Ariane 5ECA	Kourou	ESA	4
20 May 2010	(Akatsuki)-IKAROS-(Shin-en)	Venus	H-IIA 202	Tanegashima	Japan	4

Launch Date	Name	Main Target	Launcher	Launch Site	Country	Volume
01 October 2010	Chang'e 2	Moon + Asteroid	LM 3C	Xichang	China	4
05 August 2011	Juno	Jupiter	Atlas V 551	Cape Canaveral	USA	4
08 November 2011	(Fobos Grunt-Yinghuo 1)	Mars	Zenit 2-FG	Tyuratam	Russia /China	4
26 November 2011	Mars Scientific Laboratory	Mars	Atlas V 551	Cape Canaveral	USA	4
05 November 2013	Mars Orbiter Mission	Mars	PSLV	Shriharikota	India	4
18 November 2013	MAVEN	Mars	Atlas V 401	Cape Canaveral	USA	4

Missions in parentheses are missions that failed.

APPENDIX 3

Chronology of Solar System Exploration 2014–2033

20 January 2014	Rosetta	exits from hibernation
6 August 2014	Rosetta	enters orbit around Churyumov-Gerasimenko
10 August 2014	ICE	Earth return
22 September 2014	MAVEN	enters orbit around Mars
24 September 2014	MOM	enters orbit around Mars
19 October 2014	Comet Siding Spring	Very close approach to Mars
11 November 2014	Philae	lands on Churyumov-Gerasimenko
December 2014	Hayabusa 2/ PROCYON/Artsat 2	launch
12 January 2015	New Horizons	distant encounter operations begins
28 March 2015	MESSENGER	impacts Mercury
April 2015	Dawn	enters orbit around Ceres
14 July 2015	New Horizons	flyby of Pluto
21 November 2015	Akatsuki	second attempt at entering Venus orbit
December 2015	Rosetta	end of mission
December 2015	Hayabusa 2	Earth flyby
December 2015	PROCYON	Earth flyby
January 2016	PROCYON	asteroid flyby?
January 2016	Trace Gas Orbiter	launch
4 March 2016	InSIGHT	launch
5 July 2016	Juno	enters orbit around Jupiter
July 2016	Dawn	end of mission
9 July 2016	BepiColombo	launch
3 September 2016	OSIRIS-REx	launch
28 September 2016	InSIGHT	lands on Mars
19 October 2016	Trace Gas Orbiter	enters orbit around Mars
June 2017	Trace Gas Orbiter	start of the science mission
27 July 2017	Solar Orbiter	launch
September 2017	OSIRIS-REx	Earth flyby
15 September 2017	Cassini	plunges in the atmosphere of Saturn
16 October 2017	Juno	plunges in the atmosphere of Jupiter
May 2018	ExoMars rover	launch
June 2018	Hayabusa 2	reaches its target asteroid (162173) 1999JU3
16 July 2018	BepiColombo	Earth flyby
27 July 2018	Solar Orbiter	Earth flyby
31 July 2018	Solar Probe Plus	launch
September 2018	InSIGHT	primary mission ends
27 September 2018	Solar Probe Plus	Venus flyby
14 December 2018	ICE	flies by comet Wirtanen?
January 2019	ExoMars rover	lands on Mars
January 2019	OSIRIS-REx	enters orbit around asteroid (101955) Bennu
July 2019	OSIRIS-REx	samples asteroid (101955) Bennu
August 2019	Hayabusa 2	cratering experiment

22 September 2019	BepiColombo	Venus flyby
December 2019	Hayabusa 2	departs asteroid (162173) 1999JU3
21 December 2019	Solar Probe Plus	Venus flyby
4 May 2020	BepiColombo	Venus flyby
15 May 2020	Solar Orbiter	Venus flyby
25 June 2020	Solar Orbiter	Earth flyby
July 2020	NASA Mars rover	launch
5 July 2020	Solar Probe Plus	Venus flyby
23 July 2020	BepiColombo	Mercury flyby
December 2020	Hayabusa 2	returns to Earth
January 2021	NASA Mars rover	lands on Mars
5 January 2021	Solar Orbiter	Venus flyby
15 February 2021	Solar Probe Plus	Venus flyby
March 2021	OSIRIS-REx	departs asteroid (101955) Bennu
14 April 2021	BepiColombo	Mercury flyby
27 May 2021	Solar Orbiter	0.28 AU perihelion
10 October 2021	Solar Probe Plus	Venus flyby
June 2022	JUICE	launch
6 July 2022	BepiColombo	Mercury flyby
29 December 2022	BepiColombo	Mercury flyby
4 February 2023	BepiColombo	Mercury flyby
23 June 2023	Solar Orbiter	Venus flyby
15 August 2023	Solar Probe Plus	Venus flyby
24 September 2023	OSIRIS-REx	returns to Earth
1 January 2024	BepiColombo	enters orbit around Mercury
31 October 2024	Solar Probe Plus	Venus flyby
19 December 2024	Solar Probe Plus	first close perihelion
1 April 2025	BepiColombo	end of primary mission
27 April 2025	Solar Orbiter	Venus flyby
1 April 2026	BepiColombo	end of extended mission
21 July 2026	Solar Orbiter	Venus flyby
13 October 2027	Solar Orbiter	Venus flyby - end of mission
January 2030	JUICE	enters orbit around Jupiter
September 2032	JUICE	enters orbit around Ganymede
June 2033	JUICE	end of primary mission

Only those missions that are currently funded are included. ICE's 2018 Wirtanen flyby is not funded at present.

APPENDIX 4

Chronology of Mercury Exploration

Objects which flew by Mercury

Name	Country	Date	Distance (km)
Mariner 10	USA	29 March 1974	703
Mariner 10	USA	21 September 1974	48,069
Mariner 10	USA	16 March 1975	327
MESSENGER	USA	14 January 2008	201
MESSENGER	USA	6 October 2008	199
MESSENGER	USA	29 September 2009	228

Objects in orbit around Mercury

Name	Country	Entered orbit on	Parameters (km)
MESSENGER	USA	18 March 2011	Variable

APPENDIX 5

Chronology of Venus Exploration

Objects which flew by Venus

Name	Country	Date	Distance [km]	Notes
Venera	USSR	19 May 1961	100,000	Inactive
Stage L	USSR	19 May 1961	100,000?	Venera Launcher
Mariner 2	USA	14 December 1962	34,854	
Agena B	USA	14 December 1962	375,900?	Mariner 2 Launcher
Zond 1	USSR	19 July 1964	1,000,000?	Inactive
Stage L	USSR	19 July 1964	?	Zond 1 Launcher
Venera 2	USSR	27 February 1966	24,000	Inactive?
Stage L	USSR	27 February 1966	?	Venera 2 Launcher
Stage L	USSR	1 March 1966	65,500?	Venera 3 Launcher
Stage L	USSR	18 October 1967	60,000?	Venera 4 Launcher
Mariner 5	USA	19 October 1967	4,100	
Agena B	USA	19 October 1967	75,000?	Mariner 5 Launcher
Stage L	USSR	16 May 1969	25,000?	Venera 5 Launcher
Stage L	USSR	17 May 1969	150,000?	Venera 6 Launcher
Stage L	USSR	12 December 1970	?	Venera 7 Launcher
Stage L	USSR	22 July 1972	?	Venera 8 Launcher
Mariner 10	USA	5 February 1974	5,768	
Centaur	USA	5 February 1974	?	Mariner 10 Launcher
Stage D	USSR	22 October 1975	?	Venera 9 Launcher
Stage D	USSR	25 October 1975	?	Venera 10 Launcher
Centaur	USA	2 December 1978	?	Pioneer Venus Orbiter Launcher
Centaur	USA	9 December 1978	14,000?	Pioneer Venus Multiprobe Launcher
Venera 12	USSR	21 December 1978	35,000	

Stage D	USSR	21 December 1978	?	Venera 12 Launcher
Venera 11	USSR	25 December 1978	35,000	
Stage D	USSR	25 December 1978	?	Venera 11 Launcher
Venera 13	USSR	1 March 1982	36,000	
Stage D	USSR	1 March 1982	?	Venera 13 Launcher
Venera 14	USSR	5 March 1982	36,000?	
Stage D	USSR	5 March 1982	?	Venera 14 Launcher
Stage D	USSR	10 October 1983	?	Venera 15 Launcher
Stage D	USSR	14 October 1983	?	Venera 16 Launcher
Vega 1	USSR	11 June 1984	39,000	
Stage D	USSR	11 June 1984	?	Vega 1 Launcher
Vega 2	USSR	15 June 1984	39,000?	
Stage D	USSR	15 June 1984	?	Vega 2 Launcher
Galileo	USA	10 February 1990	16,106	
IUS SRM-2	USA	10 February 1990	?	Galileo Launcher
IUS SRM-2	USA	10 August 1990?	?	Magellan Launcher
Cassini	USA/ESA/Italy	26 April 1998	284	
Centaur	USA	26 April 1998	?	Cassini Launcher
Cassini	USA/ESA/Italy	24 June 1999	603	
Fregat	Russia	11 April 2006	?	Venus Express Launcher
MESSENGER	USA	24 October 2006	2,987	
MESSENGER	USA	5 June 2007	316	
Akatsuki	Japan	7 December 2010	550	
Shin'en	Japan	7 December 2010?	?	Inactive
H-IIA 2nd stage	Japan	7 December 2010?	?	Akatsuki Launcher
IKAROS	Japan	8 December 2010	80,000	
DCAM 1	Japan	8 December 2010	80,000?	Inactive
DCAM 2	Japan	8 December 2010	80,000?	Inactive

Objects in orbit around Venus

Name	Country	Parameters	Entered Orbit	In Orbit Until	Notes
Venera 9	USSR	Variable	22 October 1975		
Venera 10	USSR	Variable	25 October 1975		
Pioneer Venus Orbiter	USA	Variable	2 December 1978	8 October 1992	
Venera 15	USSR	Variable	10 October 1983		
Venera 16	USSR	Variable	14 October 1983		
Magellan	USA	Variable	10 August 1990	14 October 1994	Contacts lost 12 October 1994
STAR-48B	USA	289 × 8,458 km, 85.5°	10 August 1990		Magellan Orbit Insertion Motor
Venus Express	ESA	Variable	11 April 2006		

Other objects that entered the Venusian Atmosphere

Name	Country	Date	Notes
Venera 3	USSR	1 March 1966	near 0°N, 160°E
Venera 4 bus	USSR	18 October 1967	
Venera 5 bus	USSR	16 May 1969	
Venera 6 bus	USSR	17 May 1969	
Venera 7 bus	USSR	12 December 1970	
Venera 8 bus	USSR	22 July 1972	
Pioneer Venus Multiprobe bus	USA	9 December 1978	near 41°S, 284°E
Pioneer Venus Orbiter	USA	8 October 1992	
Magellan	USA	14 October 1994	Date extrapolated

Objects on the Venusian Surface

Name	Country	Date Landed	Longitude	Latitude	Notes
Venera 4 capsule	USSR	18 October 1967	38°E	19°N	Reached ground inactive
Venera 5 capsule	USSR	16 May 1969	18°E	3°S	Reached ground inactive
Venera 6 capsule	USSR	17 May 1969	23°E	5°S	Reached ground inactive
Venera 7 capsule	USSR	12 December 1970	9°E	5°S	
Venera 8 capsule	USSR	22 July 1972	335.25°E	10.7°S	
Venera 9 lander	USSR	22 October 1975	291.64°E	31.01°N	+ camera cover
Venera 9 heatshield sphere	USSR	22 October 1975	Close to Above	Close to Above	
Venera 9 heatshield cap	USSR	22 October 1975	Close to Above	Close to Above	
Venera 10 lander	USSR	25 October 1975	291.51°E	15.42°N	+ camera cover
Venera 10 heatshield sphere	USSR	25 October 1975	Close to Above	Close to Above	
Venera 10 heatshield cap	USSR	25 October 1975	Close to Above	Close to Above	
Pioneer Venus Large Probe	USA	9 December 1978	304°E	4°N	Failed on Impact
Pioneer Venus North Probe	USA	9 December 1978	4°E	60°N	Failed on Impact
Pioneer Venus Day Probe	USA	9 December 1978	318°E	32°S	
Pioneer Venus Night Probe	USA	9 December 1978	56°E	27°S	Failed on Impact
Venera 12 lander	USSR	21 December 1978	294°E	7°S	
Venera 12 heatshield sphere	USSR	21 December 1978	Close to Above	Close to Above	
Venera 12 heatshield cap	USSR	21 December 1978	Close to Above	Close to Above	
Venera 11 lander	USSR	25 December 1978	299°E	14°S	
Venera 11 heatshield sphere	USSR	25 December 1978	Close to Above	Close to Above	
Venera 11 heatshield cap	USSR	25 December 1978	Close to Above	Close to Above	
Venera 13 lander	USSR	1 March 1982	303.69°E	7.55°S	+ camera covers (2)
Venera 13 heatshield sphere	USSR	1 March 1982	Close to Above	Close to Above	
Venera 13 heatshield cap	USSR	1 March 1982	Close to Above	Close to Above	
Venera 14 lander	USSR	5 March 1982	310.19°E	13.055°S	+ camera covers (2)
Venera 14 heatshield sphere	USSR	5 March 1982	Close to Above	Close to Above	
Venera 14 heatshield cap	USSR	5 March 1982	Close to Above	Close to Above	
Vega 1 lander	USSR	11 June 1984	177.8°E	7.2°N	
Vega 1 heatshield sphere	USSR	11 June 1984	Close to Above	Close to Above	

Name	Country	Date Landed	Longitude	Latitude	Notes
Vega 1 heatshield cap	USSR	11 June 1984	Close to Above	Close to Above	
Vega 1 AS Upper Torus	USSR	11 June 1984	Close to Above	Close to Above	
Vega 1 AS Lower Torus	USSR	11 June 1984	Close to Above	Close to Above	
Vega 1 AS	USSR	N/A	N/A	N/A	Reached ground inactive
Vega 2 lander	USSR	15 June 1984	181.08°E	6.45°S	
Vega 2 heatshield sphere	USSR	15 June 1984	Close to Above	Close to Above	
Vega 2 heatshield cap	USSR	15 June 1984	Close to Above	Close to Above	
Vega 2 AS Upper Torus	USSR	15 June 1984	Close to Above	Close to Above	
Vega 2 AS Lower Torus	USSR	15 June 1984	Close to Above	Close to Above	
Vega 2 AS	USSR	N/A	N/A	N/A	Reached ground inactive

APPENDIX 6

Earth flybys by Deep Space Probes

Name	Country	Date	Distance (km)
Giotto	ESA	2 July 1990	22,731
Galileo	USA	8 December 1990	960
Sakigake	Japan	8 January 1992	88,997
Suisei *[Inactive]*	Japan	20 August 1992	900,000
Galileo	USA	8 December 1992	304
Sakigake	Japan	14 June 1993	255,000
Sakigake	Japan	28 October 1994	548,000
NEAR	USA	23 January 1998	540
Giotto *[Inactive]*	ESA	1 July 1999	219,000
Cassini	USA/ESA/Italy	18 August 1999	1,166
Stardust	USA	15 January 2001	6,008
Nozomi	Japan	21 December 2002	29,510
Nozomi	Japan	19 June 2003	11,023
Hayabusa	Japan	19 May 2004	3,725
Rosetta	ESA	4 March 2005	1,954
MESSENGER	USA	2 August 2005	2,347
Stardust	USA	15 January 2006	258
Rosetta	ESA	13 November 2007	5,301
Deep Impact	USA	31 December 2007	15,566
Deep Impact	USA	29 December 2008	43,000
Stardust	USA	14 January 2009	9,157
Rosetta	ESA	13 November 2009	2,480

APPENDIX 7

Chronology of Mars Exploration

Objects which flew by Mars

Name	Country	Date	Distance [km]	Notes
Mars 1	USSR	19 June 1963	193,000	Inactive
Stage L	USSR	19 June 1963	193,000?	Mars 1 Launcher
Mariner 4	USA	15 July 1965	9,846	
Agena B	USA	15 July 1965	250,000	Mariner 4 Launcher
Zond 2	USSR	6 August 1965	1,500?	Inactive
Stage L	USSR	6 August 1965	?	Zond 2 Launcher
Mariner 6	USA	31 July 1969	3,429	
Centaur	USA	31 July 1969	?	Mariner 6 Launcher
Mariner 7	USA	5 August 1969	3,430	
Centaur	USA	5 August 1969	?	Mariner 7 Launcher
Centaur	USA	14 November 1971	?	Mariner 9 Launcher
Stage D	USSR	27 November 1971	?	Mars 2 Launcher
Stage D	USSR	2 December 1971	?	Mars 3 Launcher
Mars 4	USSR	10 February 1974	1,844	Failed Orbiter
Stage D	USSR	10 February 1974	?	Mars 4 Launcher
Stage D	USSR	12 February 1974	?	Mars 5 Launcher
Mars 7	USSR	9 March 1974	1,300	
Mars 7 Lander	USSR	9 March 1974	1,300	Failed Lander
Stage D	USSR	9 March 1974	?	Mars 7 Launcher
Mars 6	USSR	12 March 1974	16,000	
Stage D	USSR	12 March 1974	?	Mars 6 Launcher
Centaur	USA	19 June 1976	80,500?	Viking 1 Launcher
Viking 1 Bioshell Base	USA	19 June 1976	80,500?	

Centaur	USA	7 August 1976	80,500?	Viking 2 Launcher
Viking 2 Bioshell Base	USA	7 August 1976	80,500?	Inactive
Fobos 1	USSR	23 January 1989	?	Inactive
Mars Observer	USA	24 August 1993	500?	Mars Observer Launcher
TOS	USA	24 August 1993	?	
Star 48B	USA	4 July 1997	?	MPF Launcher
Star 48B	USA	12 September 1997	?	MGS Launcher
Nozomi	Japan	7 September 1999	4,000,000	
Star 48B	USA	23 September 1999	?	MCO Launcher
Star 48B	USA	3 December 1999	?	MPL Launcher
Star 48B	USA	24 October 2001	?	MOd Launcher
Nozomi	Japan	13 December 2003	894?	Inactive
Fregat	Russia/ESA	25 December 2003	?	Mars Express Launcher
Star 48B	USA	4 January 2004	?	Spirit Launcher
Star 48B	USA	27 January 2004	340,000?	Opportunity Launcher
Centaur	USA	10 March 2006	?	MRO Launcher
Rosetta	ESA	25 February 2007	250	
STAR 48B	USA	25 May 2008	95,000?	Phoenix Launcher
Dawn	USA	18 February 2009	542	
Curiosity Centaur Stage	USA	8 August 2012	347,000?	Curiosity Launcher

Objects in orbit around Mars

Name	Country	Parameters	Entered Orbit	In Orbit Until	Notes
Mariner 9	USA	1,394 × 17,144 km, 64,34°	14 November 1971		
Mars 2	USSR	1,380 × 25,000 km, 48,9°	27 November 1971		
Mars 3	USSR	1,530 × 214,500 km, 60°	2 December 1971		
Mars 5	USSR	1,760 × 32,586 km, 35.33°	12 February 1974		
Viking 1	USA	Variable	19 June 1976		
Viking 1 Lander	USA	Variable	20 July 1976	20 July 1976	Landed
Viking 1 Bioshell	USA	1500 × 32800 km, 38°	20 July 1976 (?)		
Viking 2	USA	Variable	7 August 1976		
Viking 2 Lander	USA	Variable	3 September 1976	3 September 1976	Landed
Viking 2 Bioshell	USA	302 × 33240 km, 80°	3 March 1978		
Fobos 2	USSR	Variable	28 January 1989		
Fobos 2 ADU	USSR	6,145 × 6,307 km, 1°	18 February 1989		Fobos Fregat stage
Mars Global Surveyor	USA	Variable	12 September 1997		
Mars Odyssey	USA	Variable	24 October 2001		
Mars Express	ESA	Variable	25 December 2003		
Mars Reconnaissance Orbiter	USA	Variable	10 March 2006		

Other objects that entered the Mars Atmosphere

Name	Date	Notes
Mars Pathfinder Cruise Stage	4 July 1997	Entry Coordinates near that of Mars Pathfinder
Mars Climate Orbiter	23 September 1999	Failed Orbiter
Mars Polar Lander Cruise Stage	3 December 1999	Entry Coordinates near that of Mars Polar Lander
Spirit Cruise Stage	4 January 2004	Entry Coordinates near that of Spirit
Opportunity Cruise Stage	27 January 2004	Entry Coordinates near that of Opportunity
Phoenix Cruise Stage	25 May 2008	Entry Coordinates near that of Phoenix
Curiosity Cruise Stage	6 August 2012	Entry Coordinates near that of Curiosity
Curiosity Ballast Weights (2)	6 August 2012	Entry Coordinates near that of Curiosity

Objects on the Martian Surface

Name	Country	Date Landed	Longitude	Latitude	Notes
Mars 2	USSR	27 November 1971	47°E	44°S	Impact?
Mars 3	USSR	2 December 1971	158°W	45°S	Returned 20 seconds of data
Mars 3 Heatshield	USSR	2 December 1971	Close to Above	Close to Above	
Mars 3 Parachute + Container	USSR	2 December 1971	Close to Above	Close to Above	
Mars 6	USSR	12 March 1974	19.4°W	23.9°S	Failed on landing
Mars 6 Heatshield	USSR	12 March 1974	Close to Above	Close to Above	
Mars 6 Parachute + Container	USSR	12 March 1974	Close to Above	Close to Above	
Viking 1 Lander	USA	20 July 1976	47.94°W	22.48°N	+ sampler cover, debris, etc.
Viking 1 Heatshield	USA	20 July 1976	Close to Above	Close to Above	
Viking 1 Backshell + Parachute	USA	20 July 1976	Close to Above	Close to Above	
Viking 2 Lander	USA	3 September 1976	225.71°W	47.97°N	+ sampler cover, debris, etc.
Viking 2 Heatshield	USA	3 September 1976	Close to Above	Close to Above	
Viking 2 Backshell + Parachute	USA	3 September 1976	Close to Above	Close to Above	
Mars Pathfinder	USA	4 July 1997	33.52°W	19.28°N	
MPF Heatshield	USA	4 July 1997	Close to Above	Close to Above	

Name	Country	Date Landed	Longitude	Latitude	Notes
MPF Backshell + Parachute	USA	4 July 1997	Close to Above	Close to Above	
Sojourner	USA	4 July 1997	Close to Above	Close to Above	
Mars Polar Lander	USA	3 December 1999	195°W	76°S	Impact?
DS-2 Amundsen	USA	3 December 1999	195.9°W?	75.3°S?	Impact?
DS-2 Scott	USA	3 December 1999	195.9°W?	75.3°S?	Impact?
Beagle 2	UK/ESA	25 December 2003	269.7°W	11°N	Impact?
Spirit	USA	4 January 2004	175.4729°E	14.5692°S	
Spirit Base	USA	4 January 2004	Same as above	Same as above	
Spirit Heatshield	USA	4 January 2004	Close to Above	Close to Above	
Spirit Backshell + Parachute	USA	4 January 2004	Close to Above	Close to Above	
Opportunity	USA	27 January 2004	354.47417°E	1.9483°S	
Opportunity Base	USA	27 January 2004	Same as above	Same as above	
Opportunity Heatshield	USA	27 January 2004	Close to Above	Close to Above	
Oppy Backshell + Parachute	USA	27 January 2004	Close to Above	Close to Above	
Phoenix	USA	25 May 2008	234.248°E	68.219°N	
Phoenix Heatshield	USA	25 May 2008	Close to Above	Close to Above	
Phoenix Backshell + Parachute	USA	25 May 2008	Close to Above	Close to Above	
Curiosity	USA	6 August 2012	137.4417°E	4.5895°S	
Curiosity Skycrane	USA	6 August 2012	Close to Above	Close to Above	
Curiosity Heatshield	USA	6 August 2012	Close to Above	Close to Above	
Curiosity Backshell + Parachute	USA	6 August 2012	Close to Above	Close to Above	
Curiosity Ballast Weights (6)	USA	6 August 2012	Close to Above	Close to Above	

APPENDIX 8

Chronology of Asteroid Exploration

Probes that have flown by Asteroids

Name	Country	Date	Asteroid	Distance [km]	Notes
Galileo	USA	29 October 1991	951 Gaspra	1,604	
Galileo	USA	28 August 1993	243 Ida	2,410	
NEAR	USA	27 June 1997	253 Mathilde	1,212	
NEAR	USA	23 December 1998	433 Eros	3,827	
Deep Space 1	USA	29 July 1999	9969 Braille	26	
Stardust	USA	2 November 2002	5535 Annefrank	3,079	
New Horizons	USA	13 June 2006	132524 APL	101,867	
Rosetta	ESA	5 September 2008	2867 Steins	802.6	
Rosetta	ESA	10 July 2010	21 Lutetia	3,162	
Chang'e 2	China	13 December 2012	4179 Toutatis	1.564	

Probes in orbit around Asteroids

Name	Country	Asteroid	Date Entered Orbit	In Orbit Until	Notes
NEAR	USA	433 Eros	14 February 2000	12 February 2001	
Hayabusa	Japan	25143 Itokawa	28 August 2005	25 November 2005	Station Keeping in Solar Orbit
Dawn	USA	4 Vesta	16 July 2011	5 September 2012	

Objects that have landed on Asteroids and Comets

Name	Country	Object	Date	Longitude	Latitude	Notes
NEAR	USA	433 Eros	12 February 2001	279.5°W	35.7°S	passive target
Hayabusa Target	Japan	25143 Itokawa	19 November 2005	?	?	took off after 35 minutes
Hayabusa	Japan	25143 Itokawa	19 November 2005	39°E	6°S	
Hayabusa	Japan	25143 Itokawa	25 November 2005	?	?	took off after < 1 second

APPENDIX 9

Chronology of Jupiter Exploration

Objects which have flown by Jupiter

Name	Country	Date	Distance [km]	Notes
Pioneer 10	USA	4 December 1973	203,240	
TE-M-364-4	USA	4 December 1973	?	P-10 Launcher
Pioneer 11	USA	3 December 1974	42,500	
TE-M-364-4	USA	3 December 1974	?	P-11 Launcher
Voyager 1	USA	5 March 1979	348,890	
TE-M-364-4	USA	5 March 1979	?	V-1 Launcher
Voyager 2	USA	9 July 1979	721,670	
TE-M-364-4	USA	9 July 1979	?	V-2 Launcher
Ulysses	ESA	8 February 1992	379,000	
PAM-S	USA	8 February 1992	?	Ulysses Launcher
Cassini	USA/ESA	30 December 2000	9,655,000	
Ulysses	ESA	5 February 2004	120,000,000	0.8-AU flyby
New Horizons	USA	28 February 2007	2,300,000	
STAR 48	USA	28 February 2007	2,800,000?	NH Launcher

Objects in orbit around Jupiter

Name	Country	Entered orbit	In orbit until	Parameters
Galileo Orbiter	USA	7 December 1995	21 September 2003	Variable

Objects that entered the Jovian Atmosphere

Name	Country	Date	Longitude [System III]	Latitude
Galileo Probe	USA	7 December 1995	4.94°W	6.57°N
Galileo Probe heat shield	USA	7 December 1995	4.94°W	6.57°N
Galileo Orbiter	USA	21 September 2003	191.6°W	0.2°S

APPENDIX 10

Chronology of Saturn Exploration

Flybys

Name	Country	Date	Distance [km]
Pioneer 11	USA	1 September 1979	80,982
Voyager 1	USA	12 November 1980	126,000
Voyager 2	USA	26 August 1981	101,000

Orbiters

Name	Country	Entered orbit	In orbit until
Cassini	USA/Italy	1 July 2004	15 September 2017
Huygens	ESA	25 December 2004	15 January 2005

Titan landings

Name	Country	Date	Longitude	Latitude
Huygens	ESA	15 January 2005	192.3°W	10.3°S
Huygens heat shield	ESA	15 January 2005	Close to Above	Close to Above

APPENDIX 11

Chronology of Uranus and Neptune Exploration

Flybys

Planet	Name	Country	Date	Distance (km)
Uranus	Voyager 2	USA	24 January 1986	107,000
Neptune	Voyager 2	USA	25 August 1989	29,240

APPENDIX 12

Chronology of Cometary Exploration

Probes that have flown by Comets

Name	Country	Date	Comet	Distance [km]
ICE/ISEE-3	USA	11 September 1985	21P/Giacobini-Zinner	7,682
Vega 1	USSR	6 March 1986	1P/Halley	8,890
Suisei	Japan	8 March 1986	1P/Halley	151,000
Vega 2	USSR	9 March 1986	1P/Halley	8,030
Sakigake	Japan	11 March 1986	1P/Halley	6,990,000
Giotto	ESA	13 March 1986	1P/Halley	596
Giotto	ESA	10 July 1992	26P/Grigg-Skjellerup	200
Sakigake*	Japan	3 February 1996	45P/Honda-Mrkos-Pajdušáková	> 10,000
Deep Space 1	USA	22 September 2001	19P/Borrelly	2,171
Stardust	USA	2 January 2004	81P/Wild 2	236
Deep Impact	USA	4 July 2005	9P/Tempel 1	500
Deep Impact	USA	4 November 2010	103P/Hartley 2	700
Stardust	USA	15 February 2011	9P/Tempel 1	178

* probe inactive at the time of the flyby

Chapter references

[Abe-2007] Abe, M., et al., "Ground-Based Observations of Post-Hayabusa Mission Targets", paper presented at the XXXVIII Lunar and Planetary Science Conference, Houston, 2007

[Abe-2008] Abe, M., et al., "Ground-Based Observational Campaign for Asteroid 162173 1999 JU3", paper presented at the XXXIX Lunar and Planetary Science Conference, Houston, 2008

[Abilleira-2012] Abilleira, F., et al., "Entry, Descent, and Landing Communications for the 2011 Mars Science Laboratory", paper presented at the 23rd International Symposium on Space Flight Dynamics, Pasadena, October 2012

[Accomazzo-2006] Accomazzo, A. Schmitz, P, Tanco, I., "From Earth to Venus: Reaching Our Sister Planet", ESA Bulletin, 127, 2006, 38-44

[Accomazzo-2008] Accomazzo, A., "The Fly-By of Steins - Stretching Rosetta's Limits", presentation at the Rosetta Steins Fly-By Press Conference, Darmstadt, ESA/ESOC, 6 September 2008

[Adams-2010] Adams, M.L. (study lead), et al., "Chiron Orbiter Mission Study Final Report – Presented to the Planetary Decadal Survey Steering Committee and Primitive Bodies Panel", 4 May 2010

[Adler-2009] Adler, S.L., "Modeling the Flyby Anomalies with Dark Matter Scattering", arXiv astro-ph/0908.2414 preprint

[Adler-2012] Adler, M., Owen, W., Riedel, J., "Use of MRO Optical Navigation Camera to Prepare for Mars Sample Return", paper presented to the Concepts and Approaches for Mars Exploration workshop, June 2012

[Agarwal-2007] Agarwal, J., Müller, M., Grün, E., "Dust Environment Modelling of Comet 67P/Churyumov-Gerasimenko", Space Science Reviews, 128, 2007, 79-131

[Agnolon-2009] Agnolon, D., "Marco Polo: The European Contribution", presentation at the International Symposium Marco Polo and other Small Body Sample Return Missions, May 2009

[A'Hearn-2005a] A'Hearn, M.F., et al., "Deep Impact: Excavating Comet Tempel 1", Science, 310, 2005, 258-264

[A'Hearn-2005b] A'Hearn, M., Personal communication with the author, 16 July 2005

[A'Hearn-2008] A'Hearn, M.F., et al., "EPOXI's Mission to Comet 103P/Hartley 2". paper presented at the Asteroids, Comets, Meteors Meeting, 2008

[A'Hearn-2011a] A'Hearn, M.F., et al., "EPOXI at Comet Hartley 2", Science, 332, 2011, 1396-1400

[A'Hearn-2011b] A'Hearn, M.F., and the DIXI Science Team, "Comet Hartley 2: A Different

Class of Cometary Activity", paper presented at the XLII Lunar and Planetary Science Conference, Houston, 2011

[AIDA-2012] "Asteroid Impact & Deflection Assessment (AIDA) Mission: Opportunities and Tests in a US-Europe Space Mission Cooperation. Project Options", document dated December 2012

[Altenhoff-2009] Altenhoff, W.J., et al., "Why Did Comet 17P/Holmes Burst Out? Nucleus Splitting or Delayed Sublimation?", arXiv astro-ph/0901.2739 preprint

[Ammannito-2013] Ammannito, E., et al., "Olivine in an Unexpected Location on Vesta's Surface", Nature, 504, 2013, 122-125

[Anderson-2007] Anderson, M., "Don't Stop Till You Get to the Fluff", New Scientist, 6 January 2007, 26-30

[Anderson-2008] Anderson, B.J., et al., "The Structure of Mercury's Magnetic Field from MESSENGER's First Flyby", Science, 321, 2008, 82-85

[Anderson-2011] Anderson, B.J., et al., "The Global Magnetic Field of Mercury from MESSENGER Orbital Observations", Science, 333, 2011, 1859-1862

[Andrews-2010] Andrews, D., et al., "Ptolemy Operations in Anticipation of the Flyby of Asteroid 21 Lutetia", paper presented at the General Assembly of the European Geosciences Union, Vienna, May 2010

[Andrews-Hanna-2008] Andrews-Hanna, J.C., Zuber, M.T., Banerdt, B., "The Borealis Basin and the Origin of the Martian Crustal Dichotomy", Nature, 453, 2008, 1212-1215

[APL-2006a] "New Horizons Launch Press Kit", NASA; SwRI; APL, January 2006

[APL-2006b] "STEREO - The Sun in 3-D: A New Frontier in Solar Research. A Guide to STEREO's Twin Observatories", NASA; APL, 2006

[APL-2008a] "Solar Probe+ Mission Engineering Study Report", Prepared for NASA's Heliophysics Division, 10 March 2008

[APL-2008b] "Solar Sentinels: Mission Study Report", prepared for NASA by The Johns Hopkins University Applied Physics Laboratory, February 2008

[April-2008] April, R., "Where Next, Columbus?", Spaceflight, December 2008, 467-475

[April-2010] April, R., "Short Life on Hellish Planet", Spaceflight, December 2010, 424-425

[April-2011] April, R., "Five Year Focus on Planets and Moons", Spaceflight, January 2011, 16-18

[Arridge-2012] Arridge, C.R., et al., "Uranus Pathfinder: Exploring the Origins and Evolution of Ice Giant Planets", Experimental Astronomy, 33, 2012, 753-791

[Ashworth-2009] Ashworth, S., "Many Ways to Mars", Spaceflight, March 2009, 116-118

[Asker-2000] Asker, J. R., "Will Phoning Home Yield Busy Signals?", Aviation Week & Space Technology, 11 December 2000, 83-84

[ASU-2002] "SCIM Sample Collection for Investigation of Mars Fact Sheet", Arizona State University, 2002

[Atzei-2005] Atzei, A., Falkner, P., "Study Overview of the JME - Jovian Minisat Explorer TRS - An ESA Technology Reference Study", document ESA SCI-AP/2004/TN-085/AA dated 22 March 2005

[Auster-2007] Auster, H.U., et al., "ROMAP: Rosetta Magnetometer and Plasma Monitor", Space Science Reviews, 128, 2007, 221-240

[AWST-1978] "Chinese Space Plans", Aviation Week & Space Technology, 10 April 1978, 20

[AWST-2005] "Go the Extra Mile, NASA, And Fund Another Deep Impact Mission", Aviation Week & Space Technology, 11 July 2005, page unknown

[Baines-2007] Baines, K.H., et al., "Polar Lightning and Decadal-Scale Cloud Variability on Jupiter", Science, 318, 2007, 226-229

[Balint-2005] Balint, T., "Exploring Triton with Multiple Landers", Paper presented at the 56th International Astronautical Congress, Fukuoka, 2005

[Balint-2007] Balint, T., et al., "Can We Power Future Mars Missions?", Journal of the British Interplanetary Society, 60, 2007, 294-303

[Ball-1999] Ball, A.J., Lorenz, R.D., "Penetrometry of Extraterrestrial Surfaces: an Historical Overview". Paper presented at the International Workshop on Penetrometry in the Solar System, Graz, 1999

[Ball-2009] Ball, A.J., et al., "Mars Phobos and Deimos Survey (M-PADS) - A Martian Moons Orbiter and Phobos Lander", Advances in Space Research, 43, 2009, 120-127

[Ballard-2008] Ballard, S., et al., "Preliminary Results on HAT-P-4, TrES-3, XO-2, and GJ 436 from the NASA EPOXI Mission", arXiv astro-ph/0807.2803 preprint

[Balogh-2007] Balogh, A., et al., "Missions to Mercury", Space Science Reviews, 132, 2007, 611-645

[Balsiger-2007] Balsiger, H., et al., "ROSINA - Rosetta Orbiter Spectrometer for Ion and Neutral Analysis", Space Science Reviews, 128, 2007, 745-801

[Banerdt-2013] Banerdt, B., "InSight: A Geophysical Mission to a Terrestrial Planet Interior", presentation dated 7 March 2013

[Barabash-2007a] Barabash, S., et al., "The Loss of Ions from Venus Through the Plasma Wake", Nature, 450, 2007, 650-653

[Barabash-2007b] Barabash, S., "Martian Missions in Asia", undated presentation

[Barnes-2012] Barnes, J.W., et al., "AVIATR – Aerial Vehicle for In-situ and Airborne Titan Reconnaissance: A Titan airplane mission concept", Experimental Astronomy, 33, 2012, 55-127

[Barthelemy-2003] Barthélémy, P., "Rosetta, Sonde Européenne Exploratrice de Comète, Pourrait Revoir son Plan de Vol", Le Monde, 8 January 2003 (In French)

[Barucci-2007a] Barucci, M.A., Fulchignoni, M., Rossi, A., "Rosetta Asteroid Targets: 2867 Steins and 21 Lutetia", Space Science Reviews, 128, 2007, 67-78

[Barucci-2007b] Barucci, M.A., "NEO Sample Return Mission «Marco Polo»: Proposal to ESA Cosmic Vision", presentation at the Sputnik 50-Year Jubilee, Moscow, October 2007

[Barucci-2008] Barucci, M.A., et al., "Asteroids 2867 Steins and 21 Lutetia: Surface Composition from Far Infrared Observations with the Spitzer Space Telescope", Astronomy & Astrophysics, 477, 2008, 665-670

[Basilevsky-2007] Basilevsky, A.T., et al., "Landing on Venus: Past and Future", Planetary and Space Science, 55, 2007, 2097-2112

[Beebe-2010] Beebe, R., Dudzinski, L., "Mission Concept Study – Planetary Science Decadal Survey – Saturn Atmospheric Entry Probe Mission Study", April 2010

[Belbruno-1996] Belbruno, E., Genta, G., "Low Energy Comet Rendezvous Using Resonance Transitions", paper presented at the First IAA Symposium on Realistic Near-Term Advanced Scientific Space Missions, Aosta, 25-27 June 1996

[Belbruno-1997] Belbruno, E., Marsden, B.G., "Resonance Hopping in Comets", The Astronomical Journal, 113, 1997, 1433-1444

[Bell-2012] Bell, J., "Protoplanet Close-Up", Sky & Telescope, September 2012, 32-36

[Belskaya-2010] Belskaya, I.N., et al., "Puzzling asteroid 21 Lutetia: our knowledge prior to the Rosetta fly-by", arXiv astro-ph/1003.1845 preprint

[Belton-1996] Belton, M.J.S., et al., "Deep Impact: Exploration of the Mantle-Core Interface Region in a Cometary Nucleus", Bulletin of the American Astronomical Society, 28, 1996, 1088

[Belton-2005] Belton, M.J.S., et al., "Deep Impact: Working Properties for the Target Nucleus Comet 9P/Tempel 1", Space Science Reviews, 117, 2005, 137-160

[Benest-1990] Benest, D.G., "P/Ge-Wang Joins P/Slaughter-Burnham and P/Boethin in the club of comets in 1/1 Resonance with Jupiter", Celestial Mechanics and Dynamical Astronomy, 47, 1990, 361-374

[Benkhoff-2010] Benkhoff, J., et aL., "BepiColombo - Comprehensive Exploration of Mercury: Mission Overview and Science Goals", Planetary and Space Science, 58, 2010, 2-20

[Bennett-2004] Bennett, D., "Deep Impact Microlens Explorer", presentation at the Hawaiian Gravitational Microlensing Workshop, 2004

[Bensch-2006] Bensch, F., et al., "Submillimeter Wave Astronomy Satellite Observations of Comet 9/Tempel 1 and Deep Impact", arXiv astro-ph/0606045 preprint

[Berner-2002] Berner, C., et al., "Rosetta: ESA's Comet Chaser", Esa Bulletin, 112, 2002, 10-17

[Berner-2005] Berner, S., "Japan's Space Program: A Fork in the Road?", Santa Monica, RAND Corporation report TR-184, 2005

[Bertaux-2007] Bertaux, J.-L., et al., "A Warm Layer in Venus' Cryosphere and High-Altitude Measurements of HF, HCl, H2O and HDO", Nature, 450, 2007, 646-649

[Bertolami-2004] Bertolami, O., Pàramos, J., "Pioneer's Final Riddle", arXiv gr-qc/0411020 preprint

[Bibring-2007a] Bibring, J.-P., et al., "The Rosetta Lander ('Philae') Investigations", Space Science Reviews, 128, 2007, 205-220

[Bibring-2007b] Bibring, J.-P., et al., "CIVA", Space Science Reviews, 128, 2007, 397-412

[Biele-2002] Biele, J., et al., "Current Status and Scientific Capabilities of the Rosetta Lander Payload", Advances in Space Research, 29, 2002, 1199-1208

[Biele-2005] Biele, J., et al., "Philae (Rosetta Lander): Experiment Status after Commissioning", submitted to Advances in Space Research

[Binzel-2012] Binzel, R.P., "A Golden Spike for Planetary Science", Science, 338, 2012, 203-204

[Bish-2013] Bish, D.L., et al., "X-ray Diffraction Results from Mars Science Laboratory: Mineralogy of Rocknest at Gale Crater", Science, 341, 2013

[Bishop-2008] Bishop, J.L., et al., "Phyllosilicate Diversity and Past Aqueous Activity Revealed at Mawrth Vallis, Mars", Science, 321, 2008, 830-833

[Blair-2011] Blair, S., Semprimoschnig, C., van Casteren, J., "Hot Stuff: Seven Steps in Making a Mission to Mercury", ESA Bulletin, 146, 2011, 15-20

[Blake-2013] Blake, D.F., et al., "Curiosity at Gale Crater, Mars: Characterization and Analysis of the Rocknest Sand Shadow", Science, 341, 2013

[Blanc-2009] Blanc, M., et al., "LAPLACE: A mission to Europa and the Jupiter System for ESA's Cosmic Vision Programme", Experimental Astronomy, 23, 2009, 849-892

[Blewett-2009] Blewett, D.T., "Do Lunar-Like Swirls Occur on Mercury?", paper presented at the XL Lunar and Planetary Science Conference, Houston, 2009

[Blewett-2011] Blewett, D.T., et al., "Hollows on Mercury: MESSENGER Evidence for Geologically Recent Volatile-Related Activity", Science, 333, 2011, 1856-1859

[Blume-2005] Blume, W.H., "Deep Impact Mission Design", Space Science Reviews, 117, 2005, 23-42

[Bodewits-2011] Bodewits, D., et al., "Hartley-2's Puzzling Gas Anomaly", paper presented at the XLII Lunar and Planetary Science Conference, Houston, 2011

[Bondo-2004] Bondo, T., et al., "Preliminary Design of an Advanced Mission to Pluto", paper presented at the 24th International Symposium on Space Technology and Science, Miyazaki, June 2004

[Bortle-1986] Bortle, J.E., "Comet Digest", Sky & Telescope, April 1986, 426

[Bortle-2008] Bortle, J.E., "The Astounding Comet Holmes", Sky & Telescope, February 2008, 24-28

[Borucki-2009] Borucki, W.J., et al., "Kepler's Optical Phase Curve of the Exoplanet HAT-P-7b", Science, 325, 2009, 709

[Bowles-2006] Bowles, N., et al., "Venus Descent Microprobes", presentation at the Venus Entry Probe Workshop, ESTEC, Noordwijk, 19-20 January 2006

[Boynton-2009] Boynton, W.V., et al., "Evidence for Calcium Carbonate at the Mars Phoenix Landing Site", Science, 325, 2009, 61-64

[Braun-2006] Braun, R.D., Spencer, D.A., "Design of the ARES Mars Airplane and Mission Architecture", Journal of Spacecraft and Rockets, 43, 2006, 1026-1034

[Bridges-2012] Bridges, N.T., et al., "Earth-like Sand Fluxes on Mars", Nature, 485, 2012, 339-342

[Broglio-1966] Broglio, L., "Esperimento e Risultati del Satellite S. Marco I" (Experiment and Results of the San Marco 1 Satellite), Rome, Accademia Nazionale dei Lincei, 1966 (In Italian)

[Brown-2004] Brown, M.E., Trujillo, C.A., Rabinowitz, D.L., "Discovery of a Candidate Inner Oort Cloud Planetoid", arXiv astro-ph/0404456 preprint

[Brown-2005] Brown, M.E., Trujillo, C.A., Rabinowitz, D.L., "Discovery of a Planetary-Sized Object in the Scattered Kuiper Belt", arXiv astro-ph/0508633 preprint

[Brown-2013a] Brown, M.E., Hand, K.P., "Salts and Radiation Products on the Surface of Europa", arXiv astro-ph/1303.0894 preprint

[Brown-2013b] Brown, M.E., "On the Size, Shape, and Density of Dwarf Planet Makemake", arXiv astro-ph/1304.1041 preprint

[Bryant-2005] Bryant, G., "Targeting Comet Tempel 1", Sky & Telescope, June 2005, 67-69

[Buch-2009] Buch, A., et al., "Development of a Gas Chromatography Compatible Sample Processing System (SPS) for the In-Situ Analysis of Refractory Organic Matter in Martian Soil: Preliminary Results", Advances in Space Research, 43, 2009, 143-151

[Buie-2005] Buie, M.W., et al., "Orbit and Photometry of Pluto's Satellites: Charon, S/2005 P1 and S/2005 P2", arXiv astro-ph/0512491 preprint

[Buie-2010] Buie, M.W., et al., "Pluto and Charon with the Hubble Space Telescope. II. Resolving Changes on Pluto's Surface and a Map for Charon", The Astronomical Journal, 139, 2010, 1128-1143

[Buie-2012a] Buie, M.W., et al., "Searching for KBO Flyby Targets for the New Horizons Mission", paper presented at the Asteroids, Comets, Meteors meeting, 2012

[Buie-2012b] Buie, M., "The Sentinel Mission", presentation at the 7th Small Bodies Assessment Group (SBAG) meeting, Pasadena, July 2012

[Burch-2007] Burch, J.L., et al., "RPC-IES: The Ion and Electron Sensor of the Rosetta Plasma Consortium", Space Science Reviews, 128, 2007, 697-712

[Calvin-2007] Calvin, W.M., et al., "Report from the 2013 Mars Science Orbiter (MSO) Second Science Analysis Group", 29 May 2007

[Canup-2013] Canup, R., "Lunar Conspiracies", Nature, 504, 2013, 27-29

[Capuano-2012] Capuano, M., et al., "ExoMars Mission 2016, EDM Science Opportunities", paper presented at the 63rd International Astronautical Congress, Naples, 2012

[Cargill-2013] Cargill, P., "Towards ever Smaller Length Scales", Nature, 493, 2013, 485-486

[Carlisle-2009] Carlisle, C.M., "The Race to Find Alien Planets", Sky & Telescope, January 2009, 28-33

[Carnelli-2006] Carnelli, I., Gàlvez, A., Ongaro, F., "Learning to Deflect Near Earth Objects: Industrial Design of the Don Quijote Mission", paper presented at the 63rd International Astronautical Congress, Naples, 2012

[Carnelli-2013] Carnelli, I., Gàlvez, A., "Asteroid Impact Mission (AIM): ESA's NEO Exploration Precursor", presentation to the 8th Small Bodies Assessment Group (SBAG), Washington, January 2013

[Carr-2007] Carr, C., et al., "RPC: The Rosetta Plasma Consortium", Space Science Reviews, 128, 2007, 629-647

[Carry-2007] Carry, B., et al., "Near-Infrared Mapping and Physical Properties of the Dwarf-Planet Ceres", arXiv astro-ph/0711.1152 preprint

[Carry-2010] Carry, B., et al., "Physical Properties of ESA/NASA Rosetta Target Asteroid (21) Lutetia: Shape and Flyby Geometry", arXiv astro-ph/2005.5356 preprint

[Carter-2010] Carter, J., et al., "Detection of Hydrated Silicates in Crustal Outcrops in the Northern Plains of Mars", Science, 328, 2010, 1682-1686

[Carusi-1985] Carusi, A., et al., "Long-Term Evolution of Short-Period Comets", Bristol, Adam Hilger, 1985

[Carvano-2008] Carvano, J.M., et al., "Surface Properties of Rosetta's Targets (21) Lutetia and (2867) Steins from ESO Observations", Astronomy & Astrophysics, 479, 2008, 241-248

[Cassell-1998] Cassell, C.R., et al., "Asteroid Selection and Mission Design for SpaceDev's Near Earth Asteroid Prospector", paper AAS 98-183

[Cassi-2012] Cassi, C., et al., "ExoMars: One Project Two Missions", paper presented at the 63rd International Astronautical Congress, Naples, 2012

[Cecil-2007] Cecil, G., Rashkeev, D., "A Side of Mercury not Seen by Mariner 10", arXiv astro-ph/0708.0146v2 preprint

[Chandler-2008] Chandler, D., "The Burger Bar that Saved the World", Nature, 453, 2008, 1165-1168

[Chang Diaz-2000] Chang Diaz, F.R., "The VASIMIR Rocket", Scientific American, November 2000, 90-97

[Chassefière-2006] Chassefière, E., "The Lavoisier Mission Concept", presentation at the Venus Entry Probe Workshop, ESTEC, Noordwijk, 19-20 January 2006

[Chassefière-2007a] Chassefière, E., "ESA's Venus Entry Probe Workshop and Cosmic Vision Proposal", presentation at the 3rd meeting of the Venus Exploration Analysis Group (VEXAG), Crystal City, January 2007

[Chassefière-2007b] Chassefière, E., "Toward an International Venus Exploration Program", presentation at the Sputnik 50-Year Jubilee, Moscow, October 2007

[Chassefière-2007c] Chassefière, E., et al., "European Venus Explorer - A Proposed Mission for ESA's Cosmic Vision 2015-2025", presentation at the 4th meeting of the Venus Exploration Analysis Group (VEXAG), Greenbelt, November 2007

[Chen-2010] Chen, C., Hu, J., Zhu, G.,, "The Key Techniques and Design Features of YH-1 Mars Probe", Chinese Astronomy and Astrophysics, 34, 2010, 217-226

[Chen-2011a] Chen, Y., Baoyin, H.X., Li, J.F., "Design and Optimization of a Trajectory for Moon Departure Near Earth Asteroid Exploration", Science China: Physics, Mechanics & Astronomy, 54, 2011, 748-755

[Chen-2011b] Chen, Y., Baoyin H.X., Li J.F., "Target Analysis and Low-Thrust Trajectory Design of Chinese Asteroid Exploration Mission", Science China: Physics, Mechanics & Astronomics, 41, 2011, 1104-1111 (in Chinese)

[Chen-2013] Chen, Y., Baoyin, H., Li. J.-F., "Trajectory Analysis and Design for A Jupiter Exploration Mission", Chinese Astronomy and Astrophysics, 37, 2013, 77–89

[Cheng-2013] Cheng, A., et al., "AIDA: Asteroid Impact & Deflection Assessment", paper presented at the 64th International Astronautical Congress, Beijing, 2013

[Cherniy-2007] Cherniy, I., "Visokotemperaturnaya Elektronika - Klyuch k Taynam Venerii?"

(High Temperature Electronics - The Key to the Secrets of Venus?), Novosti Kosmonavtiki, November 2007 (in Russian)

[Chesley-2013a] Chesley, S., "ISIS: Impactor for Surface and Interior Science", presentation to the 8th Small Bodies Assessment Group (SBAG), Washington, January 2013

[Chesley-2013b] Chesley, S., "ISIS: Impactor for Surface and Interior Science", presentation to the 9th Small Bodies Assessment Group (SBAG), Washington, July 2013

[Christensen-2009] Christensen, P., "Science Perspective for Candidate Mars Mission Architectures for 2016-2026", presentation to the Mars Exploration Program Analysis Group (MEPAG) meeting #20, Rosslyn, Virginia, March 2009

[Christiansen-2008] Christiansen, J.L., et al., "The NASA EPOXI Mission of Opportunity to Gather Ultraprecise Photometry of Known Transiting Exoplanets", arXiv astro-ph/ 0807.2852 preprint

[Christophe-2011] Christophe, B., et al., "OSS (Outer Solar System): A Fundamental and Planetary Physics Mission to Neptune, Triton and the Kuiper Belt", arXiv astro-ph/ 1106.0132 preprint

[Cirtain-2013] Cirtain, J.W., et al., "Energy Release in the Solar Corona from Spatially Resolved Magnetic Braids", Nature, 493, 2013, 501-503

[Clark-2012] Clark, B.E., "Asteroids: Dark and Stormy Weather", Nature, 491, 2012, 45-46

[Claros-2004] Claros, V., Süss, G., Warhaut, M., "ESA Reaches Out into Space from Space - The New Cebreros Station", ESA Bulletin, 118, 2004, 17-20

[Cleave-2005] Cleave, M.L., letter to the B612 Foundation, 12 October 2005

[CLEP-2009] "Chinese Lunar Exploration Program", presentation to the Global Space Development Summit, November 2009

[Clery-2003] Clery, D., "Financial Crisis Puts Comet Mission on the Ropes", Science, 300, 2003, 1213

[Clery-2008a] Clery, D., "Cloudy Future for Europe's Space Plans", Science, 322, 2008, 1180-1181

[Clery-2008b] Clery, D., "Ministers Bankroll European Space Agency's Ambitions", Science, 322, 2008, 1447

[Cochran-2006] Cochran, A.L., et al., "Observations of Comet 9P/Tempel 1 with the Keck 1 HIRES Instrument During Deep Impact", arXiv astro-ph/0609134 preprint

[Colangeli-1999] Colangeli, L., et al., "Infrared Spectral Observations of Comet 103P/Hartley 2 by ISOPHOT", Astronomy & Astrophysics, 343, 1999, L87-L90

[Colangeli-2007] Colangeli, L., et al., "The Grain Impact Analyser and Dust Accumulator (GIADA) Experiment for the Rosetta Mission: Design, Performances and First Results", Space Science Reviews, 128, 2007, 803-821

[Colangelo-2000] Colangelo, G., et al., "Solar Orbiter: A Challenging Mission Design for Near-Sun Observations", ESA Bulletin 104, 2000, 76-85

[Colombo-1965] Colombo, G., "Rotational Period of the Planet Mercury", Nature, 208, 1965, 575

[Connors-2011] Connors, M., Wiegert, P., Veillet, C., "Earth's Trojan Asteroid", Nature, 475, 481-483

[Cooke-2011a] Cooke, B., "Orbiter Element", presentation to the Outer Planets Assessment Group, (OPAG) 19 October 2011

[Cooke-2011b] Cooke, B., "Europa Lander Study Background", presentation to the Outer Planets Assessment Group, (OPAG) November 2011

[Coradini-2007] Coradini, A., et al., "VIRTIS: An Imaging Spectrometer for the Rosetta Mission", Space Science Reviews, 128, 2007, 529-560

[Coradini-2010] Coradini, M., "The ESA/NASA ExoMars Programme", presentation to the

Mars Exploration Program Analysis Group (MEPAG) meeting #22, Monrovia, California, March 2010

[Coradini-2011] Coradini, A., et al., "The Surface Composition and Temperature of Asteroid 21 Lutetia as Observed by ROSETTA/VIRTIS", Science, 334, 2011, 492-494

[Corneille-2008] Corneille, P., "Avoiding Catastrophe", Spaceflight, October 2008, 395-399

[Coué-2007a] Coué, P., "La Chine Veut la Lune" (China Wants the Moon), Paris, A2C Medias, 2007, 159-164 (in French)

[Coué-2007b] ibid., 149-157

[Covault-2004a] Covault, C., "Hot Shot", Aviation Week & Space Technology, 26 July 2004, 58-59

[Covault-2004b] Covault, C., "Skycrane Reassessed", Aviation Week & Space Technology, 26 July 2004, 58-59

[Covault-2005a] Covault, C., "A New Mars", Aviation Week & Space Technology, 31 January 2005

[Covault-2005b] Covault, C., "Back to Mars", Aviation Week & Space Technology, 22 August 2005, 28-30

[Covault-2005c] Covault, C., "Martian Infotech", Aviation Week & Space Technology, 28 February 2005, 54-55

[Covault-2006] Covault, C., "Spying on Mars", Aviation Week & Space Technology, 23 October 2006, 24

[Covault-2007a] Covault, C., "Back to Mars", Aviation Week & Space Technology, 11 June 2007, 56-59

[Covault-2007b] Covault, C., "Bees to the Rescue", Aviation Week & Space Technology, 11 June 2007, 61

[Covault-2007c] Covault, C., "Martian Mysteries", Aviation Week & Space Technology, 11 June 2007, 60

[Covault-2007d] Covault, C., "Phoenix Streaks toward Mars", Aviation Week & Space Technology, 13 August 2007, 31

[Covault-2008a] Covault, C., "Fire and Ice", Aviation Week & Space Technology, 19 May 2008, 36

[Covault-2008b] Covault, C., "Carrying the Fire", Aviation Week & Space Technology, 2 June 2008, 24

[Covault-2008c] Covault, C., "Phoenix Delivers", Aviation Week & Space Technology, 9 June 2008, 34

[Covault-2008d] Covault, C., "Martian Arctic Revealed", Aviation Week & Space Technology, 9 June 2008, 52-55

[Covault-2008e] Covault, C., "Shake, Shake, Shake", Aviation Week & Space Technology, 16 June 2008, 41

[Covault-2008f] Covault, C., "Memory Overload", Aviation Week & Space Technology, 23 June 2008, 54

[Covault-2008g] Covault, C., "Phoenix Scores Again", Aviation Week & Space Technology, 11 August 2008, 30

[Covault-2008h] Covault, C., "Taste Test", Aviation Week & Space Technology, 7 July 2008, 28-30

[Covault-2008i] Covault, C., "Phoenix Digs Deeper", Aviation Week & Space Technology, 18 August 2008, 43

[Covault-2008j] Covault, C., "Racing the Midnight Sun", Aviation Week & Space Technology, 15 September 2008, 40

[Covault-2008k] Covault, C., "A Lander's Legacy", Aviation Week & Space Technology, 17 November 2008, 30

[Cowan-2009] Cowan, N., "Alien Maps of an Ocean-Bearing World", arXiv astro-ph/09053742 preprint

[Cowen-2012] Cowen, R., "Venus's Rare Sun Crossing May Aid Search for Exoplanets", Science, 336, 2012, 660

[Crovisier-1999] Crovisier, J., et al., "The Thermal Infrared Spectra of Comets Hale-Bopp and 103P/Hartley 2 Observed with the Infrared Space Observatory", paper presented at the Workshop on Thermal Emission Spectroscopy, Houston, 1999

[Crovisier-2005] Crovisier, J., Personal communication with the author, 2 August 2005

[Crovisier-2009] Crovisier, J., et al., "The Chemical Diversity of Comets: Synergied between Space Exploration and Ground-Based Radio Observations", arXiv astro-ph/0901.2205 preprint

[Cui-2004] Cui Pingyuan, et al., "Ivar Asteroid Exploration Mission and Trajectory Design", paper presented at the 55th International Astronautical Congress, Vancouver, 2004

[Cui-2005] Cui Pingyan, Cui Hutao, Qiao Dong, "The Scenario and Scheme of Exploring Nereus Asteroid Mission", Paper presented at the 56th International Astronautical Congress, Fukuoka, October 2005

[Cunningham-1988a] Cunningham, C.J., "Introduction to Asteroids", Richmond, Willmann-Bell, 1988, 71-76

[Cunningham-1988b] ibid., 118-119

[Cunningham-1988c] ibid., 125-128

[Cyranoski-2011] Cyranoski, D., "China forges ahead in space", Nature, 497, 2011, 276-277

[Damiani-2012] Damiani, S., Lauer, M., Müller, M., "Monitoring of Aerodynamic Pressures for Venus Express in the Upper Atmosphere During Drag Experiments Based on Telemetry", paper presented at the 23rd International Symposium on Space Flight Dynamics, Pasadena, October 2012

[David-2011] David, L., "Juno to Jupiter: Piercing the Veil", Aerospace America, July/August 2011, 40-45

[Day-2010] Day, D., "Journey to a Red Moon", Spaceflight, November 2010, 426-428

[Day-2011a] Day, D.A., "NASA's Next Discovery Class Mission", Spaceflight, July 2011, 253-254

[Day-2011b] Day, D.A., "Romancing the Stone", Spaceflight, April 2011, 134-135

[de Campos Velho-2010] de Campos Velho, H.F., Personal communication with the author, 1 November 2010

[de Groot-2012] de Groot, R., "Mars Exploration: The ESA Perspective", presentation to the Mars Exploration Program Analysis Group (MEPAG) meeting #25, Washington, DC, February 2012

[Deimos Space-2009] Deimos Space SLU "Micro/Mini-Satellite Interplanetary Mission – Proba-IP Executive Summary", document dated 23 November 2009

[de la Fuente Marcos-2013] de la Fuente Marcos, C., de la Fuente Marcos, R., "Three New Stable L5 Mars Trojans", arXiv astro-ph/1303.0124 preprint

[de León-2011] de León, J., et al., "New Observations of Asteroid (175706) 1996 FG3, Primary Target of the ESA Marco Polo-R Mission", Astronomy & Astrophysics, 530, 2011, L12

[Denevi-2009] Denevi, B.W., et al., "The Evolution of Mercury's Crust: A Global Perspective from MESSENGER", Science, 324, 2009, 613-618

[Denevi-2012] Denevi, B.W., et al., "Pitted Terrain on Vesta and Implications for the Presence of Volatiles", Science, 338, 2012, 246-249

[D'Errico-2006] D'Errico, P., Santandrea, S., "APIES: A Mission for the Exploration of the

Main Asteroid Belt Using a Swarm of Microsatellites", Acta Astronautica, 59, 2006, 689-699

[De Sanctis-2012] De Sanctis, M.C., et al., "Spectroscopic Characterization of Mineralogy and Its Diversity Across Vesta", Science, 336, 2012, 697-700

[de Selding-2011] de Selding, P., "ESA Halts ExoMars Orbiter Work to Rethink Red Planet Plans with NASA", Space News, 25 April 2011, 1, 6

[Di Pippo-1999] Di Pippo, S., Ercoli Finzi, A., Magnani, P.G., "Robotic Arms and Surface/Subsurface Sampling for Mars Exploration". Course presentation for the Summer School 1999 on Mars, Alpbach, 3-12 August 1999

[Dissly-2003] Dissly, R.W., Miller, K.L., Carlson, R.J., "Artificial Crater Formation on Satellite Surfaces Using an Orbiting Railgun", paper presented at the Forum on Jupiter Icy Moons Orbiter, Houston, 2003

[Dittus-2005] Dittus, H., et al., "A Mission to Explore the Pioneer Anomaly", arXiv gr-qc/0506139 preprint

[Dornheim-2005] Dornheim, M.A., "Crash Course", Aviation Week & Space Technology, 11 July 2005, 28-31

[Dornheim-2006a] Dornheim, M.A., "Mars Fleet Addition", Aviation Week & Space Technology, 6 March 2006, 35

[Dornheim-2006b] Dornheim, M.A., "Earning its 'O'", Aviation Week & Space Technology, 20 March 2006, 32

[Drossart-2007] Drossart, P., et al., "A Dynamic Upper Atmosphere of Venus as Revealed by VIRTIS on Venus Express", Nature, 450, 2007, 641-645

[Durda-2005] Durda, D., et al., "A Spacecraft Mission to Near-Earth Asteroid 2004 MN4: Call to Action", paper presented at the Asteroids, Comets, Meteors symposium, 2005

[Durda-2006] Durda, D., "The Most Dangerous Asteroid Ever Found", Sky & Telescope, November 2006, 29-33

[Durda-2010] Durda, D., "How to Deflect a Hazardous Asteroid", Sky & Telescope, December 2010, 22-28

[Edberg-2006] Edberg, N., "The Rosetta Mars Flyby", Diploma thesis, Master of Science Program in Engineering Physics, Uppsala University, 2006

[Edgett-2011] Edgett, K.S., "Gale Crater in Context", presentation at the Final MSL Field Site Selection, Monrovia/Arcadia, May 2011

[Edwards-2006] Edwards, C.D. Jr., et al., "Relay Communications Strategies for Mars Exploration Through 2020", Acta Astronautica, 59, 2006, 310-318

[Ehlmann-2008] Ehlmann, B.L., et al., "Orbital Identification of Carbonate-Bearing Rocks on Mars", Science, 322, 2008, 1828-1832

[Ehlmann-2011] Ehlmann, B.L., et al., "Subsurface Water and Clay Mineral Formation during the Early History of Mars", Nature, 479, 2011, 53-60

[Ehrenfreund-2012] Ehrenfreund, P., et al., "MarcoPolo-R: Near Earth Asteroid Sample Return Mission in ESA Assessment Study Phase", paper presented at the 63rd International Astronautical Congress, Naples, 2012

[Eismont-1997] Eismont, N.A., Sukhanov, A.A., "Low Cost Phobos Sample Return Mission", paper presented at the 12th International Symposium on Space Flight Dynamics, Darmstadt, 2-6 June 1997

[EJSM-2010] "Europa Jupiter System Mission (EJSM) – Exploring the Emergence of Habitable Worlds around Gas Giants", Document JPL D-67959, 15 November 2010

[Ekonomov-2008] Ekonomov, A., "How and why to Survive at Venus Surface", paper presented at the European Planetary Science Congress, Münster, 2008

[Elliott-2010] Elliott, J.O., Hunter Waite, J., "In-Situ Missions for the Exploration of Titan's Lakes", Journal of the British Interplanetary Society, 63, 2010, 376-383

[Elwood-2004] Elwood, J., et al., "Rosetta's New Target Awaits", ESA Bulletin, 117, 2004, 4-13

[Encrenaz-2005] Encrenaz, T., Personal communication with the author, 25 July 2005

[Eneev-1998] Eneev, T.M., et al., "Mission to Phobos with the Use of Electric Propulsion", paper presented at the Second IAA Symposium on Realistic Near-Term Advanced Scientific Space Missions, Aosta, Italy, June 29-July 1, 1998

[Ercoli Finzi-2007] Ercoli Finzi, A., et al., "SD2 - How to Sample a Comet", Space Science Reviews, 128, 2007, 281-299

[Eriksson-2007] Eriksson, A.I., et al., "RPC-LAP: The Rosetta Langmuir Probe Instrument", Space Science Reviews, 128, 2007, 729-744

[ESA-2000] "Solar Orbiter: A High-Resolution Mission to the Sun and Inner Heliosphere", ESA SCI(2000)6, July 2000

[ESA-2001] "Venus Express Mission Definition Report", ESA SCI(2001)6, October 2001

[ESA-2002] "CDF Study Report: ExoMars09", ESA CDF-14(A), August 2002

[ESA-2005] "Solar Orbiter Assessment Phase Final Executive Report", ESA SCI-A/2000/054/NR, 15 December 2005

[ESA-2010a] "ExoMars - EDL Demonstrator Module Surface Payload Experiment Proposal Information Package", ESA document EXM-DM-IPA-ESA-00001, 8 November 2010

[ESA-2010b] "Trojans' Odyssey: Unveiling the Early History of the Solar System", A proposal submitted as an M-class mission to ESA's Cosmic Vision Programme, December 3rd, 2010

[ESA-2011a] "JUICE: Exploring the Emergence of Habitable Worlds around Gas Giants - Assessment Study Report", ESA/SRE(2011)18, December 2011

[ESA-2011b] "Solar Orbiter: Exploring the Sun-Heliosphere Connection - Definition Study Report", ESA Document SRE(2011)14, July 2011

[Espinasse-2008] Espinasse, S., "Italian Activities and Plans in the Field of Exploration", paper presented at the first Meeting of the International Primitive Body Exploration Working Group, Okinawa, January 2008

[EST-2012a] Europa Study Team, "Europa Study 2012 Report – Introduction", JPL D-71990, 1 May 2012

[EST-2012b] Europa Study Team, "Europa Study 2012 Report – Europa Orbiter Mission", JPL D-71990, 1 May 2012

[EST-2012c] Europa Study Team, "Europa Study 2012 Report – Europa Multiple Flyby Mission", JPL D-71990, 1 May 2012

[EST-2012d] Europa Study Team, "Europa Study 2012 Report – Europa Lander Mission", JPL D-71990, 1 May 2012

[Evans-2002] Evans, N.W., Tabachnik, S.A., "Structure of Possible Long-Lived Asteroid Belts", Monthly Notices of the Royal Astronomical Society, 333, 2002, L1-L5

[Fabrega-2003] Fabrega, J., et al., "Venus Express: the First European Mission to Venus", paper presented at the 54th International Astronautical Congress, Bremen, October 2003

[Fabrega-2004] Fabrega, J., et al., "Venus Express on the Right Track", paper presented at the 55th International Astronautical Congress, Vancouver, October 2004

[Fabrega-2007] Fabrega, J., et al., "Europe Goes to Venus: The Journey of Venus Express", Journal of the British Interplanetary Society, 60, 2007, 430-438

[Farley-2014] Farley, K.A., et al., "In Situ Radiometric and Exposure Age Dating of the Martian Surface", accepted for publication in Science

[Farnham-2013] Farnham, T., "Deep Impact Continued Investigations (DI3)", presentation at the 9th Small Bodies Assessment Group (SBAG) meeting, Washington, July 2013

[Farquhar-2011] Farquhar, R.W., "Fifty Years on the Space Frontiers: Halo Orbits, Comets, Asteroids, and More", Denver, Outskirt Press, 2011, 262-263

[Farquhar-2013] Farquhar, R., et al., "A Unique Multi-Comet Mission Opportunity for China in 2018", paper presented at the 64th International Astronautical Congress, Beijing, 2013

[Feldman-2006a] Feldman, P.D., et al., "Ultraviolet Spctroscopy of Comet 9P/Tempel 1 with Alice/Rosetta during the Deep Impact Encounter", arXiv astro-ph/0608708 preprint

[Feldman-2006b] Feldman, P.D., et al., "Hubble Space Telescope Observations of Comet 9P/Tempel 1 During the Deep Impact Encounter", arXiv astro-ph/0608487 preprint

[Feldman-2006c] Feldman, P.D., et al., "Carbon Monoxide in Comet 9P/Tempel 1 Before and After the Deep Impact Encounter", arXiv astro-ph/0607185 preprint

[Ferri-2004] Ferri, P., Warhaut, M., "First In-Flight Experience with Rosetta", paper presented at the 55th International Astronautical Congress, Vancouver, October 2004

[Ferri-2005] Ferri, P., Schwehm, G., "Rosetta: ESA's Comet Chaser Already Making its Mark", ESA Bulletin, 123, 2005, 62-66

[Ferrin-2010] Ferrin, I., "Secular Light Curve of Comet 103P/Hartley 2, Target of the EPOXI Mission", arXiv astro-ph/1008.4556v1 preprint

[Fiehler-2006] Fiehler, D.I., McNutt, R.L. Jr., "Mission Design for the Innovative Interstellar Explorer Vision Mission", Journal of Spacecraft and Rockets, 43, 2006, 1239-1247

[Findlay-2012] Findlay, R., et al., "A Small Asteroid Lander Mission to Accompany Hayabusa-II", paper presented at the 63rd International Astronautical Congress, Naples, 2012

[Flight-1998] "NEAP Experiments May Fly in Discovery Mission", Flight International, 21 January 1998, 30

[Flight-1999] "Rosetta Goes on as NASA Retires", Flight International, 14 July 1999, 28

[Flight-2002] "Europe Faces new Ariane 5 Blow", Flight International, 17 December 2002, 5

[Flight-2003] "Rosetta Postponed after Ariane 5 Failure", Flight International, 21 January 2003, 7

[Fornasier-2007] Fornasier, S., et al., "Are the E-Type Asteroids (2867) Steins, a Target of the Rosetta Mission and NEA (3103) Eger Remnants of an Old Asteroid Family?", Astronomy & Astrophysics, 474, 2007, L29-L32

[Fornasier-2011] Fornasier, S., et al., "Photometric Observations of Asteroid 4 Vesta by the OSIRIS Cameras onboard the Rosetta Spacecraft", Astronomy & Astrophysics, 533, 2011, L9

[Förstner-2007] Förstner, R., Best, R., Steckling, M., "BepiColombo - Mission to Mercury", Journal of the British Interplanetary Society, 60, 2007, 314-320

[Frauenholz-2008] Frauenholz, R.B., et al., "Deep Impact Navigation System Performance", Journal of Spacecraft and Rockets, 42, 2008, 39-56

[Friedlander-1984] Friedlander, A.L., "Titan Buoyant Station", Journal of the British Interplanetary Society, 37, 1984, 381-387

[Fu-2012] Fu, R.F., et al., "An Ancient Core Dynamo in Asteroid Vesta", Science, 338, 2012, 238-241

[Fujita-2014] Fujita, K., et al., "Conceptual study and key technology development for Mars Aeroflyby sample collection", Acta Astronautica, 93, 2014, 84-93

[Fulle-2007] Fulle, M., et al., "Discovery of the Atomic Iron Tail of Comet McNaught Using the Heliospheric Imager on STEREO", The Astrophysical Journal, 661, 2007, L93-L96

[Furniss-1997] Furniss, T., "Asteroid Prospector", Flight International, 29 October 1997, 38

[Furniss-2002] Furniss, T., "Ariane 4 Puts Europe back in Space as ESA Probes 5 Loss", Flight International, 31 December 2002, 22

[Furniss-2003] Furniss, T., "Ariane 5 ECA to Try Again but Doubts over Rosetta", Flight International, 14 January 2003, 22

[Gafarov-2004] Gafarov, A.A., et al., "Conceptual Project of Interplanetary Spacecraft with Nuclear Power System and Electric Propulsion System for Radar Sounding of Ice Sheet of Europa, Jupiter Satellite", Paper presented at the 55th International Astronautical Congress, Vancouver, 2004

[Gafarov-2005] Gafarov, A.A., "Yadernaya Energiya v Kosmose" (Nuclear Energy in Space), Novosti Kosmonavtiki, January 2005 (in Russian)

[Galeev-1996] Galeev, A.A., et al., "Phobos Sample Return Mission", Advances in Space Research, 17, December 1996, 31-47

[GAO-2005] United States Government Accountability Office, "NASA's Space Vision: Business Case for Prometheus 1 Needed to Ensure Requirements Match Available Resources", GAO document 05-242, February 2005

[Gao-2012a] Gao, Y., Li, H.-N., He, S.-M., "First-Round Design of the Flight Scenario for Chang'e-2's Extended Mission: Takeoff from Lunar Orbit"; Acta Mechanica Sinica, 28, 2012, 1466-1478

[Gao-2012b] Gao, Y., "Near-Earth Asteroid Flyby Trajectories from the Sun-Earth L2 for Chang'e-2's Extended Flight", Acta Mechanica Sinica, 29, 2013, 123-131

[Garcia-2007] Garcia, M.D., Fujii, K.K., "Mission Design Overview for the Phoenix Mars Scout Mission", Paper AAS 07-247

[Garcia Yarnoz-2007] Garcia Yarnoz, D., Jehn, R., De Pascale, P., "Trajectory Design for the Bepi-Colombo Mission To Mercury", Journal of the British Interplanetary Society, 60, 2007, 202-208

[García Yárnoz-2013] García Yárnoz, D., Sanchez, J.P., McInnes, C.R., "Easily Retrievable Objects among the NEO Population", arXiv astro-ph/1304.5082 preprint

[Gardini-2003] Gardini, B., et al., "The Aurora Programme: A Stepping-Stone Path for Humans to Mars", On Station, December 2003, 10-14

[Garrick-Bethell-2005] Garrick-Bethell, I., "Artillery Based Explorers: A New Architecture for Regional Planetary Geology", Acta Astronautica, 57, 2005, 722-732

[Genta-2000] Genta, G., Vulpetti, G., "Some Consideration on Missions to the Gravitational Lens", paper presented at the Third IAA Symposium on Realistic Near-Term Advanced Scientific Space Missions, Aosta, 3-5 July 2000

[Gibney-2013] Gibney, E., "X-rays top space agenda", Nature, 503, 2013, 13-14

[Gilliland-2011] Gilliland, R.L., "Kepler Mission Stellar and Instrument Noise Properties" arXiv astro-ph/1107.5207 preprint

[Gimenez-2002] Gimenez, A., et al., "Studies on the Re-use of the Mars Express Platform", ESA Bulletin, 109, 2002, 78-86

[Gladstone-2007] Gladstone, G.R., et al., "Jupiter's Nightside Airglow and Aurora", Science, 318, 2007, 229-231

[Glassmeier-2007a] Glassmeier, K.-H., et al., "RPC-MAG: The Fluxgate Magnetometer in the Rosetta Plasma Consortium", Space Science Reviews, 128, 2007, 649-670

[Glassmeier-2007b] Glassmeier, K.-H., et al., "The Rosetta Mission: Flying Towards the Origin of the Solar System", Space Science Reviews, 128, 2007, 1-21

[Glassmeier-2009] Glassmeier, K.-H., "Magnetic Twisters on Mercury", Science, 324, 2009, 597-598

[Goesmann-2007] Goesmann, F., et al., "COSAC, the Cometary Sampling and Composition Experiment on Philae", Space Science Reviews, 128, 2007, 257-280

[Gold-2010] Gold, R.E., et al., "Uranus Mission Concept Options", Journal of the British Interplanetary Society, 63, 2010, 357-362

[González-2004] González, J.A., et al., "Don Quijote: An ESA Mission for the Assessment of the NEO Threat", paper presented at the 55th International Astronautical Congress, Vancouver, 2004

[Goswami-2013] Goswami, J.N., Radhakrishnan, K., "Indian Mission to Mars", paper presented at the XLIV Lunar and Planetary Science Conference, Houston, 2013

[Graf-2005] Graf, J.E., et al., "The Mars Reconnaissance Orbiter Mission", Acta Astronautica, 57, 2005, 566-578

[Graf-2007] Graf, J.E., et al., "Status of the Mars Reconnaissance Orbiter Mission", Acta Astronautica, 61, 2007, 44-51

[Greaves-2011] Greaves, J.S., Helling, C., Friberg, P., "Discovery of Carbon Monoxide in the Upper Atmosphere of Pluto", arXiv astro-ph/1104.3014 preprint

[Grebow-2012] Grebow, D.J., Bhaskaran, S., Chesley, S.R., "Target Search & Selection for the DI/EPOXI Spacecraft", paper presented at the AIAA/AAS Astrodynamics Specialist Conference, Minneapolis, August 2012

[Green-2004] Green, J.L., et al., "Radio Sounding Science at High Powers", paper presented at the 55th Congress of the International Astronautical Federation, Vancouver, 2004

[Grifantini-2011] Grifantini, K., "Where Did Earth's Water Come From?", Sky & Telescope, January 2011, 22-28

[Grinspoon-2010] Grinspoon, D., Tahu, G., "VCM Mission Concept Study – Planetary Science Decadal Survey Venus Climate Mission", Final Report, June 2010

[Groussin-2004] Groussin, O., et al., "The Nuclei of Comets 126P/IRAS and 103P/Hartley 2", Astronomy & Astrophysics, 419, 2004, 375-383

[Grotzinger-2013a] Grotzinger, J.P., "Analysis of Surface Materials by the Curiosity Mars Rover", Science, 341, 2013, 1475

[Grotzinger-2013b] Grotzinger, J.P. et al., "Mars Science Laboratory: First 100 Sols of Geologic and Geochemical Exploration from Bradbury Landing to Glenelg", paper presented at the XLIV Lunar and Planetary Science Conference, Houston, 2013

[Grotzinger-2014] Grotzinger, J.P., et al., "A Habitable Fluvio-Lacustrine Environment at Yellowknife Bay, Gale Crater, Mars", accepted for publication in Science

[Grover-2007] Grover, R., Desai, P., "Evolution of the Phoenix EDL System Architecture", presentation at the International Planetary Probe Workshop 5, Bordeaux, 2007

[Grover-2008] Grover, M.R., et al., "The Phoenix Mars Landing: An Initial Look", presentation at the International Planetary Probe Workshop 6, Atlanta, 2008

[Grundy-2007] Grundy, W.M., et al., "New Horizons Mapping of Europa and Ganymede", Science, 318, 2007, 234-237

[Grundy-2013] Grundy, W., "New Horizons Pluto/KBO Mission Status Report for SBAG", presentation at the 9th Small Bodies Assessment Group (SBAG), Washington, July 2013

[Gruntman-2005] Gruntman, M., et al., "Innovative Explorer Mission to Interstellar Space", paper presented at the Second IAA Symposium on Realistic Near-Term Advanced Scientific Space Missions, Aosta, 4-6 July 2005

[Gulkis-2007] Gulkis, S., et al., "MIRO: Microwave Instrument for Rosetta Orbiter", Space Science Reviews, 128, 2007, 561-597

[Guo-2002] Guo, Y., Farquhar, R.W., "New Horizons Mission Design for the Pluto-Kuiper Belt Mission", paper AIAA 2002-4722

[Guo-2004a] Guo, Y., Farquhar, R.W., "Baseline Design of the New Horizons Mission to Pluto and the Kuiper Belt", paper presented at the 55th International Astronautical Congress, Vancouver, October 2004

[Guo-2004b] Guo, Y., "New Horizons II Mission Design", presentation dated 16 June 2004

[Hand-2007] Hand, E., "The Girl Next Door", Science, 450, 2007, 606-608

[Hand-2008] Hand, E., "Phoenix: A Race Against Time", Nature, 456, 2008, 690-695

[Hand-2011a] Hand, E., "NASA Picks Mars Landing Site", Nature, 475, 2011, 433

[Hand-2011b] Hand, E., "Dragon offers ticket to Mars", Nature, 479, 2011, 162

[Hand-2012a] Hand, E., "Space Missions Trigger Map Wars", Nature, 488, 2012, 442-443

[Hand-2012b] Hand, E., "The Time Machine", Nature 487, 2012, 422-425

[Hand-2012c] Hand, E., "NASA Set to Choose Low-Cost Solar System Mission", Nature, 487, 2012, 150-151

[Hansen-2007] Hansen, K.C., et al., "The Plasma Environment of Comet 67P/Churyumov-Gerasimenko Throughout the Rosetta Main Mission", Space Science Reviews, 128, 2007, 133-166

[Hansen-2009] Hansen, C., Hammel, H.B., "Argo: Voyage Through the Outer Solar System", presentation to the Small Bodies Assessment Group (SBAG), January 2009

[Hansen-2011] Hansen, C.J., et al., "Seasonal Erosion and Restoration of Mars' Northern Polar Dunes", Science, 331, 2011, 575-578

[Hansen-2013] Hansen, C.J., "Juno Status and Earth Flyby Plans", presentation at the NASA Outer Planets Assessment Group (OPAG), July 2013

[Harmon-1992] Harmon, J.K., Slade, M.A., "Radar Mapping of Mercury: Full-Disk Images and Polar Anomalies", Science, 258, 1992, 640-643

[Harmon-2001] Harmon, J.K., Perillat, P.J., Slade, M.A., "High-Resolution Radar Imaging of Mercury's North Pole", Icarus, 149, 2001, 1-17

[Harmon-2002] Harmon, J.K., Campbell, D.B., "Mercury Radar Imaging at Arecibo in 2001", paper presented at the XXXIII Lunar and Planetary Science Conference, Houston, 2002

[Harri-2003] Harri, A.-M., et al., "MetNet - The Next Generation Lander for Martian Atmospheric Science", paper presented at the 54th International Astronautical Congress, Bremen, 2003

[Harri-2007a] Harri, A.-M., et al., "MetNet - In Situ Observational Network and Orbital Platform to Investigate the Martian Environment", Finnish Meteorological Institute, Helsinki, 2007

[Harri-2007b] Harri, A.-M., et al., "MetNet - Atmospheric Science Network for Mars", presentation at the Sputnik 50-Year Jubilee, Moscow, October 2007

[Harri-2008] Harri, A.-M., et al., "MMPM - Mars MetNet Precursor Mission", paper presented at the European Planetary Science Congress, Münster, 2008

[Harris-2008] Harris, A., "What Spaceguard Did", Nature, 453, 2008, 1178-1179

[Hartogh-2011] Hartogh, P., et al., "Ocean-like Water in the Jupiter-Family Comet 103P/Hartley 2", Nature, 478, 218-220

[Hassler-2014] Hassler, D.M., et al., "Mars' Surface Radiation Environment Measured with the Mars Science Laboratory's Curiosity Rover", accepted for publication in Science

[Hauck-2010] Hauck, S.A., Eng, D.A., Tahu, G.J., "Mercury Lander Mission Concept Study", NASA Mission Concept Study, revision 2 April 2010

[Hayati-2009] Hayati, S., "Technology Planning for Future Mars Missions", presentation to the Mars Exploration Program Analysis Group (MEPAG) meeting #21, Providence, Rhode Island, July 2009

[Head-2008] Head, J.W., et al., "Volcanism on Mercury: Evidence from the First MESSENGER Flyby", Science, 321, 2008, 69-72

[Head-2011] Head, J.W., et al., "Flood Volcanism in the Northern High Latitudes of Mercury Revealed by MESSENGER", Science, 333, 2011, 1853-1856

[Heaton-2001] Heaton, A.F., Longuski, J.M., "The Feasibility of a Galileo-Style Tour of the Uranian Satellites", paper AAS 01-464

[Hecht-2009] Hecht, M.H., et al., "Detection of Perchlorate and the Soluble Chemistry of Martian Soil at the Phoenix Lander Site", Science, 325, 2009, 64-67

[Heidemann-1998] Heidemann, J., "Imaging of Extrasolar Advanced Terrestrial Planets", paper presented at the Second IAA Symposium on Realistic Near-Term Advanced Scientific Space Missions, Aosta, 29 June-1 July 1998

[Herkenhoff-2007] Herkenhoff, K.E., et al., "Meter-Scale Morphology of the North Polar Region of Mars", Science, 317, 2007, 1711-1715

[Hermalyn-2011] Hermalyn, B., et al., "The Detection and Location of Icy Particles Surrounding Hartley 2", paper presented at the XLII Lunar and Planetary Science Conference, Houston, 2011

[Heyman-2008] Heyman, J., "Australia's Plan for Sundiver Spacecraft", Spaceflight, January 2008, 5

[Hibbard-2010] Hibbard, K., et al., "Trojan Tour Mission Concepts Provide Several Options for Cost-Effective Break-Through Science", Journal of the British Interplanetary Society, 63, 2010, 351-356

[Hibbard-2011] Hibbard, K., "Flyby Element", presentation to the Outer Planets Assessment Group, (OPAG), 19 October 2011

[Hill-2000] Hill, W., et al., "Using Microtechnologies to Build Micro-Robot Systems", paper presented at the 6th ESA Workshop on Advanced Space Technologies for Robotics and Automation 'ASTRA 2000', Noordwjik, December 2000

[Hirose-2012] Hirose, C., et al., "The Trajectory Control Strategies for Akatsuki Re-Insertion into the Venus Orbit", paper presented at the 23rd International Symposium on Space Flight Dynamics, Pasadena, October 2012

[Ho-2011] Ho, G.C., et al., "MESSENGER Observations of Transient Bursts of Energetic Electrons in Mercury's Magnetosphere", Science, 333, 2011, 1865-1868

[Holt-2008a] Holt, J.W., et al., "Radar Sounding Evidence for Buried Glaciers in the Southern Mid-Latitudes of Mars", Science, 322, 2008, 1235-1238

[Holt-2008b] Holt, R., "Plan to Set Up Martian Weather Station Network", Spaceflight, October 2008, 367

[Holt-2010] Holt, J.W., et al., "The Construction of Chasma Boreale on Mars", Nature, 465, 2010, 446-449

[Hoofs-2005] Hoofs, R.M.T., et al., "Venus Express - Initial Science Observations at Venus", paper presented at the 56th International Astronautical Congress, Fukuoka, October 2005

[Hou-2013] Hou, J., et al., "Joint Mars Exploration with Main-Sub Satellites in Group", paper presented at the 64rd International Astronautical Congress, Beijing, 2013

[Hsieh-2004] Hsieh, H.H., Jewitt, D.C., Fernàndez, Y.R., "The Strange Case of 133P/Elst-Pizarro: A Comet Among the Asteroids", The Astronomical Journal, 127, 2004, 2997-3017

[Hu-2010] Hu, X., et al., "An Emulation Research on the Radio Occultation Exploration of Martian Ionosphere", Chinese Astronomy and Astrophysics, 34, 2010, 100-112

[Hu-2013] Hu, S., et al., "Combined Orbit Determination for CE-2 and Toutatis Based on Optical Data at Fly-by", paper presented at the 64th International Astronautical Congress, Beijing, 2013

[Huang-2011] Huang, H., "Chinese Mars Exploration Mission Analysis", paper presented at the 7th UK-China Workshop on Space Science and Technology, August 2011

[Huang-2012a] Huang, J., et al., "Research and Development of Chang'e-2 Satellite", paper presented at the 63rd International Astronautical Congress, Naples, 2012

[Huang-2012b] Huang, H., et al., "Chang'E-2 Satellite Lagrange L2 Point Mission", paper presented at the 63rd International Astronautical Congress, Naples, 2012

[Huang-2013a] Huang, J., et al., "The Engineering Parameters Analysis of 4179 Toutatis

Flyby Mission of Chang'e-2", Science China Technological Sciences. 43, 2013, 596-601 (in Chinese)

[Huang-2013b] Huang, J., et al., "The Ginger-shaped Asteroid 4179 Toutatis: New Observations from a Successful Flyby of Chang'e-2", published online by Nature Scientific Reports, 12 December 2013

[Iannotta-2006] Iannotta, B., "A New Day for Dawn", Aerospace America, June 2006, 26-30

[IAUC-8315] "International Astronomical Unit Circular No. 8315", 4 April 2004

[IKI-2008] "Nauchnaya I Nauchno-Organizatsionnaya Deyatelnost' - Otchet Za 2008 G.", (Scientific and Scientific-Organizational Activity - 2008 Report), Russian Academy of Sciences' Institut Kosmicheskikh Isledovanii, 2008

[Imamura-2006] Imamura, T., "Planet-C: Venus Climate Orbiter from Japan & Technologies for Future Missions", presentation at the Venus Entry Probe Workshop, ESTEC, Noordwijk, 19-20 January 2006

[Imamura-2007] Imamura, T., et al., "Planet-C: Venus Climate Orbiter mission of Japan", Planetary and Space Science, 55, 2007, 1831-1842

[Imamura-2013] Imamura, T., "Akatsuki Mission Update", presentation at the 10th meeting of the Venus Exploration and Analysis Group (VEXAG), Washington, November 2013

[Ingersoll-2007] Ingersoll, A.P., "Express Dispatches", Nature, 450, 2007, 617-618

[INSIDER-2013] "INSIDER Interior of Primordial Asteroids and the Origin of Earth's Water – A White Paper Submitted in Response to ESA's L2/L3 Call for Ideas", 24 May 2013

[ISAS-2001] Venus Exploration Working Group, "Japanese Venus Mission Proposal", ISAS, January 2001

[Ishii-2009] Ishii, N., et al., "System Analysis and Orbit Design for PLANET-C Venus Climate Orbiter", paper presente at the 27th International Symposium on Space Technology and Science, 2009

[Izenberg-2007] Izenberg, N.R., et al., "The MESSENGER 2007 Venus Flyby: Peeking Through Atmospheric Windows with MASCS, MDIS and Venus Express' VIRTIS", paper presented at the Lunar and Planetary Science Conference XXXVIII, Houston, 2007

[Izenberg-2009a] Izenberg, N.R., et al., "MESSENGER Views of Crater Rays on Mercury", paper presented at the XL Lunar and Planetary Science Conference, Houston, 2009

[Izenberg-2009b] Izenberg, N.R., et al., "Resolved Ultraviolet to Infrared Reflectance Spectroscopy of Mercury from the Second MESSENGER Flyby", paper presented at the XL Lunar and Planetary Science Conference, Houston, 2009

[Jaeger-2007] Jaeger, W.L., et al., "Athabasca Valles, Mars: A Lava-Draped Channel System", Science, 317, 2007, 1709-1711

[Jafry-1994] Jafry, Y.R., Cornelisse, J., Reinhard, R., "LISA – A Laser Interferometer Space Antenna for Gravitational-Wave Measurements", ESA Journal, 18, 1994, 219-228

[Jäger-2004] Jäger, M., et al., "Launching Rosetta - The Demonstration of Ariane 5 Upper Stage Versatile Capabilities", paper presented at the 55th International Astronautical Congress, Vancouver, October 2004

[Jakosky-2010] Jakosky, B., Lin, B., "MAVEN: Mars Atmosphere and Volatile Evolution Mission", presentation at the Planet Mars III Workshop, Les Houches, April 2010

[Jakosky-2012] Jakosky, B., Grebowsky, J., Mitchell, D., "The 2013 Mars Atmosphere and Volatile EvolutioN (MAVEN) Mission", presentation to the Mars Exploration Program Analysis Group (MEPAG) meeting #25, Washington, DC, February 2012

[Jansen-2013] Jansen, F., et al., "Waking Rosetta", ESA Bulletin, 156, 2013, 19-27

[Jaumann-2012] Jaumann, R., et al., "Vesta's Shape and Morphology", Sicence, 336, 2012, 687-690

[JAXA-2009] "Small Solar Power Sail Demonstrator 'IKAROS'", JAXA undated leaflet

[JAXA-2011] JAXA report 4-2 dated 26 January 2011

[Jenniskens-2004] Jenniskens, P., "2003 EH1 is the Quadrantid Shower Parent Comet", The Astronomical Journal, 137, 2004, 3018-3022

[Jenniskens-2009] Jenniskens, P., et al., "The Impact and Recovery of Asteroid 2008 TC3", Nature, 458, 2009, 485-488

[Jerolmack-2013] Jerolmack, D.J., "Pebbles on Mars", Science, 340, 2013, 1055-1056

[Jewitt-1990] Jewitt, D.C., Luu, J.X., "CCD Spectra of Asteroids. II. The Trojans as Spectral Analogs of Cometary Nuclei", The Astronomical Journal, 100, 1990, 933-944

[Jewitt-2010] Jewitt, D., Jing, L., "Activity in Geminid Parent (3200) Phaethon", arXiv astro-ph/1009.2710 preprint

[Jonaitis-2003] Jonaitis, J., et al., "A Solar Powered Spacecraft for the INSIDE Jupiter Mission", Acta Astronautica, 52, 2003, 237-244

[Jones-1999] Jones, J.A., et al., "Balloons for Controlled Roving/landing on Mars", Acta Astronautica, 45, 1999, 293-300

[Jones-2003] Jones, R.M., "Surface and atmosphere geochemical explorer (SAGE) baseline design from March 2003 Team X studies", presentation at the 38th Vernadsky/Brown Microsymposium on Comparative Planetology, Moscow, Russia, 27-29 October 2003

[Jones-2013] Jones, M.H., Bewsher, D., Brown, D. S., "Imaging of a Circumsolar Dust Ring Near the Orbit of Venus", Science, 342, 2013, 960-963

[Jorda-2008] Jorda, L., et al., "Asteroid 2867 Steins: I. Photometric Properties from OSIRIS/Rosetta and Ground-Based Visible Observations", Astronomy & Astrophysics, 487, 2008, 1171-1178

[JPL-2005] "Prometheus Project: Final Report", JPL document 982-R120461, 1 October 2005

[JPL-2010a] "Mission Concept Study - Planetary Science Decadal Survey Mars Polar Climate Concepts", report dated May 2010

[JPL-2010b] "Mission Concept Study - Planetary Science Decadal Survey Mars Geophysical Network", report dated June 2010

[JPL-2010c] "Mission Concept Study - Planetary Science Decadal Survey Enceladus Orbiter", report dated May 2010

[JPL-2010d] "Mission Concept Study - Planetary Science Decadal Survey JPL Team X Titan Lake Probe Study", report dated April 2010

[Ju-2008] Ju, G., et al., "A Feasibility Study on Korean Lunar Exploration Mission", paper presented at the 2008 KSAS-JSASS Joint International Symposium

[Jurewicz-2002] Jurewicz, A.J.G., et al., "Investigating the use of Aerogel Collectors for the SCIM Martian-Dust Sample Return", paper presented at the XXXIII Lunar and Planetary Science Conference, Houston, 2002

[Jutzi-2010] Jutzi, M., Michel, P., Benz, W., "A large crater as a probe of the internal structure of the E-type asteroid Steins", Astronomy & Astrophysics, 509, 2010, L2

[Jutzi-2013] Jutzi, M., et al., "The Structure of the Asteroid 4 Vesta as Revealed by Models of Planet-Scale Collisions", Nature, 494, 2013, 207-210

[Karkoschka-1998] Karkoschka, E., "Clouds of High Contrast on Uranus", Science, 280, 1998, 570-572

[Kaspi-2013] Kaspi, Y., et al., "Atmospheric Confinement of Jet Streams on Uranus and Neptune", Nature, 497, 2013, 344-347

[Kawaguchi-2004] Kawaguchi, J., "A Solar Power Sail Mission for a Jovian Orbiter and Trojan Asteroid Flybys", paper presented at the 55th International Astronautical Congress, Vancouver, 2004

[Kawaguchi-2010] Kawaguchi, J., "Solar Power Sail - Hybrid Propulsion and its Applications

- A Jovian Orbiter and Trojan Asteroid Flybys", presentation at the 2nd International Symposium on Solar Sailing, 2010

[Kayali-2008] Kayali, S., "Juno Project: Challenges for a Jupiter Mission", presentation at the Microelectronics Reliability and Qualification Workshop, Manhattan Beach, California, December 2008

[Kayali-2010] Kayali, S., "Juno Project Overview and Challenges for a Jupiter Mission", presentation at the NASA Project Management Challenge, February 2010

[Kean-2010a] Kean, S., "Making Smarter, Savvier Robots", Science, 329, 2010, 508-509

[Kean-2010b] Kean, S., "Forbidden Planet", Air & Space Smithsonian, November 2010

[Keck-2012] "Asteroid Retrieval Feasibility Study", report prepared by the Jet Propulsion Laboratory for the Keck Institute for Space Studies, 2 April 2012

[Keller-2005] Keller, H.U., et al., "Deep Impact Observations by OSIRIS Onboard the Rosetta Spacecraft", Science, 310, 2005, 281-283

[Keller-2007] Keller, H.U., et al., "OSIRIS - The Scientific Camera System Onboard Rosetta", Space Science Reviews, 128, 2007, 433-506

[Keller-2008] Keller, H.U., and the OSIRIS Team, "Asteroid (2867) Steins", presentation at the Rosetta Steins Fly-By Press Conference, Darmstadt, ESA/ESOC, 6 September 2008

[Keller-2010] Keller, H.U., et al., "E-Type Asteroid (2867) Steins as Imaged by OSIRIS on Board Rosetta", Science, 327, 2010, 190-193

[Kelley-2009] Kelley, M.S., et al., "Spitzer Observations of Comet 67P/Churyumov-Gerasimenko at 5.5-4.3 AU From the Sun", arXiv astro-ph/0903.4187 preprint

[Kelly Beatty-2009] Kelly Beatty, J., "Mercury Gets a Second Look", Sky & Telescope, March 2009, 26-28

[Kelly Beatty-2012] Kelly Beatty, J., "Mercury's Marvels", Sky & Telescope, April 2012, 27-33

[Kelly Beatty-2013] Kelly Beatty, J., "Curiosity Hits the Road", Sky & Telescope, January 2013, 22-25

[Keppler-2012] Keppler, F., et al., "Ultraviolet-Radiation-Induced Methane Emissions from Meteorites and the Martian Atmosphere", Nature, 486, 2012, 93-96

[Kerr-1988] Kerr, R.A., "Another Asteroid Has Turned Comet", Science, 241, 1988, 1161

[Kerr-2006] Kerr, R.A., "In Search of the Red Planet's Sweet Spot", Science, 312, 2006, 1588-1590

[Kerr-2007] Kerr, R.A., "Is Mars Looking Drier and Drier for Longer and Longer?", Science, 317, 2007, 1673

[Kerr-2008a] Kerr, R.A., "Layers Within Layers Hint at a Wobbly Martian Climate", Science, 320, 2008, 867

[Kerr-2008b] Kerr, R.A., "Water Everywhere on Early Mars But Only for a Geologic Moment?", Science, 321, 2008, 484-485

[Kerr-2008c] Kerr, R.A., "To Touch the Water of Mars and Search for Life's Abode", Science, 320, 2008, 738-739

[Kerr-2008d] Kerr, R.A., "Culture Wars Over How to Find an Ancient Niche for Life on Mars", Science, 322, 2008, 39

[Kerr-2009a] Kerr, R.A., "Phoenix Rose Again, but not All Worked out as Planned", Science, 323, 2009, 872-873

[Kerr-2009b] Kerr, R.A., "Europa vs. Titan", Science, 322, 2008, 1780-1781

[Kerr-2009c] Kerr, R.A.,"Priorities Nearer to Home in Need of Better Cost Estimates", Science, 323, 2009, 579

[Kerr-2010a] Kerr, R.A., "Iceball Mars Proving a Tough Place to Find Liquid Water", Science, 327, 2010, 1075

[Kerr-2010b] Kerr, R.A., "Phoenix Lander Revealing a Younger, Livelier Mars", Science, 329, 2010, 1267-1269

[Kerr-2010c] Kerr, R.A., "Liquid Water Found on Mars, But It's Still a Hard Road for Life", Science, 330, 2010, 571

[Kerr-2011a] Kerr, R.A., "Mercury Looking Less Exotic, More a Member of the Family", Science, 333, 2011, 1812

[Kerr-2011b] Kerr, R.A., "How an Alluring Geologic Enigma Won the Mars Rover Sweepstakes", Science, 333, 2011, 508-509

[Kerr-2011c] Kerr, R.A., "Price Tags for Planet Missions Force NASA to Lower Its Sights", Science, 331, 2011, 1254-1255

[Kerr-2012] Kerr, R.A., "Could a Whiff of Methane Revive The Exploration of Mars?", Science, 336, 2012, 1500-1503

[Kerr-2013a] Kerr, R.A., "Radiation Will Make Astronauts' Trip to Mars Even Riskier", Science, 340, 2013, 1031

[Kerr-2013b] Kerr, R.A., "Life Could Have Thrived on Mars, but Did It? Curiosity Still Has No Clue", Science, 339, 2013, 1373

[Kerr-2013c] Kerr, R.A., "New Results Send Mars Rover on a Quest for Ancient Life", Science, 342, 2013, 1300-1301

[Kerr-2013d] Kerr, R.A., "Planetary Scientists Casting Doubt on Feasibility of Plan to Corral Asteroid", Science, 340, 2013, 668-669

[Keszthelyi-2011] Keszthelyi, L.P., "Europa Awakening", Nature, 479, 2011, 485

[Kiefer-2008] Kiefer, W.S., "Forming the Martian Great Divide", Nature, 453, 2008, 1191-1192

[Kissel-2007] Kissel, J., et al., "COSIMA - High Resolution Time-of-Flight Secondary Ion Mass Spectrometer for the Analysis of Cometary Dust Particles Onboard Rosetta", Space Science Reviews, 128, 2007, 823-867

[Klaasen-2003] Klaasen, K.P., Greeley, R., "VEVA Discovery Mission to Venus: Exploration of Volcanoes and Atmosphere", Acta Astronautica, 52, 2003, 151-158

[Klesh-2013] Klesh, A., "INSPIRE Interplanetary NanoSpacecraft Pathfinder In a Relevant Environment", presentation at the 10th Low-Cost Planetary Missions Conference, Pasadena, June 2013

[Kletzkine-2014] Kletzkine, P., Personal communication with the author, 8 January 2014

[Klingelhöfer-2007] Klingelhöfer, G., et al., "The Rosetta Alpha Particle X-Ray Spectrometer (APXS)", Space Science Reviews, 128, 2007, 383-396

[Kofman-2007] Kofman, W., et al., "The Comet Nucleus Sounding Experiment by Radiowave Transmission (CONSERT): A Short Description of the Instrument and of the Commissioning Stages", Space Science Reviews, 128, 2007, 413-432

[Kok-2012] Kok, J., "Martian Sand Blowing in the Wind", Nature, 485, 2012, 312-313

[Kolyuka-2012] Kolyuka, Yu.F., et al., "Arrangement and Results of the Phobos-Grunt Emergency Flight Monitoring and its Re-Entry Impact Window Estimation in Russian Conrol[sic] Center", paper presented at the 23rd International Symposium on Space Flight Dynamics, Pasadena, October 2012

[Konstantinov-2004] Konstantinov, M.S., "Missions to Asteroids Approaching with the Earth on Very Close Distances", paper presented at the 55th International Astronautical Congress, Vancouver, 2004

[Kopik-2003] Kopik, A., "Rossyiskiye Meshplanyetniye Planiy" (Russian Planetary Plans), Novosti Kosmonavtiki, No. 11, 2003, page unknown (in Russian)

[Kopik-2004] Kopik, A., "Mars v Oblasty Nashikh Interesov", (Mars as Part of Our Interests) Novosti Kosmonavtiki, No. 5, 2004, page unknown (in Russian)

[Korablev-2003] Korablev, O., "Russian Programme for Deep Space Exploration", presentation at the IAA-ESA Workshop "The Next Steps in Exploring Deep Space", Noordwijk, September 2003

[Korablev-2006] Korablev, O., et al., "Venera-D: Russian mission for complex investigation of Venus", presentation at the Venus Entry Probe Workshop, ESTEC, Noordwijk, 19-20 January 2006

[Korablev-2009] Korablev, O.I., Martinov, M.B., "Russian Plans for Mars and Venus", presentation at the International Conference on Comparative Planetology: Venus-Earth-Mars, ESTEC, Noordwjik, May 2009

[Korpenko-2000] Korpenko, S., "Nasha Mesplanetnaya Stantsiya (Proyekt Rossiskoy AMS 'Fobos-Grunt')" (Our Planetary Probe, the Russian Project 'Fobos-Grunt'), Novosti Kosmonavtiki, No. 3, 2000, 28-32 (in Russian)

[Koschny-2007] Koschny, D., et al., "Scientific Planning and Commanding of the Rosetta Payload", Space Science Reviews, 128, 2007, 167-188

[Koschny-2008] Koschny, D., "Marco Polo – A sample return mission from a Near Earth Object (part of ESA's Cosmic Vision programme studies)", presentation at the Marco Polo Cannes Workshop, June 2008

[Kozyrev-2009] Kozyrev, A.S., et al., "Studying Mercury Surface Composition by Mercury Gamma-Rays and Neutrons Spectrometer (MGNS) from BepiColombo Spacecraft", paper presented at the XL Lunar and Planetary Science Conference, Houston, 2009

[Kremer-2008] Kremer, K., "Phoenix Hits Mars Jackpot", Spaceflight, October 2008, 378-385

[Kremer-2009a] Kremer, K., "Science from Arctic Mars", Spaceflight, September 2009, 349-353

[Kremer-2009b] Kremer, K., "NASA Delays Mars Science Laboratory to 2011", Spaceflight, February 2009, 44-45

[Kresak-1984] Kresak, L., Pittich, E.M., "Opportunities of Ballistic Missions to Long-Period Comets", Bulletin of the Astronomical Institutes of Czechoslovakia, 35, 1984, 364-375

[Kronk-1984a] Kronk, G.W., "Comets: A Descriptive Catalog", Hillside, Henslow, 1984, ., 324-325

[Kronk-1984b] ibid., 232

[Kronk-1984c] ibid., 306-308

[Kronk-1984d] ibid., 225

[Kronk-2005] "85P/Boethin", G.W. Kronk internet website

[Krupp-2007] Krupp, N., "New Surprises in the Largest Magnetosphere of Our Solar System", Science, 318, 2007, 216-217

[Ksanfomality-2003] Ksanfomality, L.V., "Mercury: The Image of the Planet in the 210°-285° W Longitude Range Obtained by the Short-Exposure Method", Solar System Research, 37, 2003, 469-479

[Kumar-2006] Kumar, V., "Indo-US Cooperation in Civil Space", presentation at the 2nd Space Exploration Conference, Houston, December 2006

[Küppers-2005] Küppers, M., et al., "A large dust/ice ratio in the nucleus of comet 9P/Tempel 1", Nature, 437, 2005, 987-990

[Küppers-2007a] Küppers, M., et al., "Determination of the Light Curve of the Rosetta Target Asteroid (2867) Steins by the OSIRIS Cameras Onboard Rosetta", Astronomy & Astrophysics, 462, 2007, L13-L16

[Küppers-2007b] Küppers, M., et al., "Observations of Gravitational Microlensing Events with OSIRIS: A Proposal for a Cruise Science Observation", ESA proposal dated 2007

[Kusnierkiewicz-2005] Kusnierkiewicz, D.Y., et al., "A Description of the Pluto-Bound New Horizons Spacecraft", Acta Astronautica, 57, 2005, 135-144

[Kuzmin-2003] Kuzmin, R.O., Zabalueva, E.V., "The Temperature Regime of the Surface Layer of Phobos Regolith in the Region of Potential the Fobos-Grunt Space Station Landing Site", Solar System Research, 37, 2003, 480-488

[Lai-2012] Lai, H.R., et al., "The Return of Asteroid 2201 Oljato to Venus Conjunction: New IFEs?", paper presented at the European Planetary Science Congress, Madrid, 2012

[Lakdawalla-2012] Lakdawalla, E., "Touchdown on the Red Planet", Sky & Telescope, November 2012, 20-27

[Lämmerzahl-2006] Lämmerzahl, C., Preuss, O., Dittus, H., "Is the Physics Within the Solar System Really Understood?" arXiv gr-qc/0604052 preprint

[Lamy-1998] Lamy, P.L., et al., "The Nucleus and Inner Coma of Comet 46P/Wirtanen", Astronomy & Astrophysics, 335, 1998, L25-L29

[Lamy-2007] Lamy, P.L., "A Portrait of the Nucleus of Comet 67P/Churyumov-Gerasimenko", Space Science Reviews, 128, 2007, 23-66

[Lamy-2008a] Lamy, P.L., et al., "Asteroid 2867 Steins: II. Multi-Telescope Visible Observations, Shape Reconstruction, and Rotational State", Astronomy & Astrophysics, 487, 2008, 1179-1185

[Lamy-2008b] Lamy, P.L., et al., "Asteroid 2867 Steins: III. Spitzer Space Telescope Observations, Size Determination, and Thermal Properties", Astronomy & Astrophysics, 487, 2008, 1187-1193

[Lamy-2008c] Lamy, P.L., et al., "Spitzer Space Telescope Observations of the Nucleus of Comet 67P/Churyumov-Gerasimenko", Astronomy & Astrophysics, 489, 2008, 777-785

[Landis-2001] Landis, G.A., "Exploring Venus by Solar Airplane", paper presented at the STAIF Conference on Space Exploration Technology, Albuquerque 11-15 February 2001

[Landis-2002a] Landis, G.A., LaMarre, C., Colozza, A., "Atmospheric Flight on Venus", paper AIAA-2002-0819

[Landis-2002b] Landis, G.A., LaMarre, C., Colozza, A., "Venus Atmospheric Exploration by Solar Aircraft", paper presented at the 53rd International Astronautical Congress, Houston, 2002

[Landis-2004] Landis, G.A., "Robotic Exploration of the Surface and Atmosphere of Venus", paper presented at the 55th International Astronautical Congress, Vancouver, 2004

[Landis-2013] Landis, G.A., "Venus Landsailer: A New Approach to Exploring Our Neighbor Planet", undated presentation

[Lange-2010] Lange, C., et al., "Baseline Design of a Mobile Asteroid Surface Scout (MASCOT) for the Hayabusa-2 Mission", paper presented at the 7th International Planetary Probe Workshop, Barcelona, June 2010

[Lara-2007] Lara, L.M., et al., "Behavior of Comet 9P/Tempel 1 Around the Deep Impact Event", Astronomy & Astrophysics, 465, 2007, 1061-1067

[Lardier-2009] Lardier, C., "Le Nouveau Scénario d'ExoMars" (The New ExoMars Scenario), Air & Cosmos, 23 October 2009, 37 (in French)

[Lardier-2010] Lardier, C., "Les Futures Evolutions de la Propulsion Spatiale" (The Future Evolutions of Space Propulsion), Air & Cosmos, 14 May 2010, 50-52 (in French)

[Larson-2013] Larson, T., A'Hearn, M., Chesley, S., "Deep Impact/EPOXI Status and Plans", presentation to the 8th Small Bodies Assessment Group (SBAG), Washington, January 2013

[Lauretta-2008] Lauretta, D., "OSIRIS: Regolith Explorer", paper presented at the first open Workshop on the Marco Polo Mission, Cannes, June 2008

[Lawler-2006] Lawler, A., "Long-Term Mars Exploration Under Threat, Panel Warns", Science, 313, 2006, 157

[Lawler-2008] Lawler, A., "Rising Costs Could Delay NASA's Next Mission to Mars and Future Launches", Science, 321, 2008, 1754

[Lawler-2009] Lawler, A., "Can a Shotgun Wedding Help NASA and ESA Explore the Red Planet?", Science, 323, 2009, 1666-1667

[Lawrence-2009] Lawrence, D.J., et al., "Identification of Neutron Absorbing Elements on Mercury's Surface Using MESSENGER Neutron Data", paper presented at the XL Lunar and Planetary Science Conference, Houston, 2009

[Lawrence-2012] Lawrence, D.J., et al., "Hydrogen at Mercury's North Pole? Update on MESSENGER Neutron Measurements", paper presented at the XLIII Lunar and Planetary Science Conference, Houston, 2012

[Lazzarin-2009] Lazzarin, M., et al., "New Visible Spectra and Mineralogical Assessment of (21) Lutetia, a Target of the Rosetta Mission", Astronomy & Astrophysics, 498, 2009, 307-311

[Lefèvre-2009] Lefèvre, F., Forget, F., "Observed Variations of Methane on Mars Unexplained by Known Atmospheric Chemistry and Physics", Nature, 460, 2009, 720-723

[Leipold-1999] Leipold, M., et al., "Solar Sails for Space Exploration-The Development and Demonstration of Critical Technologies in Partnership", ESA Bulletin, 98, 1999, 102-107

[Leipold-2010] Leipold, M., "Interstellar Heliopause Probe (IHP) - System Design of a Challenging Mission to 200 AU", presentation at the 2nd International Symposium on Solar Sailing, 2010

[Lele-2013] Lele, A., "Mission Mars: India's Quest for the Red Planet", Springer, Berlin, Heidelberg, 2013, 39-69

[Lellouch-2009] Lellouch, E., et al., "Pluto's Lower Atmosphere Structure and Methane Abundance from High-Resolution Spectroscopy and Stellar Occultations", Astronomy & Astrophysics, 495, 2009, L17-L21

[Lenard-2000] Lenard, R.X., "NEP for a Kuiper Belt Object Sample Return Mission", paper presented at the Third IAA Symposium on Realistic Near-Term Advanced Scientific Space Missions, Aosta, 3-5 July 2000

[Leshin-2002] Leshin, L.A., et al., "Sample Collection for Investigation of Mars (SCIM): An Early Mars Sample Return Mission through the Mars Scout Program", paper presented at the XXXIII Lunar and Planetary Science Conference, Houston, 2002

[Leshin-2013] Leshin, L.A., et al., "Volatile, Isotope, and Organic Analysis of Martian Fines with the Mars Curiosity Rover", Science, 341, 2013

[Lewicki-2012] Lewicki, C., "Planetary Resources", presentation at the 7th Small Bodies Assessment Group (SBAG) meeting, Pasadena, July 2012

[Lewis-2008] Lewis, K.W:, "Quasi-Periodic Bedding in the Sedimentary Rock Record of Mars", Science, 322, 1532-1535

[Li-2010] Li, F., "Mars Sample Return Discussions", presentation to the Mars Exploration Program Analysis Group (MEPAG) meeting #22, Monrovia, California, March 2010

[Li-2011] Li, X., et al., "Sun Polar Probe Trajectory Design Based on Multi-objective Genetic Algorithm", Chinese Journal of Space Science, 31, 2011, 653-658

[Li-2012a] Li, M., Zheng, J., "Low Energy Trajectory Optimization for CE-2's Extended Mission After 2012", paper presented at the 63rd International Astronautical Congress, Naples, 2012

[Li-2012b] Li, M., Personal communication with the author, 10 October 2012

[Li-2013] Li, C., Li., H., "Chang'e 2 Flyby of Toutatis", presentation at the 8th Small Bodies Assessment Group (SBAG) meeting, Washington, January 2013

[Lim-2012] Lim, L.F., "OSIRIS-REx Asteroid Sample Return Mission", at the 7th Small Bodies Assessment Group (SBAG) meeting, Pasadena, July 2012

[Lipinski-1999] Lipinski, R.J., et al., "NEP for a Kuiper Belt Object Rendezvous Mission", paper presented at the Space Technology and Applications International Forum 2000, Albuquerque, 3 Nov 1999

[Lisse-2006] Lisse, C.M., et al., "Spitzer Spectral Observations of the Deep Impact Ejecta", Science, 313, 2006, 635-640

[Lisse-2009] Lisse, C.M., et al., "Spitzer Space Telescope Observations of the Nucleus of Comet 103P/Hartley 2", arXiv astro-ph/0906.4733 preprint

[Liu-2012] Liu, L., Personal communication with the author, 18 August 2012

[Lockwood-2012] Lockwood, M.K., et al., "Solar Probe Plus Mission Definition", paper presented at the 63rd International Astronautical Congress, Naples, 2012

[Lodiot-2009] Lodiot, S., et al., "The First European Asteroid 'Flyby'", ESA Bulletin, 137, 2009, 69-74

[Lognonné-2010] Lognonné, P., "Network Science on Mars", presentation at the Planet Mars III Workshop, Les Houches, April 2010

[Lorenz-2000] Lorenz, R.D., "Post-Cassini Exploration of Titan: Science Rationale and Mission Concepts", Journal of the British Interplanetary Society, 53, 2000, 218-234

[Lorenz-2001] Lorenz, R.D., "Flight Power Scaling of Airplanes, Airships, and Helicopters: Application to Planetary Exploration", Journal of Aircraft, 38, 2001, 208-214

[Lorenz-2008a] Lorenz, R.D., "Titan Bumblebee: a 1 kg Lander-Launched UAV Concept", Journal of the British Interplanetary Society, 61, 2008, 118-124

[Lorenz-2008b] Lorenz, R.D., "A Review of Balloon Concepts for Titan", Journal of the British Interplanetary Society, 61, 2008, 2-13

[Lorenz-2009] Lorenz, R.D., "Titan Mission Studies - A Historical Review" Journal of the British Interplanetary Society, 62, 2009, 162-174

[Lorenzini-1990] Lorenzini, E.C., Grossi, M.D., Cosmo, M., "Low Altitude Tethered Mars Probe", Acta Astronautica, 21, 1990, 1-12

[Lowry-2001] Lowry, S.C., Fitzsimmons, A., "CCD Photometry of Distant Comets II", Astronomy & Astrophysics, 365, 2001, 204-213

[Lowry-2012] Lowry, S., et al., "The Nucleus of Comet 67P/Churyumov-Gerasimenko: A New Shape Model and Thermophysical Analysis", Astronomy & Astrophysics, 548, 2012, A12

[Lu-2005] Lu, E.T., Love, S.G., "A Gravitational Tractor for Towing Asteroids", preprint astro-ph/0509595, 2005

[Luz-2011] Luz, D., et al., "Venus's Southern Polar Vortex Reveals Precessing Circulation", Science, 332, 2011, 577-580

[Lyngvi-2004] Lyngvi, A., et al., "Technology Reference Studies", paper presented at the 55th International Astronautical Congress, Vancouver, 2004

[Lyngvi-2005a] Lyngvi, A., et al., "The Solar Orbiter", paper presented at the 56th International Astronautical Federation Congress, Fukuoka, 2005

[Lyngvi-2005b] Lyngvi, A., et al., "The Solar Orbiter Thermal Design", paper presented at the 56th International Astronautical Federation Congress, Fukuoka, 2005

[Lyngvi-2007] Lyngvi, A.E., van den Berg, M.L., Falkner, P., "Study Overview of the Interstellar Heliopause Probe – An ESA Technology Reference Study", SCI-A/2006/114/IHP, 17 April 2007

[Maccone-1998] Maccone, C., "The Science Payload and Antenna of the 'Focal' Space Mission to 550 A.U.", paper presented at the Second IAA Symposium on Realistic Near-Term Advanced Scientific Space Missions, Aosta, 29 June-1 July 1998

[Maccone-2000a] Maccone, C., Bussolino, L., "The Trojan Asteroids as Bases to Monitor Other Asteroids Potentially Dangerous for Earth", paper presented at the Third IAA Symposium on Realistic Near-Term Advanced Scientific Space Missions, Aosta, 3-5 July 2000

[Maccone-2000b] Maccone, C., "Sunlensing the Cosmic Microwave Background from 763 AU by Virtue of NASA's Interstellar Probe", paper presented at the Third IAA

Symposium on Realistic Near-Term Advanced Scientific Space Missions, Aosta, 3-5 July 2000

[MacRobert-2005] MacRobert, A., "Asteroid 2004 MN4: A Really Near Miss", Sky & Telescope, May 2005, 16-17

[Maddé-2006] Maddé, R., et al., "Delta-DOR: A New Technique for ESA's Deep Space Navigation", ESA Bulletin, 128, 2006, 69-74

[Mahieux-2012] Mahieux, A., et al., "Densities and Temperatures in the Venus Mesosphere and Lower Thermosphere Retrieved from SOIR on Board Venus Express. Carbon Dioxide Measurements at the Venus Terminator", Journal of Geophysical Research, 117, 2012, E07001-E07015

[Mainzer-2012] Mainzer, A., "Near-Earth Object Camera NEOCam", presentation at the Small Bodies Assessment Group Meeting, 10 July 2012

[Marchi-2010] Marchi, A., et al., "The Cratering History of Asteroid (2867) Steins", arXiv astro-ph/1003.5655v1 preprint

[Marchi-2012] Marchi, S., et al., "The Violent Collisional History of Asteroid 4 Vesta", Science, 336, 2012, 690-694

[Marchis-2006] Marchis, F., et al., "A Low Density of 0.8 g cm-3 for the Trojan Binary Asteroid 617 Patroclus", Nature, 439, 2006, 565-567

[Margot-2007] Margot, J.L., et al., "Large Longitude Libration of Mercury Reveals a Molten Core", Science, 316, 2007, 710-714

[Marinova-2008] Marinova, M.M., Aharonson, O., Asphaug, E., "Mega-Impact Formation of the Mars Hemispheric Dichotomy", Nature, 453, 2008, 1216-1219

[Markiewicz-2007] Markiewicz, W.J., et al., "Morphology and Dynamics of the Upper Cloud Layer of Venus", Nature, 450, 2007, 633-636

[Marlow-2009] Marlow, J., "Seeking ET", Spaceflight, April 2009, 140-146

[Marraffa-2000] Marraffa, L., et al., "Inflatable Re-Entry Technologies: Flight Demonstration and Future Prospects", ESA Bulletin, 103, 2000, 78-85

[Martin-Mur-2012] Martin-Mur, T.J., et al., "Mars Science Laboratory Navigation Results", paper presented at the 23rd International Symposium on Space Flight Dynamics, Pasadena, October 2012

[Martynov-2009] Martynov, M.B., et al., "The Concept of Expedition to Europa, the Jupiter's Satellite", presented at the International Workshop Europa Lander: Science Goals and Experiments, IKI, Moscow, February 2009

[Martynov-2010] Martynov, M., "'Phobos-Grount' Project - Mission Concept & Current Status of Development". Presentation at the Moscow Solar Syatem Symposium, IKI, 13 October 2010

[Mason-2011] Mason, L., et al., "A Small Fission Power System for NASA Planetary Science Missions", Journal of the British Interplanetary Society, 64, 2011, 76-87

[Matousek-2005] Matousek, S., "The Juno New Frontiers Mission", paper presented at the 56th International Astronautical Congress, Fukuoka, 2005

[Mattingly-2008] Mattingly, R., "A Constellation-Enabled Mars Sample Return (MSR) and Preparation for Humans-to-Mars aka CEMMENT (Constellation-Enabled Mars Mission Exhibiting New Technology)", presentation at the Ares-V Utilization Workshop, NASA Ames, 16 June 2008

[Mattingly-2010a] Mattingly, R., "Mission Concept Study - Planetary Science Decadal Survey MSR Orbiter Mission (Including Mars Returned Sample Handling)", report dated March 2010

[Mattingly-2010b] Mattingly, R., "Mission Concept Study - Planetary Science Decadal Survey MSR Lander Mission", report dated April 2010

[Maue-2008] Maue, T., "Mars Flyby with the Dawn Framing Camera", paper presented at the European Planetary Science Congress, Münster, 2008

[MAVEN-2008] "MAVEN, Mars Atmosphere and Volatile Evolution Mission Fact Sheet", University of Colorado brochure, 2008

[MAVEN-2013] "MAVEN - Mars Atmosphere and Volatile Evolution Mission Press Kit", November 2013

[McAdams-2006] McAdams, J.V., et al., "Trajectory Design and Maneuver Strategy for the MESSENGER Mission to Mercury", Journal of Spacecraft and Rockets, 43, 2006, 1054-1064

[McAdams-2012] McAdams, J.V., et al., "MESSENGER at Mercury: from Orbit Insertion to First Extended Mission", paper presented at the 63rd International Astronautical Congress, Naples, 2012

[McAdams-2013] McAdams, J.V., Personal communication with the author, 2 December 2013

[McClintock-2008a] McClintock, W.E., et al., "Spectroscopic Observations of Mercury's Surface Reflectance During MESSENGER's First Mercury Flyby", Science, 321, 2008, 62-65

[McClintock-2008b] McClintock, W.E., et al., "Mercury's Exosphere: Observations During MESSENGER's First Mercury Flyby", Science, 321, 2008, 92-94

[McClintock-2009] McClintock, W.E., et al., "MESSENGER Observations of Mercury's Exosphere: Detection of Magnesium and Distribution of Consituents", Science, 324, 2009, 610-613

[McComas-2007] McComas, D.J., et al., "Diverse Plasma Populations and Structures in Jupiter's Magnetotail", Science, 318, 2007, 217-220

[McComas-2012] McComas, D.J., et al., "The Heliosphere's Interstellar Interaction: No Bow Shock", Science, 336, 2012, 1291-1293

[McCord-2012] McCord, T.B., et al., "Dark Material on Vesta from the Infall of Carbonaceous Volatile-Rich Material", Nature, 491, 2012, 83-86

[McCoy-2005] McCoy, D., Siwitza, T., Gouka, R., "The Venus Express Mission", ESA Bulletin, 124, 2005, 10-15

[McEwen-2007a] McEwen, A.S., et al., "Mars Reconnaissance Orbiter's High Resolution Imaging Science Experiment (HiRISE)", Journal of Geophysical Research, 112, 2007, E05502

[McEwen-2007b] McEwen, A.S., et al., "A Closer Look at Water-Related Geologic Activity on Mars", Science, 317, 2007, 1706-1709

[McEwen-2010] McEwen, A.S., et al., "The High Resolution Imaging Science Experiment (HiRISE) during MRO's Primary Science Phase", Icarus, 205, 2010, 2-37

[McEwen-2011] McEwen, A.S., et al., "Seasonal Flows on Warm Martian Slopes", Science, 333, 2011, 740-743

[McGranaghan-2011] McGranaghan, R., et al., "A Survey of Mission Opportunities to Trans-Neptunian Objects", Journal of the British Interplanetary Society, 64. 2011, 296-303

[McLennan-2014] McLennan, S.M., et al., "Elemental Geochemistry of Sedimentary Rocks at Yellowknife Bay, Gale Crater, Mars", accepted for publication in Science

[McNutt-2007] McNutt, R.L. Jr., et al., "Energetic Particles in the Jovian Magnetotail", Science, 318, 2007, 220-222

[McNutt-2012] McNutt, R.L. Jr., et al., "The MESSENGER Mission Continues: Transition to the Extended Mission", paper presented at the 63rd International Astronautical Congress, Naples, 2012

[Meech-2005] Meech, K.J., et al., "Deep Impact: Observations from a Worldwide Earth-Based Campaign", Science, 310, 2005, 265-269

[Meech-2011] Meech, K.J., the EPOXI Earth-based observing team and the EPOXI/DIXI Science Team, "The EPOXI Earth-Based Observing Campaign", paper presented at the XLII Lunar and Planetary Science Conference, Houston, 2011

[Menon-2006] Menon, C., Ayre, M., Ellery, A., "Biomimetics: A New Approach for Space System Design", ESA Bulletin, 125, 2006, 21-26

[Meslin-2013] Meslin, P.-Y., et al., "Soil Diversity and Hydration as Observed by ChemCam at Gale Crater, Mars", Science, 341, 2013

[Messidoro-2014] Messidoro, P., "Dai Disegni di Marte al Design per Marte" (From Drawing Mars to Designing for Mars), le Stelle, January 2014, 64-67 (in Italian)

[Messina-2003] Messina, P., Ongaro, F., "Aurora: The European Space Exploration Programme", ESA Bulletin, 115, 2003, 34-39

[Messina-2006] Messina, P., et al., "The Aurora Programme: Europe's Framework for Space Exploration", ESA Bulletin, 126, 2006, 11-15

[Michel-2012] Michel, P., et al., "MarcoPolo-R: Near Earth Asteroid Sample Return Mission in Assessment Study Phase of ESA M3-Class Missions", paper presented at the Asteroids, Comets, Meteors conference, 2012

[Michel-2013] Michel, P., et al., "Marco Polo-R", presentation at the 9th Small Bodies Assessment Group (SBAG) meeting, Washington, July 2013

[Milani-2006] Milani, G.A., et al., "Photometry of Comet 9P/Tempel 1 During the 2004/2005 Approach and the Deep Impact Module Impact", arXiv astro-ph/0608180 preprint

[Ming-2013] Ming, X., et al., "Deimos Encounter Trajectories Design for Piggyback Spacecraft Launched for Martian Surface Reconnaissance", paper presented at the 64rd International Astronautical Congress, Beijing, 2013

[Ming-2014] Ming, D.W., et al., "Volatile and Organic Compositions of Sedimentary Rocks in Yellowknife Bay, Gale Crater, Mars", accepted for publication in Science

[Montagnon-2005] Montagnon, W., Ferri, P., "Rosetta on Its Way to the Outer Solar System", paper presented at the 56th International Astronautical Congress, Fukuoka, October 2005

[Morgan-2013] Morgan, G.A., et al., "3D Reconstruction of the Source and Scale of Buried Young Flood Channels on Mars", Science, 340, 2013, 607-610

[Morimoto-2008] Morimoto, M.Y., Kawakatsu, Y., Kawaguchi, J., "Trajectory Options of the Planet-C Auxiliary Payload after the Venus Swing-by", paper presented at the 26th International Symposium on Space Technology and Science, 2008

[Morin-2011] Morin, H., Jégo, M., "Espoir pour la Sonde Russe Phobos-Grunt" (Hope for the Russian probe Fobos-Grunt), Le Monde, 26 November 2011 (in French)

[Morley-2009] Morley, T., Budnik, F., "Rosetta Navigation for the Fly-by of Asteroid 2867 teins", paper presented at the 21st International Symposium on Space Flight Dynamics, Toulouse, October 2009

[Morley-2012] Morley, T., "Rosetta Navigation for the Fly-by of Asteroid 21 Luteti"a, paper presented at the 23rd International Symposium on Space Flight Dynamics, Pasadena, October 2012

[Morring-2002] Morring, F. Jr, "Nuclear-Powered Mars Rover Planned in '09", Aviation Week & Space Technology, 20 May 2002, 64-65

[Morring-2003a] Morring, F. Jr, "Prometheus Bound", Aviation Week & Space Technology, 11 August 2003, 63-64

[Morring-2003b] Morring, F. Jr, "Data Dump", Aviation Week & Space Technology, 23 June 2003, 53-55

[Morring-2003c] Morring, F. Jr, "Exploration Plans", Aviation Week & Space Technology, 16 June 2003, 181-182

[Morring-2004a] Morring, F. Jr, "Distant Destinations", Aviation Week & Space Technology, 13 December 2004, 56

[Morring-2004b] Morring, F. Jr., "Looking Ahead", Aviation Week & Space Technology, 5 July 2004, 25-27

[Morring-2005a] Morring, F. Jr, Taverna, M.A., "Tightening Focus", Aviation Week & Space Technology, 23 May 2005, 42

[Morring-2005b] Morring, F. Jr, "Solar Studies", Aviation Week & Space Technology, 5 September 2005, 62-64

[Morring-2005c] Morring, F. Jr, "Crunch Time", Aviation Week & Space Technology, 14 February 2005, 28

[Morring-2006] Morring, F. Jr, "Cash-Poor Astrobiologists Hope for Gold in Human Exploration", Aviation Week & Space Technology, 10 April 2006, 48

[Morring-2007] Morring, F. Jr., "New Horizons Returns a Treasure Trove of Jupiter Data", Aviation Week and Space Technology, 7 May 2007, 80

[Morring-2008] Morring, F. Jr., "Beyond Crew Launch", Aviation Week & Space Technology, 14 April 2008, 34-35

[Morring-2010] Morring, F. Jr., Mathews, N., "Spacefaring Nation", Aviation Week & Space Technology, 14 June 2010, 62-64

[Morring-2011a] Morring, F. Jr., "Back to Jupiter", Aviation Week & Space Technology, 21 March 2011, 50-53

[Morring-2011b] Morring, F. Jr., "Parting the Veil", Aviation Week & Space Technology, 21 March 2011, 54

[Morring-2011c] Morring, F. Jr., "Sky Crane", Aviation Week & Space Technology, 1 August 2011, 38-42

[Morring-2012a] Morring, F. Jr., Norris, G., "Advancing the Art", Aviation Week & Space Technology, 13 August 2012, 24-27

[Morring-2012b] Morring, F. Jr., Norris, G., "Curiosity's Next Moves", Aviation Week & Space Technology, 20 August 2012, 30

[Morring-2012c] Morring, F. Jr., "Losing Thrust", Aviation Week & Space Technology, 20 February 2012, 33-34

[Morring-2012d] Morring, F. Jr., "Next Steps", Aviation Week & Space Technology, 5 March 2012, 38-39

[Morring-2012e] Morring, F. Jr., Svitak, A., "Working Together", Aviation Week & Space Technology, 1 October 2012, 36-37

[Morring-2013a] Morring, F. Jr., "Closing In", Aviation Week & Space Technology, 29 July 2013, 21-23

[Morring-2013b] Morring, F. Jr., Norris, G., "Poor Protection", Aviation Week & Space Technology, 21 January 2013, 31-32

[Morring-2013c] Morring, F. Jr., "Climate Change", Aviation Week & Space Technology, 26 August 2013, 40-42

[Morring-2013d] Morring, F. Jr., "Red Planet Payoff?", Aviation Week & Space Technology, 16 December 2013, 24-26

[Morris-2009] Morris, J., Taverna, M.A., "Slimming Down", Aviation Week & Space Technology, 16 March 2009, 33-35

[Morrow-2006] Morrow, M.T., Woolsey, C.A., Hagerman, G.M., "Exploring Titan with Autonomous, Buoyancy Driven Gliders", JBIS, 59, 2006, 27-34

[Mottola-2007] Mottola, S., et al., "The ROLIS Experiment on the Rosetta Lander", Space Science Reviews, 128, 2007, 241-255

[MPEC-2007a] Minor Planet Electronic Circular 2007-V69, "2007 VN84", 8 November 2007

[MPEC-2007b] Minor Planet Electronic Circular 2007-V70, "Editorial Notice", 9 November 2007

[MPPG-2012] Mars Program Planning Group, "Summary of the Final Report", presentation dated 25 September 2012

[Mueller-2008] Mueller, N., et al., "Correlation Between Venus Nightside Near Infrared Emissions Measured by VIRTIS/Venus Express and Magellan Radar Data", paper presented at the European Planetary Science Congress, Münster, 2008

[Muirhead-1999] Muirhead, B., Kerridge, S., "The Deep Space 4/Champollion Mission", Acta Astronautica, 45, 1999, 407-414

[Müller-2012] Müller, T.G., et al., "Physical Properties of OSIRIS-REx Target Asteroid (101955) 1999 RQ36 derived from Herschel, ESO-VISIR and Spitzer observations", arXiv astro-ph/1210.5370 preprint

[Mumma-2009] Mumma, M.J., et al., "Strong Release of Methane on Mars in Northern Summer 2003", Science, 323, 2009, 1041-1045

[Muñoz-2012] Muñoz, P., et al., "Preparations and Strategy for Navigation During Rosetta Comet Phase", paper presented at the 23rd International Symposium on Space Flight Dynamics, Pasadena, October 2012

[Murchie-2008] Murchie, S.L., "Geology of the Caloris Basin, Mercury: A View from MESSENGER", Science, 321, 2008, 73-76

[Mustard-2008] Mustard, J.F., et al., "Hydrated Silicate Minerals on Mars Observed by the Mars Reconnaissance Orbiter CRISM Instrument", Nature, 454, 2008, 305-309

[Mustard-2013] Mustard, J.F., et al., " Report of the Mars 2020 Science Definition Team", posted July 2013, by the Mars Exploration Program Analysis Group (MEPAG)

[Nakamura-2007] Nakamura, M., et al., "Planet-C: Venus Climate Orbiter Mission of Japan", Planetary and Space Science, 55, 2007, 1831-1842

[Nakamura-2008] Nakamura, M., Imamura, T., "Present Status of Japanese Venus Climate Orbiter", presentation the 5th Meeting of the Venus Exploration Analysis Group (VEXAG), Greenbelt, May 2008

[Nakamura-2012] Nakamura, M. et al., "Return to Venus of the Japanese Venus Climate Orbiter Akatsuki", paper presented at the 63rd International Astronautical Congress, Naples, 2012

[NAS-2006] National Academy of Sciences, "Priorities in Space Science Enabled by Nuclear Power and Propulsion", Washington, the National Academies Press, 2006, 17

[NASA-2005a] "Deep Impact Launch Press Kit", NASA, January 2005

[NASA-2005b] "Final Report of the New Horizons II Review Report", Washington, NASA, report dated 31 March 2005

[NASA-2006] Mars Advance Planning Group, "2006 Update to Robotic Mars Exploration Strategy 2007-2016", document dated 28 November 2006

[NASA-2008a] "MESSENGER - Mercury Flyby 1 Press Kit", NASA, January 2008

[NASA-2008b] "Phoenix Landing - Mission to the Martian Polar North", NASA, May 2008

[NASA-2009a] "MESSENGER - Mercury Flyby 3 Press Kit", NASA, September 2009

[NASA-2009b] "Kepler: NASA's First Mission Capable of Finding Earth-Size Planets - Press Kit", NASA, February 2009

[NASA-2011] "Juno Launch Press Kit", NASA, August 2011

[Nature-2008] "What Next for Mars?", Nature, 456, 2008, 675

[Nature-2010] "Galileo's Send-Off", Nature, 468, 2010, 6

[Nedelcu-2007] Nedelcu, D.A.., et al., "E-Type Asteroid (2867) Steins: Flyby Target for Rosetta", Astronomy & Astrophysics, 473, 2007, L33-L36

[Neumann-2006] Neumann, G.A., et al., "Laser Ranging at Interplanetary Distances", paper presented at the 15th International Workshop in Laser Ranging, Canberra, 2006

[Neumann-2012] Neumann, G.A., et al., "Dark Material at the Surface of Polar Crater

Deposits on Mercury", paper presented at the XLIII Lunar and Planetary Science Conference, Houston, 2012

[Ni-2006] Ni, W.-T., et al., "ASTROD I: Mission Concept and Venus Flyby", Acta Astronautica, 59, 2006, 598-607

[Nielsen-2001] Nielsen, E., et al., "Antennas for Sounding of a Cometary Nucleus in the Rosetta Mission". Paper presented at the 11th International Conference on Antennas and Propagation, Manchester, 17-20 April 2001

[Niles-2010] Niles, P.B., et al., "Stable Isotope Measurements of Martian Atmospheric CO2 at the Phoenix Landing Site", Science, 329, 2010, 1334-1337

[Nilsson-2007] Nilsson, H., et al., "RPC-ICA: The Ion Composition Analyzer of the Rosetta Plasma Consortium", Space Science Reviews, 128, 2007, 671-695

[Nimmo-2008] Nimmo, F., et al., "Implications of an Impact Origin for the Martian Hemispheric Dichotomy", Nature, 453, 2008, 1220-1223

[Nittler-2011] Nittler, L.R., et al., "The Major-Element Composition of Mercury's Surface from MESSENGER X-ray Spectrometry", Science, 333, 2011, 1847-1850

[Noble-1998] Noble, R.J., "Radioisotope Electric Propulsion of Sciencecraft to the Outer Solar System and Near-Interstellar Space", paper presented at the Second IAA Symposium on Realistic Near-Term Advanced Scientific Space Missions, Aosta, 29 June-1 July 1998

[Noll-2005] Noll, K.S., "Solar System Binaries", paper presented at the Asteroid, Comets, Meteors 2005 conference

[Norris-2012a] Norris, G., Morring, F. Jr., "Commissioning", Aviation Week & Space Technology, 20 August 2012, 28-29

[Norris-2012b] Norris, G., "Early Promise", Aviation Week & Space Technology, 3 September 2012, 36-38

[NRC-2002] Solar System Exploration Survey, Space Studies Board, National Research Council, "New Frontiers in the Solar System - An Integrated Exploration Strategy", July 2002

[NRC-2003] National Research Council, "New Frontiers in Solar System Exploration", Washington, the National Academies Press, 2003

[NRC-2008a] Space Studies Board, "Opening New Frontiers in Space - Choices for the Next New Frontiers Announcement of Opportunity", Washington, the National Academies Press, 2008

[NRC-2008b] Space Studies Board, "Science Opportunities Enabled by NASA's Constellation System - Interim Report", Washington, the National Academies Press, 2008

[NRC-2011] Space Studies Board, "Vision and Voyages for Planetary Science in the Decade 2013-2022", Washington, the National Academies Press, 2011

[Oberst-2012] Oberst, J., et al., "Dynamic and Morphologic Studies of the MarcoPolo-R Binary Asteroid System by Laser Altimetry", paper presented at the 63rd International Astronautical Congress, Naples, 2012

[Okubo-2007] Okubo, C.H., et al., "Fracture-Controlled Paleo-Fluid Flow in Candor Chasma, Mars", Science, 315, 2007, 983-985

[Oleson-2011] Oleson, S.R., et al., "Kuiper Belt Object Orbiter Using Advanced Radioisotope Power Sources and Electric Propulsion", Journal of the British Interplanetary Society, 64, 2011, 63-69

[Olkin-2006] Olkin, C.B., et al., "The New Horizons Distant Flyby of Asteroid 2002 JF56", presentation at the American Astronomical Society, DPS meeting No.38, 2006

[Ortiz-2012] Ortiz, J.L., et al., "Albedo and Atmospheric Constraints on Dwarf Planet Makemake from a Stellar Occultation", Nature, 491, 2012, 566-569

[Paige-1992] Paige, D.A., Wood, S.E., Vasavada, A.R., "The Thermal Stability of Water Ice at the Poles of Mercury", Science, 258, 1992, 643-646

[Paige-2012] Paige, D.A., et al., "Thermal Stability of Frozen Volatiles in the North Polar Region of Mercury", paper presented at the XLIII Lunar and Planetary Science Conference, Houston, 2012

[Pappalardo-2011] Pappalardo, B., "Europa Study Update", presentation to the Outer Planets Assessment Group, (OPAG), 19 October 2011

[Pappalardo-2013] Pappalardo, R., et al., "The Europa Clipper – OPAG Update", presentation to the Outer Planets Assessment Group, (OPAG), 15-16 July 2013

[Parker-2007] Parker, T., Manning, R., "Mars Litter Inventory: Using HiRISE to Find out Stuff", presentation dated 28 February 2007

[Pätzold-2007a] Pätzold, M., et al., "Rosetta Radio Science Investigations (RSI)", Space Science Reviews, 128, 2007, 599-627

[Pätzold-2007b] Pätzold, M., et al., "The Structure of Venus' Middle Atmosphere and Ionosphere", Nature, 450, 2007, 657-660

[Pätzold-2011] Pätzold, M., et al., "Asteroid (21) Lutetia - Low Mass, High Density", Science, 334, 2011, 491-492

[Peacock-2005] Peacock, A., "On the Feasibility of a Fast Track Return to Mars: Mars Lander(s) 2011 Mars Demonstration Landers (MDL)", presentation to the 1st Mars Express Science Conference, Noordwijk, 2005

[People's Daily-2009] "China's First Mars Probe Set for Launch in Oct.", People's Daily Online, 9 June 2009

[Peplowski-2011] Peplowski, P.N., et al., "Radioactive Elements on Mercury's Surface from MESSENGER: Implications for the Planet's Formation and Evolution", Science, 333, 2011, 1850-1852

[Pergola-2013] Pergola, P., "Small Satellite Survey Mission to the Second Earth Moon", Advances in Space Research, 53, 2013, 1622-1633

[Perozzi-1993] Perozzi. E., Pittich, E.M., "Small Satellite Missions to Long-Period Comets". In: "Systèmes et Services à Petits Satellites", Toulouse, Cépaduès, 1993, 181-184

[Perozzi-1996] Perozzi, E., et al., "Small Satellite Missions to Long-Period Comets: the Hale-Bopp Opportunity", Acta Astronautica, 39, 1996, 45-50

[Phillips-2008] Phillips, R.J., et al., "Mars North Polar Deposits: Stratigraphy, Age, and Geodynamical Response", Science, 320, 2008, 1182-1185

[Phillips-2011] Phillips, R.J., et al., "Massive CO2 Ice Deposits Sequestered in the South Polar Layered Deposits of Mars", Science, 332, 2011, 838-841

[Phipps-2004] Phipps, A., et al., "Venus Orbiter and Entry Probe: An ESA Technology Reference Study", paper presented at the 55th International Astronautical Congress, Vancouver, 2004

[Phipps-2005] Phipps, A., et al., "Mission and System Design of a Venus Entry Probe and Aerobot", paper presented at the 56th International Astronautical Congress, Fukuoka, 2005

[Piccioni-2007] Piccioni, G., et al., "South-Polar Features on Venus Similar to Those Near the North Pole", Nature, 450, 2007, 637-640

[Piccioni-2008] Piccioni, G., et al., "First Detection of Hydroxyl in the Atmosphere of Venus", Astronomy & Astrophysics, 483, 2008, L29-L33

[Pichkhadze-1996] Pichkhadze, K.M., et al., "Mission to the Sun with Low Thrust", paper presented at the First IAA Symposium on Realistic Near-Term Advanced Scientific Space Missions, Aosta, 25-27 June 1996

[Pieters-2012] Pieters, C.M., et al., "Distinctive Space Weathering on Vesta from Regolith Mixing Processes", Nature, 491, 2012, 79-82

[Pillinger-2007] Pillinger, C., "Space is a Funny Place", London, Barnstorm, 2007, 197

[Plaut-2010] Plaut, J., Smrekar, S., Zurek, R., "Mars Reconnaissance Orbiter: Progress, Status, Plans", presentation at the Planet Mars III Workshop, Les Houches, April 2010

[Polischuk-2005a] Polischuk, G.M., et al., "Perspektivniy Avtomaticheskiy Kosmicheskiy Kompleks Dliya Issledovaniya Marsa" (Perspective Automatic Spacecraft for the Exploration of Mars), Polyet, 6, 2005, 7-11 (in Russian)

[Polischuk-2005b] Polischuk, G.M., et al., "Perspektivniye Proyekti Avtomaticheskikh Kosmicheskikh Kompleksov Dliya Issledovaniya Planet-Gigantov i Ikh Sputnikov" (Perspective Projects of Automatic Spacecraft for the Exploration of the Giant Planets and Their Satellites), Polyet, 7, 2005, 12-15 (in Russian)

[Polischuk-2006] Polischuk, G., et al., "Proposal on Application of Russian Technical Facilities for International Mars Research Program for 2009-2015", Acta Astronautica, 59, 2006, 113-118

[Portree-2001] Portree D.S.F., "Humans to Mars: Fifty Years of Mission Planning 1950-2000", Washington, NASA, 2001

[Potter-1985] Potter, A., Morgan, T., "Discovery of Sodium in the Atmosphere of Mercury", Science, 229, 1985, 651-653

[Potter-1990] Potter, A., Morgan, T., "Evidence for Magnetospheric Effects on the Sodium Atmosphere of Mercury", Science, 248, 1990, 835-838

[Powell-2009] Powell, J.W., "Seven Minutes of Terror", Spaceflight, September 2009, 338-348

[Prado-2008] Prado, J.-Y., "International Campaign for the Improvement of the APOPHIS Ephemeris", presentation at the United Nations Office for Outer Space Affairs Scientific and Technical Subcommittee Forty-fifth Session, February 2008

[Prado-2010] Prado, J.-Y., "Apophis 2029 a Unique Mission Opportunity", presentation at the United Nations Office for Outer Space Affairs Scientific and Technical Subcommittee Forty-seventh Session, February 2010

[Pratt-2009] Pratt, L.M., et al., "Mars Astrobiology Explorer-Cacher (MAX-C): A Potential Rover Mission for 2018", Final Report of the Mars Mid-Range Rover Science Analysis Group (MRR-SAG), October 14, 2009

[Prettyman-2012] Prettyman, T.H., et al., "Elemental Mapping by Dawn Reveals Exogenic H in Vesta's Regolith", Science, 338, 2012, 242-246

[Prockter-2004] Prockter, L., and the Jupiter Icy Moons Orbiter Science Definition Team, "The Jupiter Icy Moons Orbiter: An Opportunity for Unprecedented Exploration of the Galilean Satellites", paper presented at the 55th Congress of the International Astronautical Federation, Vancouver, 2004

[Prockter-2006] Prockter, L.M., et al., "Enabling Decadal Survey Science Goals for Primitive Bodies Using Radioisotope Electric Propulsion", paper presented at the XXXVII Lunar and Planetary Science Conference, Houston, 2006

[Prockter-2009] Prockter, L.M., et al., "The Curious Case of Raditladi Basin", paper presented at the XL Lunar and Planetary Science Conference, Houston, 2009

[Prockter-2010] Prockter, L.M., at al., "Evidence for Young Volcanism on Mercury from the Third MESSENGER Flyby", Science, 329, 2010, 668-671

[Randolph-2003] Randolph, J., et al., "Urey: To Measure the Absolute Age of Mars", paper presented at the 2003 IEE Aerospace Conference

[Rao-2009] Rao, R., "Beyond the Moon", Flight International, 3 February 2009, 37-38

[Rathke-2004] Rathke, A., "Testing for the Pioneer Anomaly on a Pluto Exploration Mission", arXiv astro-ph/0409373 preprint

[Rayman-2004a] Rayman, M.D., "Why no Magnetometer on Dawn?", posting to the FPSpace discussion group, 28 April 2004

[Rayman-2004b] Rayman, M.D., et al., "Dawn: A Mission in Development for Exploration of

Main Belt Asteroids Vesta and Ceres", paper presented at the 55th Congress of the International Astronautical Federation, Vancouver, 2004

[Rayman-2005] Rayman, M.D., et al., "Preparing for the Dawn Mission to Vesta and Ceres", paper presented at the 56th Congress of the International Astronautical Federation, Fukuoka, 2005

[Rayman-2012a] Rayman, M.D., Mase, R.A., "Dawn's Exploration of Vesta", paper presented at the 63rd International Astronautical Congress, Naples, 2012

[Rayman-2012b] Rayman, M.D., Personal communication with the author, 24 May 2012

[Rayman-2013] Rayman, M.D., Mase, R.A., "Dawn's Operations in Cruise from Vesta to Ceres", paper presented at the 64th International Astronautical Congress, Beijing, 2013

[Raymond-2013] Raymond, C.A., Russell, C.T., "Dawn Update", presentation at the 8th Small Bodies Assessment Group (SBAG) meeting, Washington, January 2013

[RDIME-2008] R&D Institute of Mechanical Engineering brochure, 2008

[Read-2013] Read, P., "Plumbing the Depths of Uranus and Neptune", Nature, 497, 2013, 323-324

[Reddy-2012a] Reddy, V., et al., "Color and Albedo Heterogeneity of Vesta from Dawn", Science, 336, 2012, 700-704

[Reddy-2012b] Reddy, V., et al., "Composition of Near-Earth Asteroid (4179) Toutatis", arXiv astro-ph/1210.2853 preprint

[Reich-2010] Reich, E.S., "NASA Panel Weighs Asteroid Danger", Nature, 467, 2010, 140-141

[Reichhardt-2005] Reichhardt, T., "Mars Orbiter Ready to Scout for Future Landing Sites as NASA Looks Ahead", Nature, 436, 2005, 613

[Reichhardt-2010] Reichhardt, T., "Titan Air", Air & Space Smithsonian, June/July 2010, 20-23

[Reinhardt-1999] Reinhardt, R., "Ten Years of Fundamental Physics in ESA's Space Science Programme", ESA Bulletin, 98, 1999, 121-132

[Reitsema-2013] Reitsema, H., "The Sentinel Mission", presentation at the 9th Small Bodies Assessment Group (SBAG) meeting, Washington, July 2013

[Rengel-2008] Rengel, M., Hartogh, P., Jarchow, C., "HHSMT Observations of the Venusian Mesospheric Temperature, Winds, and CO Abundance around the MESSENGER Flyby",arXiv astro-ph/0810.2899 preprint

[Renton-2006] Renton, D., "Deimos Sample Return Technology Reference Study, Executive Summary", ESA SCI-A/2006/010/DSR, 8 February 2006

[Retherford-2007] Retherford, K.D., et al., "Io's Atmospheric Response to Eclipse: UV Aurorae Observations", Science, 318, 2007, 237-240

[Reuter-2007] Reuter, D.C., et al., "Jupiter Cloud Composition, Stratification, Convection and Wave Motion: A View from New Horizons", Science, 318, 2007, 223-225

[Rieber-2009] Rieber, R., "The Contingency of Success: Deep Impact's Planet Hunt", paper presented at the 2009 IEEE Aerospace Conference

[Riedler-2007] Riedler, W., et al., "MIDAS-The Micro-Imaging Dust Analysis System for the Rosetta Mission", Space Science Reviews, 128, 2007, 869-904

[Robertson-2008] Robertson, D.F., "Parched Planet", Sky & Telescope, April 2008, 26-30

[Robinson-2008] Robinson, M.S., et al., "Reflectance and Color Variations on Mercury: Regolith Processes and Compositional Heterogeneity", Science, 321, 2008, 66-69

[Roelof-2004] Roelof, E.C., et al., "Telemachus: a mission for a polar view of solar activity", Advances in Space Research, 34, 2004, 467-471

[Romstedt-2001] Romstedt, J., Novara, M., "Master", ESA Bulletin 105, 2001, 52-53

[Roth-2014] Roth, L., et al., "Transient Water Vapor at Europa's South Pole", Science, 343, 2014, 171-174

[Rouméas-2003] Rouméas, R., "Aurora Exploration Programme: ExoMars Mission", presentation to the Aurora Industry Day, Noordwijk, ESTEC, 6 February 2003

[Russell-2004] Russell, C.T., et al., "Dawn: A Journey in Space and Time", Planetary and Space Science, 52, 2004, 465-489

[Russell-2007] Russell, C.T., et al., "Lightning on Venus Inferred from Whistler-Mode Waves in the Ionosphere", Nature, 450, 2007, 661-662

[Russell-2012] Russell, C.T., et al., "Dawn at Vesta: Testing the Protoplanetary Paradigm", Science, 336, 2012, 684-686

[Ryan-2012] Ryan, A.J., Christensen, P.R., "Coils and Polygonal Crust in the Athabasca Valles Region, Mars, as Evidence for a Volcanic History", Science, 336, 2012, 449-452

[SAGE-2010] "NASA Facts: Surface and Atmosphere Geochemical Explorer", May 2010

[Saki-2012] Saki, T. et al., "Development Status of Small Carry-on Impactor for Hayabusa-2 Mission", paper presented at the 63rd International Astronautical Congress, Naples, 2012

[Sánchez Pérez-2012] Sánchez Pérez, J.M., "Trajectory Design of Solar Orbiter", paper presented at the 23rd International Symposium on Space Flight Dynamics, Pasadena, October 2012

[Sasaki-2009] Sasaki, S., "Japan's Mars Exploration Plan: MELOS", presentation to the Mars Exploration Program Analysis Group (MEPAG) meeting #20, Rosslyn, Virginia, March 2009

[Satoh-2011] Satoh, T., et al., "In-flight observations performed by Akatsuki/IR2", paper presented at the European Planetary Science Congress, Nantes, 2011

[Sawada-2010] Sawada, H., et al., "Report on Deployment Solar Power Sail Mission of IKAROS", presentation at the 2nd International Symposium on Solar Sailing, 2010

[Sawada-2012] Sawada, H., et al., "The Sampling System of Hayabusa 2 Missions", paper presented at the 63rd International Astronautical Congress, Naples, 2012

[Schaefer-2008] Schaefer, B.E., Buie, M.W., Smith, L.T., "Pluto's Light Curve in 1933-1934", arXiv astro-ph/0805.2097 preprint

[Scheeres-2004] Scheeres, D.J., Schweickart, R.L., "The Mechanics of Moving Asteroids", Paper AIAA 2004-1446

[Scheeres-2005] Scheeres, D.J., et al., "Abrupt Alteration of Asteroid 2004 MN4's Spin State During its 2029 Earth Flyby", accepted for publication in Icarus

[Schenk-2008] Schenk, P., Matsuyama, I., Nimmo, F., "True Polar Wander on Europa from Global-Scale Small-Circle Depressions", Nature, 453, 2008, 368-371

[Schenk-2012] Schenk, P., et al., "The Geologically Recent Giant Impact Basins at Vesta's South Pole", Science, 336, 2012, 694-697

[Schilling-2003] Schilling, G., "Star-Crossed Comet Chaser Eyes New Target", Science, 299, 2003, 1638

[Schmidt-2009] Schmidt, B.E., et al., "The 3D Figure and Surface of Pallas from HST", paper presented at the XL Lunar and Planetary Science Conference, Houston, 2009

[Schmidt-2011] Schmidt, B.E., et al., "Active Formation of 'Chaos Terrain' over Shallow Subsurface Water on Europa", Nature, 479, 2011, 502-505

[Schoenmaekers-2008] Schoenmaekers, J. et al., "Mission Analysis: Towards a European Harmonization", ESA Bulletin, 134, 2008, 10-19

[Schröder-2008] Schröder, S.E., et al., "In-Flight Calibration of the Dawn Framing Camera", paper presented at the European Planetary Science Congress, Münster, 2008

[Schultz-2011] Schultz, P.H., et al., "Geology of 103P/Hartley 2 and Nature of Source Regions for Jet-Like Outflows", paper presented at the XLII Lunar and Planetary Science Conference, Houston, 2011

[Schultze-2002] Schultze, R., "Aurora: A European Roadmap for Solar System Exploration", On Station, March 2002, 4-5

[Schumacher-2001] Schumacher, G., Gay, J., "An Attempt to Detect Vulcanoids with SOHO/ LASCO Images", Astronomy & Astrophysics, 368, 2001, 1108-114

[Schwehm-1994] Schwehm, G., Hechler, M., "Rosetta ESA's Planetary Cornerstone Mission", ESA bulletin, 77, 1994, 7-18

[Schweickart-2003] Schweickart, R.L., et al., "The Asteroid Tugboat", Scientific American, November 2003, 54-61

[Schweickart-2005] Schweickart, R.L., "A Call to (Considered) Action", B612 Foundation Occasional Paper 0501, May 2005

[Scoon-1998] Scoon, G., et al., "The Venus Sample Return Mission", paper presented at the 5th ESA workshop on Advanced Space Technologies for Robotics and Automation, Noordwjik, 1-3 December 1998

[Scott-2000] Scott, W.B., "NASA Revisits Nuclear Propulsion for Long Space Missions", Aviation Week & Space Technology, 30 October 2000, 72-74

[Seidensticker-2007] Seidensticker, K.J., et al., "SESAME-An Experiment of the Rosetta Lander Philae: Objectives and General Design", Space Science Reviews, 128, 2007, 281-299

[Seu-2007] Seu, R., et al., "Accumulation and Erosion of Mars' South Polar Layered Deposits", Science, 317, 2007, 1715-1718

[Sheppard-2010] Sheppard, S.S., Trujillo, C.A., "Detection of a Trailing (L5) Neptune Trojan", Science, 329, 2010, 1304

[Shinbrot-2004] Shinbrot, T., et al., "Dry Granular Flows Can Generate Surface Features Resembling Those Seen in Martian Gullies", Proceedings of the National Academy of Sciences, 101, 2004, 8542-8546

[Shirley-2003] Shirley, J.H., et al., "Icy Satellites Impactor Probes for the Jovian Icy Moons Orbiter", paper presented at the Forum on Jupiter Icy Moons Orbiter, Houston, 2003

[Showalter-2006] Showalter, M.R., Lissauer, J.J., "The Second Ring-Moon System of Uranus: Discovery and Dynamics", Science, 311, 2006, 973-977

[Showalter-2007] Showalter, M.R., et al., "Clump Detection and Limits on Moons in Jupiter's Ring System", Science, 318, 2007, 232-234

[Showalter-2011] Showalter, M.R., et al., "The Impact of Comet Shoemaker-Levy 9 Sends Ripples through the Rings of Jupiter", Science, 332, 2011, 711-713

[Sicardy-2011] Sicardy, B., et al. "A Pluto-like Radius and a High Albedo for the Dwarf Planet Eris from an Occultation", Nature, 478, 2011, 493-496]

[Siddiqi-2000] Siddiqi, A. A., "Challenge to Apollo", Washington, NASA, 2000, 334-337 and 745-746

[Siersk-2011] Siersk, H., et al., "Images of Asteroid 21 Lutetia: A Remnant Planetesimal from the Early Solar System", Science, 334, 2011, 487-490

[Sietzen-2009] Sietzen, F., "Mars Laboratory Lands on Red Ink", Aerospace America, October 2009, 24-28

[Siili-2011] Siili, T., et al., "On Mars and Moon Mission Activities at the Finnish Meteorological Institute (FMI)", paper presented at the XIII meeting of Finnish Space Researchers FinCOSPAR 2011

[Slade-1992] Slade, M.A., Butler, B.J., Muhleman, D.O., "Mercury Radar Imaging: Evidence for Polar Ice", Science, 258, 1992, 635-640

[Slavin-2008] Slavin, J.A., et al., "Mercury's Magnetosphere After MESSENGER's First Flyby", Science, 321, 2008, 85-89

[Slavin-2009] Slavin, J.A., et al., "MESSENGER Observations of Magnetic Reconnection in Mercury's Magnetosphere", Science, 324, 2009, 606-610

[Slavin-2010] Slavin, J.A., at al., "MESSENGER Observations of Extreme Loading and Unloading of Mercury's Magnetic Tail", Science, 329, 2010, 665-668

[Smith-2000a] Smith, B.A., "NASA Weighs Mission Options", Aviation Week & Space Technology, 11 December 2000, 54-59

[Smith-2000b] Smith, B.A., "NASA Invests Heavily in New Technology", Aviation Week & Space Technology, 11 December 2000, 63-67

[Smith-2006] Smith, D.E., et al., "Two-Way Laser Link over Interplanetary Distances", Science, 311, 2006, 53

[Smith-2009a] Smith, D.E., et al., "Does Mercury have Lunar-Like Mascons?", paper presented at the XL Lunar and Planetary Science Conference, Houston, 2009

[Smith-2009b] Smith, P., et al., "H2O at the Phoenix Landing Site", Science, 325, 2009, 58-61

[Smith-2010] Smith, I.B., Holt, J.W., "Onset and Migration of Spiral Troughs on Mars Revealed by Orbital Radar", Nature, 465, 2010, 450-453

[Smith-2012] Smith, D.E., et al., "Gravity Field and Internal Structure of Mercury from MESSENGER", Science, 336, 2012, 214-217

[Smrekar-2010] Smrekar, S.E., et al., "Recent Hot-Spot Volcanism on Venus from VIRTIS Emissivity Data", Science, 328, 2010, 605-608

[Snodgrass-2010] Snodgrass, C., et al., "A collision in 2009 as the origin of the debris trail of asteroid P/2010 A2", Nature, 467, 2010, 814-816

[Solomon-2007] Solomon, S.C., "The MESSENGER Venus Flybys", presentation at the 3rd meeting of the Venus Exploration Analysis Group (VEXAG), Crystal City, January 2007

[Spencer-2003] Spencer, J., et al., "Finding KBO Flyby Targets for New Horizons", Earth, Moon and Planets, 92, 2003, 483-491

[Spencer-2007] Spencer, J.R., et al., "Io Volcanism Seen by New Horizons: A Major Eruption of the Tvashtar Volcano", Science, 318, 2007, 240-243

[Spencer-2009] Spencer, D.A., et al., "Phoenix Landing Site Hazard Assessment and Selection", Journal of Spacecraft and Rockets, 46, 2009, 1196-1201

[Spilker-2003a] Spilker, T.R., Young, R.W., "JIMO Delivery and Support of a Jupiter Deep Entry Probe", paper presented at the Forum on Jupiter Icy Moons Orbiter, Houston, 2003

[Spilker-2003b] Spilker, T.R., "Saturn Ring Observer", Acta Astronautica, 52, 2003, 259-265

[Spilker-2010] Spilker, T.R., et al., "Saturn Ring Observer Concept Architecture Options", Journal of the British Interplanetary Society, 63, 2010, 345-350

[Spohn-2007] Spohn, T., et al., "MUPUS-A Thermal and Mechanical Properties Probe for the Rosetta Lander Philae", Space Science Reviews, 128, 2007, 339-362

[Squyres-2010] Squyres, S., "Solar System 2012: The Planetary Science Decadal Survey", presentation to the Mars Exploration Program Analysis Group (MEPAG) meeting #23, Monrovia, California, September 2010

[STDT-2008] "Solar Probe Plus: Report of the Science and Technology Definition Team (STDT)", 14 February 2008

[Steffl-2013] Steffl, A.J., et al., "A Search for Vulcanoids with the STEREO Heliospheric Imager", arXiv astro-ph/1301.3804 preprint

[Steltzner-2008] Steltzner, A.D., et al., "Mars Science Laboratory Entry, Descent, and Landing System Overview", paper presented at the IEEE Aerospace Conference, 2008

[Stern-1995a] Stern, A., "The Chiron Perihelion Campaign", Sky & Telescope, March 1995, 32-34

[Stern-1995b] Ster, S.A., et al., "Future Neptune and Triton Missions". In: Cruikshank, D.P. (ed.), "Neptune and Triton", University of Arizona Press, 1995, 1151-1178

[Stern-2000] Stern, A., Durda, D.D., Tomlinson, B., "Low-Cost Airborne Astronomy Imager to Begin Research Phase", Eos Transactions, 7 March 2000, 101-105

[Stern-2002] Stern, S.A., "Journey to the Farthest Planet", Scientific American, May 2002, 56-63

[Stern-2004] Stern, S.A., Spencer, J., "New Horizons: The First Reconnaissance Mission to Bodies in the Kuiper Belt", 2004

[Stern-2005a] Stern, S.A., et al., "Characteristics and Origin of the Quadruple System at Pluto", arXiv preprint astro-ph/0512599

[Stern-2005b] Stern, S.A., et al., "New Horizons 2", document dated 2005

[Stern-2007a] Stern, S.A., et al., "ALICE: The Rosetta Ultraviolet Imaging Spectrograph", Space Science Reviews, 128, 2007, 507-527

[Stern-2007b] Stern, S.A., "The New Horizons Pluto Kuiper Belt Mission: An Overview with Historical Context", arXiv astro-ph/0709.4417 preprint

[Stofan-2009] Stofan, E., "Titan Mare Explorer (TiME): The First Exploration of an Extra-Terrestrial Sea", presentation to the Decadal Survey, 25 August 2009

[Stolper-2013] Stolper, E.M., et al. "The Petrochemistry of Jake_M: A Martian Mugearite", Science, 341, 2013

[Stone-2009] Stone, R., "Mars Mission Delayed as Mad Dash to Prep Probe Falls Short", Science, 326, 2009, 27

[Strom-2008] Strom, R.G., et al., "Mercury Cratering Record Viewed from MESSENGER's First Flyby", Science, 321, 2008, 79-81

[Sukhanov-2010] Sukhanov, A.A., et al., "The Aster Project: Flight to a Near Earth Asteroid", Cosmic Research, 48, 2010, 443-450.

[Sunshine-2006] Sunshine, J.M., et al., "Exposed Water Ice Deposits on the Surface of Comet Tempel 1", Science, 311, 2006, 1453-1455

[Sunshine-2009a] Sunshine, J.M., et al., "Temporal and Spatial Variability of Lunar Hydration as Observed by the Deep Impact Spacecraft", Science, 326, 2009, 565-568

[Sunshine-2009b] Sunshine, J., "Comet Hopper (CHopper)", presentation dated 28 May 2009

[Svedhem-2005] Svedhem, H., Witasse, O., Titov, D.V., "The Science Return from Venus Express", ESA Bulletin, 125, 2005, 24-32

[Svedhem-2007a] Svedhem, H., et al.., "Venus as a More Earth-Like Planet", Nature, 450, 2007, 629-632

[Svedhem-2007b] Svedhem, H., "Venus Express During MESSENGER Venus Fly-By", presentation at the 4th meeting of the Venus Exploration Analysis Group (VEXAG), Greenbelt, November 2007

[Svedhem-2008a] Svedhem, H., Witasse, O., Titov, D.V., "Exploring Venus: Answering the Big Questions with Venus Express", ESA Bulletin, 135, 2008, 3-9

[Svedhem-2008b] Svedhem, H., "Status and a Selection of Results from Venus Express", presentation the 5th Meeting of the Venus Exploration Analysis Group (VEXAG), Greenbelt, May 2008

[Svedhem-2013] Svedhem, H., Titov, D., "Venus Express Status, Results and Future plans - The Variable Character of Venus", presentation the 10th Meeting of the Venus Exploration Analysis Group (VEXAG), Washington, November 2013

[Svitak-2011] Svitak, A., "Drifting Apart", Aviation Week and Space Technology, 8 August 2011, 24-26

[Svitak-2012] Svitak, A., Morring, F. Jr., "Changing Partners", Aviation Week & Space Technology, 19-26 March 2012, 35

[Sweetser-2003] Sweetser, T., et al., "Venus sample return missions – A range of science, a range of costs", Acta Astronautica, 52, 2003, 165-172

[Syvertson-2010] Syvertson, M., "Mission Concept Study - Planetary Science Decadal Survey Mars 2018 MAX-C Caching Rover", report dated March 2010

[Tabachnik-1999] Tabachnik, S., Evans, N.W., "Cartography for Martian Trojans", The Astrophysical Journal, 517, 1999, L63-L66

[Taguchi-2012] Taguchi, M., et al., "Characteristic Features in Venus' Nightside Cloud-Top Remperature Obtained by Akatsuki/LIR", Icarus, 219, 2012, 502-504

[Tan-1999] Tan, Z., et al., "Hard Landing Impact of Planet Probe", Missiles and Space Vehicles, 4, 1999 (in Chinese)

[Tang-2013] Tang, G., et al., "Optical-Image-Based Precise Estimation of Chang'E II Fly-By Distance to Toutatis", paper presented at the 64th International Astronautical Congress, Beijing, 2013

[Tan-Wang-1998] Tan-Wang, G.H., Sims, J.A., "Mission Design for the Deep Space 4/ Champollion Comet Sample Return Mission", paper AAS 98-187

[Taverna-1999] Taverna, M.A., "Mercury and Venus Sample Returns Eyed", Aviation Week & Space Technology, 15 February 1999, 23-24

[Taverna-2000] Taverna, M.A., "Europe to Have Major Sample Return Role", Aviation Week & Space Technology, 11 December 2000, 60-63

[Taverna-2002] Taverna, M.A., "Venus Express Go-Ahead Reflects End of Italian Space Policy Review", Aviation Week and Space Technology, 18 November 2002, 46

[Taverna-2005] Taverna, M.A., "Back Down to Earth", Aviation Week and Space Technology, 18 April 2005, 30-31

[Taverna-2006] Taverna, M.A., Morring, F. Jr., "Robotic Reconnaissance", Aviation Week and Space Technology, 23 October 2006, 66

[Taverna-2008a] Taverna, M.A., "Spoiler", Aviation Week and Space Technology, 6 October 2008, 36

[Taverna-2008b] Taverna, M.A., "New Programs Will Move ESA Into New Areas of Activity", Aviation Week and Space Technology, 1 December 2008, 39

[Taverna-2008c] Taverna, M.A., "Bringing It Home", Aviation Week & Space Technology, 11 August 2008, 62

[Taverna-2009a] Taverna, M.A., Morring, F. Jr., "Exploring Together", Aviation Week and Space Technology, 29 June 2009, 54

[Taverna-2009b] Taverna, M.A., "Split Decision", Aviation Week & Space Technology, 19 October 2009, 34

[Taverna-2009c] Taverna, M.A., "Fighting Inflation", Aviation Week & Space Technology, 7 December 2009, 46

[Taverna-2010] Taverna, M.A., "Then There Were 3", Aviation Week & Space Technology, 1 March 2010, 26

[Taylor-2001] Taylor, B.G., "The Selection of New Science Missions", ESA Bulletin, 2001, 44-45

[Thomas-2011a] Thomas,P.C., et al., "The Shape and Geological Features of Comet 103P/ Hartley 2", paper presented at the XLII Lunar and Planetary Science Conference, Houston, 2011

[Thomas-2011b] Thomas, P.C., "Cold-Trapping Mars' Atmosphere", Science, 332, 2011, 797-798

[Titov-2008] Titov, D.V., et al., "Atmospheric Structure and Dynamics as the Cause of Ultraviolet Markings in the Clouds of Venus", Nature, 456, 2008, 620-623

[Tolson-2008] Tolson, R., et al., "Atmospheric Modeling Using Accelerometer Data During Mars Reconnaissance Orbiter Aerobraking Operations", Journal of Spacecraft and Rockets, 45, 2008, 511-518

[Trotignon-2007] Trotignon, J.G., et al., "RPC-MIP: The Mutual Impedance Probe of the Rosetta Plasma Consortium", Space Science Reviews, 128, 2007, 713-728

[Trujillo-1998] Trujillo, C., Jewitt, D., "A Semiautomated Sky Survey for Slow-Moving Objects Suitable for a Pluto Kuiper Express Encounter", The Astronomical Journal, 115, 1998, 1680-1687

[TSSM-2009] "TSSM – Titan Saturn System Mission", NASA/ESA Joint Summary Report, 16 January 2009

[Tsuda-2012] Tsuda, Y., et al., "System Design of Hayabusa 2 - Asteroid Sample Return Mission to 1999JU3", paper presented at the 63rd International Astronautical Congress, Naples, 2012

[Tubiana-2007] Tubiana, C., et al., "Photometric and Spectroscopic Observations of (132526) 2002 JF56: Fly-by Target of the New Horizons Mission", Astronomy & Astrophysics, 463, 2007, 1197-1199

[Tubiana-2008] Tubiana, C., et al., "Comet 67P/Churyumov-Gerasimenko at a large Heliocentric Distance" Astronomy & Astrophysics, 490, 2008, 377-386

[Turrini-2009] Turrini, D., Magni, G., Coradini, A., "Probing the History of the Solar System through the Cratering Records on Vesta and Ceres", arXiv astro-ph/0902.3579 preprint

[Turtle-2010] Turtle, E., Niebur, C., "Mission Concept Study - Planetary Science Decadal Survey - Io Observer", May 2010

[Turyshev-2004] Turyshev, S.G., Nieto, M.M., Anderson, J.D., "Lessons Learned from the Pioneers 10/11 for a Mission to Test the Pioneer Anomaly", arXiv gr-qc/0409117 preprint

[Tytell-2005a] Tytell, D., "Deep Impact's Hammer Throw", Sky & Telescope, October 2005, 34-39

[Tytell-2005b] Tytell, D., "Deep Impact Revisited", Sky & Telescope, December 2005, 16-17

[Tytell-2005c] Tytell, D., "The Nightmare Before Christmas", Sky & Telescope, April 2005, 20

[Ulamec-2002] Ulamec, S., Biele, J. and the Rosetta Lander Team, "Rosetta Lander - Overview". In: Proceedings of the Second European Workshop on Exo/Astrobiology, Graz, Austria, 16-19 September 2002

[Ulamec-2003] Ulamec, S., et al., "Rosetta Lander: Implications of an Alternative Mission", paper presented at the 54th International Astronautical Congress, Bremen, October 2003

[Ulivi-2004] Ulivi, P., with Harland, D.M., "Lunar Exploration: Human Pioneers and Robotic Surveyors", Chichester, Springer-Praxis, 2004, 297-303

[UNITEC-2009] University Space Engineering Consortium, "Call for Support on Tracking and Receiving RF Signal for First Interplanetary University Satellite UNITEC-1", 29 May 2009

[Vago-2006] Vago, J., et al., "ExoMars: Searching for Life on the Red Planet", ESA Bulletin, 126, 2006, 17-23

[Vago-2013] Vago, K., et al., "ExoMars: ESA's Next Step in Mars Exploration", ESA Bulletin, 155, 2013, 12-23

[van de Haar-2004] van de Haar, G., "Messenger on Its Way to Mercury", Spaceflight, October 2004, 382-384

[van de Haar-2006] van de Haar, G., Corneille, P., "And Finally - Our First Mission to the Last Planet", Spaceflight, March 2006, 93-96

[Van den Broecke-2011] Van den Broecke, J., et al., "Landing Site Selection and Characterization for the ExoMars 2016 Mission", undated paper

[Vaniman-2014] Vaniman, D.T., et al., "Mineralogy of a Mudstone at Yellowknife Bay, Gale Crater, Mars", accepted for publication in Science

[Van Winnendael-1999] Van Winnendael, M., et al., "Nanokhod Microrover Heading Towards Mars", paper presented at the 5th International Symposium, ISAIRAS '99, Noordwijk, 1-3 June 1999

[Vaubaillon-2006] Vaubaillon, J., Christou, A.A., "Encounters of the Dust Trails of Comet 45P/Honda-Mrkos-Pajdusakova with Venus in 2006", Astronomy & Astrophysics, 451, 2006, L5-L8

[Verdant-1998] Verdant, M., Schwehm, G.H., "The International Rosetta Mission", ESA Bulletin, 93, 1998, 38-50

[Vervack-2010] Vervack, R.J., at al., "Mercury's Complex Exosphere: Results from MESSENGERis Third Flyby", Science, 329, 2010, 672-675

[VITaL-2010] "VITaL: Venus Intrepid Tessera Lander - Mission Concept Study Report to the NRC Decadal Survey Inner Planets Panel", Final Report, 6 April 2010

[VME-2009] "Venus Mobile Explorer - Mission Concept Study Report to the NRC Decadal Survey Inner Planets Panel", Final Report, 18 December 2009

[Vokrouhlick-2000] Vokrouhlick, D., Farinella, P., Bottke, W.F. Jr., "The Depletion of the Putative Vulcanoid Population via the Yarkovsky Effect", Icarus, 148, 2000, 147-152

[von Eshleman-1979] von Eshleman, R., "Gravitational Lens of the Sun: Its Potential for Observations and Communications over Interstellar Distances", Science, 205, 1979, 1133-1135

[Vorontsov-2010] Vorontsov, V.A., et al., "Perspektivnyi Kosmicheskyi Apparat Dlya Issledovaniya Veneri. Proyekt 'Venera-D'" (Perspective Spacecraft for Venus Research. The 'Venera-D' Project), Vestnik FGUP "NPO im. S.A. Lavochkina", 4, 2010, 70-80 (in Russian)

[Vorontsov-2012] Vorontsov, V.A., et al., "Perdlosheniya pa Rastschireniyu Programmi Issledovaniya Veneri s Ychetom Opita Proyektnikh Rasrabotok NPO im. S.A. Lavochkina" (Proposals for Augmenting the Exploration of Venus based on the Experience of Projects of the NPO named after S.A. Lavochnkin"), Elektronniy Jurnal "Trudy MAI", 52, 2012 (in Russian)

[Vourlidas-2007] Vourlidas, A., et al., "First Direct Observation of the Interaction between a Comet and a Coronal Mass Ejection Leading to a Complete Plasma Tail Disconnection", The Astrophysical Journal, 668, 2007, L79-L82

[Walker-2006] Walker, R.J., et al., "Concepts for a Low-Cost Mars Micro Mission", Acta Astronautica, 59, 2006, 617-626

[Wall-2005] Wall, R., Taverna, M.A., "Mars Muddle", Aviation Week & Space Technology, 14 March 2005, 88-89

[Wallace-1995] Wallace, R.A., et al., "Measure-Jupiter: Low-Cost Missions to Explore Jupiter in the post-Galileo Era", Acta Astronautica, 35, 1995, 277-286

[Wang-2012] Wang, G., Ni, W.-T., "Time-delay Interferometry for ASTROD-GW", Chinese Astronomy and Astrophysics, 36, 2012, 211-228

[Wargo-2010] Wargo, M.J., "Exploration Precursor Robotic Missions (xPRM) Point of Departure Plans", presentation to the Mars Exploration Program Analysis Group (MEPAG) meeting #23, Monrovia, California, September 2010

[Warhaut-2003] Warhaut, M., Martin, R., "Talking to Satellites in Deep Space from New Norcia", ESA Bulletin, 114, 2003, 38-41

[Warhaut-2005] Warhaut, M., Marin, R., Claros, V., "ESA's New Cebreros Station Ready to Support Venus Express", ESA Bulletin, 124, 2005, 38-41

[Warner-2005a] Warner, E.M., Redfern, G., "Deep Impact: Our First Look Inside a Comet", Sky & Telescope, June 2005, 40-44

[Warner-2005b] Warner, E.M., Redfern, G., "Amateurs and the Deep Impact Mission", Sky & Telescope, June 2005, 70-71

[Watters-2009] Watters, T.R., et al., "Evolution of the Rembrandt Impact Basin on Mercury", Science, 324, 2009, 618-621

[Way-2006] Way, D.W., et al., "Mars Science Laboratory: Entry, Descent, and Landing System Performance", paper presented at the 2006 IEEE Aerospace Conference

[Weaver-2009] Weaver, H.A., "Ultraviolet and Visible Photometry of Asteroid (21) Lutetia Using the Hubble Space Telescope", arXiv astro-ph/0912.4572 preprint

[Weaver-2012] Weaver, H., "New Horizons Pluto/KBO Mission Status Report for SBAG", presentation at the 7th Small Bodies Assessment Group (SBAG) meeting, Pasadena, July 2012

[Weaver-2013] Weaver, H., "New Horizons Pluto/KBO Mission: Status Report for SBAG", presentation at the 8th Small Bodies Assessment Group (SBAG) meeting, Washington, January 2013

[Webster-2013a] Webster, C.R., et al., "Isotope Ratios of H, C, and O in CO2 and H2O of the Martian Atmosphere", Science, 341, 2013, 260-263

[Webster-2013b] Webster, C.R., et al., "Low Upper Limit to Methane Abundance on Mars", Science, 342, 2013, 355-357

[Weiss-2011] Weiss, B.P., et al., "Evidence for Thermal Metamorphism or Partial Differentiation of Asteroid 21 Lutetia from Rosetta", paper presented at the XLII Lunar and Planetary Science Conference, Houston, 2011

[Weissman-1999] Weissman, P.R., et al., "The Deep Space 4/Champollion Comet Rendezvous and Lander Technology Demonstration Mission", paper presented at the Lunar and Planetary Science Conference XXX, Houston, 1999

[Weissman-2007] Weissman, P.R., Lowry, S.C., Choi, Y.-J., "Photometric Observations of Rosetta Target Asteroid 2867 Steins", Astronomy & Astrophysics, 466, 2007, 737-742

[Weissman-2012] Weissman, P., "Rosetta Update", presentation at the Small Bodies Assessment Group Meeting, 10 July 2012

[Wellnitz-2009] Wellnitz, D., "EPOXI Status", presentation at the Small Bodies Assessment Group Meeting, 18 November 2009

[Werner-2011a] Werner, D., "Juno May Be Last Chance to Obtain Jupiter Data for a Decade", Space News, 4 April 2011, 12

[Werner-2011b] Werner, D., "Rising Costs Cast Shadow on NASA Planetary Program", Space News, 10 January 2011, 13

[Whipple-1980] Whipple, F.L., "Rotation and Outbursts of Comet P/Schwassmann-Wachmann 1", The Astronomical Journal, 85, 1980, 305-313

[Whiteway-2009] Whiteway, J.A., et al., "Mars Water-Ice Clouds and Precipitation", Science, 325, 2009, 68-70

[Williams-2013] Williams, R.M.E., et al., "Martian Fluvial Conglomerates at Gale Crater", Science, 340, 2013, 1068-1072

[Williamsen-2004] Williamsen, J., Evans, H., Strober, J., "Surviving a Comet: Shielding the Deep Impact Spacecraft", Aerospace America, November 2004, 18-21

[Wilson-2011] Wilson, H.F., Militzer, B., "Rocky Core Solubility in Jupiter and Giant Exoplanets", arXiv astro-ph/1111.6309 preprint

[Winton-2005] Winton, A.J., et al., "Venus Express: the Spacecraft", ESA Bulletin, 124, 2005, 16-22

[Witze-2013] Witze, A., "Mars Mission Set for Launch", Nature, 503, 2013, 178

[Woerner-1998] Woerner, D.F., "Revolutionary Systems and Technologies for Missions to the Outer Planets", paper presented at the Second IAA Symposium on Realistic Near-Term Advanced Scientific Space Missions, Aosta, 29 June-1 July 1998

[Wood-2008] Wood, B.E., et al., "Comprehensive Observations of a Solar Minimum CME with STEREO", arXiv astro-ph/0811.3226v1 preprint

[Worstall-2013] Worstall, T., "Asteroid Miners Hunt for Platinum... Leave Common Sense in the Glovebox", Spaceflight, February 2013, 72-73

[Wright-2007] Wright, I.P., et al., "Ptolemy - An Instrument to Measure Stable Isotopic Ratios of Key Volatiles on a Cometary Nucleus", Space Science Reviews, 128, 2007, 363-381

[Wu-2010] Wu, J., et al., "Scientific Objectives of China-Russia Joint Mars Exploration Program YH-1", Chinese Astronomy Astrophysics, 34, 2010, 163-173

[Wu-2011] Wu, J., et al., "Imaging Interplanetary CMEs at Radio Frequency from Solar Polar Orbit", Advances in Space Research, 48, 2011, 943-954

[Wu-2012] Wu, W.R., et al., "Pre-LOI Trajectory Maneuvers of the CHANG'E-2 Libration Point Mission", Science China: Information Sciences, 55, 2012, 1249-1258

[Yamada-2007] Yamada, T., et al., "Venus Atmosphere Observation Mission by Venus Entry Probe and Water-Vapor Balloon", presentation at the 2007 VEP Meeting, Oxford, January 2007

[Yamakawa-1999] Yamakawa, H., et al., "Preliminary ISAS Mercury Orbiter Mission Design", Acta Astronautica, 45, 1999, 187-195

[Yamakawa-2001] Yamakawa, H., Kimura, M., "Orbit Synthesis of the ISAS Venus Climate Orbiter Mission", paper AAS 01-460

[Yamamoto-2007] Yamamoto, S., et al., "Comet 9P/Tempel 1: Interpretation with the Deep Impact Results", arXiv astro-ph/0712.1858 preprint

[Yang-2012] Yang, G., Heng-Nian, L., Sheng-Mao, H., "First-Round Design of the Flight Scenario for Chang'e-2's Extended Mission: Takeoff from Lunar Orbit", published in Acta Mechanica Sinica, 2012

[Yano-2008] Yano, H., "Science, Technology and Programmatic Progress of the Japanese Team toward the Joint Marco Polo Phase-A Study", presentation at the Marco Polo Cannes Workshop, June 2008

[Yen-1989] Yen, C.-W. L., "Ballistic Mercury Orbiter Mission via Venus and Mercury Gravity Assists", The Journal of the Astronautical Sciences, 37, 1989, 417-432

[Ying-2013] Ying, C., et al., "Design for Mars Plural Mode Combination Exploration Mission", paper presented at the 64rd International Astronautical Congress, Beijing, 2013

[Yoshikawa-2008] Yoshikawa, M., "JAXA's Primitive Body Exploration Program", paper presented at the first Meeting of the International Primitive Body Exploration Working Group, Okinawa, January 2008

[Young-2003] Young, R.W., Spilker, T.R., "Science Rationale for Jupiter Entry Probe as Part of JIMO", paper presented at the Forum on Jupiter Icy Moons Orbiter, Houston, 2003

[Young-2004] Young, L.A., et al., "Rotary-Wing Decelerators for Probe Descent through the Atmosphere of Venus", paper presented at the 2nd International Planetary Probe Workshop, NASA Ames Conference Center, 23-26 August 2004

[Yuan-2011] Yuan, Y., Yu, Z., Hu, Z., "Concept Research of Mars Penetrator" paper presented at the 7th UK-China Workshop on Space Science and Technology, August 2011

[Yue-2011] Yue, T., et al., "The Application of Chang'E-2 CMOS Camera Technologies", Spacecraft Recovery & Remote Sensing, 32, 2011, 12-17 (in Chinese)

[Zahnle-2011] Zahnle, K., Freedman, R.S., Catling, D.C., "Is There Methane on Mars?", Icarus, 212, 2011, 493-503

[Zak-2008] Zak, A., "Mission Possible", Air & Space, August/September 2008, 60-63

[Zander-2011] Zander, J., "Peering Beneath Jupiter's Clouds", Sky & Telescope, September 2011, 19-23

[Zarka-2008] Zarka, P., et al., "Ground-Based and Space-Based Radio Observations of Planetary Lightning", Space Science Reviews, 137, 2008, 257-269

[Zasova-2011] Zasova, L.V., et al., "Russian Mission Venera-D - New Conception", paper presented at the European Planetary Science Congress, Nantes, 2011

[Zeitlin-2013] Zeitlin, C., et al., "Measurements of Energetic Particle Radiation in Transit to Mars on the Mars Science Laboratory", Science, 340, 2013, 1080-1084

[Zelenyi-2004] Zelenyi, L.M., et al., "Russian Space Program: Experiments in Solar-Terrestrial Physics". In: Proceedings of IAU Symposium No. 223 on Multi-Wavelength Investigations of Solar Activity, 2004, 573-580

[Zelenyi-2005] Zelenyi, L.M., Petrukovich, A.A., "Prospects of Russian Participation in the International LWS Program", Advances in Space Research, 35, 2005, 44-50

[Zelenyi-2007] Zelenyi, L.M., et al., "Phobos Sample Return Project", presentation at the Sputnik 50-Year Jubilee, Moscow, October 2007

[Zhang-2007] Zhang, T.L., et al., "Little or No Solar Wind Enters Venus' Atmosphere at Solar Minimum", Nature, 450, 2007, 654-656

[Zhang-2012] Zhang, T.L., et al., "Magnetic Reconnection in the Near Venusian Magneto-tail", Science, 336, 2012, 567-570

[Zhao-2008] Zhao, H., "YingHuo-1 - Martian Space Environment Exploration Orbiter", Chinese Journal of Space Science, 2008, 28, 395-401

[Zhao-2011] Zhao, B.C., et al., "Overall Scheme and On-Orbit Images of Chang'E-2 Lunar Satellite CCD Stereo Camera", Science China: Technological Sciences, 54, 2011, 2237-2242

[Zimmerman-2009] Zimmerman, R., "Cosmic Cataclysms", Sky & Telescope, April 2009, 26-32

[Zou-2014] Zou, X., et al., "The Preliminary Analysis of the 4179 Toutatis Snapshots of the Chang'e-2 Flyby", Icarus, 229, 2014, 348–354

[Zuber-2007] Zuber, M.T., "Density of Mars' South Polar Layered Deposits", Science, 317, 2007, 1718-1719

[Zuber-2008] Zuber, M.T., et al., "Laser Altimeter Observations from MESSENGER's First Mercury Flyby", Science, 321, 2008, 77-79

[Zuber-2009] Zuber, M.T., et al., "Observations of Ridges and Lobate Scarps on Mercury From MESSENGER Altimetry and Imaging, and Implications for Lithospheric Strain Accommodation", paper presented at the XL Lunar and Planetary Science Conference, Houston, 2009

[Zuber-2012] Zuber, M.T., et al., "Topography of the Northern Hemisphere of Mercury from MESSENGER Laser Altimetry", Science, 336, 2012, 217-220

[Zurbuchen-2008] Zurbuchen, T.H., et al., "MESSENGER Observations of the Composition of Mercury's Ionized Exosphere and Plasma Environment", Science, 321, 2008, 90-92

[Zurbuchen-2011] Zurbuchen, T.H., et al., "MESSENGER Observations of the Spatial Distribution of Planetary Ions Near Mercury", Science, 333, 2011, 1862-1865

[Zurek-2009] Zurek, R., Chicarro, A., "Final Report of the 2016 Mars Orbiter Bus Joint Instrument Definition Team", 10 November 2009

[Zurek-2012] Zurek, R., "Mars Reconnaissance Orbiter Status", presentation to the Mars Exploration Program Analysis Group (MEPAG) meeting #25, Washington, February 2012

[Zurek-2013] Zurek, R., Diniega, S., Madsen, S., "Comets at Mars", presentation to the Mars Exploration Program Analysis Group (MEPAG), July 2013

Further reading

BOOKS

Godwin, R., (editor), "Mars: The NASA Mission Reports Volume 2", Burlington, Apogee, 2004

Kelly Beatty, J., Collins Petersen, C., Chaikin, A. (editors), "The New Solar System", 4th edition, Cambridge University Press, 1999

MAGAZINES

Aerospace America
Aviation Week & Space Technology
ESA Bulletin
Espace Magazine (in French)
Flight International
Nature
Novosti Kosmonavtiki (in Russian)
Science
Scientific American
Sky & Telescope
Spaceflight

INTERNET SITES

ESA (www.esa.int)
Jonathan's Space Home Page (planet4589.org/space/space.html)
JPL (www.jpl.nasa.gov)
Novosti Kosmonavtiki (www.novosti-kosmonavtiki.ru)
Spaceflight Now (www.spaceflightnow.com)
The Planetary Society (planetary.org)

Previous volumes in this series

Part 1: The golden age 1957–1982

Part 2: Hiatus and renewal 1983–1996

Part 3: Wows and Woes 1997–2003

About the authors

PAOLO ULIVI

I was born in 1971, at a time when the Apollo lunar exploration program was slowing down. In spite of this, like many kids of the 1970s, I grew up when space was still cool and we had something 'spacey' everywhere: my father had kept some magazines with pictures of the Apollo 8 and 11 missions, we had truly awful shows on TV such as *Space 1999*, we had space-themed disco music, and we had some of the greatest sci-fi movies in theaters. It is no surprise, then, that many of us were fascinated with space and its technology. I vaguely remember TV images of the Apollo-Soyuz mission, and magazines with pictures from the Viking landers on Mars, but the events that had me enamored with space and solar system exploration were the first launch of the Space Shuttle (I remember where I was when I watched it live on TV) and, most of all, the magnificent, oddly colored images of Saturn from Voyager 2. While in high school, I developed a new interest in aeronautics, which led to my decision to study aerospace engineering at the Milan Politecnico university. The discovery of the Internet in 1991, and the time spent downloading Voyager images from slow FTP servers, led me back to space and astronomy. The realization, as Carl Sagan once explained, that "planets are places", shaped my preference for solar system exploration. After spending much time on other things, it was finally time to choose a subject for my thesis. I went to speak with a professor at my university, Amalia Ercoli Finzi, who is also the Principal Investigator for the drill instrument of the Rosetta cometary mission. She suggested, rather nonchalantly, that I may be interested in working on the DeeDri, an Italian drill which was to fly to Mars in 2003 as part of NASA's "faster, cheaper, better" sample-return mission. So that was the subject of my thesis, and when in 1999 the Mars Polar Lander crashed on the planet I still remember the frustration of not knowing what the future would bring for our drill. In the end, I graduated, but the DeeDri and the 2003 mission didn't make it; hopefully a relative of that drill will fly on ExoMars in 2018. I then went to work, initially in railway engineering and then in aerospace. I began on the 'space' side of aerospace, with racks for the European Columbus module of the International Space Station, but since then have spent most of my time working

David Harland and Paolo Ulivi.

in the 'aero' side. If you fly on a Boeing 787, or an Airbus 380 or 350, you will be flying on many hours of my work. Alongside my day job, I began to write articles and books on space flight and astronomy. After Springer-Praxis published my *Lunar Exploration* in 2004, the logical follow up was a history of the robotic missions which explored the solar system. It seemed like an easy task. Ten years and four volumes later, the "easy task" has finally been completed!

DAVID M. HARLAND

Having been born in 1955, I was just old enough to be enthralled a decade later when the first robotic probes landed on the Moon. I was amazed by the picture taken by an orbiter peering over the rim of the crater Copernicus. The newspapers called this the 'Picture of the Century'. Instead of being an astronomer's perspective, looking down from above and at great distance, it depicted this vast crater, with its complex central peak and terraced walls, from the point of view of an astronaut in lunar orbit. Patrick Moore's television programme *Sky at Night*, then approaching its tenth anniversary, broadened my horizons. I decided that when I grew up I would be an astronomer. For some reason, most of my friends wanted to become footballers, bank managers or engine mechanics. In 1969 I sat up through the night to watch Neil Armstrong and Buzz Aldrin walk on the Sea of Tranquility. The next big event for me was when a mobile camera let us follow Dave Scott and Jim Irwin as they explored a valley in the Apennine mountains. I think I was the only child in my school to stay home to watch these moonwalks. I was so happy that, when asked afterwards by a teacher whether I had been ill, I was truthful: no, I was watching television! I got away with it because I was the 'weird kid' who wanted to be an astronomer. I was the first in the school to sign up for astronomy O-level. No one was teaching this subject, but I had the course book by Patrick Moore and knew I could study it on my own; I passed. Shortly after the final Apollo lunar mission, I headed to university to study astronomy. By the time I graduated with a degree in that subject, I was of a mind to follow with a master's in computer science. After getting my doctorate in that subject, the obvious career was a university lecturer. That opened the door to consulting work, and after several years I left academia to work in industry. In 1990 I returned to academia, this time to pursue research management. By 1995 I had resurrected my interest in space and decided to 'retire' in order to write full time. Since then, I have written a fair few books on my own, coauthored a number with other people and, in recent years, have edited dozens of books by fellow space enthusiasts, most notably this series with Paolo Ulivi on the robotic exploration of the solar system.

Index